Unifying Themes in Complex Systems

VOLUME II

New England Complex Systems Institute Series on Complexity

Yaneer Bar-Yam, Editor-in-Chief

Unifying Themes in Complex Systems, Volume I
Proceedings of the First International Conference on Complex Systems
Edited by Yaneer Bar-Yam

Unifying Themes in Complex Systems, Volume II
Proceedings of the Second International Conference on Complex Systems
Edited by Yaneer Bar-Yam and Ali Minai

Virtual Worlds: Synthetic Universes, Digital Life, and Complexity
Edited by Jean-Claude Heudin

Unifying Themes in Complex Systems

VOLUME II

Proceedings of the First International Conference on Complex Systems

Edited by Yaneer Bar-Yam and Ali Minai

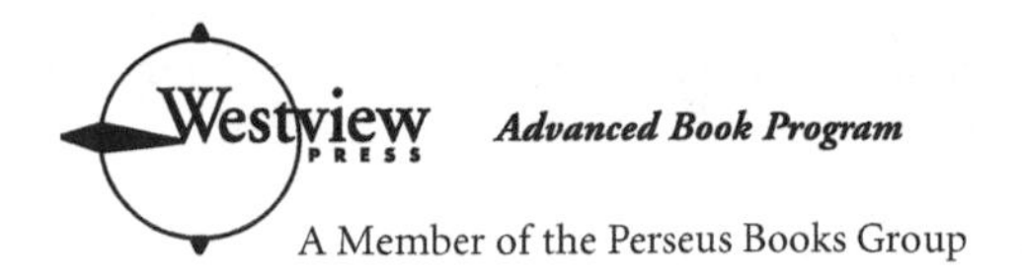

Advanced Book Program

A Member of the Perseus Books Group

Published in the United States of America by Westview Press, A Member of the Perseus Books Group, 5500 Central Avenue, Boulder, Colorado 80301–2877, and in the United Kingdom by Westview Press, 12 Hid's Copse Road, Cumnor Hill, Oxford OX2 9JJ.

Find us on the world wide web at www.westviewpress.com

Westview Press books are available at special discounts for bulk purchases in the United States by corporations, institutions, and other organizations. For more information, please contact the Special Markets Department at the Perseus Books Group, 11 Cambridge Center, Cambridge, MA 02142, or call (617) 252–5298, (800) 255–1514 or email j.mccrary@perseus
books.com.

A Cataloging-in-Publication data record for this book is available from the Library of Congress.
ISBN 0-8133-4124-8 (hardcover) / 0-8133-4123-X (pbk.)

The paper used in this publication meets the requirements of the American National Standard for Permanence of Paper for Printed Library Materials Z39.48–1984.

10 9 8 7 6 5 4 3 2 1

Contents

Introduction

The *International Conference on Complex Systems* provides a unique opportunity for scientists to rise above disciplinary boundaries and explore unity in complex systems. This volume contains the proceedings of the *Second International Conference on Complex Systems.* The sessions ranged from the opening talk by Nobel laureate Philip Anderson on emergence and the relevance of complex systems to physics to the final panel that included many of the great management gurus of this generation. The plenary session topics for the second conference were emergence, description and modeling, self-organization, networks, time series, agents in action, and complexity and management. Talks in these sessions often focused on contributions in specific disciplines but each session contained a diverse set of disciplines to show the strategies and concepts that can be applied generally to complex systems. Overall, the conference presentations dealt with issues of significant public interest including individual psychology, dynamics of social and economic change, the human genome, and ecology.

The International Conference on Complex Systems was the first conference to explicitly encompass general issues of complex systems and research and applications related to complex systems sub-areas. This conference is hosted by the New England Complex Systems Institute as an effort to build the international community devoted to the study of complex systems. The study of complex systems in a unified framework is perhaps the ultimate interdisciplinary field. It is strongly rooted in the advances that have been made in diverse fields ranging from physics to anthropology, from which it draws inspiration, and to which it is relevant. The study of complex systems is recognized by faculty and students as a growing and vibrant area. It has attracted significant attention throughout the scientific community.

Complex systems are systems with multiple interacting components whose behavior cannot be simply inferred from the behavior of the components. While two of the traditional foci of this field are in non-linear dynamics, and fractal models, this has broadened in recent years. Concepts such as emergence, complexity, adaptation, networks, evolution, and self-organization play an important role in our understanding of these systems. The properties of descriptions / representations of complex systems and their effective utilization, including information and computation theory play an important role. Methods for structuring, accessing and characterizing large bodies of information describing complex systems (informatics) are also central.

Among the motivations for progress in this field is a growing recognition that novel quantitative analytic and simulation tools can be applied to biological, economic and social systems. There are two primary thrusts of these developments. The first arises from the development of analytic tools in statistical physics that can be applied, with care, to understanding various aspects of the behavior of complex systems. Among the early applications should be included the study of polymers and complex fluids that has had considerable success. More recent applications are to the dynamics of stochastic non-equilibrium systems using dynamic scaling theory, and to the analysis of simple models of networks (including biochemical networks, neural networks and socioeconomic networks), proteins, pattern formation, etc. The second thrust arises from the tools of simulations, such as Monte Carlo simulations, Cellular Automata and Agent Based models that may be applied to a wide variety of complex systems.

Another new area of exploration is the application of ideas from quantum physics to computer science – reciprocally advancing our understanding of quantum phenomena. Researchers have made important progress in exploring connections between quantum physics and information and computation theory: quantum cryptography and quantum computation. There are also important studies of the connection between computer science and the properties of non-equilibrium physical systems: information mechanics.

More generally, each field has made significant contributions to the study of complex systems. It should be apparent that as systems become more complex, many of the same issues must arise in each discipline in one form or another. Thus, cross-fertilization and utilization of diverse advances in each discipline should advance all fields. A specific example may be given from the field of linguistics where the development of an approach to representation of tree structures describing the decomposition of sentences has been applied to the dynamics of biological development through the formalism of L-systems.

1 New England Complex Systems Institute

The development of interdisciplinary fields, where progress requires close interactions between members of widely different disciplines, is not well served by the present structure of departmental and institutional boundaries. The development of efforts on complex systems requires building a community of researchers, a common vocabulary, and questions that can guide the development of knowledge. To this end a collaboration of faculty from MIT, Harvard, Brandeis, Boston University, Tufts University, Northeastern University and other institutions from around New England have formed the New England Complex Systems Institute (NECSI). Together these individuals will address problems ranging from the integration of information theory with physics, to understanding the rapidly changing structure of human organizations.

NECSI is an educational and research institute dedicated to the advancement of interdisciplinary communication and collaboration, and the rapid dissemination of results of quantitative research in the behavior of complex systems. The board of directors of the institute is drawn from the faculty of major academic institutions in New England.

The targeted areas of research include theoretical and experimental work in a wide diversity of areas spanning the range of human knowledge in the physical, chemical, biological and social sciences, mathematics and engineering. Many areas represent independent fields. Targeted research, however, will represent work that has cross-disciplinary relevance to these areas. By bringing investigators from different disciplines together to explore these areas, NECSI strives to stimulate research collaborations for scientific and educational projects.

To promote communication and interaction between researchers, NECSI is organizing working groups and topical workshops focusing on interdisciplinary subjects. It has established an Internet based peer-reviewed journal for rapid dissemination of research results in all areas of science and engineering.

2 NECSI Charter

The New England Complex Systems Institute is an independent research and educational institute dedicated to the advancement of knowledge about the behavior and universal properties of complex systems. A complex system is a system formed out of many components whose behavior is emergent, that is, cannot be simply inferred from the behavior of its components. The amount of information necessary to describe the behavior of such a system is a measure of its complexity.

The study of complex systems spans systems on all scales, from particle fields to the universe. Among the most complex systems with which we are familiar are biological and social systems including biological macromolecules, biological organisms, ecosystems, human social and economic structures. As a unified endeavor the field of complex systems is new and must itself be promoted. This will increase the ability of researchers to make contributions through timely publication, and to obtain funding and institutional support.

The field of complex systems research includes a number of subfields. Information Mechanics is dedicated to exposing the fundamental connection between information and physical law, measurement and observation. Artificial Life explores the processes of evolution, reproduction and development. Adaptive Systems investigates adaptation, neural systems, learning and consciousness. In every traditional field, extensions toward the study or design of ever more complex phenomena lead to overlap with the field of complex systems. For example, statistical physics traditionally investigated the emergence of simple rather than complex collective phenomena; however in recent years the emergence of complex phenomena occupies increasing attention. Throughout, the field of complex systems places emphasis on applying quantitative analytic and simulation tools to the study of problems previously treated largely through qualitative analogies. The quantitative tools provide an ability to predict the behavior of complex systems, and the limit of the accuracy of such predictions.

Applications of the study of complex systems are varied and include: Drug design through applied molecular evolution; design of software agents and self-organizing software systems; and the design of corporate structures for networked organizations. One system particularly important for the field of complex systems is human civilization - the history of social and economic structures and the emergence of an interconnected global civilization. Applying principles of complex systems to enable us to gain an understanding of its past and future course is ultimately an important objective of this field. We can anticipate a time when the implications of economic and social developments for individual human beings and civilization will become an important application of the study of complex systems.

The merging of disciplines in the field of complex systems runs counter to the increasing specialization in science and engineering. It provides many opportunities for synergies and the recognition of general principles that can form a basis for education and understanding in all fields.

Yaneer Bar-Yam
Cambridge, Massachusetts

Second International Conference on Complex Systems: Organization and Program

Organization:

Host:

New England Complex Systems Institute

Partial financial support:

Perseus Books
Harvard University Press

Conference Chair:

* Yaneer Bar-Yam - NECSI

Executive Committee:

* Larry Rudolph - MIT
† Ali Minai - University of Cincinnati

Organizing Committee:

Philip W. Anderson - Princeton University
Kenneth J. Arrow - Stanford University
* Michel Baranger - MIT
Per Bak - Niels Bohr Institute
Charles H. Bennett - IBM
William A. Brock - University of Wisconsin
* Charles R. Cantor - Boston University
Noam A. Chomsky - MIT
Leon Cooper - Brown University
Daniel Dennett - Tufts University
* Irving Epstein - Brandeis University
Michael S. Gazzaniga - Dartmouth College
* William Gelbart - Harvard University
Murray Gell-Mann - CalTech/Santa Fe Institute
Pierre-Gilles de Gennes - ESPCI
Stephen Grossberg - Boston University
Michael Hammer - Hammer & Co
John Holland - University of Michigan
John Hopfield - Princeton University
* Jerome Kagan - Harvard University
Stuart A. Kauffman - Santa Fe Institute
Chris Langton - Santa Fe Institute
Roger Lewin - Harvard University
Richard C. Lewontin - Harvard University
Albert J. Libchaber - Rockefeller University
* Seth Lloyd - MIT
Andrew W. Lo - MIT
Daniel W. McShea - Duke University
Marvin Minsky - MIT
Harold J. Morowitz - George Mason University
Alan Perelson - Los Alamos National Lab
Claudio Rebbi - Boston University
Herbert A. Simon - Carnegie-Mellon University
* Temple F. Smith - Boston University
H. Eugene Stanley - Boston University
* John Sterman - MIT
* James H. Stock - Harvard University
Gerald J. Sussman - MIT
Edward O. Wilson - Harvard University
Shuguang Zhang - MIT

Session Chairs:

Dan Stein - University of Arizona
Jeffrey Robbins - Addison-Wesley
* Yaneer Bar-Yam - NECSI
Steve Lansing - University of Arizona
David Litster - MIT
* Irving Epstein - Brandeis University
Richard Bagley - Digital Equipment Corporation
Yasha Kresh - Drexel University
Tim Keitt - SUNY Stony Brook
* Les Kaufman - Boston University
Mark Bedau - Reed College
† Dan Braha - Ben-Gurion University
Dan Frey - MIT
Sean Rice - Yale University
Max Garzon - University of Memphis
Bob Savit - University of Michigan
* Larry Rudolph - MIT
Jerry Chandler - George Mason University
Richard Cohen - MIT
Kosta Tsipis - MIT
Walter Willinger - AT&T Bell Laboratories
† Helen Harte - NECSI
Farrell Jorgensen - Kaiser Permanente
Joel MacAuslan - Speech Technology and Applied Research
Anjeli Sastry - MIT
Walter Freeman - UC Berkeley
† Ali Minai - University of Cincinnati
Michael Jacobson - University of Georgia
William Fulkerson - Deere & Company
* Tom Petzinger - Wall Street Journal

* NECSI Co-faculty
† NECSI Affiliate

Subject areas: Unifying themes in complex systems

The themes are:

EMERGENCE, STRUCTURE AND FUNCTION: substructure, the relationship of component to collective behavior, the relationship of internal structure to external influence.

INFORMATICS: structuring, storing, accessing, and distributing information describing complex systems.

COMPLEXITY: characterizing the amount of information necessary to describe complex systems, and the dynamics of this information.

DYNAMICS: time series analysis and prediction, chaos, temporal correlations, the time scale of dynamic processes.

SELF-ORGANIZATION: pattern formation, evolution, development and adaptation.

The system categories are:

FUNDAMENTALS, PHYSICAL & CHEMICAL SYSTEMS: spatio-temporal patterns and chaos, fractals, dynamic scaling, non-equilibrium processes, hydrodynamics, glasses, non-linear chemical dynamics, complex fluids, molecular self-organization, information and computation in physical systems.

BIO-MOLECULAR & CELLULAR SYSTEMS: protein and DNA folding, bio-molecular informatics, membranes, cellular response and communication, genetic regulation, gene-cytoplasm interactions, development, cellular differentiation, primitive multicellular organisms, the immune system.

PHYSIOLOGICAL SYSTEMS: nervous system, neuro-muscular control, neural network models of brain, cognition, psychofunction, pattern recognition, man-machine interactions.

ORGANISMS AND POPULATIONS: population biology, ecosystems, ecology.

HUMAN SOCIAL AND ECONOMIC SYSTEMS: corporate and social structures, markets, the global economy, the Internet.

ENGINEERED SYSTEMS: product and product manufacturing, nano-technology, modified and hybrid biological organisms, computer based interactive systems, agents, artificial life, artificial intelligence, and robots.

Program:

Sunday, October 25

PEDAGOGICAL SESSION - Dan Stein session chair

Alfred Hubler - overview

Gerard Vichniac - cellular automata

Charlie Doering - convection, stability & turbulence

Geoffrey West - scaling

Cris Moore - computational complexity and statistical physics

Matt Wilson - neural systems

Fred Nijhout - patterns in development

RECEPTION SESSION - Jeffrey Robbins - session chair

Richard Wrangham - primate behavior

Monday, October 26

Yaneer Bar-Yam - welcome

EMERGENCE - Yaneer Bar-Yam - session chair

Philip Anderson - emergence, hierarchy, complexity

Stephen Kosslyn - the visual mind

Norman I. Badler - virtual humans

David Sloan Wilson - altruism in nature and human nature

DESCRIPTION AND MODELING - Steve Lansing - session chair

John Sterman - system dynamics

David Harel - statecharts

Richard Palmer - statistical strategies

Gerald Sussman - AI perspectives

FUNDING - David Litster - session chair

James Anderson - NIH (NIGMS)

Mike McCloskey - NSF

Tuesday, October 27

SELF-ORGANIZATION - Irving Epstein - session chair

David Campbell - physical systems

Leo Liu - development of C. elegans

Jerry Lettvin - vision

J. A. Scott Kelso - patterns in the brain

AFTERNOON PARALLEL BREAKOUT SESSIONS

Richard Bagley - quantum information and computation: session chair

David Divincenzo, Seth Lloyd, David Meyer, Tim Havel

Yasha Kresh - medical complexity: session chair

Jason Haugh, Steven Kleinstein/Jaswinder Pal Singh /Martin Weigert, Andre Van Der Kouwe/Richard C. Burgess/Jeffrey I. Frank, Ary Goldberger, Donald Coffey

Tim Keitt/Les Kaufman - ecology: session chairs

Nic Gotelli, Jim Drake/Craig Zimmermann, Jay Bancroft, Frencesco Santini, Jordi Bascompte, Brian Enquist/Geoffrey B. West/James H. Brown

Mark Bedau - artificial and natural evolution: session chair

Lynn Caporale, Dimitris Stassinopoulos, Hiroki Sayama, Wolfgang Banzhaf, Carlo Maley

EVENING PARALLEL BREAKOUT SESSIONS

Dan Braha/Dan Frey - modeling the design process: session chairs

Bill Birmingham, David Brown, Pradeep Khosla, James Rinderle, Seth Lloyd

Sean Rice - evolution: multi level selection and the structure of biological systems: session chair

Leticia Aviles, Charles Goodnight, David Sloan Wilson

Max Garzon - complex systems aspects of DNA computing: session chair

Toni Kozic, Laura Landweber, Erik Winfree, Natasha Jonoska/Max Garzon

Bob Savit - modeling social systems: session chair

Bennett Levitan/Stuart Kauffman/Jose Lobo/Richard Schuler, Marco Propato/James Uber/Ali Minai, Thad Brown, David Meyer, Josh Epstein

Wednesday, October 28

NETWORKS - **Larry Rudolph** - session chair

Eli Upfal - structure and analysis of networks

Kai Nagel - transportation network simulations

Walter Willinger - Internet traffic analysis

AFTERNOON PARALLEL SESSIONS - **Jerry Chandler** - session chair

Thread A

Temple Smith - Nature's legos

William Gelbart - genomes proteomes and phenomes

Irving Epstein - waves in inhomogeneous biological media

Yasha Kresh - heart-rate dynamic adaptation after transplantation

Steve Lansing - emergence of cooperation

Sean Rice - evolution of structure at different scales

Eric Beinhocker - complexity consulting

Thread B

Ernest Hartmann - sleep and dreaming

Terrence Deacon - consciousness

Michael Jacobson - learning the sciences of complexity

Mark Bedau - toward a zeroth law for evolving systems

Helen Harte - medical management

Susanne Kelly - evolving the enterprise

Juergen Kluever - modeling social differentiation

Thread C

Jim Stock - economic predictions

Robert Savit - adaptive competition, market efficiency & phase transitions

Philip Auerswald - industry dynamics with learning

Michael Caramanis - multi-scale manufacturing

Atlee Jackson - evolution of Nature's dynamics
Max Garzon - DNA computing
Edward Lowry - optimum data objects

BANQUET SESSION

Dean LeBaron - complex understandings of global economics today

Thursday, October 29

TIME SERIES - Richard Cohen - session chair

Doyne Farmer - market force, ecology and evolution

Andrew Lo - time series models in economics and finance

Jim Collins - biological time series

MANAGING COMPLEXITY

Harris Berman - medical management and Tufts Healthcare

AFTERNOON PARALLEL BREAKOUT SESSIONS

Kosta Tsipis - global systems: session chair

Frank Schweitzer, Manjula Waldron, Susan Aaron, Roberto Serra

Walter Willinger - networks: session chair

Bruce Sawhill, Daniel Coore, Duncan Watts, Sergey Buldyrev

Helen Harte/Farrell Jorgensen - health care management: session chairs

Mark Van Kooy, Martin Merry, William Braswell, Mark Smith, Jeff Goldstein

Joel MacAuslan - dynamics and complexity of physical systems: session chair

Hagai Arbell, Jerzy Maselko, Jay Palmer, Richard Squire/Mariusz H. Jakubowski/Ken Steiglitz, Martin Zwick

EVENING PARALLEL BREAKOUT SESSIONS

Anjeli Sastry - system dynamics: session chair

Chris White, Jim Hines/Jody House

Walter Freeman/Ali Minai - neural computation: session chairs

Steve Bressler, John Lisman, Randy Mcintosh, Mike Hasselmo, Jorge Jose, John Symons

Michael Jacobson - education in complex systems: session chair

Win Farrell, Marshall Clemens, Uri Wilensky/Walter Stroup/Kenneth Reisman, Kenneth Brecher, Jim Kaput

Michael Lissack - new concepts in human organizations: session chair

Jeho Lee, Kurt Richardson, Steve Maguire, Chang-Hyeon Choi/E. Andres Garcia, Vasant Honavar, Hernan Lopez-Garay/Jose Joaquin Contreras Garcia

Friday, October 30

AGENTS IN ACTION - **William Fulkerson** - session chair

Rob Axtell - emergence of multi-agent institutions and organizations

Abhijit Deshmukh - caution! agents-based systems in operation

Van Parunak - how to model a supply chain

Win Farrell (discussant)

COMPLEXITY AND MANAGEMENT (EXTENDED SESSION) - **Tom Petzinger** - session chair

Michael Hammer

William McKelvey

Henry Mintzberg

Laurence Prusak

Peter Senge

Ron Schultz (discussant)

Poster session

Iqbal Adjali, **D. Collings**, **A. A. Reeder**, **A. Karelas** and **M. H. Lyons** Analysing the behaviour of an evolutionary macro-economic model of innovation and imitation

Prateek S. Aggarwal A recursive stochastic learning model for efficient Bayesian estimation of discrete stochastic variables

Mary Ann Allison Relationships among communications media, transportation methods, and human group self organization

Jay Bancroft A model system and individual-based simulation for developing statistical techniques and hypotheses for population dynamics in fragmented populations

Wojciech Burkot Feeding the Schroedinger cat: On the power of structural approach in solving hard problems with quantum computers

Val Bykoski Computer-generated physical reality and its applications

Dmitriy Chistilin The self-organizing of USA economy system as the basis for leading position in the world economy

Marshall Clemens Working towards an integrated picture of CSS

Brian Clouse and **Manjula B. Waldron** Human performance technology from a complex adaptive systems view

Raja Laifa Clouse and **Manjula B. Waldron** The effect of history to achieve self adaptation

Ron Cottam, **Willy Ranson** and **Roger Vounckx** Diffuse rationality

Gavin Crooks Excursions in Statistical Dynamics

Robert Cutler Analyzing complexity in international relations: A theoretical synthesis of existing systemic approaches

Somalee Datta Gradient clogging in depth filtration

Alice Davidson, **Yaneer Bar-Yam** and **Martin H. Teicher** Measuring environmental complexity as related to the well-being of the elderly

Tanya de Araujo Conceptual models as emerging complex systems

Karl C. Diller ¡title unknown¿

John C. Gallagher Does evolution make sense? Finding order in a class of evolved continuous time recurrent neural network central pattern generators

Ben Goertzel Continuous learning in sparsely connected ANN's, with application to Hopfield nets and the Webmind Architecture

Paul Halpern Evolutionary algorithms on a self-organized dynamic lattice

Karthik Balakrishnan, **Rushi Bhatt** and **Vasant Honavar** Animal spatial learning: A computational model and behavioral experiments

Dirk Holste Optimization and redundancy of composition measures: Implications for recognizing coding DNA

Sui Huang and **Don Ingber** Regulation of gene activation patterns by cell shape

Stuart Kauffman, **Jose Lobo** and **William Macready** Optimal search on a technology landscape

Joel MacAuslan and **Juan Sanchez** The dimensionality and tangent space of experimental data

Ali Minai Synchronization in non-homogeneous arrays of chaotic maps

Paul W. Morris Organizational intelligence

Erik Rauch Pattern formation and functionality in swarms

Jonathan A. Rees A time-symmetric interaction calculus

Jeff Schank A new paradigm for investigating animal social systems

Rebecca Schulman and **Jeremy Zucker** Towards an understanding of the principles of pattern formation

Mark Smith, **Yaneer Bar-Yam** and **William Gelbart** Towards quantitative languages for complex systems

Vassilios Spyropoulos and **G. Papagounos** Some aspects of decision making in the complex system "hospital"

Christina Stoica SOZAIN - Social cellular automaton with an interactive network

Hideaki Suzuki One-dimensional unicellular creatures evolved with genetic algorithms

Yasuhiro Suzuki and **Hiroshi Tanaka** Degree of halting computation and complexity – Halting property and behavior of abstract chemical system

Andrew Tait, **Chris Casement** and **Fran Ackermann** Creating "real" knowledge in virtual communities

Wilson A. Truccolo-Filho Dynamics and information processing in a generic cortical area model

Burton Voorhees Complexity, complementarity, and consciousness

Shi-ching (Billy) Wu Chaos theory may provide an answer to the number of hexagrams in I-ching

Tomohiko Yamaguchi, **H. Kuriyama**, **D. Murray**, **T. Ohmori**, **T. Amemiya**, **T. Kanamori**, **M. Tanaka** and **Y. Mitsui** Respiration dynamics in yeast culture

Martin Zwick Structure and dynamics of elementary cellular automata

NOTE: Uri Wilensky provided an interactive demonstration throughout the conference.

Publications:

Proceedings:

Conference proceedings (this volume)
Video proceedings are available to be ordered through the New England Complex Systems Institute.

Journal articles:

Individual conference articles were published through the refereed on-line journal InterJournal and are available on-line (http://interjournal.org/) as manuscripts numbered 217-271.

Other products:

An active email discussion group has resulted from the conference. Access and archives are available through links from http://necsi.org/.

Web pages:

http://necsi.org/
Home page of the New England Complex Systems Institute with links to the conference pages.
http://necsi.org/html/iccs2.html
Second International Conference on Complex Systems (this volume).
http://necsi.org/events/iccs/iccs2program.html
Conference program.
http://necsi.org/html/iccs.html
First International Conference.
http://necsi.org/html/iccs3.html
Third International Conference.
http://interjournal.org/
InterJournal: refereed papers from the conference are published here.

Part I

Transcripts

Pedagogical Sessions

Dan Stein
University of Arizona
Charlie Doering
University of Michigan
Geoffrey West
Los Alamos National Laboratory
Chris Moore
Santa Fe Institute
Matt Wilson
MIT
Fred Nijhout
Duke University

Mr. Dan Stein: Yaneer has asked each of the session chairs to say a few words at the beginning of the session to capture the essence of the session. And so as I was flying here yesterday I tried to think of what that might be. And the first thing I came up with was sorry to make you all come on a Sunday; but let's look on the bright side. I think the bright side is that a few years ago we had very few conferences of this kind, and now there's an increasing number. And I think that is a clear sign of the increasing success of this field.

I think that complex systems has become a growing discipline, or at least a quasi-discipline. It's been fueled, in large part, by tremendous success, and problems that twenty, thirty years ago had long been considered intractable. Problems ranging from cognition to fluid turbulence, to the brain, to what have you. And the – rather than try to list these subjects, since everybody has their own favorite, maybe I should say instead that even though systems are very different, they do share a number of common traits. Typically they share traits like strong nonlinearities. Seemingly random or unpredictable behavior, and wildly fluctuating regions in space and/or time. And it's difficult to try to attack these problems using the traditional tools of mathematical analysis.

Now, the other aspect of these systems is that the problems themselves often cross – or transcend is maybe a better word – disciplinary boundaries. Problems that people deal with in this area are pattern formation in biology or chemistry, anything having to do with the brain, fluid turbulence, and so on. A lot of these also involve heavy computation. But the basic point is is that because these problems do cross disciplinary boundaries so often, and because the representation at a meeting like this is from many different areas, this conference is quite a bit different from a lot of the other ones that generally are on the area of complexity, in the sense that a lot of conferences tend to be more specialized. You know, you tend to have conferences on artificial life, or adaptive computation, or rugged landscape, or what have you. And this is really an attempt to try to cover, as much as possible, as many different areas of complexity. And that means that we're going to have participants from a very wide range of different fields. And that's the need that was felt for this pedagogical session. It was felt that it would be a good idea to have a number of speakers from different areas. Obviously, we can't cover every area that's going to be covered in the conference, but just to give people sort of a broad overview of the kinds of things that people look at in complexity, and to give an introduction to some of the main themes of the conference that you'll be attending over the following week.

1 Convection, Stability, and Turbulence

Mr. Charlie Doering: It's a pleasure to be here this morning, and I'm impressed by the turnout early this morning. I was expecting sort of a dozen hardy souls. More than that I see. I'm going to talk about a specific physical system which provides examples of some of the very specific and quantitatively measurable, and modelable examples of paradigms, several of them, that Alfred Hubler talked about this morning. The title says it all "Convection, Stability, and Turbulence". I will go backwards, and I will try to give you some indication of what I mean by turbulence, what I mean by stability, and what I mean by convection.

The outline of the talk, which I'll run through quickly, is I will tell you what it is, and I'll tell you why this is a very important problem from the point of view of both basic physics, and applied physics. Then I will talk about how one can mathematically model such systems, and I'll talk about the phenomenology. By that I mean I'll describe the kinds of phenomena that are observed in these systems. Both experimentally, and in simulations. Then I'll talk about some stability notions in convection, and talk about the connection between stability and turbulence, which is the self-contradiction part of the talk. Stability and turbulence really shouldn't have anything to do with one another. And then I'll talk about how we can use notions of stability, and make rigorous a hand-waving argument which sounds a whole lot like self- organized criticality, and get actual limits on how much heat can be transported by turbulent convection. And then I'll conclude.

Let me first say what is stability. I could do like Alfred, and make a list. Ask people what they think stability means. But I think we can agree that stability is the property of a system such that it's impervious to perturbations. You can kick it, or knock it, and it relaxes back to its previous state. Imperturbable might be a good synonym for that. So stability is something that we're all familiar with. Perhaps we would like a little bit more of it in our lives. But I think we might all agree that stability – at least in terms of dynamics – could be considered to be the opposite of turbulence. By turbulence I mean a very wild, chaotic motion, of a fluid in particular, on many space and time scales. And here are some snapshots from some experiments. I'll describe what these are later in the talk. But what these are are snapshots of an experimental system that is being stressed at a higher and higher rate. And in Alfred's language, I've got increasing amounts of flux of something going through the system. For this particular example it's a flux of momentum that's being forced through the system. And as the flux increases, you can see that the fine scaled structure increases here, and the system becomes more and more active on smaller and smaller time scales – I mean space scales there. And actually things go faster also. So that's what I mean by turbulence.

And convection. Now what do I mean by convection? Well, I'll spend a few minutes on this. By convection I'm talking about a buoyancy driven fluid flow that transports heat. So again the basic idea here is that we're taking a system, and we're forcing heat to flow through it. And we're going to see how the system responds to this kind of an imposed flux. And the basic idea is that heat rises. Of course heat only rises as long as gravity is pointing downward. But in most of our experiences, gravity does point downward.

Convection is what drives many of the phenomena that affect your daily life, and your life on many scales. For example, meteorology. It's convection. It's the sun shining down on the earth making the hot air near the surface, which then rises. And if you're anywhere near a lake or body of water, you'll almost always get air rushing in. Cool air rushing in over there as the hot air rises generates the wind, not to mention things like rain, and transporting other things besides just heat and momentum there.

Another application is in groundwater flows. Here's the surface of the earth, and here is some dirt, or some porous medium which we mentioned before this morning. And fluid can flow, can seep through a porous medium. And very far underground often there are sources of heat which heat up the water, which then tends to rise. And if there's some hot stuff rising, then since, generally speaking, these fluids are incompressible, or nearly so, then something cold has to drop, and you get induced motions there. And this is used for geothermal energy tapping, that kind of thing, for the transport of pollutants underground. Those are the things people think of.

Going from the land to the sea in oceanography. A lot of physical oceanography is exactly concerned with the transport of two major quantities, heat and salt, due to motions of the ocean. Here the situation is the sun shines on the surface of the ocean. It heats up the surface. You get some evaporation of the water, but the salt remains. So that generally speaking, if everything was static, you would have saltier water on the surface of the ocean. But it turns out salty water is heavier than fresh water. You know that because you float better in salty water than you do in fresh water. And so the salty heavy water tends to fall. So here you have a competition between warm stuff wanting to rise, and salty stuff wanting to fall. And you get all sorts of wild dynamics going on there.

On a larger scale, the earth itself is a big example of convection. Starting from the center we have a solid core there some place, but there's a liquid core. At least this is the standard model of the earth here. And that liquid core is actually made up of liquid metal, which is being heated from the center. As the heat flows out, there is some expansion, and some buoyancy forces which induce currents here. It's the motion of those conducting fluids down there which is presumed to be the source of the magnetic field of the earth.

And then going up another level, there is the mantle, which is rock. To us it seems like rock. It's the farthest thing there is from a fluid. But it actually – over geological time scales – it behaves just like a fluid. It's heated from below, the hot stuff is slightly less dense. It moves up. It has to push cold stuff down. And it's exactly this convection in the mantle which drives the plates on the surface. So this is the source of plate tectonics on the surface. What we're witnessing when we look at the continents, and continental drift, and so on on the surface is just the edge of a convection layer there.

Then there's at least one more scale up I can go, which is into astrophysics. And what one of the central challenges of astrophysics is to explain the structure of astronomical objects, like stars. And a star, in the first approximation, is a big ball of fluid. Well, it's a big ball of fluid with a lot of things going on. In the center, there's all sorts of nuclear reactions which are occurring, which are generating tremendous amounts of heat. That heat has to get out somehow in order to sustain some kind of relatively stable and steady structure there. One of the main mechanisms of the transport of heat is actually by the fluid of the star carrying it out to the surface, and then you get the large scale motions of the body of the stars going on simultaneously. This is complicated in astrophysics by the fact that most stars are rotating, so now you have complicated fluid motions with rotation, which adds extra what are called fictitious forces, but they have very real effects there. And there's also magnetic fields floating around.

So that this is a somewhat fancied up version of the problem. But in fact what astronomers think of, and astrophysicists think of very much is trying to take a very complicated phenomenon like a star, and break it down into its elements in order to try to understand one element at a time. That's the reductionist point of view. Now it turns out that for an astrophysicist, there's a quote here that sort of hits the nail on the head.

This is from a review in the Annual Reviews of Astronomy and Astrophysics some years ago by a well-known astronomer, Ed Spiegel. This is a whole review about convection in stars. He talks about the basic theoretical techniques of modern convection theory have been developed and tested on the basic problem of convection in a thin plain layer of fluid. So the point here is that the first step is to simplify it from some ball of rotating, conducting gas with magnetic fields, and all sorts of things going on, to "let's just consider a layer of fluid". Let's just look at the simplest possible system that may have some of the same physical mechanisms. And that's what I'm going to concentrate on here.

There's another comment here that addresses the relative complexity of the simplest looking phenomena to the whole system. Spiegel claims that however the astrophysicist, interested in a simple recipe for calculating stellar models, will be disappointed to find that instead the stellar structure calculation in these approaches constitutes a subroutine in the convection program. So that it's harder to figure out what's going on with convection than it is to put convection into some larger model, and figure out what the star is doing.

So there's been a lot of attention focused on the simplest set up here. I'll have some equations, but I won't do any calculations in public here today. But I'm going to define some quantities, and say what the variables are, and so on; but I'll stick with a lot of pictures.

When somebody talks about convection, like myself, or theorists, or to some extent experimentalists as best as they can, they're trying to consider just a fluid layer contained between two boundaries which just limit the motion of the fluid. So we don't want things spilling all over the place. And we're going to heat it from below. We can impose a boundary condition, and heat it from below. We're going to cool it from above. And the idea is that gravity is pulling down, and the hot stuff is slightly less dense, slightly more buoyant, and will tend to rise, perhaps.

Now the variables in this simplest possible scenario here are the temperature, as a function of position and time in the fluid, the velocity vector field of the fluid. So each point of the fluid has a velocity which depends upon where it is, and what time it is. And that will have components (U,V,W) and a pressure field. There will be a pressure field in there, which will depend upon space and time. And the simplest physics approximation, or the simplest approximation to the essence of the phenomena, which we can apply is to neglect everywhere the compressibility of the fluid, except in the one term where it counts, which is in the buoyancy force. And we're going to assume that the change of density, the fractional change of density of the fluid is just simply proportional to the deviation of the temperature from some reference temperature. So this is a linear approximation to the thermal expansion of the fluid there.

And then once one does that, there are the three partial differential equations which we think we know, and we think we understand. The first one is just the heat equation. It says that the rate of change with respect to time of the temperature of a little object is just a thermal diffusion coefficient times the Laplacian of the temperatures. So that's just (Fourier's) law of heat conduction. And I remind you that when we have a fluid around here that the derivatives of a material point are the partial derivative with respect to time, plus this convective term.

So what this says is that the rate of change of a little piece of fluid if I move with the fluid is just heat diffusion. But if I hold my focus fixed on a certain point – of course hot stuff and cold stuff can blow by – and that's what this term describes there.

Now the second equation is the physicist's favorite equation: F = MA. This is the density of the fluid, which is the M, and that's the acceleration of an element of fluid, that's the A. And that's equal to the force that's applied on that piece of fluid, which is minus the gradient of the pressure. Plus there's viscosity, so there's the different pieces of the fluid rubbing against one another, which essentially looks like a diffusion of the momentum, or the velocity. And then there's the buoyancy force. And this says that the change in density times the acceleration of gravity is giving me a force in the vertical direction. K is the vertical unit vector there.

And then the last one is incompressibility. It says that other than that little term there, we're going to assume that this fluid has to get out of its own way. It can concentrate anyplace. So all this says is that it's the free motion of material which rubs against itself, which tries to get out of its own way, and when it's hot, it wants to go up. The quantity of interest here – I mean, there's a million questions you could ask – but the quantity of interest is how much heat is transported. If we impose some temperature boundary conditions there. If we say how much the temperature drop is across the bottom and the top, how much flux of heat, or how much flux of energy is going through the system. Now of course if we have a bigger temperature drop, we expect a bigger flux of heat through the system. And one would like to predict that. One would like to have a quantitative relationship between the boundary conditions, the imposed temperature drop, and the heat current. Now we can identify the heat current by looking at the temperature equation. Writing it as a continuity equation. Multiplying by the specific heat of the material. Anyway, there's an explicit expression for the heat current, in terms of the solutions of the PDE, if we could solve them.

And let me just mention what some of these other parameters are. Kappa is the thermal diffusion coefficient. I want to make the point there's a lot of parameters here, but life does get a little simpler for us. There is the viscosity of the fluid, there is the average density of the fluid, there is the acceleration of gravity G, there is this reference temperature, and there is this alpha, which is the thermal expansion coefficient. So you can see this system, even in its simplest guise, already has a lot of parameters, and a lot of variables there.

Now it turns out that those parameters aren't at all all independent, that they can be reduced down to actually two numbers which completely characterize the system, and which completely characterize all of the possible dynamics of the system. And the way one does that is to use the time-honored method of going to non-dimensional variables, which is often a very good way to start analyzing any problem. If we measure length in units of the height of the gap, H. And we measure time in units of H squared over (kappa), which is the time it would take for heat to just diffuse across the gap of height H. And we measure temperature in a temperature scale, which is T hot minus T cold, then our domain and boundary conditions become very simple. We look at a gap of height one, and we have temperature one on the bottom, zero on the top. And our equations, when we do the rescaling in the variables, turn into something relatively simple. We have a heat equation with unit thermal diffusion coefficient. In the F = MA equation, when the smoke clears, we only have two parameters that show up. There's this parameter Sigma, and this parameter Rayleigh. Now the parameter Rayleigh is multiplying this buoyancy force term. So this is going to be a thing which tells us basically how strong we're stressing the system. How much we're trying to pump through the system. Sigma changes my time scale, the time scale of the fluid motion there. It's multiplying the time derivative there, and then the incompressibility condition. These two numbers. Prandtl number, Sigma, and the Rayleigh number are given in terms of the original variables as follows. Sigma is just the ratio of the viscosity to the density times the thermal diffusion coefficient. It only depends upon the fluid, it doesn't depend upon how big the system is. It doesn't depend upon the geometry, it doesn't depend upon how much temperature I put across it. For a given fluid, there's a given Prandtl number, and that's fixed for that system. The Rayleigh number, though, the term that appears right here is what we think of as the control parameter for the system. It's basically proportional to the applied temperature drop. So you should think of the temperature drop is like an applied voltage, and the heat current is as a current. And that's what we want to get out of the system. Now so it boils down from having six, or seven, or eight, or nine parameters down to having two parameters, one of which is a material parameter, one of which is a natural control parameter.

And the thing that we want to measure is how much heat flux there is in the system. How much current results from the applied stress? In non-dimensional units this is measured in terms of what's called the Nusselt number. And that's the total vertical heat flux on average. The total average vertical heat flux, divided by the purely conductive heat flux. The denominator here is just given by Fourier's law of heat conduction. It's the overall temperature gradient times the thermal diffusion coefficient. And we divide that into the long time average of the vertical component of the heat flux vector. And there's many ways of expressing that in terms of the variables in the system. But one of them is quite simply that the Nusselt number is one, plus the long time average of the vertical component of velocity times temperature. That's just an identity.

Now we've got a non-dimensional variable, the Nusselt number tells me how much heat is flowing through the system. That's going to be a function of the Rayleigh number, which is the applied stress, in this material parameter Sigma. And there's no further I can go by dimensional analysis, because now I'm asking for a dimensionless variable as a function of two other dimensionless variables. Now one has to look at physics there.

Well, it's exactly these variables that experimentalists look at. And here's a picture from a very influential paper from 1987 by a group that used to all be at Chicago, and I think none of whom is anymore. And they were doing laboratory experiments on convection, a fluid heated from below. And they were doing the experiments in gaseous helium at low temperature, at a few degrees Kelvin. The reason they were doing that was because they had very tight experimental control both over the details of the boundary conditions that they were imposing on the system, and also on the material parameters there. They're measuring the Nusselt number, that's how much heat is flowing through the system relative to how much we just conduct through, as a function of the Rayleigh number, which you can think of as the temperature drop across the system. And what they find are various regimes of behavior. At very low Rayleigh number they find that the Nusselt number is one. That means that the heat is just conducting, that the fluid is not flowing at all through the system. The fluid is not moving. And then there's a bifurcation at a critical value of the Rayleigh number. And there is some convection, which I'll draw you a picture of in a moment. Then they get all sorts of intermediate regimes; oscillations, chaos, and transition they call it. And then they go into a regime that they coined the name of soft turbulence for. And in that regime the Nusselt number goes like the Rayleigh number to the one third in this regime over something that was several orders of magnitude for their experiment. And then they went and crossed over to a new regime at a very high Rayleigh number here where the scaling changed from a one- third exponent to a two-sevenths exponent, which is not dramatic. One-third is .33, two-sevenths is .28. So experimentally they had to get good clean data to see this. And you can see the range of Rayleigh number. In this experiment they were able to vary their Rayleigh number over eight orders of magnitude, which is pretty good for an experimentalist. Nature does it all the time.

I just say up here there's another regime which was predicted theoretically back in the early sixties by a fellow named Robert Kraichman that predicted that the Nusselt number ought to go like the Rayleigh number to the one-half power; which this experiment did not see. But let me give you an idea what is going on in those different regimes of behavior. It really goes from the simplest to the most complex that one can have in fluid dynamics.

So what happens at a low Rayleigh number when I don't stress the system very hard – I'm just applying a temperature difference, but not too much of a temperature difference – is that because the fluid is constrained by these plates, the plates are holding the fluid at rest there. And if there's enough damping in the system, if the viscosity is large enough, or the thermal diffusion is large enough, then what happens is the fluid just wants to stay at rest, and it's just going to conduct fluid up there so that there is no fluid motion. And this state, this pure conduction state is stable, in every sense of the word. And there's various senses which I (won't) mention. In that case, the Nusselt number is one, as I said. And if I looked at the temperature profile across the layer going from top to bottom, I would get a function that looked like this. The temperature at the bottom is one, the temperature at the top is zero, and there's just a linear temperature profile across there. What happens above that critical value of the Rayleigh number, where the convection sets in? Well at first, you start to get rolls. You get convection rolls. You get regions where the hot stuff rises, and the cold stuff drops. And if you have a nice symmetric system, and very carefully prepared boundary conditions, and so on, you'll get a nice long range pattern of rolls. So you get a spontaneous breaking of the translation invariance in the system. The top system pure conduction state is translation invariant. And you get a spontaneous symmetry breaking into the convection state. The motion of the fluid is now helping to conduct the heat. The Nusselt number increases. You get a change in the temperature, the average temperature profile across the system.

And this kind of regime is where you see patterns, convection patterns, convection rolls. You can get oscillations of these things. You can have chaotic dynamics of these rolls. There's a whole industry in physics and mathematics that is concerned with what's going on right here in that dynamic regime.

When you push the system too hard, what happens eventually, the rolls break up, and you no longer have a spatial coherence. In these systems I think you lose temporal coherence first. You go chaotic temporally. But then eventually you're going to go chaotic spatially, where you don't see the structure of the rolls anymore. That's this soft turbulence regime. What happens then is you get a layer of hot fluid down at the bottom, a layer of cold fluid at the top, and occasionally a plume. A plume of hot stuff will break off the bottom, and float up to the top. A plume of cold stuff will break off the top, and go down to the bottom. They'll mix in, and it will start to look turbulent there. In this case, the observation is that the Nusselt number scales pretty much like a third experimentally. And I'll give you a theoretical argument why that's the case. And the temperature profile starts to look quite interesting. It looks like in the center of the layer, if I took a horizontal average, the fluid is about temperature one-half. It's about exactly halfway between the two. It's like a thermal short circuit in here that it's isothermal. It's got infinite conductivity there. But then all of the temperature drop occurs across these boundary layers at the top and the bottom.

What happens at an even higher Rayleigh number is there's that cross over to that new regime called hard turbulence. And there it's observed that spontaneously in any given experimental system you'll start to get a large scale flow, a flow in the cell which is as large as the cell is, however big your cell is. And what happens there is that's going to start to sweep these plumes. The way I've got this thing oriented it's blowing the hot plumes to the left here, and dragging them up here, and pulling the cold ones down here. Experimentally it's observed that the scaling changes. And there's a theoretical argument for why this number should do what it does, which I'll give you in a moment. And the temperature profile again, it just looks an even more extreme version of this thing. Like a thermal short circuit across the center, and the temperature drop across the boundaries there.

And then as the Rayleigh number increases further, the idea in Kraichman's 1962 model is that in this situation we have two kinds of boundary layers. There's a boundary layer for the fluid motion, and a boundary layer for the thermal structure there. And if the fluid boundary layer crosses over, and becomes thinner than the thermal boundary layer, then we'll get to this what's called the ultimate regime. Because that's presumed never to change furthermore. At least in the model. And where the scaling of one-half may appear.

Now what does all this have to do with stability? Well, let me first say something about a couple of ideas of stability as they're applied to convection theory. The simplest notion of stability that anybody with a technical background is familiar with is linear stability theory. Linear stability theory says we take a system at a fixed point, or at a stationary point, and we kick it only a little bit; and we look at the linearized evolution of the perturbation. And for convection the idea is that if there's a solution, a temperature field, velocity field, and a pressure field which is a stationary solution of the equations of motion, then any solution, of course, can be written in terms of the stationary solution, plus a perturbation. The perturbations get plugged into the equations. For linearized stability theory, you look at the equations of motion, and any terms which are higher than linear in the perturbation you drop.

And then you have linear equations for the perturbations. Usually they're complicated linear equations, but they're linear. And then you can do a spectral analysis. You look at the Eigenvalues of the operators that are in there, which means you're assuming an e to the minus lambda t kind of time dependence. And then what you find is that when you study what are the possible values of lambda for this system – of course I'm talking about, you know, several hundred years of mathematics condensed into three lines here! But the bottom line is that if there's any Eigenvalues in my notation which have a negative real part, then that means that some of those perturbations, if lambda's negative, then some of those perturbations will grow with time. That means an infinitesimal perturbation will grow until it becomes noninfinitesimal. We say the solution is unstable in that case.

For the pure conduction state in convection, it's a classical calculation. Which actually was set up by Rayleigh, but not solved until the Forties. That the critical value of the Rayleigh number is 1708. And it's interesting that it doesn't depend upon the Prandtl number.

Now linear stability theory only tells me if I'm stable to infinitesimal perturbations. Really the only information it gives me is a sufficient criterion for instability. It really does not tell me whether something is stable. It doesn't give me a sufficient condition for stability. Because it doesn't say if I kicked it maybe a little bit harder so that the linearization was not valid. It's silent about whether or not that's going to elicit a response. But there is another notion of stability theory, which generalizes to many other applications in physics. And the idea here is not to linearize the equations of motion. Try to find some criterion that comes directly from the evolution equations without any apologies for assumptions which may not hold. And the idea for convection, it turns out, is to identify a certain quantity. What one does for convection, in fact, is you look at the kinetic energy in the perturbation. V is the perturbation velocity field. You look at the total kinetic energy there, which is the integral of that over the space. Plus the integral of the square of the temperature perturbation. You add them up in this particular combination, and you look at the time evolution of that. Now, what you find from the equations of motion by doing the most fundamental kind of mathematical analysis – which is integration by parts, as we learned earlier, and which remains true – is that this is equal to just the minus (sum) in a quadratic form, which depends upon the perturbations, and depends parametrically on the solution I'm working with here. Now the observation is that this quadratic form, that emerges from the equations has a property that it may or may not be a positive quadratic form. A positive quadratic form is like a matrix that has positive Eigenvalues. This is something that can be checked mathematically. We can check whether the particular exact stationary solution generates a quadratic form that has this positivity property, or not. If it does have this positivity property, then that will tell me that any perturbation is always going to have a negative time derivative, or this functional of it. It's always going to decay. In fact, it will decay exponentially. That condition will give us an absolute stability condition for the system.

Now it turns out for convection, one can apply this condition to the pure conduction solution. And one finds that the pure conduction solution is absolutely stable for Rayleigh less than a critical Rayleigh number which turns out to be exactly the same as the critical Rayleigh number of linear stability. That gives us an absolutely complete understanding of this transition right here, which doesn't take us a lot of way into turbulence, but it tells us that prior to that critical Rayleigh number, the system is absolutely stable, and above that Rayleigh number, the conduction solution is unstable, so it must have bifurcated.

But what about stability in the turbulent regime? There's an argument that goes back to the fifties by a fellow named Will Malkus, who's at MIT, and Lou Howard helped quantify this in the sixties. I'll make it in the context of stability. And those of you in the know will say this sounds a lot like self-organized criticality. In fact, this is self-organized criticality pre-Per Bak. Now the idea is the following: in the turbulent state, it's observed experimentally that the temperature profile has this boundary layer structure that looks like a thermal short in the middle; and that's where all the fluid is flowing. And basically the fluid is flowing so well, it's conducting the heat as though it was a thermal short circuit there. But we have these boundary layers at the bottom and the top, where basically the fluid is at rest, because it's pinned there by the walls. And it turns out that if we could figure out how thick this boundary layer was, then I could compute the heat flux. It turns out to be just one half of the gap height divided by delta. Just that non-dimensional ratio. That's not hard to see, because basically it says that the heat is just conducting across here like pure conduction, because the fluid's not flowing. And whatever heat is flowing into or out of one layer has to flow in and out of another layer so that there's the same amount of heat flowing through each layer in a steady state.

Well, the hypothesis, or this heuristic marginal stability argument, is that the boundary layer adjusts itself to be exactly the correct thickness so that as a convection system unto itself, the boundary layer is just marginally stable. Now the argument is that the Rayleigh number defined for the variables in the boundary layer ought to be about equal to the critical Rayleigh number value, which is around seventeen hundred, or some number around a thousand or so. Now we can apply that quantitatively. If we write down the Rayleigh number for the boundary layer, we use the G row alpha – I don't expect you to remember the definition of the Rayleigh number – but those were in there. And we take the temperature drop across the boundary layer; which is half of the total temperature drop, so I have a half there. We multiply by the gap height cubed. Well, here the gap for the boundary layer is just delta cubed, divided by the viscosity, and the diffusion coefficient. Then if we multiply and divide by H cubed, and pull out the one half to rewrite this in terms of the real Rayleigh number, what happens is we find that this critical value –that's seventeen hundred, or a thousand, or so – must be equal to the product of the Rayleigh number times delta cubed over H cubed.

Now if I solve that for delta, and then convert that to the Nusselt number, then you can probably do the algebra in your head. I just invert this, take the cube root. Then it tells me that the Nusselt number ought to be some prefactor which is just a numerical value, which I could put numbers in there, and compute, times Rayleigh to the one-third. And this exactly gives the scaling that's observed in experiments there.

The onset of scaling tells us several things. Well, it tells us that this argument gives a quantitatively correct answer. But it also – whenever you see scaling – and maybe we'll hear about this more in the next talk, that usually means that some certain aspect of the problem is not relevant to the problem. In this case, it turns out that when we have this scaling, the Nusselt number goes like Rayleigh to the one-third. That the actual heat conduction in the layer now depends upon how thick the layer is. That now you could just consider the heat conduction into a semi-infinite layer. And if that was well-defined, this would be the scaling that would appear there.

Well, it turns out, then, that – for the other regime, this hard turbulent regime, one can generalize this argument by observing that with that extra shear flow that sets up – the large scale flow that sets up in a cell there, that shears these boundary layers. That changes their stability properties. And it's a very nontrivial calculation, which is not at all rigorous. But one can make the argument that the shearing of the boundary layers stabilizes them. Allows them to be a little thicker, which actually results in a slight decrease in the scaling from the one-third to the two-sevenths exponent. That says slightly less heat is transported in the hard turbulent regime than in the soft turbulent regime.

Finally, I guess I will quickly say that there is a rigorous way to impose this. This idea of marginal stability – which is one of these paradigms which has come up in the last few years by myself, and Peter Constantine, who's a mathematician at the University of Chicago – is to do analysis rather than making models. But the statement of the theorem is very simple. It says let's consider any temperature profile that you might want to consider for the system. It doesn't have to be a solution of the equations. And the statement is that if that temperature profile is nonlinearly stable as if it was a solution, which is a mathematical criterion that I can check. At a slightly rescaled Rayleigh number. At twice the actual Rayleigh number. Then the convective heat transport is rigorously not more than half of what it would be if that was a solution. So there's a formula.

So this says that there's a certain stability notion, which is this nonlinear stability notion, not linearized stability. And if you give me a profile, and it's nonlinearly stable, which is something we can check. Then we can plug it into this formula, and prove that the heat conduction is not bounded or is bounded by that amount.

What this theorem gives, at large Rayleigh numbers, is the fact that the Nusselt number can never exceed the Rayleigh to the one half scaling with specific pre-factors that we could compute. When you have the Nusselt going like Rayleigh to the one half scaling, it turns out that the physical heat conduction then becomes independent of the material parameters, the viscosity of the fluid, and the heat conduction of the fluid.

And then in this ultimate regime, this hard turbulent regime, you have this self-contradictory statement where you have heat conduction through a layer, but that would persist, even if there were no heat conduction in the fluid, which is pretty wild.

Well, it turns out that last year, November of 1997, there was an experimental group in France who was still working on gaseous helium. In fact, Castaing – he's the head of the group – was one of the experimentalists in Chicago ten years prior. And they pushed their experiment out to a higher Rayleigh number, and started to see the scaling with this ultimate regime. So it seems that nature does indeed try to saturate these things, and nature does indeed try to realize these scalings.

So there are many other problems that this can go onto, but the bottom line here is that this marginal stability notion, which was up on one of Alfred's slides this morning, is something which can be applied in a physical system to give quantitative predictions. It can also be implemented in mathematically rigorous ways to give rigorous limits on the behavior of complex systems. That's if we all agree that turbulence is an example of complexity.

Q: Are there are ways that these results could have more general applications?

Mr. Doering: Well, see, now that I'm in a mathematics department, I can't do that. But let me take off that hat, and just make some comments. This idea of self-organized criticality, or marginal stability, or self-realized marginal stability, which is something that has come up over and over again in physics. It's not original to Per Bak. You know, he didn't disagree with me after I showed him these examples. These are people in fluid dynamics in the fifties, and even prior to that, who were applying exactly this idea. That somehow an ignorant system – which is not even agents that are trying to agree with one another, or trying to compete with one another – but somehow it will drive itself to a limit of stability, or a boundary of stability. And these systems apply physical examples, and physical realizations of that.

There are examples where this is important. Where it has quantitative and computable implications which can be experimentally verified. And of course now, the physicist in me says well, you know, that must be true for other things if it's true for this. There is another problem for the modeling - to see where the model breaks down. For the modelers, or for physics. To see when it breaks down. It's believed, actually, that this simple model – even though there's other quantitative corrections which are going to come in, and maybe change prefactors, or change things a little bit. It's believed that this model holds even over ten, or twelve, or fifteen orders of magnitude. And there are many examples where it doesn't hold, and those are well understood, and well studied.

2 Scaling

Mr. Geoffrey West: Dan had asked me to talk about scaling, and I suppose like most of the speakers I don't know quite who the audience is. It's, I'm sure, a very mixed one. And so I put together a talk – bits and pieces of talks, actually, that I think people from many different backgrounds, and all levels of sophistication of their mathematics and physics I think might get a little bit out of. Which means that it may end up being a total disaster, but I've tried to make it sort of seamless, and tried to – part of the idea is to explain why some of the ideas that are associated with scaling actually work. Like the one Charlie referred to, the classic dimensional analysis. I want to talk a little bit about fractals, I want to talk about something called the renormalization group, and I want to show you those in some sort of unified picture.

I want to start out, first of all, at the most elementary level, and talk first of all in a slightly historical mode. Because the first person to think about scaling was the first person that started us on this great enterprise of mathematical physics, in a sense, and that was Galileo. Because Galileo was the first to think about what happens when you do the most primitive form of scaling. And that is, you take a system, and you just look at its geometrical expansion. You simply scale all lengths of the system by the same amount.

And the question he asked was why is there a limit to the height of buildings and trees? Why is it that you can't have infinitely tall buildings? And the answer he gave was a classic scaling argument. He said look, if you increase all the lengths similarly, the weight, obviously, increases like the length cubed. Whereas the strength of that pillar over there, for example, holding up this room, only increases like its cross-sectional area. The length of that pillar plays almost no role, it is just the cross-sectional area. And that only increases like a length squared. Therefore, the ratio of strength to weight supported that ratio decreases like one over length. Or put the other way round, the weight to be supported divided by the strength to support it increases like a length. And so ultimately, some length is big enough that the building would collapse under its own weight. This is a very elegant and simple argument, but exceedingly profound; because it says that there's limits to growth, and something has to change when you start so-called scaling. So either what has to change is the material – so therefore, at some sense the force laws governing this – or the architectural design. And I will come back to this at the end.

One nice little example might be to think of the concept of bridge. Where if you go to look at a little creek, you could cross it. The back of your garden you take a little stream, a little creek; you just put a piece of wood down. It's the simplest form of a bridge. But of course if you think of a little river, you have to actually build a structure. You could still build it out of wood. But eventually when the river gets big enough, you have to make it out of stone, or concrete, or something, and you have to maybe put an arch in it. To make it bigger, you have to put several arches. And then eventually you have to make it out of steel, and then ultimately you come to the Golden Gate Bridge, where you not only make it out of a wholly different new material, but you have a whole different design, you have a suspension bridge. And you've gone from a very simple system to a highly complex system. And what you see that has happened are two things. One is the concept of bridge-ness is still there that you can see in the piece of wood, and you can see it in the Golden Gate Bridge. And the Golden Gate Bridge is built out of little pieces of wood. It's built out of – I mean, pieces of steel, but they're the same. But you have implicitly, and explicitly changed the architecture, and the dynamics. So that was Galileo, and I summarized it just here. Galileo drew this picture of how bones have to change. I won't go through that.

And here's a picture I found in an anthropology book. My mother, who left school at age eleven, could tell me immediately that this animal is bigger than that animal, even though they're scaled to the same size. And she could also tell me, incidentally, that gravity points downwards from this picture. That is, if that picture were drawn that way, gravity would be pointing to the right. All from scaling.

Now here is the surface area of salamanders, versus their mass. To show you that Galileo's argument is right that it scales like mass to the two-thirds. The two from the area, the three from the weight, the volume. And here it is even more explicitly. Here is the strength of weight-lifters versus their body weight; Olympic weight-lifters. And it's plotted on a log log scale. And what you see are two things. One is that it's exactly the same as the area of salamanders. And also you see that it's very important to realize that my mother, who thinks that this is the strongest man in the world, is wrong from a physicist's viewpoint. This in fact is the weakest, because he lies below the normalization curve, which is this two-thirds law; and this guy here, who lies a little bit above, is actually the strongest.

Okay. So these are some elementary ideas in scaling. There are these very simple geometric ones. But life, which is the most complex of all complex systems, satisfies remarkably simple scaling laws. I mean, it's amazing it satisfies any scaling laws, but it satisfies remarkably simple ones. And here's what I consider the mother of all scaling laws. I mean, it's the greatest scaling law, I think, except for one other, which I will come to at the end, which is: how much energy do you need to stay alive, versus your weight? And we're here somewhere – because it's all mammals. It actually goes up to the whale, which sits there. But it begins with your cell, but here's your mitochondrion, which is where your energy comes from. But where your energy actually comes from is a bunch of molecules here. And they all scale on one curve, and they all scale with a slope that is not two-thirds, but is actually three-quarters. And the origin of that three-quarters is, we believe – we meaning myself, James Brown, and a student Brian Enquist – to do with the hierarchical branching networks that exist in your body. That's a whole other subject, which I will not discuss here.

The metabolic power was measured originally by just measuring the heat that came out. But the way they do it is mostly measuring the amount of oxygen you use. And at the cellular level you just simply measure the energy. And at the molecular level, which that went to, it's a chemical reaction. And that graph went over twenty-seven orders of magnitude. That's quite remarkable.

Now, there's lots of scaling laws in biology. I'll just show you a couple. That's your heart rate, which also has this curious one-quarter. That's heart rate versus mass. Your lifetime increases like mass to the one-quarter. Which means that the number of heartbeats in a lifetime is an invariant, is the same for all mammals. Which means that if you understood that number, which is about one point five billion. If you understood that number, you would understand something very profound about why many of us in this room are getting old, and decrepit, and are going to die within the next few years. That number determines that. And if you understood it, it would be extremely important. But you see it comes from very simple scaling.

So this is by way of introduction of just telling you a few things about scaling. Now let's look at some elementary ideas in an area that used to be called similitude, invented by Lord Rayleigh. That is usually called dimensional analysis, and it's a tool that every scientist either explicitly or implicitly uses in his or her work. And it was implied in the equations that were written down. And then you remember the reduction of parameters from eight or nine to two. And I want to talk about that, and in a slightly different way. And I want to do it in the following way, by way of a story.

There was a great hydrodynamicist, one of the great theoretical physicists, named G.I. Taylor, an Englishman who worked on the Manhattan Project at Los Alamos. And he went back to England after the war, and stopped working on weapons type things. But then the hydrogen bomb went off in whenever it was. 1950? I don't remember the dates. And he was in England, and he got Life Magazine. And in Life they published pictures of the bomb. But they didn't just publish one great big picture, they published lots of pictures with a time sequence. And Taylor, being a very smart man, realized immediately by looking at those pictures he could determine the energy yield of the bomb, which was a secret. And he calculated it from those pictures, and told people the number. In fact, he wrote a little paper. But he told people the number, and it created a huge stir, because there had been assumed that there had been some terrible leak, and he was investigated, and so on. So I'm going to now tell you what he did, how he calculated that. And it's quite elegant, and simple.

What he did was the following. He said there's the bomb, and he wrote down the variables of the problem, which were the radius of a shock wave after this thing went off at some time, T, E is the energy yield, and rho is the density of air against which the shock wave is pushing. Whatever the radius of this shock wave is, it's some function of these variables is what he says. And one of the points about this is having the right physical intuition to choose the right set of variables. So these are the important set of variables. And then the method that is used, which was invented by Rayleigh, is to say well, let's first assume this is some power law. And then what you do is you write down the dimensions of each of these objects on both sides of these equations. The dimension of radius is length, and density is mass per unit volume, and so on. You write them down, and then you just equate the exponents, and you get a bunch of simultaneous equations. I won't do a hell of a lot of mathematics. I'm just writing them down. It's trivial to solve. And you end up with this curious law that the radius depends upon T, the time, to the two-fifths; with an arbitrary coefficient outside. And Rayleigh knows the density of air. He knows R versus T, because Time had published the pictures with the time sequence. And so he can, if he assumes C is of order one, he can estimate what E is. And this is actually R versus T, and it in fact agrees. The way it's been drawn is a funny way, but it agrees precisely with this two-fifths power law. And he determined it, and got into deep doo-doo. So it's a very simple method. It's one we're all familiar with, and use.

I will go through one other example of dimensional analysis, because it's related to hydrodynamics, and it's the classic one that Rayleigh started the whole process with. And that's the drag force on a ship. I'm not going to drag you through this, I'm just going to tell you the result. I'll throw this up. But the idea is, the drag force depends upon the velocity of the ship, the length of the ship, the density of the ship, and the viscosity of water. For the moment I will assume this. And then if you go through the same thing, what you realize immediately is that there's a slightly different problem here. That is that there are now one, two, three, four of these exponents. And before, if I were to bring it back up, you would see that there were just three arbitrary exponents. But the way the method went was that you were assuming that there were basic three independent basic units like mass, length, and time. That's how one got the equations. And before there were three independent exponents with three fundamental units. So there was a unique solution. That was the two-fifth. Here there are now four, so there's no unique solution. And the way the solution comes out now is that there's some arbitrary power which, according to Rayleigh, one just raises to some arbitrary function. And it's a function of this combination of variables, so that the force is equal to the density times the velocity squared, times the length squared, times a function of this set of dimensionless variables; meaning that these don't change if you change scale. And this number is called the Reynolds number, and characterizes something about the flow. Now, there are many ways of thinking about this. But one nice way is that what this says is that no matter how you make the measurements. That is, if you were to plot, for example, the velocity versus length, normally you would get a whole bunch of different curves, like this. But if you plot the data in a special way, namely, if you were to plot it this ratio, which has no dimensions, versus this ratio, then you would get a universal curve. And that's what's been done here, but this is an ancient classic experiment of a little bowl falling through fluids. And so what it is, this is F divided by V squared L rho, and this is the Reynolds number. And this is a universal curve. Meaning that what has been done is that data on all different balls falling through all different liquids all fall on one curve – so you have a universal curve. So scaling has involved in it some implicit idea, which here is made explicit, that of universality.

The other aspect of it is that notice that something that could have been an arbitrary function of these things separately is a function of this combination. Which means that if you wanted to model this system of a real ship that satisfied only these variables. A real ship, which would have a very large L moving slowly, you could get exactly the same solution. You would see exactly the same as what's going on if you made a small model of it, but made it move very fast. But this is done in such a way that this Reynolds number was the same. So that's sort of the primitive idea of modeling in engineering.

So these are some examples. And now what I want to talk about is why in the hell does all this work? What is going on? Why can we use power laws? Why do we solve this set of simultaneous equations according to what Rayleigh told us to do? And so on. That's what I want to talk about now. So the first thing I want to do is prove to you that dimensions of physical objects have to be powers. Now everybody just takes them to be powers. Maybe they could have been inverse tangents, or exponentials, or whatever. Why are they powers? Why is it that when you measure units of things it's always something to the centimeters per second, or centimeters to the one half? Why isn't it log of centimeters, or whatever? So that's the first thing. But we have to do a little bit of mathematics. These dimensions should occur as power laws. The idea is very simple, as I said. So let's consider some parameter, some quantity that I will label little X, that is a function, or depends on certain independent units. Now normally you choose mass, length, and time. But of course there's nothing fundamental about mass, length, and time. It could be velocity, it could be Planck's constant. It could be, I don't know, your father's weight, whatever. It doesn't matter. It could be anything. So it could be any set. But you have to assume that they're somehow independent.

Now, the first thing you realize is that the very quantity that you're considering you could have chosen as a fundamental unit itself. There's nothing special about any particular unit. And in fact, what I'm going to now show you is that this theorem follows from the fact that there is no unit that is more fundamental than another. And how do you do that? This is the standard argument of scaling. That's why I want to go through it, and I will use it two or three more times in this talk.

What you do is you take the original unit, and you scale it by some scale factor lambda. So you take Xi, and you scale it by the quantity lambda i, you produce a new value. So you go from centimeters to inches, whatever. Now what happens, of course, the quantity that you're looking at, which depends on these various Xi. When I write Xi, I mean that I can go from one, two, three, four. And here I mean that X depends upon several of these Xs. X1, X2, X3. Whatever there are. So this changes by some scale factor. Of course the Xs change accordingly to here. You just shift each one of these. But I said that this original X itself could have been used as a fundamental unit. So there's two different ways of getting the same answer. I could either do what I did here. That is change, for example, mass, length, and time. Or I could have changed the dependent quantity X. I could have changed by some scale factor, which for various reasons I call Z. I label it with a sub X because it's connected to this X. But it of course depends upon the scale factors that I've used to make the scaling for each of the ones that I originally called the fundamental units. And any of these could be used. So here I've written down the two different ways of doing it. Either I scaled the individual units that I'd written this in terms of, or I make an overall scaling of this from here to here with some factor, Z, which depends upon the scaling that I've chosen. So that's the constraint on every physical quantity that we ever use.

Well, how do we exploit that? We exploit that by a trick first invented by Euler, which is that you differentiate this with respect to these arbitrary scale factors, and then set them equal to one. And you do it for each one of these i's. And it's very simple. If you do that, you see from here you'll have to know what dz, the lambda i is. You need to know how this changes the lambda i, and set it equal to one. But that's just some number, which I will call alpha i. And if you look at that equation, and you differentiate it, you end up with this very simple equation. I'm not going to bother you with details. But the answer is that you immediately determine that X has to have this power law behavior. And then you realize that this alpha i here, which we call the dimension of the physical object, is just this quantity, the Z, the lambda i which I had for this equation.

Now, so that proves that you have to have powers. Now I want to try to give you an indication as to why you can do what Lord Rayleigh did, and what we all do. Why that works, and why you get the sort of this set of simultaneous equations, and dimension analysis works from this more formal viewpoint. And the idea is to do pretty much what we just did, but do it for something, some function of various quantities. This is like we had before for the explosion. We had time, density, the energy, and something else. We have a series of dependent variables. And the idea is exactly the same.

We make a scale transformation on the so-called fundamental units which we choose; whatever they are. We just choose some, we then make a scale transformation. We know that each one of these responds in this way where these are powers. That we just proved. But now we want to know: how does this respond? Well, it responds by some scale factor which depends upon the scale factors used to make the transformation. That's what's been done here. So we end up with a very similar situation that we can go two ways for getting the answer if we make the scale transformation at the top here. That is, either we can change each one of the variables Xi. We can just make a simple transformation of each one of those where these are powers; or we can make an overall scaling on F itself. So it's very similar to the equation we had before where we proved that there must be powers. And this has to be true for each one of these lambda i's. And we do exactly the same as we did before. We simply use Euler's trick. And I'm not going to bore you with the details. I've written them there for those interested. But you go through, and you end up with a whole series of equations, which are of this form. Which are just a bunch of simultaneous equations. And if you look at them carefully, what you see is that if the number of independent variables – like mass, length, and time – is three, and you have three dependent variables only, there's a unique solution. Because the determinant of this object here, which is the analog of the exponents, just tells you how you differentiate; how these scale factors change with the arbitrary scale one has done. Then you get a unique solution. And in fact, if you look at this carefully, you see that it exactly reproduces the idea that Rayleigh introduced, but in a formal way, and shows you why you have to always have solutions of forms where you have dimensionless pieces.

So I've sort of skipped through that quickly. But I just want to give you a flavor of the technique. Because what I now want to do is change gears, and talk about fractals. And then I want to relate it back to this idea. Because fractals are often disconnected from this, even though the same word scaling are used. And I want to show you that connection.

So I want to spend a few minutes reviewing fractals. And many of you are familiar, some not so familiar. And I'm going to do it on a sort of standard text book way at the very beginning. And then I'm going to move into something that relates it back to this. And the way I would like to introduce it is to think of the idea of measurement. You take a yardstick, a ruler, a piece of wood, and you lay it down so many times across this room, and the length of this room is just defined by the number of times you lay this down. Now of course it doesn't quite fit. So what you do is the perfect thing. What you do is you take this, and you make it smaller, and you lay it down again. And of course you count up again the number of times it fits, but there's still a little bit left over. And you just keep shrinking it down. And there's a limiting process in which the length of this ruler gets smaller and smaller, and the number of steps that it goes in gets bigger and bigger. And of course as it gets smaller and smaller, this thing gets to where it effectively does fit. And that limiting process is such that the number of times that you lay it down times the length of the ruler is a constant. And that constant defines the length of this room. That's sort of the measuring procedure, and that thing converges quite quickly if you measure this room.

It converges even quite quickly if you go to something that's slightly more complicated. That is, you measure the circumference of a circle. That is, here you just lay it inside. You try to approximate by trying to fit the best sort of polygon you can inside the circle. It fits in N times, so a good approximation to the circumference is N times the length of the ruler. And then, say, here's what I did is I took the ruler that was only one third of the size. And this now fits in maybe three, almost three times as many. And you keep doing it. And if you go through that procedure, and I guess the Greeks were the first to do this, you get two pi times the radius of that circle. So what's plotted here is the length, versus the length of the ruler or yardstick. And it gradually converges – well quickly, actually – converges to two pi times the radius. So that's great. And that's how we think of measurement.

Even though I just said all those words, no one had actually carried out that experiment, as far as I know, until about 1960. It's amazing. From Greek times to now no one had actually done it. Everybody assumed that was the way you did things. And it was this wonderful man named Richardson, a geographer in England, who actually carried it out by measuring the coastline of Britain. The famous measurement of the coastline of Britain, which I will now go through with you. Except I will do a better example, which is the coastline of Norway, because it illustrates it more. And here's how he did it.

So what you do is you lay down a grid of a given size, and you start counting the number of boxes. You start at the top. One, two, three, four, five, six. That gives you a very gross course measurement. And then you halve the size of the boxes, and of course you get a better estimate, and so on. And you keep doing that. You do exactly what I just said, but the laying down of a ruler is replaced just by using a grid in some way. And he did this for the British coastline. And I think, to his amazement, so I am told, he got not what we saw here for a circle, but he got this. Namely, it started to diverge. And then when it was plotted on a log-log scale – this is not his data, this is for Norway – he got a nice power law. And it was a very simple relationship, in fact, but it's a power law. And that power law says, if you extrapolate it – what I showed you, that picture – that in fact, the length of the coastline of Norway is infinite (or of Britain, the original one). And that introduced the idea of fractals, because it was Benoit Mandelbrot who realized the profound implications of this, and realized that it had manifestations in many different aspects of nature.

So I want to talk a little bit about this, in terms of the things that I have said about scaling, and relate it back to it.

So, here is the reason, if you think about it for a moment, why this thing is diverging. Why? Because when you look at this measurement, you miss all the little fjords, and so on. As you start shrinking down, you start measuring. Imagine doing this thing with a ruler. You go down the coastline, and with a big ruler you miss a fjord. And then you make the ruler smaller, and you realize my God, I have to now go down the fjord. And I go down, and I come back, and I get a new measurement. And then I shrink the ruler more, and I go back through. And then I shrink it enough where a fjord coming off a fjord starts to be seen. You have to go down that. So I started, came down here, I have to go down, and I sort of see this, but I didn't see this. And then I come here, I saw this thing, and I saw this, but I didn't see this until I get a small enough resolution.

So, the fundamental moral of this story is: a measurement of distance in general, including the width of this room, is a meaningless statement unless you tell me the scale of the ruler you used to measure it with. It is absolutely meaningless, from a conservative viewpoint, to say the length of this room is whatever it is. Fifty feet. That's meaningless, unless you told me the scale of that ruler was a hundred feet, or whatever, or an inch, or whatever. You have to give me a scale. So that's the critical thing. And I wasn't going to show this, but I will since you raised it. The textbook way of explaining it is with something called a Koch Curve. It just gives you an idea of what's going on, if you didn't get that picture. That is, you take a line of length D, and you raise an isosceles triangle of one-third of the length. So each one of these are equal. One-third. And then you just keep repeating. You raise another one here of one-third, another one here, one here. That produces this. And then you keep doing it. And what you see is that you have this so-called self-similarity as you go down and down, and this thing is supposed to go on ad infinitum. And what you also see is that each piece of it is a contracted version of this sort of fundamental element. So that's sort of what's going on.

When it comes to measurement and scales, the statement is simply that you need – except for most manmade objects, most objects in nature you need to specify the scale associated with the measurement. That's the lesson I want to leave you with, because what I want to finish this up with.

And what I want to talk about now, briefly, is how to formalize that. I'd like to spend a few minutes just relating this to the dimensional analysis of scaling, the simple scaling. Even though I erected this what looks maybe a little bit complicated apparatus to show why this hundred year old method works that we all use. But I now want to relate this to that in a similar way. So here's the idea. Here's some curve. Here's some curve, and this one, the black one; and one now does this thing of laying down the ruler. These are supposed to be equal length, Lo. And the curve goes from this point to this point. And the distance between these is like the curve you're trying to get. That is, that I've measured its length S, where the ruler is commensurate with this size, and doesn't have this problem of either the atoms, or the nature of the crinkliness of the wall, or whatever.

Now, if there were no other scale in the problem. If there weren't a ruler scale, if there weren't this, then since there's only one scale in this problem, namely this distance, it must be that L is equal to some number times S. In fact, that's true in general. But if there is no other scale in the problem, Alpha, which is dimensionless, must be just some number, and that's it. It can't depend upon anything. It's just some number defining how many S's there are in L, that's all. So there's a very simple relationship between L and S, and that doesn't have the character that we saw when Richardson made his measurements, and this idea of fractals, which has funny powers. And so inevitably it means that since Alpha is dimensionless, there has to be another dimension for quantity. Another quantity with size of length in the problem, which we can identify with the ruler that is being used to make the measurement, or the scale at which you are talking about the measurement of that curve. So therefore, in that case, what I should write – this is the most general case – is that you should think of this L not just as a function of this simple distance, I call it S, but as a function also of the ruler length. In which case, this proportionality constant is now actually a function of S over Lo. It's dimensionless, and it must be a function of dimensionless quantities. And one could rewrite this as the number of times you use Lo to make the measurement. But this number, the number of times you do it, depends in some way upon S over Lo.

Now, if you do what Galileo talked about, which is you scale all lengths equally. You just multiply all lengths by the same quantity, lambda. Then clearly if I chose to change S by a factor lambda, and L naught by a factor lambda this doesn't change. So since Lo is changing by factor lambda, L just changes by a factor lambda. And so it's very simple. It's just like Galileo said. But there's now a different possibility for scaling. A different possibility, which is that we take this curve.

So what Galileo was doing, and the scaling we were thinking of, was scaling all lengths. So this thing just sort of scaled isometrically. But now we can imagine keeping this fixed, and doing the thing implicit in what Richardson and Mandelbrot were doing. That is, ask what happens if I change the length of the ruler. So that's a different kind of scaling. And I've written that down here in this hieroglyphics down here. But I'm going to just give you the answer as to what happens when you ask that. And you get what is in some ways the most general statement about scaling. I've actually kept the length of the ruler fixed, and I've scaled the curve. But it could be either way round. And what you find is that there is a very similar equation to what we had before. That this thing just scales by a scale factor that depends upon this scale parameter, say, the length scale, any particular length scale in the problem. And also this ratio of S over Lo.

Now, we just follow our nose as before. Differentiate this equation with respect to lambda, and set it equal to one. Do what Euler said. And if you do that, you end up with a generalization of the equations that gave you all of dimensional analysis. The generalization is extremely simple. It is that when you differentiate with respect to lambda, and set it equal to one, you have to know how this changes with lambda. Here it is here: dz / d lambda. And before, this was not there. And that was just a constant. Now it is not a constant, it depends upon this ratio, and you have this equation, which I've written down here, and this quantity gamma I will call the generalized fractal or anomalous dimension. And it's given by just this quantity. And you can actually solve this equation. And I've written down the general solution there. It's a simple differential equation to solve; and there it is.

I want to talk about one special case of this. And that is the special case when you do this scaling, when S over Lo does not occur in this. That is the Z here, the Z depends only upon lambda, and not upon this ratio. So in the picture, the scaling that one is doing is keeping the shape the same. The ratio of S to L naught is kept the same. So in some sense the shape is kept the same as you scale. And if you do that, this gamma is now just a number. And what you find is that you get a power law. And so in that case it defines something called self-similarity, and it gives rise to a power law, which is the fractal.

Now I want to finish up by connecting it briefly to something called the renormalization group, and the evolution of the universe. This is not a theory of everything, but a lot. But based on the same ideas. That's why I went through all this junk. Whether you know quantum mechanics or not is not going to be crucial to this, but whether you know it or not, you all know that the most fundamental statement about quantum mechanics is the uncertainty principle of Heisenberg. And the uncertainty principle says the following: when you make a theory of measurement, and you have a resolution of a given time, say with a certain time resolution. If it's a very narrow resolution in time, it's a very big resolution in energy. So there's a lot of energy available during that time resolution for all sorts of processes to happen that can violate things like the conservation of energy. They're allowed to violate it within this time resolution. And that's the uncertainty principle.

And what that says, specifically, this idea that the finer resolution that you have, the more energy is available for things to happen that would not normally be allowed to happen. That gives rise to the following phenomena. That is, that if you think of that as, say, an electron; a nice, well-defined electron moving along, that's a coarse resolution that you look at sort of almost macroscopically. But if you go to a fine resolution, and you took a picture of it. Imagine taking a photograph, what you actually would see at a finer resolution is an electron with a photon. That's a photon, a piece of radiation associated with it. And if you go to a finer resolution there may be two pieces of radiation, or maybe three as you go finer, and finer. This is getting finer and finer time resolution allowing more and more energy to produce more and more radiation until you get to situations like this, which you see have this sort of branching, hierarchical network structure to them.

And incidentally, the energy here is so powerful that it can produce antiparticles in this small period of time. But what you're seeing is that the very nature of an electron is scale-dependent. The very nature of something we think is very well-defined, and so on, is dependent upon the resolution of time. That means that all things are dependent on the resolution, from the most elementary level. And if you have a theory, you can calculate those.

So in particular, you can calculate how the nature of that electron, or any elementary particle, changes with the scale you use to look at it with. And the equations, in a certain sense now not surprisingly, the equations that describe that have identical character to the fractal equations, and to the equations of dimension analysis. That is the equation. That is what are the deviations from a simple, elementary particle are given by this where this is the momentum of the particle, and this is the strength of interaction of the particle, and this quantity mu is just some arbitrary scale used to tell you what the strength of the interaction is, and so on.

But the important point of this equation – you don't have to understand its details – is that it has exactly the same character. That is, if you make a change of scale, there is an overall factor, like the ones we've talked about, like we had in the ordinary dimension analysis, and we had in fractals times the original. This statement is a highly nontrivial statement in quantum field theory. This is what really, in a sense, got Feynman and Shringer, and Tomonaga the Nobel Prize.

But you can use that, and you do exactly the same trick of differentiating that equation with respect to that arbitrary scale change. Setting that lambda equal to one. That equation is the renormalization group equation, and its solutions have the same characteristics as the equations for the fractal solution.

And why is it connected to the evolution of the universe? It's because it tells you how the character of elementary particles, and their interactions change with scale from the smallest scales to the largest scales. And that's what happens in the expansion of the universe. Because as the universe expands, it cools. And so the average temperature available for the formation of any object is decreasing. So that means energy is going down as time increases. And that just traces out a scale change as you go. And that is given. That scale change is governed by precisely these same equations.

So to summarize quickly, in some ways it's sort of remarkable that from the very microscopic to the very origin of the universe, the same sorts of pictures emerge. This sort of hierarchical set of pictures that have implicit scaling with the same formalism, exactly the same formalism all come all the way through even to determine how you sustain yourself through, say, the hierarchical structure of your cardiovascular system, respiratory system. They're identical in their structure to the evolution of the universe. It's all scaling.

3 Computational Complexity and Statistical Physics

Chris Moore: This is going to be half a pedagogical session, and half a presentation of some work. I know it's supposed to be mainly a pedagogical session, so I'm trying to figure out how to fulfill the obligations of both the abstract, and the claim that it's pedagogical; so I'll try to do that.

My degree is in physics, but I've become increasingly interested in using ideas from computer science as a way of talking about complex systems. I think of all the definitions of complexity, I think that the approach of computational complexity is, in my opinion, the best – the most rigorous, and best – defined. That doesn't mean it has relevance as broad as we might want for a definition of complexity, though. Maybe that's a tradeoff that we just have to accept. That the more rigorous and well-defined the mathematical idea is, that may narrow down the set of things that it's relevant to. But for some of the things that I work on, computational complexity theory works very well.

Some of the complexity classes that probably a lot of you have already heard of are NP, where N stands for non-deterministic and P stands for polynomial time. So this is the set of things, like the traveling salesman problem, where you might not be able to find the solution in a polynomial amount of time. But once you guess a solution, you can check to see if it is the solution in a polynomial amount of time. That being what nondeterministic means in this case. Then there are the things that we can do in deterministic polynomial time. So for instance, in graph theory, if you have a bunch of points, and they're connected by lines, and the lines have numbers on them indicating their length, finding the shortest path between two points is very easy, and can be done in polynomial time. Finding a path that visits all of them of the shortest length is much harder, we believe. And that is in NP.

I'm interested in some subsets of P which are less well-known outside the computer science community. One of them is NC. N here does not stand for nondeterministic; it stands for Nick, because this was a class defined and studied by Nick Pittinger. And NC is the subset of P, of problems which can be efficiently parallelized. So, to use this big O notation, P is the set of things that can be done in order of N to the k time for sum K. NC is the set of things that can be done in order log to the k of N, by which I mean log N to some power. Parallel time with N to some other constant, a polynomial number of processors.

So the idea here is that suppose you have something which can be solved in polynomial time on a serial computer. So there's a polynomial number of computation steps that you can solve it with. The question is if you had many, many processors, could you do a lot of those steps at once, or is the problem inherently sequential? Meaning that you have to do the steps in order, because, for instance, you might not know what thing you need to try in step number two until you first find out the outcome of step number one. That would be a problem which would be harder to parallelize. So NC is the subset of P for which you can gain a lot by parallelization.

Now our model of parallel computation here is unfortunately not as physical as it might be. We're assuming that this polynomial number of processors can speak to each other instantly. We're not paying attention to communication delays, routing problems, or any of the other issues that arise when people actually try build highly parallel computers. But it still might be a good theoretical class to think about, because it allows us to de-emphasize the communication questions, and emphasize the fundamental question of parallelization. Can we do a lot of the work at once? If you think about the processors as nodes in a circuit with time proceeding downward, then parallel time can also be thought of as the depth of the circuit. So that each layer of the circuit corresponds to an additional step, and then the polynomial number of processors corresponds to a polynomial width of the circuit.

So another way to put this is this is the set of problems which can be solved by circuits of log or polylogarithmic depth. Log into some power with polynomial width. Now there are a variety of subsets below that of even smaller kinds of circuits. First of all, we can ask what is the power of the logarithm here? And a lot of interesting problems are down at NC two, or NC one, namely log, or log squared depth circuits. And one interesting point that Christos Papadymetrio makes in his excellent 1994 book on computational complexity from Addison Wellesley, which is very readable and will help bring you up to date. If the only computational theory text you have on your shelf is Hopcroft and Olman's old book, this will help bring you up to date on a lot of more recent developments. He points out that as actual computer technology has gotten more powerful, we've actually gotten more conservative in our definition of what kind of computation is feasible.

So that back in the 1950s, people were making distinctions between whether you have exponential time, or exponential of an exponential, and so on. How many layers of exponentials you had. Whereas then in the 1970s people said no, the real question is: can you do it in polynomial time, or not? Now people are saying well, N to the tenth is a polynomial, but it's still a pain in the butt. And so perhaps it's because we're interested in solving larger problems. The N here is meant to be the size of the input, the number of cities in the traveling salesman problem, or the number of variables in a Boolean satisfy-ibility problem, or whatever. And then as computers improve, we care about bigger N, because we're able to do bigger N. And as we care about bigger N, we want a function that increases more slowly.

Papadymetrio, for instance, argues that NC two is kind of a good definition of feasible computation now. And that's a more conservative definition than P.

Down here there are all kinds of fun classes, even dealing with circuits whose depth is actually a constant, but where we allow some fancier kinds of gates. Gates that combine an arbitrary number of things into an AND or an OR gate at once. And there are some beautiful algebraic theorems down here. And down here is one of the few places in this entire hierarchy where we can actually say that one class is smaller than another; that one computational class is less powerful than another. As you all probably know, everyone believes that P is strictly contained in NP, but no one has been able to prove that. We all feel intuitively that checking whether a solution is right is a lot easier than finding a solution in an exponentially large space.

But no one has been able to prove that. In the same way, it is believed that NC is strictly contained in P. It's believed that there are problems which can be solved in polynomial time, which cannot be efficiently parallelized. But again, no one has been able to prove this. Then it's also nice to remember, if you're interested in sort of learning about computational complexity, that all of this is only the beginning of various infinite hierarchies. NP is like adding a single existential quantifier, as the logicians say. A there exists to P. So that NP is sort of P where you're making guesses about one object. Is there a path that visits all the cities, and that makes the salesmen happy within a certain length? And if there is, I'll guess it, and then check for it. We can add additional quantifiers, and look at properties of, you know, saying like for all X there exists a Y such that this property is true, or there exists an X such that for all Y there exists a Z, and so on. Those things all add up into something called the polynomial hierarchy, sometimes called PH. That is contained in P space, which is the set of things which can be solved with a polynomial amount of memory, or space in the computer. That is contained in things that can be done with exponential time, which is contained in the things that can be done in exponential space, and so on. All of these are contained in things which have some sensible bound on them, on their time or space, in terms of their elementary functions. This thing up here is called elementary. And then this is contained in the recursive things. And then, of course, above the recursive, or computable things, we have

domain boundaries diffuse around in a pretty random-seeming way. For people interested in actually proving things like this, this has only been proven for up to two particles; domain boundaries. It's been proved that for two that they move in an uncorrelated way. But it seems from a numerical experiment that you can actually have a lot of them moving around in an uncorrelated way.

Now when two of these meet, like this, they annihilate, and the domain in between the two of them go away. And so over time, the number of these particles goes down to zero, at a certain rate in the system; roughly one over the square root of T, because they move in a random walk.

The only reason I'm putting this up is that this is the most interesting cellular automaton I've found where you can actually predict it on a highly parallel computer much more quickly than by simulating it. And the way you do that is first ignore the color. From a coarse grain point of view, the whole system is just (addition mod two). And a cellular automaton, which is just addition mod two, is very easy to predict quickly, because it obeys a nice scaling rule. If this is A, B, C, this is A plus B, B plus C. Then since B plus B is zero, mod two, this is A plus C. So doing the rule for two time steps is equivalent to doing it once on a sort of dilated set of initial conditions. The same is true for all powers of two. And that means that if here are your initial conditions, and you want to know this state here, you can jump back by the biggest power of two up here, and then by the next biggest, and so on, until you get back to your initial conditions. And in the worst case, where the time here is a power of two minus one, the depth of this circuit is just log of T.

So that you can do very quickly, and on a parallel computer you can actually find the zeroness or oneness, again we're ignoring the color, of every state below our initial conditions for as far out as we like, quickly. Once we know the zeroness or oneness, we just use the fact that the boundaries travel. But whether or not the boundaries get to you, you get to a particular site here. And which domain it maps back onto in the initial conditions becomes a kind of problem called reachability. And reachability is just where I give you a map of countries, and I tell you which countries are bordered on which other ones. And then I ask you: can you get here from there – as they say in New England – or there from here? And the answer is yes or no. That problem can be solved very quickly by a parallel machine.

So that's an example of a CA where you can do it quickly. But note that these particles, by diffusing around, and annihilating in pairs, they don't get to carry very much information around. They don't get to do that much computation. You certainly couldn't use this cellular automaton to build a computer; because as soon as two of these particles touch, they both disappear. So it's not as if you can get them to carry information around from place to place, and interact, and then carry the result of that interaction somewhere else. Because that's what you would need to build a computer.

So Jonathan Matta, who spoke at the satellite meeting yesterday on statistical mechanics, was doing some similar work, and so I'm going to describe some of that. But I want to point out one other thing before I do that, which is that this whole question of whether we can simulate something quickly is a special case of a question about when a logical circuit is equivalent to a much shallower logical circuit. Because after all, if we have our sites of our cellular automaton here, you can think of them being combined as we go forward into time. We have some set of wires which carry what these states are with truth values, with bits of information. Then those wires are combined with some circuit of AND and OR gates, and then the output comes out here, and this happens again here.

And so we can think of the process of simulation as going through a circuit whose depth is equal to the number of time steps, and which has this nice simple periodic structure. Now in general, there is the problem of circuit value, and this is the canonical P-complete problem. Just as there are NP complete problems, like the traveling salesman problem, such that if any one of them could be done in polynomial time, then in fact, every other problem in NP also could be. There are also P complete problems. The P complete problems are the hardest problems in the class P. And those are the ones that most computer scientists believe, or probably all, cannot be parallelized. Those problems are the ones that are believed to be inherently sequential. You have to do your work step by step. Even if you have a lot of processors, you don't gain that much.

So the circuit value problem, we take our inputs, and I give you some circuit with AND gates, and OR gates, and maybe NOT gates. And I give you the truth values here, true, false, true; and I ask you: is the output true or false? So like the problem of simulating our cellular automaton, we can't always just go through the circuit level by level until we get to the bottom, and then you can tell me the answer. The question of whether P equals NC is equivalent to the question of can I always take a logical circuit, of any depth, and come up with an equivalent circuit, which does exactly the same thing, but which is far shallower? In fact exponentially shallower, because I'm going from a polynomial to a logarithm, or a power of the logarithm. And no one believes this, but no one has been able to show it can't be done. I think the intuition is that since the levels of a logical circuit can affect the truth values on the next level in an almost arbitrary way, how can you do any better than go through it step by step? How could you possibly skip over a lot of it, or do a lot of levels at once? But again, no one has been able to prove this. And this circuit I've showed you here is a circuit of logarithmic depth. Because each time I add a level, I get twice as much time that I'm predicting it for. So this is a case where there is an equivalent circuit which is shallower.

Now so the problem is that in computer science we can hardly ever prove that something can't be done. There are some specific cases where for specific reasons we can do that. But in general, lower bounds are very hard to prove. Usually the best we can do is say that well, I've proved that this problem is NP complete, or I've proved that this problem is P complete. And that means that unless everybody is wrong that searching is harder than checking, and P and NP are actually the same. Then, unless that's wrong, then there's no polynomial time algorithm. In the same way, if I prove something is P complete, then unless P and NC are the same, and it turns out that everything is parallelizeable, which would really change the world views of a lot of computer scientists, then it cannot be efficiently parallelized.

The way to show that a problem is P complete is to show that another P complete problem can be mapped onto it; or as the parlance goes, reduced to it. If one problem can be reduced to another, if problem A can be reduced to problem B, then that means that problem B is at least as hard as problem A. But the reason that is is that if you can reduce problem A to problem B, if you had a fast trick for solving problem B, you could use that as a fast trick for solving problem A, by first turning the problem A into the problem B, and then solving it there. And then taking your answer back.

So let me show one way to prove how a system is P complete. One system that I'm interested in, that lots of people are interested in is the Ising model of magnetism. In the Ising model of magnetism, you have a lattice of atoms. And either because they're quantum objects, or just to make things easy on ourselves, each one is a little magnet which only points up or down. And then whenever two neighbors point in the same direction they're happy, their energy is lower. Whenever they point in opposite directions they're unhappy, and their energy is higher. And the system wants to find a low energy state. If it is at some non-zero temperature, then there is a chance, according to a formula of Boltzman, that we will make transitions that make things temporarily worse, that increase the energy. Like flipping an atom up even though all of its neighbors are down. That would increase the energy in a square lattice by four, because it has four neighbors that it's now disagreeing with.

Now we've gone from four happy neighborhood relationships to four unhappy ones. And then in a cubic lattice it would be six, or whatever. But at zero temperature, we would presumably only make moves that keep the energy the same, or that reduce it – in other words, that make the system happier. We're not willing to make things temporarily worse in the hope that they'll get better later. So that means that the atom will do things like flip up if a majority of its neighbors are up. And if its neighbors are tied so there are an equal number up and down, then it might flip up or down with probability a half.

So this is kind of like a cellular automaton. There's a cellular automaton you can look at where you do a majority vote. You take, on a square lattice, yourself, and your four neighbors. That's a five-site neighborhood. Or on a cubic lattice, yourself and your six neighbors: up, down, left, right, and in front and behind of you. And then that and you make seven, and then you take a majority vote to see who's up or down. And then whatever the majority is, you conform to the majority, and become that. This is fairly close to the Ising model at zero temperature. So let me prove to you that, unless literally hundreds of computer scientists are wrong (since I don't know how many computer scientists there are in the world), this system cannot be parallelized quickly. And one way to do that, or pretty much the only way we have to do that, is to show that it's possible to build gadgets into this system which actually implement arbitrary logical gates.

So here are our gates. This thing here is a wire, or sort of a fuse, because it can only be used once. Each of these blocks which is occupied. You can say up is occupied, and down is unoccupied. Each one is happy, because this one has itself, and three of its neighbors. And that's four out of seven, and that's a majority. The things in the trough have just three neighbors, which is one short of a majority. So they will stay unoccupied if not otherwise disturbed. However, if we fill this block in, then this one will now have a majority of occupied neighbors, and it will fill in; and then this one will fill in, and so on. So they will carry these things around. They can't get unfilled later with this particular shape, but they will fill in. So if getting filled in is true, and not getting filled in is false, then this here is an AND gate. So what happens here is that if we fill in the bottom wire, it will come up to the center, and then stop, because the central site in the center there only has two occupied neighbors, which is two below the threshold. But if we now fill in the top one also, it will come down, and then fill out the one on the side. So if we turn any two of the inputs on, the third one will turn on.

This is an OR gate, and it took me some time to find it. It's hard to see, but there's a notch taken out the back. Here the central site has three neighbors. So if we fill in any one of the three inputs, then the other three will get filled in. So this is an OR gate. This is an elbow because you need to be able to bend wires around, and plug them into other things. This is another kind of elbow, in case your wires are entered that way.

So what this means is that if you give me any Boolean circuit of AND's and OR's, I can construct for you an initial condition in the cellular automaton such that a particular site will turn on if, and only if, the output of the circuit is true. I can do the same for the Ising model I can give you. We'll set the atoms up just this way, and then there will be a non-zero chance of this atom flipping on, if, and only if, the circuit has a true output. And that means that this problem is P complete, and that means that unless P equals NC there is no way to predict this that much faster than by simulating it. There's a caveat there. There might be some way to gain a power law or a polynomial improvement. You just have to simulate it step by step. But you probably won't be able to parallelize it to logarithmic depth.

M: You divide your lattice into two separate lattices.

Mr. Moore: This will still work. From a mathematical point of view, what I care about is not: can we do it twice as fast, or even ten times as fast? I want to know if I need a hundred million time steps, can I do it in something proportional to the logarithm of a hundred million, or log squared of a hundred million? Or even a hundred million to some power less than one. I care about sort of the functional dependence of how hard the problem gets as I increase the size of the problem, or the amount of time I want.

I mean, once you give me your circuit, then the question of the truth values of the inputs of the circuit is just a question of whether we fill in certain blocks or not in the initial condition. I don't think this is profound, but I think it is cute. Now let me tell you why I think it's not profound; because maybe that's a good thing to focus on. Before I step away from this, I want to mention a cute open problem. I've only shown you how to do what are called monotone Boolean circuits. Monotone Boolean circuits are the ones without negation. So that they're built out of just AND's and OR's. And they have the property that turning an input on will never have the effect of turning the output off. Okay. So the output is a non-decreasing function of the inputs.

So in three dimensions, this is okay. You may be aware that any logical circuit can be made into a planar logical circuit; one where the wires don't cross. It is also true that any logical circuit can be made into a monotone logical circuit where you just negate some of the inputs, and where everything after the input level is monotone. But it is not the case that every logical circuit can be made both planar and monotone. And in fact, the question of circuit value for circuits that are both planar and monotone is in NC. It can't be efficiently parallelized. So you can no longer do all the computations you want.

So the reason I bring this up is that in two dimensions this proof fails, because in two dimensions, we can't make the wires cross over each other – at least not any way that I can see. So the two-dimensional case of this majority voting CA, or zero temperature using model dynamics, is an open question. I haven't been able to prove it's P complete, but I have not been able to find a fast parallel algorithm for it either.

The same is true for another system which you've probably heard of. This is the Ebelian sand pile where you have a lattice. Let's say it's a square lattice. And these are popular models of self-organized criticality, and each lattice has a certain number of grains of sand on it, and we dribble sand in random places. And whenever your site has four or more grains of sand, it loses them, and sends one each to its neighbors. This is called toppling. But one of your neighbors may now be over the threshold, and it might topple, and that might set one of its neighbors off. And so you get these avalanches that go around the system. And if you measure how often avalanches of different sizes occur, you actually get a very nice power law. It's a pretty straight line on a log log plot, and then people say oh, these sand piles, they're like earthquakes, and they're like extinctions, and they're like everything else in the world. So it is a nice model of self-organized criticality.

In three dimensions we can build circuits in the same way. A wire is simply a row of threes. Because if you add one to this, and then it has four, and it topples, and then its neighbor has four, and so on. And it goes like that. And then whenever you have a branching in the wires, then if there's a three there, it's an OR gate, because if any one of them gets burned up, it acts just like a fuse, and burns up the other two. If the thing in the middle is a two, then it acts like an AND gate, because you have to burn up this one, and then this one. And then this one reaches its threshold, and it goes like that. But again, in two dimensions it's an open question. I cannot prove that predicting the final outcome of the sand pile is P complete. But I've not been able to find a fast parallel algorithm for it either. There are actually people who do large numerical experiments on the two-dimensional sand pile. So if there were a fast parallel algorithm, it would be of some practical use to them. At least once we actually build highly parallel computers. Which isn't something we've done to the extent that this theoretical model of parallel computation imagines.

And I just want to mention, in case you missed his talk yesterday, that Jonathan Machta has done similar kinds of thinking for other physical systems, like diffusion-limited aggregation, which builds these large fractal blobs. Let's suppose I wanted to build a really big fractal blob that was generated by diffusion-limited aggregation, I want it to have a billion particles, but I don't want to add each particle one at a time. You know, diffusion-limited aggregation means you take a particle, and it wiggles around until it hits your cluster, and then it sticks. And suppose you don't want to go wiggle, wiggle, stick, wiggle, wiggle, stick a billion times; you want a shortcut. He proved that unless P equals NC there isn't one.

But getting back to the point I made at the beginning about relevance. What do I like about this, and what don't I like about it? Well, first of all, from the physical point of view, we really don't give a hoot about whether a particular site in, you know, a crystal, or a large lattice system turns on most of the time. Most of the time, all we really care about is bulk averages of the properties of large sections of the lattice. In the Ising model, we care about the overall magnetization, but it doesn't really matter to us very much about whether a single thing flips on or off.

Now there are some tricks you can do to try to magnify the effect of these circuits so that it's not just one site, that it does something macroscopic. But that would be the first criticism that I would raise is that we're talking about microscopic things here.

Criticism number two, which applies to a lot of computational complexity theory in general, is that all of these are worst-case results. I am setting up very carefully constructed initial conditions that are hard to predict. That does not mean that a typical initial condition, which is drawn randomly by randomly initializing your set, or whatever other place you might get an initial condition from. It doesn't mean that a typical one will be hard to predict. It might not have structures in it that do these complicated things, and take a long time to do them. There are people in computer science who are working very hard on defining average case complexity.

It was recently pointed out by a lot of people, that lots of NP complete problems are easy most of the time. The traveling salesman problem is easy most of the time. It just is problems that are kind of precisely tuned that are very hard. So now people are trying to discuss the idea of average case NP completeness as perhaps more relevant to the real world definition of NP completeness. And the same thing could be done here.

Another good example is that Jonathan Machta, and his student Ken Moriarty showed that while it's true that diffusion-limited aggregation is P complete in the worst case, most of the time you can gain a pretty significant advantage from parallelization. Not down to these logarithmic depth things, but you can do a cluster of N things in N to the A time, where A is something less than one, which is a big improvement for a large N.

And then my last criticism would be that again lots of problems that we encounter are simply not encodeable in this nice, digital way. Or if you do encode them in a nice, digital way, you're probably missing the point about the problem. So problems that involve lots of continuous degrees of freedom, problems that involve, for instance, biological objects that are sort of discrete, but not really, totally discrete. You know, I don't necessarily claim that this approach will work for those. But I do think that the overall philosophy of this approach is a very good one. And that gets back to the hierarchy I drew before. That it is a good thing to try to make qualitative distinctions between different kinds of systems, based on how much information processing they can do, how different parts of them can communicate with each other, how much memory they can have, in what way they can draw on that memory, or manipulate it. And this is the kind of qualitative hierarchy that people have done in the digital realm. A lot of people, including me, are interested in trying to do the same thing for analog computation, where the degrees of freedom are continuous, rather than discrete. So it might be that we can build a similar hierarchy for analog computation, and use it as a vocabulary. Even if we can never actually build analog computers, there might be a good way to talk about the complexity of analog systems; just as this is, I think, a good way to talk about digital ones.

Mr. Moore: I didn't mean to say that that literature doesn't exist yet, or that it's original with me. I mean, there's Blumen, Schubb, and Smale's machines. The difference, though, is that I don't think that any one of those models has reached the same level of acceptance that the Turing machine has in the discrete area. They're still rather different from each other, depending on what your basic operations are. But I agree with you, yes. That's good work.

M: I have one short question. I was wondering if there's anything that you're interested in that caused you to omit from your survey.

Mr. Moore: Well, mainly the limited time. But for instance, Jonathan Machta's result about diffusion-limited aggregation assumes a non-deterministic system, because you have random diffusion. And so he had to phrase the question fairly carefully in that he asked you have a bunch of random bits which are going to drive the choices of direction, the diffusion. Let's say we make those part of the input. Then can we go from there to the result quickly? Actually, I think there might be a nice mapping between the fact that in general in physics we might have random initial conditions, and the fact that we want to have these randomized algorithms.

RNC is probably a bigger class than NC. RP is probably a bigger class than P, and so on. So yes, that's a good direction to look at.

Yes, I'd be happy to talk with you guys more about that. My only knowledge of randomized algorithms comes from reading Papadymetrio's book, and trying to read journal articles. And I'm just catching up on that myself now.

4 Neural Systems

Mr. Matt Wilson: I was charged with delivering a largely pedagogical lecture on neural systems. Now that's a fairly broad topic, and I don't know that I'm quite up to the task of explaining how the brain works in an hour in an instructive, and enlightening way. But I think that I can – by focusing on a specific area that is both of interest to me, and one that I think will be of general interest to you – give you an idea of the approaches that are being used (in contemporary) neuroscience to understand basic problems of neural systems, neural coding, and neural function.

And so what I'm going to describe are some of our efforts to understand the structure of memory formation. And I think that this is important, because it points out how the study of neural systems really breaks down into two primary areas. One looking at the structure, structure of neural coding, as an example; and the other is process. And that is looking at the manner in which interactions between neural coding systems gives rise to function. So this is a bit more specific than I will get into. And that is looking at the formation of memories. But the basic concepts that are embodied in that can really be elaborated.

I will go through my description of the system that I've shown here. And what we have here is a picture of a rodent brain. Thank you. A picture of a rodent brain with overlying neocortex removed to reveal the structure known as the hippocampus. Now, the hippocampus is, again, I think of general pedagogical interest because of its historical background. And that is that back in the early fifties, there was a very important clinical observation that was made in a human subject. The human subject is known as the case of HM. And the patient HM underwent a surgical procedure in which the hippocampus was bilaterally removed to treat intractable epilepsy. This was at a point where our knowledge of how neural function was distributed was fairly limited, and so the consequences were not fully anticipated. But the consequence of this manipulation was the total and complete loss of new memory formation – the capability of forming long-term memories, episodic memories in particular (that is, memory of people, places, things, and events). So quite literally, this person in his early twenties ceased to accrue memory. He's now in his seventies. To this day, his life begins as of about five minutes ago. He doesn't recognize himself in the mirror. He doesn't recognize investigators who have been working with him for the past few decades.

Now, the fact that a single structure could be so critically responsible for maintaining what we think of as a basic property of our nervous system – that is, the ability to form memories – really pointed out that it is possible to focus on a specific structure, understand how it processes information, and come to some enlightenment about how memory itself is formed, a very complex system.

Now this is a timeline which again begins back in the fifties with the clinical observation, and then proceeds through several stages, which are relevant, because they track the progress of neuroscience. In the early seventies, the result of work looking at recording of events in the brain using microelectrodes, which had begun, actually, in the forties, fifties, sixties, but refined itself to the point where individual neurons in a behaving animal could actually be monitored and tracked. Now John O'Keefe, in 1971, carried out such a recording. Placing microelectrodes into the structure of the hippocampus that I showed you. And discovered that individual cells would discharge when an animal is in certain locations in its environment; and termed these cells place cells in the regions of space-placed fields. What he literally did was had an animal on a table. Had a wire in the animal's brain. Was listening to individual neurons. And when the animal would move, let's say to the corner of the table, the cell would start to discharge. Pah, pah, pah, pah, pah, pah, pah. The animal would move off, and the cell would become silent. If he went back, pah, pah, pah, pah, pah. The animal would move away, the cell was silent. This could be repeated over and over during a session, across days, even over weeks. This cell would continue to discharge in the same location. Suggesting that it was maintaining information about complex information within its environment. Information that allowed it to determine its spatial location, independent of manipulations of cues. Changes in cues, time of day, where the experimenter happened to be, the behavioral state or general state of arousal of the animal. All were minimized relative to this information about absolute spatial location. Now this was very convenient for studying the structure and function of the hippocampus, because here was a very powerful behavioral correlate. One that could be repeated, one that could be easily elicited simply by allowing the animal to move about. And it turns out that on the order of sixty, seventy, eighty percent of all the neurons within the hippocampus exhibit this characteristic property of spatial firing.

Now, it was at that time convenient to think of this, a behavioral correlate of experience to somehow be related to memory. But how could memory, demonstrated clinically, be tied mechanistically to these neurons of the hippocampus? And that's when, in 1973, Tim Bliss and Travo Lomo identified the first mechanism that could actually form the basis for memory within the brain. And that was the phenomenon of long-term potentiation. An activity-dependent change in the strength of connections between neurons. And this was quite striking in that simply delivering a very brief tetannic stimulus, a brief high frequency burst of activity to the input pathway of the hippocampus could produce a long-lasting change; a strengthening of synapses.

Now, this subsequently has been found throughout the synaptic pathways within the hippocampus. Pathways outside of the hippocampus in the neocortex. In fact, virtually everywhere within the brain we see long term changes which can be elicited by very brief periods of electrical stimulation.

Now, given these two or three facts –hippocampus involved in memory; the identification of a strong behavioral correlate; and a mechanism of plasticity which could relate all of these things to long term memory itself – John O'Keefe and Lynn Nadel came up with a theory for how the hippocampus could bring these together to form actual representations of space, of a cognitive map. Subsequent experimental work tested the relationship between the mechanisms of plasticity, and learning, and memory at the spatial level. By blocking plasticity within the hippocampus, synaptic plasticity, and asking the question: if the synapses can't change, can the animal still learn? And the answers to that were somewhat ambiguous. In 1986, the answer was no. You block synapses, animals can't learn. I'm reminded of a comment by Mark Twain regarding his assessment of his father's intellectual prowess. And that is, in these ten years in 1996, we've been surprised at how much the hippocampus has changed in its ability to control memory. Because in 1986, plasticity in the hippocampus was necessary for memory; 1996, it appeared that it was not. Now the question is has the hippocampus itself changed, or has our ability to study it changed?

Now, what I'm going to try to describe to you is an experiment involving manipulation at three primary levels. The molecular, the systems, and the behavioral level. We're trying to understand these basic findings of plasticity within the hippocampus, neural correlates, place cells, and involvement in learning and memory at the behavioral level. And to do that, I'm just going to point out the hippocampus lies at the apex of a convergent pathway of associative information within the brain. And that is information from basic sensory systems – we can think of starting at the back of the brain – moves forward along the temporal pathway, in which we find increasingly specific object-related visual information, as an example. What has been termed the "what" pathway.

Simultaneously along this parietal pathway, or the dorsal pathway, we find that there is a preponderance of spatial information. So we can think of sensory information moving from the back of the brain, across the upper parietal or dorsal pathway, carrying spatial information. Converging with the hippocampus, which is located sort of medial temporally here at the center of the brain. The same time, information about objects is passing along this temporal pathway, the ventral pathway, and again, converging within the hippocampus. So by the time we reach the hippocampus, we have a convergence of spatial and object information; which could give us the ability to combine this information to form complex representations of specific experience.

So just to break this down, we're going to try to look at four basic levels of organization within these neural systems. The molecular, or sort of protein level; cellular, the single neuron level; to the network group level; to the behavioral level, which involves multiple brain regions. And we're going to look specifically at the function of a particular receptor, the NMDA receptor, which is known to regulate plasticity. And look at a specific form of plasticity, synaptic plasticity; that of LTP. Look at it in the context of this hippocampal place code. And then examine how this hippocampal place code changes with experience, and is translated between different regions of the brain.

So now the first technological development that was required to move us beyond the approaches that had been employed in the early 1970s, which allowed us to look at individual neurons. As I said, the technological development involved modifying our single wire electrodes to allow monitoring of large collections. Simultaneous monitoring of individual neurons in different regions of the brain. And what we have here is a schematic diagram showing a cutaway view of a multielectrode microdriver A; which here we see mounted on an animal. And the principle is that we can independently manipulate very fine microwires.

Now these microwires are made out of insulated wire that has been bound together in a very fine bundle. The dimensions of this bundle are on the order of thirty to forty microns; so about a fraction of a human hair. And the principle is this. And that is that the tips of the electrode are exposed, and allowed to monitor the local electrical field in the vicinity of a population of cells. Now when a neuron produces (an action) potential, it will produce a field, an electric field, which is attenuated with distance. So the farther we are away from a neuron, the smaller the amplitude of this electrical event.

Now we can see the challenge in this. This is a cross-section here through the brain at the level of the hippocampus. And we can see the various areas, many areas that are involved. Note the dimensions here – this is on the order of a centimeter. We want to put on the order of a dozen electrodes, drive them down such that the tips fall here within this very small layer, about a hundred microns thick. And place them such that this electrode does not move more than ten to twenty microns. While an animal is running about, the brain, of course, is just encased in sort of a fluid-filled bag which is jostling around. But in order to achieve stable recording, we have to maintain stability, at the order of ten microns, over days to weeks.

And this is just the basic principle in which we have three neurons, shown here, each producing a pattern of action potentials of fixed amplitude. And what we want to do is we want to place a wire such that we can record, and disambiguate the activity of those three neurons. So what we're going to do is place four wires in close juxtaposition. Each of the wires will see the signal from each of these neurons, but the amplitude will be graded based on its distance. So for instance, wire number one will see the purple source number three with a high amplitude because it's close, and the green one with a low amplitude, because it's distant. Conversely, wire number four will see the green with high amplitude, and the purple with low amplitude.

And so we can construct a distribution here of individual events, based on their relative amplitude, across these individual wires. So here we see a pattern in which the amplitude on sensor one relative to sensor four will give us a cluster of points corresponding to the action potentials arriving from cell number one. Now using this technique, we can identify and separate the contributions of on the order of ten to twenty individual neurons within the vicinity of about a fifty micron sphere.

There is no feedback stabilization. That would be nice if we could actually take the wires, and adjust them such that they maintain the integrity of the signal. But we're relying upon the inherent stability of the recording wires. And physically what occurs is the wires are implanted very gradually. And over a period of time, they actually anchor themselves to the local tissue through glial adhesion. So they become a part of the brain, quite literally.

Here is about an hour's worth of data. Each point represents a single discharge from an individual neuron, and we can see these clusterings in this amplitude space demonstrating the contributions of multiple neurons in a local vicinity. But now, with the ability to monitor, simultaneously, multiple neurons, how can we use that to examine the structure of these neural representations during the formation of let's say spatial memories, or during behavior in which we would like to examine neural function? So here is a schematic just showing, diagrammatically, how we're going to approach this. In the rodent, the rodent's plugged in with a tether which allows us to monitor activity. He's moving about in a small, enclosed arena, and we're recording the activity of these place cells. Different cells will fire in different locations. And so as the animal moves about, we're going to see a shifting pattern of activity, which will be consistent. Different cells will fire. Cell one when the animal moves through this location, two when he moves through here, cell three when he moves through here. And if he revisits these, those cells will also discharge at their respective locations. This doesn't really bother the animal at all. This is chronically implanted. So once this goes on the animal's head, it never comes off; so they will live with this for the rest of their short lives.

This pink stuff, by the way, is dental acrylic; it's a plastic material. So the actual surgical invasion is very minimal. It's just small hole about three millimeters that's drilled in his skull. And then this is glued on top. So this is literally stuck on top of his head, and then through that small hole these electrodes are advanced.

Well, so what would distinguish an optical method from a direct measurement of electrical activity? Current optical methods can be divided into two camps. One which looks at general metabolic activity. And that is, they examine general qualitative activity within a region. The other involve – the introduction of specific molecules that act as voltage sensors. And so we're replacing an electrical signal with an optical signal. And of course the drawbacks with that have been that we have to alter the system. We actually have to introduce a signaling molecule into the cell, which did not formerly exist. Now here we're introducing a wire which did not formerly exist. But what is key is that we be able to identify the activity of individual neurons.

Now as an example: within the hippocampus, two neurons that are sitting directly adjacent to one another will not necessarily respond in a similar manner. They may respond to completely disparate regions of space. So it is not generally possible to lump together regions of neural tissue, and treat them as though this were simply a population with some general distribution. It is really essential to be able to resolve the activity of individual cells. So we have no optical technique, at this moment, which will allow us to resolve the activity of individual neurons at that large scale.

So here's the environment that we're going to be testing this animal. Just allowing it to explore briefly. And this is the map, within the animal brain, of that small box. Each of these panels represents the activity of an individual neuron, individual hippocampal brain cell, as the animal explores in space. The color indicates the firing rate of that cell. And so here we can see a cell that fires in the center of that box. So here again as the animal walks around in this box, as he's walking around the edge, the blue indicates that the cell is silent. As soon as he moves to the center of the box, the cell starts to fire vigorously. Silent, vigorous firing. Different cells. Here's one that fires on the left-hand wall. Here's one that fires on the right-hand corner over here. And we see a number of them that are silent. So this was generated by averaging activity over about ten minutes. But if this animal were taken out, replaced in this box days or weeks later, we would see exactly the same pattern. And that is about thirty to forty percent of all the neurons in the hippocampus, firing with preferred spatial locations. And a small fraction of them here showing very broad distribution of activity. Firing all over the environment. And it turns out that these represent a different class of cell; the inhibitory interneurons.

So we can broadly separate neurons within the hippocampus, and generally within the brain, into these two camps. One the principal parametal cells, which tend to be projection neurons, sending information out of a particular area; and the inhibitory interneurons. These interneurons tend to be local. They tend to discharge at relatively high rate, and are believed to modulate the general excitability of local regions of neural tissue. So this neuron exists at a ratio of about one to twenty, relative to parametal cells. That is, for every twenty parametal cells, you'll find a single inhibitory interneuron.

And it turns out that these inhibitory interneurons are also responsible for regulating and coordinating the general synchronized oscillatory behavior of these networks. So that this activity in the hippocampus is organized into this rhythmic firing pattern. And so for instance, if we're listening to a cell here, a collection of cells, as the animal moves through space, you can hear this 8 to 10 hertz rhythm. It sounds something like this: ch, ch, ch, ch, ch, ch, etc. That rhythm is related to locomotion. It's known as the theta rhythm. It's present throughout the limbic system. And is present in all mammals. And is believed to be related to the processing of sensory information and arousal. So that we can see that when an animal is attentive, and acquiring information, cells are modulated in this rhythmic way.

Now what's interesting is that that tends to be coordinated in the rodent with whisking and sniffing. So as the animal is sampling its environment at about 8 to 10 hertz – that is, it's sniffing and whisking at about 8 to 10 hertz – these regions, which are receiving sensory information, if you remember that flow diagram, are oscillating at about that same frequency.

These rhythms are generally broken down by the frequency bands. One thing about human rhythms, the various rhythms, is that they are cortical rhythms. So here this is a limbic theta rhythm, which is similar, but not identical. But it's in the same general frequency range; 5 to 12 hertz.

So here we have individual cells. So now we're going to try to determine how does that pattern change? We can look at the activity of a lot of cells. We can ask: how does that pattern change with experience?

And so here is a manipulation in which we're going to allow behavior to change, and see how it correlates with the neural representation. That is, simply introduce an animal into a new region of space, and ask: how does that pattern change?

Now here we have an example of an individual cell – this is now over time. We're going to track a single neuron here up at the top. You can see one cell. During a period where the animal is being exposed to a familiar box, that same old box. The cell fires in a particular location. Now this manipulation. We're going to remove a partition within this box to reveal a new space. The animal will enter that new space, so we can ask: what do the cells begin to do? Now, we're looking for changes in the dynamic patterns of activity within this population of neurons. And what we see is that over time the robustness of firing increases. That is, cells will initially fire sounds something like this: [imitates sound] as I move through this location in space. The second time the animal goes through: [imitates sound]. The third time the animal goes through will now have a very robust pattern: [imitates sound]. Which is consistent both in its timing, and in its covariance with other neurons that are firing at about the same time. So if we can think of two cells that are active in the same location, we can think of both of them firing. During those initial periods, we'll get activity that sounds something like this: [imitates sound]. After the animals had a bit of experience, these cells will discharge in a covariant fashion: [imitates sound]. And we can speculate that this change is a result of some form of learning. And that that learning has, at its mechanistic root, some form of synaptic plasticity.

How can that be tested? Now one thing that we can do is to look at population covariants. That is, how is not just an individual cell changing its firing pattern, but how is a whole population changing in its robustness of representation? Well, we can try to decode the activity of ensembles. And that is, guess the location in space where the animal is, based on ongoing neural activity. And that's what we've done here. Here we can see our efforts to decode the neural ensemble at one second intervals. The red is the estimate, blue is the actual location of the animal. And you can do pretty well with this hippocampal activity coming to within about a centimeter every second. Now what we can see is when the animal enters this new region of space there's a period of about five minutes, five to ten minutes, when the error of that estimation is high. And that is, we're unable to decode accurately this neural ensemble. But that rapidly improves such that the robustness of the ensemble in the new space is comparable to that in the familiar space. Now what we would like to be able to do is to tie this to some mechanism of plasticity. But how can we manipulate plasticity in a very specific way?

Now again, historically, efforts to alter mechanisms, functional mechanisms within the brain, fell into two camps. One was the physical lesion; actually removing structures from the brain. The other was a pharmacological manipulation. And that is blocking the action of certain molecules. Now the problem with these two approaches, the physical lesion had regional specificity. You could actually target a specific region of the brain, but it had no functional specificity. You just scoop out tissue. You can't target synapses over cell bodies. You can't target specific pathways, i.e., fibers passing through a region, as opposed to the cells within that region itself. So an additional refinement was required.

And so what we see here is the result of new techniques involving genetic manipulation in which we use the genetic signature of individual neurons to achieve both the regional or tissue specificity, as well as the functional specificity. And this is the result of application, in the lab of Sisuma Tonagawa, of a new genetic technique which allows the knockout, or the lesion of a specific receptor in cells in this part of the brain. This is the hippocampus. A particular subregion of the hippocampus. And the genes encoding, the NMDA receptor, which was responsible for a certain form of plasticity. Those genes have been eliminated, but only from the cells in that region of the brain.

Now, if I can just quickly go through, and compare two approaches for carrying out this kind of molecular genetic manipulation. The first is, as I described, the genotypical manipulation. That is, you target a gene. You know the function you want to eliminate. You create a genetic construct. That is, you design the gene such that it does not have that, or is able to be manipulated such that that desired gene is eliminated. Now the advantage is, it gives you great specificity. The disadvantage is you have to know what you're looking for, to begin with.

Now, an alternate approach – which has been employed in invertebrates primarily, in Drosophila flies, to great advantage, and now in vertebrates, in fish – is that of the phenotypical manipulation. And that is you put in some kind of mutagen. You randomly perturb the gene. You create many different mutants. And you screen them, you test them to see whether or not they carry the characteristic that you're interested in. If you want to identify a mutant that has poor vision, or a mutant that has exceptional hearing, or that walks with a limp, or has, you know, one leg longer than the other, or learns better than another, has better spatial memory. All of these things can be tested, and screened; selected for. And over multiple generations, one can generate an animal that has manipulation of its genome that will give you this desired functional or phenotypical characteristic.

Now, the cons are that carrying out such a manipulation is massive in its scale, typically requiring on the order of tens to hundreds of thousands of animals to achieve even a single, targeted phenotype. As opposed to the genotypical approach, which can be done in a small number of generations.

This is just a, again, a very brief rundown. Right now, we have the ability to achieve tissue specificity with some limited induce-ability. That is, temporal control, as well as regional control. At the moment, this manipulation is one way. That is, once the gene has been modified, it's not reversible. But that itself is also being addressed. And, of course, the benefits are that if we have the ability to control specific mechanisms within a neural system, coupled with the means to monitor, very specifically, the fundamental elements, the individual neurons, we can both manipulate and monitor at a large scale, system scale, with the requisite specificity to look at function. So we can see things at the necessary resolution, and manipulate them at the necessary targets. It's actually quite simple. And it relies upon the existence of exogenous transcriptional factors. That is, you have a gene, and the expression of a gene is under the control of a specific promoter. So you have a gene, and there's a little sequence here that requires, that's looking for some factor. When it sees a factor, it will allow transcription of the rest of the gene. It will allow the rest of that sequence to be read out, to be transcribed.

Now, what we can do is to introduce what's known as a recombinase, (a cree recombinase). And what this does is when that recominase is present, it will cause sequences here in another part of the genome to come together. So we can bring together or pinch out certain regions of the genome. So if you have a gene, you can target it. Place these two little factors, these little elements on either side of it. And then trigger the expression of this novel substance, recominase. So when the recominase is present, it causes that portion of the genome to be deleted. And the trick is, can we only get this stuff to be expressed in a particular region of the brain? What that involves is identifying a promoter which is sensitive to a factor; is looking for a substance that's only present in a particular region of the brain. Once you find this promoter, you can put it upstream of this recominase. And now you have the ability to knock out any gene of interest. So you can create a gene that has been flanked in this way, cross it with another animal that has this cree. So it's known as the cree lock system, or I think of it as the key and lock system. One animal has the key, one animal has the lock. When you bring them together, you get tissue-specific deletion of a specific gene. Specific gene, specific region.

Now, we can take an animal that has been targeted for deletion of that specific receptor, and test it behaviorally. Now, there's a simple spatial test that involves putting an animal in a pool of water that we saw here, and asking: can it find a hidden platform submerged under the water? The only way you can find it is if you know where it is. Now it turns out that animals which have had deletions of that specific gene only in that specific area of the brain here are unable to find that platform –this is the search time – in that pool. You can see that normal animals go right to that platform. They know where it is, and they find it. These animals do not. Further, this just shows that electrophysiologically we can look at the currents that are elicited following synaptic activation. And in this animal, that current, carried by the NMDA receptor, is absent; and plasticity – this looks at a measure of plasticity, LTP – is blocked completely in this region of the brain, not in another region, in this animal, and is normal in other controls. So with plasticity blocked, with a specific learning impairment, we can implant one of these animals with these electrodes, and try to tie together the mechanism, the NMDA receptor, with the behavior through this neural representation at the level of hippocampal place cells.

And this is a mouse that's been implanted. It's quite a bit smaller than the rat. And we can look at a few things. So one property that you might expect could change would be the stability. And that is, if the animal has difficulty remembering, perhaps it's because the representation is unstable. Maybe that's what plasticity does. Well, it doesn't appear to be the case. The patterns are quite stable over time. But if we do look at these, and compare normal animals above, with animals for which this plasticity has been deleted, we see that the pattern is simply more diffuse. That the firing is not as robust. It doesn't fire in the same small region of space. It's really more spread out. And here we see an animal running around in a large circular platform. These are top down views of a little track. The animals running back and forth on the track. Here you see it fires right in one spot. Here it fires at several spots. So that deleting plasticity leads to a change in the robustness of firing. So if we think of how this cell is firing as this animal walks along this track, it sounds something like this: [imitates sound]. Versus this animal; he walks through that spot. [Imitates sound]. Here it's [imitates sound]. Now, if you compare those, [imitates sound], versus the very robust [imitates sound], what does that sound like? It's very much like the difference between that first pass in that novel environment, and the second or third pass. Now a normal animal, after three or four passes in a new region of space, makes a transition from here, this variable pattern, to a robust pattern. But it appears that these animals, with the deletion of plasticity, are unable to make that transition. So what it appears is that by making targeted manipulation of plasticity, you can see the attribute that changes as a result of learning, which may be related to the formation, or familiarization with a particular environment.

And one thing that's interesting is that amnesiacs that have damage to the hippocampus report their experience as one of perpetual novelty. That is one of the consequences of disrupting this long term, episodic memory, as I mentioned with the subject HM. That his life begins as of five minutes ago. So this quite literally may be what familiarity looks like within the brain, and that we've identified it by combining our observations of normal animals with experience, with manipulation of specific mechanisms within that region, correlating with behavior, and then identifying the structure of neural activity that corresponds to it. And then this is just showing that in fact that covariance, the correlated firing between cells is completely disrupted simply by disturbing plasticity within that particular subregion of the brain.

Last, I just want to show you this. It's a simple model which we can actually put together with a random network with the known convergence and divergence that we have within this particular region of the brain, which tries to determine whether or not a simple mechanism of Hebbian-like plasticity – and that is a correlation rule for changing synapses – is capable of capturing all of the phenomenology that I described to you. And it turns out that it can. That here is a simple model that takes a place field, a single cell. There's the animal running back and forth on a track. Now, with experience as we strengthen synapses, our fields go from this multiple-peaked variable pattern, to a very robust pattern. And the only principle that's really being – that's being exploited is one of enhanced signal to noise. That we're increasing the relative signal to noise of these individual cells to incoming spatial information by selectively and competitively enhancing response to a particular subset of its inputs. And at the same time, adjusting its threshold such that it maintains a fixed level of sparsity. And that is the relative fraction of time for which the cell is active remains constant. In the hippocampus, this is roughly two to five percent.

So by attempting to maintain a relative constant fraction of output, while at the same time enhancing the selectivity – that is, regions of space. So you can imagine here if I want to get five percent – five percent on the red trace corresponds to, well, a very restricted region; whereas five percent here, on the purple trace, corresponds to multiple locations. And because we're very close to the threshold, this is going to be highly variable, whereas this will not. So both the variability, and specificity comes through enhanced signal to noise.

And then the very final thing. I'll just say that we can now take this one step further, and ask what does coordinated firing within the hippocampus do to the other structures downstream? And this is just an example of simultaneous recordings of collections of neurons, but in separate regions of the brain. Here in the hippocampus, and the pre-frontal cortex; which we know behaviorally is involved in decision making, planning. And by carrying this out, we can see that these coordinated firings of cells within the hippocampus leads to coordinated activity within this region of the pre-frontal cortex. So coordinated discharge within the hippocampus propagates to the pre-frontal cortex, and triggers an oscillatory, coordinated discharge. Here we see it in the EEG, coordinated oscillations in different parts of the brain.

So that's the final, which is we can see that coordinated activity can be maintained through activation of specific mechanisms, can be expressed as a result of experience, can be measured by looking at covariance of activity across ensembles, and has, as its consequence, the coordination of activity not just within an area, but across brain areas during a period which we suspect may be involved in the long term consolidation of memory. And that is that this is not occurring when the animal is running around, as you saw in the previous slides. This is occurring while the animal is asleep, and presumably reactivating information related to recent experience.

5 Patterns in Development

Fred Nijhout: I am doing this on the assumption that there are very few biologists in this audience, and that probably far fewer geneticists than that. So I'll take it a step at a time. And what I want to do is let me start by outlining what the problem of development is. The general problem of development is that you start with a homogeneous undifferentiated, usually spherical cell. And in a period of a few days or a few weeks, this thing becomes differentiated, and it develops into something that can look as different as these different objects on here. The remarkable thing about this is it does it a million times, with relatively little variation, and relatively little error – if this is a giraffe egg, it will likely turn into a giraffe. And there are also, starting from roughly similar-looking eggs, more than a million alternative end points to this development. And so the problem of how this is regulated, how you get the stability within a lineage, this diversity between lineages is sort of the big problem in developmental biology. And one could arguably say that this is the big problem in biology in general. And so I'm going to walk us through how people have thought about these problems.

First of all, defining development is a little bit difficult in its own right; depending on whom you ask. You can say well, development is a kind of process that takes us from the simple to the complex, from the general to the special, from eggs to embryos, or from embryos to adults. They're kind of vague definitions. In order to make some progress in development, we tend to specify much more narrowly what it is that we're talking about, and what we're worrying about.

So most of the time, and particularly today – and I'll be exclusively referring to this today – we're talking about problems of pattern formation. In which we're basically asking the question how do you start with a homogeneous field, like an egg, and how do you end up, after a brief period of time, with a spatially patterned field that is patterned in some predictable and useful way? And the control of pattern formation, we know from experimental work, is underlain by some spatially patterned gene expression. And what that means is that in different locations of the field, different genes turn on. And that this bears some relationship to what will subsequently happen in development. So another sort of the general problem that I will address is how do you get from genes to form? At least that's what this is in service of.

And just to give you some general sense of what do we mean by patterns: there are the prosaic patterns of pigments on the backs of zebras; where you've got cells that make no melanin next to cells that do make melanin in a fairly well-patterned way. There are patterns like these in sea urchins of gene expression where these dark points indicate gene expression that determine where particular movements will take place during embryo genesis. Other patterns are determined a bit later on. Where the five arms will form, and within the arms some other patterns of gene expression, again indicated in black here, which will tell you exactly where the little tube feet, and so forth, will form. So the question then is: why should these cells express this particular gene that is important for arm formation, and not the other cells?

And then some other things that I've done a lot of my own research on, which is the color patterns of butterflies. Again asking the same question. Why is it that some of these cells make a blue pigment, others a yellow, other an orange in this patterned way that serves some kind of a communication mechanism, in turn, for the butterfly.

Genes are involved in all these processes in some way. And the reason we know that is because when something goes wrong with those genes, when there is a mutation, something goes wrong with the process. And in humans the observation is you have a normal gene, normal function. You have a defective gene, we call that a disease; a genetic disease. In animals, if it's a defective gene, we call that a research opportunity, or a terribly interesting system to analyze, and we don't worry about how the animal feels about it, typically.

The conclusion of this one is that a defective gene causes the disease. Which seems on the surface like a reasonable conclusion. I'll show you as we go along that it might not be quite as straightforward as this. In terms of pattern formation, we can say: what are the causes of form, or of pattern? Whenever I say "form", you can read "pattern". I'll use those interchangeably. For biologists they pretty much mean the same thing. The observation, again, is that a change in a gene obscures the whole process. A change in a gene causes a change in form, of some sort. And by change in gene we typically mean mutation. A mutation, or something that we can do experimentally to cause that gene not to turn on, or to turn on abnormally. And from that, because a change in gene causes a change in form, the conclusion typically is that that gene somehow causes or controls form. That's a bit more of a tenuous conclusion, as you can almost see from the way I've phrased it here.

What's a gene? Now most of you probably have some concept of what a gene is. I'm not going to go through, and ask you each, and make a list like Alfred did over there, because we'd come up with far more definitions of a gene than this. But here are some ways in which some of my colleagues have thought about it. You know, genes, or DNA, provides a program that controls the development of the embryo, by Lewis Wolpert. Or a quotation by DeLisi, "This collection of chromosomes constitutes the complete set of instructions for development, determining the timing and details of the formation of the heart, the central nervous system, the immune system, and every other organ and tissue required for life." So that's what genes as an ensemble do. The instructions for how the egg develops are written in the linear sequences of bases along the DNA, and we all recognize that in the fertilized egg there is a set of genes, and these genes give the instructions that ultimately produce a complex adult. You can pick up any textbook, any article, particularly the more popular articles on genetics, and you will find some kind of a statement about genes like this.

That makes you wonder: what do genes do? These are all metaphorical attributes of genes. This is not what genes do; what these statements say. These are political statements, largely; they're metaphorical. Genes provide, in those metaphors, instructions, programs, controlled blueprints. You've probably heard one or more of these terms used before. Those metaphorical attributes of genes stand in very direct contrast to what we actually know genes do. And these very same people that I have just quoted are actually the ones that have been involved in discovering what genes really do, which is only one thing. Genes are sequences of DNA that code for the sequence of amino acids in a protein, period. That's all. That is the only thing that genes are known to do. That is their only function. What happens subsequent to that actually has relatively little to do with that property of the genes. Although if you change genes, you change some of those consequences in interesting ways; which leads to the metaphors which I'll try to point out.

This is how we teach genetics in Introductory Biology. Here we have two genes which in different variants, in different permutations, for instance, can cause these different forms of the comb of a cock. And we can say well, these genes control the shape of the comb of a rooster. Here's what genes really do. So that sequence of genes controls the primary structure of a protein. What we call the primary structure of a protein is a sequence of amino acids in a protein. So a protein is just a linear chain of amino acids. Twenty different kinds of amino acids. They bear different charges, and different active groups; so that different charges either attract each other, or repel each other, and cause that chain of amino acids to fold in particular ways so that it forms a dynamically stable ensemble, which is the shape of the protein we call the tertiary structure. And that's the important thing about a protein. The linear sequence of amino acids really is not what makes a protein. A protein is all shape. It's because the shape causes the right kinds of nooks and crannies to be formed, the right kinds of charge appositions that allow that protein to interact with other things in its environment. Get the shape wrong, get one of these loops wrong, and the protein is totally functionless.

Now, what does the linear sequence of amino acids have to do with this shape? It turns out almost nothing. There is not even enough information in the linear sequence of amino acids to specify this shape correctly, except for the very simplest of proteins. And the reason for it is that for the larger proteins there are an indefinitely large number, or very often a very large number, of equally stable states in which this thing could fold at some minimal energy level. The reason it folds in life in only one way, has to do with the fact that while it is being formed, there are lots of other proteins involved in this process that are called chaperones, or chaperoning proteins, or chaperonins that guide as this protein is being synthesized. Guide its various parts through folding just to write one single, unique way that is required for this thing to end up a functional protein rather than a functionless protein. And a mutation in one of those chaperonins would have just as dramatic an effect on malforming that protein as a mutation in the protein itself would have.

So now we have already lost a whole bunch of information. DNA genes code for the sequence of amino acids in a protein, but they don't even code quite yet for what is functional about a protein, what is important about a protein. So we're some steps removed already from sort of a very simple cause and effect relationship between genes and form. The slide is a bit off the top of the board, but it doesn't matter. Here is a sort of an historical sequence of how people have thought about the relationship between genes and form, or genes and characters, or genes and patterns. Initially because we knew that many mutations in genes had very definite and unique effects on particular characters. It was thought that one gene by and large coded for one character. It was before we knew really much about anything about genes.

As we've begun to learn something about biochemistry, it became clear that a great many genes code for functional proteins that are enzymes that catalyze reactions. Hence the hypothesis of one gene really codes for one enzyme, and that's what's the important part. So that became the definition of the gene. With the elucidation of the central dogma. The fact that DNA codes for RNA, which codes for proteins. That we sort of relaxed the restriction that proteins can only be enzymes, and now proteins became any kind of a polypeptide, any kind of a protein that has some kind of a function. And now we know that one gene can actually code for a great many different kinds of proteins, depending exactly on how it's sliced and spliced later on in development.

And so this has removed us pretty far away from how genes control characters. It's pretty clear that you don't have a one to one relationship between a gene and a character. So what's the alternative? Well, lots of genes control a character. So there's polygenic control of characters. And since we know that a great many characters share the input for many different kinds of genes, we get this network of effects hypothesis, where a number of genes control character. Presumably the weighting of those arrows is not quite equal so that some permutational contribution that is unique specifies different kinds of characters. So there's no mechanism here, but there's just a correlational analysis.

Character here means anything that changes when a gene changes. So a character can be your fingernail, the color of your eyes, the length of your – well, not the length of your hair after a haircut. The length of your body, the color of your skin, those kinds of things, the number of toes on your foot.

There are actually two kinds of characters. Characters and phenotypes, I think, we generally call the same thing. There are some characters that are fairly simple, we call those protein phenotypes. They are the direct consequence of gene products. And when the gene is altered, those characters are altered in a very defined way. You know, blood groups, and biochemical deficiencies. A lot of metabolic diseases are controlled by single genes, or affected by single genes, because they're directly attributed to a defective single protein. A lot of the modern genetic diseases that people are looking at are directly ascribable to a single protein. So those do occur, but they're in the minority. Most of the things that we're really interested in are what you might call complex phenotypes, or complex characters such as size, shape, pattern, form where the mapping of gene to character is indirect because genes don't directly lead to form, but something called development, some black box called development intervenes where lots of things are going on that produce something that has got five fingers at the end, or something like that. That's the sort of a thing we like to look at.

So then how do genes control development, or what is the role of genes in development? There are a whole variety of ways in which developmental biologists have thought about this; and they've been reasonably vague about this. First of all, there's this thing called the sequential gene expression hypothesis where you turn on a particular gene that somehow affects, or changes, or produces – it depends on who you're reading – a particular character. That character has as a consequence that a second gene is caused to turn on. You get some cascade of events in which progressively more genes turn on. And the structure in which this occurs progressively gets more and more complex, or acquires more and more attributes. That's not particularly interesting, probably completely incorrect; but it was alive for quite awhile.

The gene interaction hypothesis basically takes a bunch of genes, inserts them into a black box, and out come a bunch of characters.

The master regulatory gene hypothesis, which is still pretty active today, you know, pretty current today, I would say, is basically a variant of this really old sequential gene expression hypothesis; except that we now call these genes regulatory genes, and they have a whole bunch of subservient genes that they turn on that then interact in some, you know, through this set of complex interactions to produce a particular character set. So those are sort of conceptual ways in which people have thought, up until maybe ten years ago, about how genes and characters might be related to each other.

Here's the reality of it. You know, this is a black box of development. This is sort of a minimal series of things that have to happen before you can get from a gene to a functional character structure, phenotype, pattern, whatever you want to call it.

Genes code for proteins. We know that's the only thing they do. Proteins can be enzymes, they can be structural. You know, things like collagen, and whatnot. You know, things that actually make things. They can feed back to produce regulatory proteins which can turn other genes on or off. They can come together to form larger molecular structures. Some of those can be membranes. Membranes and matrix you need in order to make cells.

Then we get into a whole different regime here; a higher structural regime. Once we have cells, we know they can interact with each other to form sheets or clumps which are tissues. Tissues and cells can be compartmentalized for different behaviors. Behavior suddenly becomes an emergent property of this system. Cell behaviors, together with regional specialization, can cause morphogenetic movement. Things can start migrating, and all of the sudden you start to get form, shape emerging out of those what was initially an amorphous mass. Morphogenetic movements bring very distant cell groups together. You can get induction, you get whole new things to happen that would normally not have happened. And eventually you end up with some sort of an organ, or a tissue, or whatever.

All of these steps, almost every one of these arrows in here involves a protein, or a series of protein; is mediated by proteins in some way. So that genes not only provide proteins directly, but the activity of proteins mediate a lot of these arrows in this process. And that's why genes are as important as they are in development; because their products are involved in behaviors, basically.

So what genes do, then, is genes code for proteins, period, nothing more; but what proteins do is truly magnificent. I mean, they do lots of different kinds of things. They control gene expression. In other words, genes are kind of stupid. They don't even know when to turn themselves on. They need something else to turn them on, which is typically a protein. They control the rates of things, chemical reactions that transform one thing into another. Synthesize and degrade substances. They provide structural elements for support, for movement, for the maintenance of shape of things. They serve for energy transfer between energy producing, and energy requiring reactions. They generate signals, they can talk to each other, they can cause one part of the organism to talk to another part of the organism, and coordinate things.

Let me just back up for a second – How then do we get from these more sort of static, this sort of static view of what proteins do, to a more dynamic view of what proteins do as they work and develop? And one way that we've been exploring is starting from sort of the one really good dynamic thing that we know that proteins are involved in, which is a catalysis of biochemical reactions. We have an enzyme that acts on a substrate that produces an enzyme substrate complex that then breaks down into that enzyme, that protein again, but the substrate has now been altered into this product.

The rates of those reactions, these rate constants, are the properties of the enzyme, or the properties of the protein. They're captured by, you know, sort of in the dynamic equation for this reaction, by this KP, and KM, which are defined below. And what it basically is is that each of these forward and reverse reactions are properties of a different part of this protein molecule that is the enzyme that catalyzes this reaction. So you can have independent mutations, independent alterations in an enzyme that alter one of these arrows. And of course since the Michaelis constant is a function of the combination of those arrows, a mutation in any one of those parts of a protein can alter the dynamics of that reaction. So store that away. I mean, we'll come back to this as sort of a dynamic. You know, as a way of controlling dynamics in development.

So how do we now start thinking about how processes can control phenotypes? And again, phenotypes are the same thing as characters, or the same thing as structures or patterns; and we use all those terms interchangeably.

There are a number of phenomena for which the processes that underlie them are understandable, or at least are beginning to be understood, and how genes might control those processes, or might be involved in those processes are beginning to be understood. Allometry, I'll give you some examples of this. I won't say much about cellular automata, we've heard about those off and on; but cells are, you know, do in many cases behave as cellular automata do. They exchange signals with each other, and alter their state, depending on what their neighbors are doing. Lateral inhibition processes, particularly as mediated by reaction diffusion, have been important models, and diffusion gradient threshold models; which I'll spend a great deal of time on when we get there.

Let's talk about allometry first. What do we mean by that? Allometry means heterogonic growth. When some part of an organism grows at a different rate than another part of an organism. So that where we have these beetles that for fairly small increases in body size, there are disproportionately big increases in some organ; as the legs in this case. And we know from cases that's probably due to the fact that some time early in development, these legs, or the tissue that forms these legs develop more receptors for a particular hormone that caused them to grow more in slightly larger individuals, than in smaller ones. And again, the hormone receptors are proteins, which are directly gene products. You can see how genes might be involved in that process.

If you look at them in multiple dimensions, you get these things that were originally described, by Darcy Thompson, as Cartesian transformations. And what that really implies, it's a very fundamental observation, is that lineages of organisms that are reasonably closely related to each other are all plastic deformations of each other. In other words, they're within what we call a phylum, for instance; which is a major lineage of organisms. There's by and large no novelty. No new things happen. What happens in the course of the evolution and the radiation of these things, or in the course of the ontogeny in the development of the organism, are basically rate changes of growth in particular directions, and kind of very profound effects. [This is] not to imply that this is ancestral, and these are deformations from an ancestor, but purely a mapping function where you can see that the differences between, you know, baboon, human, and chimpanzee skulls, although they look fundamentally different really differ only sort of quantitatively, and where you have focused your growth. If you put most of your growth up front, and very little of it in the back, you get these big jawed, big snouted kinds of things. Again, we know how to do that mechanistically by with the use of growth hormones, and their receptors.

Reaction diffusion has been a very important theoretical mechanism for pattern formation in biological systems. Reaction diffusion, by the way, is as a rule just a way of implementing lateral inhibition. And what that means is in lateral inhibition is that if you have a particular event, the expression of a gene, or a cell division, or something happening in one location, it tends to inhibit that same event from occurring in its immediate vicinity so that there's some regular spacing of patterns (or pairs). In reaction diffusion you typically have some activating substance, and some inhibiting substance that interact, who affect each other's synthesis by these terms. And a diffusion term, which simply means that these things move around from cell to cell. That material activated that is produced in one cell can be transmitted to another cell. And under the right kinds of conditions – and they're often very, very restrictive – you can get diffusive instability out of these kinds of coupled differential equations. And you can get the emergence of pattern if you let these things run with the right kind of initial conditions, or right kind of boundary conditions, and if you specify all your reaction constants just right.

These reaction constants are again like these "k"s in the Michaelis Menton equation that I just showed you. These are the kinds of things that genes could affect. So these are the activities. The activities of genes are your "k"s, in this case. Let me show you what some of the consequences of that are. This is some of Jim Murray's work. He was interested in spotting patterns in vertebrates. By altering the ratios of these "k"s to each other you can get standing wave patterns for pigment synthesis produced. With different scaling, these produce a variety of different spotting patterns that do resemble those that we see on real animals. And again, the variations that you see between these different panels, although they look qualitatively very different from each other, are really only very small quantitative variations in this generating formula.

Something that we have done a lot of work on is to wonder how butterflies develop their patterns. As you can see from this thing, the butterfly color pattern typically is just a repetition of fairly similar color patterns; where we can reduce this problem to trying to answer exactly, you know, how the pattern in one of these little compartments originates. And so we do this as follows. As we take a butterfly wing we model one of these compartments as a rectangular thing. We know we have a venation system on the outside, which is the circulatory system basically, through which you can supply material. If through that venation system you supply a little bit of that activator from that reaction diffusion equation that I just showed you, then as that activator starts diffusing in, it starts inhibiting its own production, you know, within this field. You get a fairly complex dynamic of activator and inhibitor synthesis that evolves; and it evolves as follows. These are from a computer simulation where the numbers are iterations. But again, we feed it in initially from the side, we get an area of inhibition around that side, and a slight simulation of synthesis in the midplane. After a period of time a very strong autonomasynthesis of this same material begins to appear on the midline. At the free end here, you get a very sharp peak that now begins to inhibit synthesis around it, and this bar eventually declines, retreats, and you've got a permanently stable spot of synthesis of this activator in the middle. So all that we've done with this is we've started the system off by feeding a little bit of activator in here, and we've stably turned on a spot of activator synthesis in the center here.

What does that have to do with genetics and development? These are some results of Sean Carroll, who's been looking at the expression of particular genes in the wings of some butterflies. And what he has found is that this particular gene called distalis again, if this is one of those compartments on an early developing wing, initially has a fairly broad distribution. There is a little streak of strong synthesis that occurs in the middle, as you can see here, with a head at the tip. Eventually that head begins to inhibit synthesis of that material of that particular gene product. And at the end you get the single spot of activation of this gene product that is perfectly stable thereafter.

And where do these things occur on the wing? They occur in all the locations where the future butterfly will develop eyespots. So these centers of gene synthesis or gene expression define the centers of these eyespots; these white spots in the color pattern of these eyespots. And what these cells are actually doing: as they get turned on in a second round in development, they begin to produce a diffusable substance, we believe – it's not been identified yet. But the dynamics seem to be consistent with a simple diffusion of material out of that eyespot whose contours of the diffusion gradient that form are these circular rings of pigment synthesis. So with distance, and with decreasing concentration of this diffusing signal, the thing begins to synthesize different pigments.

What I just showed you was this end point here. That reaction diffusion equation I showed you it had four "k"s in it, you know, four reaction constants in it. By twiddling those reaction constants, you can actually stabilize this point, either high, low, or in the middle. You can cause a bifurcation to occur so that it actually stabilizes as lines along the wing veins. You can stabilize two points in the corners. You can stabilize points over here. You can actually also create a pretty long term stability of some of these long midline sources as well, by just changing the values of these reaction constants. Okay?

So here we have a way of linking enzyme activity, which we understand, to spatial pattern.

Q: In that slide, are the eyes on the veins themselves?

Mr. Nijhout: No, they're exactly in between the veins. The veins are actually right here, but there is a little fold that runs about this far right in the middle between the veins. It looks like a vein, but it isn't.

So here we have then a way in which we can get quantitative variation in genetics. And just quantitative variation in those "k"s. The kinds of things that could be achieved by a very small mutation that just alters the activity of an enzyme just a little bit. And how that can result in what appear to be qualitative changes in a spatial pattern. This is the stuff of development. This is what we want to do, what we want to be able to do. And so we've developed, and proposed – and this is largely yet unproven – a model for how you can get this diversity of wing patterns in butterflies to occur. Where you start with this pre-pattern of the wing venation. I mean that's critically important. If you don't have that, the system doesn't work. But you use those as source material for this activator. You can stabilize that activator production, depending on circumstances in a variety of patterns. These then become the sources of a new set of materials that start diffusing out, and whose gradients that are formed then define these number of different shapes of patterns. So you can get from quantitative changes in genetic variables to qualitative changes in source positions, to truly visually dramatically different changes in the final pattern that results.

And just to give you a sense of what these things look like, there are some of these arcs, and dashes, and whatnot. There's a different animal with different shapes of arcs and dashes. But in any event, so these patterns can get relatively complicated.

So now this is what we were heading for. Drosophila is the organism on which virtually all genetics of development have been worked out. Ninety-nine point five percent of all the genetics of development has been done on drosophila. And this system in which these different colors of bands in a drosophila egg now indicate different genes, different patterns of gene expression. So you can stain the gene product, the protein that a gene produces with particular dyes. You can demonstrate that some of these genes have these patterns that run in perfectly fine stripes along the middle of these eggs. And it turns out that each pair of these stripes corresponds to what in the adult fly will be a segment. So we have three segments in the thorax, a bunch of segments in the abdomen. So the longitudinal organization of the fly is critically dependent on getting these stripes of pattern to occur. And if you disrupt this, either mechanically or genetically, the striping pattern, or the segmental pattern of the fly gets screwed up. So we know that there's a connection there.

How these stripes originate has been the subject of a great deal of research, and some fairly complex and elaborate models have been written for that. A lot of them depending on using reaction diffusion variants of that, where you start with an initially homogeneous egg, and you try to set up conditions under which you get a stable wave, you know, some sort of a stable wave form to be produced that has this frequency that you see here. Enormous amount computational and intellectual modeling effort has gone into this. It turns out what is really going on is that at the beginning of development, there is actually originally a gradient in a protein called bicoid; which is highly concentrated on one side of the egg, and gradually peeters out towards the other side of the egg. And this gradient of bicoid protein, at various thresholds of its concentration, causes different other genes to be turned on. And by turned on mean that those genes now begin to synthesize the protein that they make. That they code for.

So the thing that evolved from this initial simple gradient is a local synthesis of a number of different kinds of proteins. And so that's what you see in this bottom picture. So there's your bicoid gradient. A short time later, you begin to see bands of other kinds of proteins that are being synthesized. But they're all being cued by different concentrations of this bicoid gradient. And then as the second set of gradients gets set up, they in turn begin to activate other genes, and new gradients get set up. And you get these bunches of these intersecting gradients. These little circles have pluses and minuses in them. The little arrows indicate activation and inhibition. But basically what you get is a sequence of activation, of genes being turned on, their protein diffusing away, at particular critical concentrations turning on other genes. Their product begins to diffuse away from that location, and they turn, at particular thresholds, yet other genes on. So it's a sequential diffusion gradient threshold model.

And so what was wrong with the initial attempt at creating the stripes is it was trying to generate them all at once out of some homogeneous, or randomly perturbed field. In reality, what is going on, is that the whole system starts with a pre-pattern, starts with a gradient. Just like the wing pattern started with a pre-pattern of wing veins, and out of that emerged a color pattern.

There was something I was going to say about thresholds. There are ways of making thresholds very sharp through a process called cooperativity. When Matthew was talking about this little region in front of a gene that controls whether that gene will be turned on, whether it will be transcribed. Well, it turns out there are all these things that we call transcription factors that can bind to that region that can determine whether in fact this molecule called RNA polymerase, which actually transcribes the DNA, and will make the precursor for the proteins, can actually figure out where this thing is. And so these things compete with each other. They're activators and inhibitors. They interact with each other in some cooperative fashion, it is believed. So that if one has bound probability, that the next one will bind is enhanced. And all together, this could cause a very sharp change in whether this gene is on or off, depending on a very smoothly graded distribution in the concentration of these different transcription factors. That's it in short.

The longer part of it is that that's all hypothetical. This is a place that almost no proper modeling has been done. For any of you who are interested in looking at this stuff quantitatively, this is a virgin field. This kind of a quantitative regulation of gene transcription that, you know, you could employ thirty of you profitably for quite a long time. Nobody's doing this. Molecular geneticists aren't interested in this stuff. They're interested in sequencing genes, but they're clueless about the dynamics of this process. And they're very, very bad about collaborating with theoreticians, which is sort of unfortunate. No, theory is a really bad name in molecular genetics. In evolutionary biology, you know, theory drives the whole thing. In molecular genetics, if it's not ignored, it's actually ridiculed. It's pretty awful.

So we have these previous theories for pattern formation where the mechanisms were largely some version, some implementation of lateral inhibition, usually reaction to fusion, where the initial conditions were at least to be homogeneous steady states randomly perturbed. But the empirical findings over and over and over showing us – not just in the example I just showed you but in bunches of other examples, in maginal disk development, and so forth – is that the mechanisms in the vast majority of cases are diffusion gradient threshold mechanisms. And the initial condition is there is always an antecedent pattern. You're always building on something that is already there, no matter how far back you go. You never start with initial homogeneous conditions. Biology just doesn't work that way.

And this is just a graphical way of saying that random to pattern doesn't occur. Pre-pattern to pattern: that's what biological systems do. In development. In ecology it might be something else.

Diffusion gradient thresholds. This is just a long list of genes. If you know anything about genetics there are lots of friends in here. Things that form diffusion gradients, and other genes that respond to it. This is simply a support for the fact that the diffusion gradient threshold is an important mechanism in development. At the moment it seems like it's one of the most important mechanisms in development.

And so it's worth asking the question what the effect of genetic variation is. In other words, what is the effect of mutation, or of normal genetic heterogeneity as you find in most animal populations on this kind of a system? And what we mean now by the effect of genetic variation is that a lot of these things are subject to sort of genetic deterministic noise. The rate at which you synthesize this stuff is determined by an enzyme, by a protein. You can have different mutations in that protein, so you get different rates at which you synthesize this stuff. The way you measure the threshold is the property of a protein. The rate at which this stuff diffuses, the diffusion coefficient, is the property of a protein. The rate at which it's broken down enzymatically is the function of yet another protein. And all of these things are variable. So that if you think of the character now as the distance from the point where the source is to the point where this white circle is, you can have variation in all these different parameters, genetic variation, which would alter the position of this phenotype, this character.

Now from an evolutionary point this is a very, very exciting kind of thing, because the only way you can get evolution is if you have variation. And so what an evolutionary biologist would ask about this. And again, the interaction between evolution and development is a very hot subject these days. An evolutionary biologist would ask well, how stable is this system, really? What is the evolutionary tendency of this system? If by some small amount of variation – which we know there exists in genes that control all these various factors – how would such a system respond to natural selection, for instance, what would these phenotypes really look like in a population? We know that no two individuals look quite alike in any of their characters. Can this kind of genetic variation account for the normal variation that we see in the size of this character, for instance, in natural populations?

And so what we did is we looked at this. I'm going to show you the effect of variation of this function source on the position of this vertical line. And that's what is shown here by these three curves where we make three different assumptions about what all the other variables, what all the other parameters, all the other genes are doing that affect all the other. Things like threshold, and diffusion coefficient, and so forth. So if we assume that these are variable, then if they all have a particularly high value, the relationship between variation in the source, and variation in the phenotype is this red line. If they all have very low values it's this purple line. If they have some intermediate value, it's this green line.

The point of all of this is that for any value, any genetic value of this source, you can create any range of phenotypes, any range of forms, depending entirely what all the other genes are doing. Likewise, you can get any value for the phenotype, like this value ten, for any of these values of this gene. Depending entirely on what all the other genes are doing. So no gene by itself determines the value of the phenotype; it's a property of the ensemble. It seems kind of logical. The six variables that we thought were involved in the standard diffusion threshold problem are source value, diffusion coefficient, a timing constant, the value of the threshold, decay of the material. So this is synthesis here, this is decay, and there is always some background value, some background rate of synthesis. The point is that all of these have a space that defines the relationship between its value, and the value of the phenotype that is very, very broad. So the variation in any one of these by itself has very, very little variation on the phenotype.

Now that sounds like a very bothersome kind of a thing, if we think that genes control characters. Here is a character that is completed determined by these six genes, yet none of them has a particularly big effect on it. A surprise to me, maybe not to many of you; but there we are.

So we ask the question: suppose now we do selection? Suppose we do to these things what happens in nature; we do a selection experiment. Selection only happens on phenotypes. That's the only thing that natural selection can see. So we created a computer population of organisms that created this phenotype, this character by means of these six genes. You had some variation in those genes. So now we create a big population that's a normal distribution of this phenotype. And we said okay, now let's select for small values of the phenotype. Everybody that has a large value of the phenotype doesn't get to breed. Everybody with a small value contributes to the next a small value, or a large value that will contribute to the next generation. What distribution of phenotypes do they now generate? Now because there is such a poor relationship between genotype, and phenotype, and because a lot of these things compensate for each other; so a large value in, you know, threshold can be offset by a low value in source number, in source value, it looks like this kind of a system would be fairly nondeterministic, would be a bit of a mess, and a selection experiment wouldn't get you very far. But this is what happens. This is an upward selection. So you only allow breeding of those individuals in a population that have a particularly large phenotype.

This is the response to selection of the average phenotypic value. If we start with the value down here. How we get it we can talk about later. Over a number of generations of selection, we get a beautiful response to selection. That phenotype does everything that you would expect a beautifully behaved phenotype under selection to do; like you get out of laboratory selection experiments.

When we looked, however, at what the genes were doing, at what the frequencies of the different alleles, the different mutant forms of these genes that we had in this simulation were doing, what we saw is that most of them, during the early parts of the simulation, didn't change at all. But there were one or two that very rapidly responded to the selection by changing in frequency. When those started to level off, the next gene began to respond. When that leveled off, the next set of genes began to respond. So underlying a very smooth phenotypic character response to selection is a fairly messy, though highly repeatable, genetic response to selection. So we begin to wonder how can you explain this? What is accounting for the fact that some of these genes don't seem to follow the phenotype very much? And, if we simply ignore this lower panel for a second, what we did is we calculated the correlation between the variation in each of these genes, again, that are involved with making this pattern. The correlation between each of those genes, and the phenotype. So this is essentially asking the question if I made a mutation in that gene, what effect would that mutation have on the phenotype? A low correlation means the mutation in the gene has no effect. It doesn't change the phenotype at all. A high correlation means it does change the phenotype by quite a lot. So what we have is at the beginning of this experiment, genes just differed very greatly in their correlation with the phenotype. There was one gene that was highly correlated, and everybody else was sort of intermediate, or low. As selection proceeded, progressed, this gene became progressively less correlated with the phenotype, and a different gene became highly correlated with the phenotype. Then this guy gradually declined, and a different gene became highly correlated with the phenotype. Each of these genes took their turn being the most important gene.

Now, if you had been an experimentalist, and you did not know this, and you had had a population of organisms at about this point in their evolutionary time, and you had a mutation in this gene, you would say well, this gene, the mutation has no effect on the phenotype. That gene is not involved. You had a mutation in this gene, all of the sudden the phenotype, bang, disappears. You say ah, this is the gene for that character we're studying.

If you had taken the population at some other time, this gene would have had no effect. It would be this other gene that you would have said oh yes, that's the gene for, you know, long tails, or whatever this character is. The point is that which gene you identify as being the most important gene for the character depends entirely on what all the other genes are doing. And that is the complexity of development. And it isn't that genes are themselves very complicated. They're kind of stupid little things. It's all in the interactions among them. And this is the kind of thing for which there's very, very little theory. Likewise there are ways of explaining how this happens after the fact. At the moment there are yet no ways of predicting this.

I told you I was going to come back to this slide, and I'm just about done. That this one observation seemed to be really kind of obvious, where you had this defective gene that caused the disease. So therefore a defective gene is correlated with, in fact causes, disease. That is okay for a simple disease. But for complex diseases that are controlled where the character is the property of the interaction of many genes, you cannot make that statement; even in genetic diseases. A good example would be cancer genes, you know, breast cancer gene, for instance, which we know is associated with breast cancer only a very small percentage of the time. And only in certain populations. So in other words, it's already telling us that this gene, whether it causes, whether it's associated with breast cancer depends on what all the other genes that this individual happens to carry are doing. So we can very well there have a situation that is very much like this one. I mean, okay, this is a breast cancer gene for this population, but for a population with a slightly different genetic makeup, it's all right. You can still have breast cancer, but it will be for a completely different reason. We know this in cancers in general. That they have many, many contributing factors; out of which any two or three will get you over the hump, and cause trouble, but no single one is particularly necessary.

Q: ???

Mr. Nijhout: Oh, that's my personal prejudice. I have nothing against plants, I just haven't thought about them. Plants grow in a somewhat different way in the sense that they don't have morphogenetic movement, so they tend to have cell division patterns are much more important than cell pushing patterns. But it ought to behave in just these ways, with some plantlike variance on it. But nothing really that would make them in principle different.

Q: You mentioned that there are these pre-formed gradients as far back as you go; which says that life comes from life.

Mr. Nijhout: Precisely.

Q: And the question is: what do you think is the source? How is that transmitted?

Mr. Nijhout: You've got to go back to the pre-Cambrian. In development, the mother puts that gradient. When the mother makes this egg, for instance, as she is packing that egg with yolk, at one end of the egg she packs a whole bunch of RNA, for instance, that codes for this bicoid protein that's going to make that gradient. So she sets up that asymmetry in making the egg, and then she actually puts some other proteins in on the other side. So, I mean, there are some intersecting gradients.

Q: Don't you think there are other influences?

Mr. Nijhout: I think there are a lot of other influences; for instance, that gradient. You know, I mean, it is a gradient of a gene product. But the important thing is the gradient, you know, which is a physical process that produces that.

M: I'm speaking more specifically. In mechanistic terms, maybe there's an effect of the membrane.

Mr. Nijhout: Yes. But again, the membrane is fifty percent protein, so fifty percent direct gene product. So you don't know whether genes are involved in the process. The best way to think about this, and about to what degree genes are involved in these processes, are again: genes only code for proteins. If a protein is instrumental in something, it's a direct gene effect. If it is not, then it's an indirect gene effect. If a protein is somewhere in that chain of causation. If you can develop a chain of causation that ends not at a protein, but at an environmental signal. You know, photons of blue light, or something like that. Then that's a cause, and there's no genetic effect.

Q: The constituency in this room is concerned with biology, but also other social practices. Two authors you didn't mention at the beginning. One guy – Steven Weiss, I think – has written a book quite recently. He actually works on the brain. And he's tried to emphasize the disintegration of the nature/nurture divide very strongly, which is more or less where you ended up. But he would have liked, I think, to put a bigger loop in there between environment and genetics. Could you put more loops in there?

Mr. Nijhout: Yes. Well, if I may just take the liberty of going back two steps here I'll tell you why I – in the time allowed – the six panels where the genetic variation lead the phenotypic variation. Each one of those scales for genetic variation could be an environmental variable just as easily. For example, he right diffusion coefficient, temperature, you know – you know so they're completely translatable. And they're all effective. I mean you could do a normive reaction, for instance, if you're familiar with that term, on that system, just as you could do a genetic effect, and get fairly similar results.

It so happens in the drosophila egg, there are no cell divisions until fairly late. There are a bunch of nuclear divisions. The nucleus is in the middle of the egg. It divides seven or eight times until there are a bunch – several hundred nuclei. By that time a lot of these gradients have been set up. Then these nuclei migrate to the surface, and become cellularized. And now all of the sudden you freeze in that pattern in cells.

Mr. Dan Stein: We started with a progression, after our overview, of some physical sciences, to computational, to biological. The computational was sort of a nice bridge. And there were a number of themes that were common through a lot of these. In the morning, I think pattern formation really was the overarching theme; as it is, I think, in much of complexity anyway. From some set of fairly simple rules, you end up with rather complicated patterns. And how you go from one to the other is a mystery. And this happens in physics, chemistry, biology, economics, sociology, and so on. And in the morning, a lot of the themes that cropped up again and again, of course, were scaling, stability, and chaos. In the afternoon it was adaptation, plasticity, and so on. In some of the cases, particularly the physical cases, the dynamical rules were static. They don't modify as time goes on. Often in biological cases they do. But both kinds of adaptive and nonadaptive systems do lead to interesting complex behaviors. We'll see a lot more of that during the next week. I hope that during the week, a lot of those themes are tied together. And I think that this session made a very good start.

Reception Session

Jeffrey Robbins
Perseus Books
Richard Wrangham
Harvard University

Good evening. I'm Jeffrey Robbins, executive editor for the advanced book program at Perseus Books. I'm very happy to be here at NECSI's second international conference on complex systems. Those of you who were here last year for the first of NECSI's international conferences pretty much know what to expect. An active, productive week of expanding your horizons, and those of your discipline, in the talks and discussions with researchers in a variety of fields of science and scholarship. Those of you who are newcomers, get set for the most stimulating conference you have attended in years. This is a very exciting moment for the advanced book program, because as NECSI's chosen book publisher, we are about to publish the first two books of the NECSI/Perseus partnership: Unifying Themes in Complex Systems. Proceedings of last year's conference. They're edited by Yaneer Bar-Yam, founder and editor-in-chief of the series. Actually, it's called the NECSI series on complexity. And "Virtual World", edited by Jean Claude Heudin of the International Institute of Multimedia of Leonardo da Vinci University, Paris.

The NECSI series on complexity is intended as a reference series that reflects both the diversity, and the unified nature of the study of complex systems. Those first two volumes I mentioned will be out in January. And you can get preliminary information on them at the Perseus booth in the exhibit area, which will open tomorrow, and will be open every afternoon from Monday to Friday.

Tonight I have a special pleasure. A graduate of both Oxford and Cambridge Universities, Richard Wrangham is presently Professor of Anthropology at Harvard. He is the recipient of many awards, including a MacArthur Fellowship. He serves on the editorial board of four journals. He is a trustee of the Jane Goodall Foundation, and is on the scientific board of the Gorilla Foundation. For you radicals in the audience, please note that that's gorilla spelled "G-O-R-I".

Recently he published a landmark popular science book on the biological origins of male violent behavior entitled "Demonic Males". Which one reviewer called "One of the most important books of the decade. Read it, learn, and if you're male, resolve to fight the urges that demean you, and hurt others." Richard's current fieldwork takes him to Western Uganda for from two to seven months out of every year as director of the Kibali chimpanzee project. I'm very glad to say that this is not one of those months, and so I have the pleasure of introducing Richard Wrangham, who will speak on his theory of the roots of violence as adaptive survival behavior, and the similarities between chimpanzee, and human patterns of violence. Richard Wrangham.

Mr. Richard Wrangham: Thank you for that introduction. I am duly daunted by being faced with a lot of complex systems people. I am the light entertainment, and I will stick to it in that way. But of course it's odd to be doing the light entertainment on a subject that is fairly heavy; namely: why are human males so violent? That's the topic that I've become fascinated by since being exposed to the fact that chimpanzees are rather human-like in their patterns of violence. And given that fact is obviously, rightly, that there is some kind of evolutionary similarity between chimpanzees and humans, and maybe we follow the same rules as they do. In which case, it's rather an engaging thought to explore whether or not in the much simpler system of chimpanzee violence, we can see some of the same causes of violence as may be applicable to humans.

So what I'm going to do is talk about three kinds of human violence, if time doesn't run out; and two if it does. We'll be getting increasingly towards wild speculation as we move onwards. It won't be just the fact that you're falling asleep that you don't follow the argument. And the essence of what I want to do is to look for simple explanatory systems. I am aware that human violence is a complex system. And the principle that I take is that what we need to do is to start out with some very simple patterns in order to answer some of the big questions. And the sort of basic big question that I begin with is this. We only know of two species of animals in which males live in groups with other individuals with whom they form coalitions. And those coalitions sometimes go to neighboring ranges, and look for other individuals, and hunt, and kill them. And those two species are: humans, and chimpanzees.

So the sort of premise that I start with is it seems likely that there is a rich vein of explanation to be had in that comparison.

So what I want to do is to introduce you a little bit to one of our close relatives, mainly. "Demonic Males" was a book that reviewed humans in comparison to other apes. And part of the premise is that there is a special demoniacal quality to some males. And I think of this exaggerated pattern of coalitional aggression as being part of that demonic quality.

We've been running into just a little bit of flack in a couple of reviews, because it turns out that this picture, which is on the cover of the book, has offended some people. Not because they come from the Deep South, and are puritanical, but because the figure on the right turns out to have been made of an African. Something of which I was ignorant at the time that I chose this.

I thought that what we were doing was choosing just a representative human. And the fact that his skin is the same color as the skin of the apes suggested to me that it would not be a black person, because apes don't have black skins. However, if you look very carefully, apparently you can detect something in the physiognomy which tells this. This is ironic, because the whole point of the book is that Dostoyevsky was right when he said that this is not something about different kinds of humans. The demoniacal qualities are something that I attribute to everybody. So when he says in every man, a demon lies hidden; and then goes on with the kinds of demons, that's the attitude I'm taking. Not that I am seriously worried that anybody here is going to leap out of their chairs and attack me with a meat axe. But that, under appropriate circumstances, there is more than just the capacity to do so. There will be, in some cases, a tendency to do so.

And this is very different from the politically correct view, which was nicely summarized in a Seville statement of violence in 1986, where on the right we see the claim that violence is neither in our evolutionary legacy, nor in our genes. And this is very much part of the social science program; which I think is an absolutely admirable concept. The only question is whether or not it is accurate. I think it comes from the fact that people are keen on seeing that the same species who invented war is capable of inventing peace. In other words, that there is no tendency pushing us one way or another.

Now that sort of criticism I think leads to some odd perspectives. Here is Stephen Jay Gould saying why it is that the kind of thing I do is so terrible. He says if both darkness and light lie within our capacities, which of course is clear, and if both tendencies operate at high frequency in human history, then we learn nothing by speculating that either or probably both lie within the evolutionary adaptive Darwinian heritage. Nothing. However, he does go on and say at the very most, biology might help us to delimit the environmental circumstances that tend to elicit one behavior rather than the other. Which would seem to me to be rather a triumph if we can do that in a predictable way.

So I just want to get my position clear. I would love it if it were the case that we could say with a magic wand that there is no evolutionary heritage of aggression in us. It doesn't seem to me to be a likely position, but I want to just explore it.

I'm going to be doing these three parts. The first part, I'm going to be thinking about the evolution of raiding into neighboring ranges by males. The second I'm going to be thinking about battles. And the third, if we get to it, I'll be thinking about male/female relationships.

The overall sort of thought here is that humans are really a very odd species. Not just in the fact that there is this extraordinary tendency for coalitional aggression to be expressed against each other in the form of lethal violence; but also that there are tremendous amounts of violence which is created by male/male competition for females, which very often takes the form of males locking up females, guarding females, fighting over females. And that much of this is related to the fact that females are continuously sexually attractive.

These are odd things for a species. And they're unsolved problems as to where these patterns come from in humans; because they don't happen in our close relatives. So I want to just think about that stuff too.

So for those of you who haven't seen it recently, the way it happens in chimps is that you have a small party of chimpanzees lying within their territory. The way chimpanzee social system is organized is that a group of fifty to a hundred individuals share a large-ish home range, a territory, within which they break up into small parties, and over a twenty year period it may well be that all members of the community never come together. They only just associate in their small parties. And the males within the party form alliances with each other. And on a particular day the males of several adjacent parties might shout at each other, and charge among each other; particularly fired up by one individual, who will tend to be the alpha, or the highest ranking male. And then he will lead them off to the edge of their range. And this is a peculiar expedition, these boundary patrols, because instead of foraging on the way, all they do is go and sit, and listen, and look into the neighboring territory. Now that's particularly striking, because they spend over half their time foraging. So this is an odd thing that they shouldn't be doing that. And if they hear a large party of chimpanzees in the neighboring range, they will shout back at them, and then run back into their range. But what they're really looking for is this, a lone individual. And if they see one, and detect one by hearing, then they will hunt them down. And the hunting is very subtle, very careful. It can be slow and quiet until they get to within a few meters, and then make a rush. And in the course of the rush, if they are able to get there before the other one has escaped, someone grabs a foot, someone else grabs an arm, and then they do very nasty things. This is not just escalated fighting in contest over a limited resource. There is no resource at stake. They are very deliberately going after this individual, and they do very deliberately nasty things. Such as taking a limb, and starting to twist it until it's three hundred and sixty degrees, and then the limb breaks. The muscles tear, the bone can break. They're very strong. They can put a bit of skin in their teeth of the victim, and then rear the head back, and skin strips all the way down the arm. In my study site in Uganda last month we had a case where it was our chimps rather than the strangers killed a stranger in which they ripped his trachea out, and took his testicles, and laid them a couple of meters away. This is not chance occurrence. So the result is that sort of thing. This is Ruin Zorie, who I knew as a fine young male; and this is he after he was found by the neighbors.

Now these are individuals that expect to live a long time. They've got a life expectancy of maybe fifty years maximum life span. And so it's a bad idea for them to have this happen, and of course they try to avoid it.

In many ways it is very similar to the kinds of patterns of intergroup aggression that occur in many peoples living in isolated villages. Politically isolated camps. Similar in the sense that a small group of males go off on a deliberate expedition in search of a potential victim, and then kill him by any of a variety of means. By plunking spears into him, or setting fire to a hut, or whatever it happens to be.

We don't know too much about the rates of death from these things, but where it has been possible to get data from societies that were still fighting with each other, independent of any colonizing power, or powerful neighboring tribe, what we find for thirty non-state societies is that for humans the median society had a percentage of males dying from violence between twenty and thirty percent. Dying from inter-village violence. Going up sometimes very high. Females also suffering. And then here is a chimpanzee pattern. Coalitional aggression, remember, is virtually unknown as being responsible for violence in any other mammals.

The raiding. Here is Lawrence Keeley's compilation of data on the rate of death from raiding in pre-state societies. Lots of variation, but a high median. Here it's expressed differently as percentage kill per year. And what is striking is that this little sample of state societies in battles; that includes Germany and Russia during this century. So what we think of as a simply appalling century, if we had the rate of death that has been estimated from raiding in politically independent villages, we would be talking about billions rather than millions dead.

So the problem of human violence used to be thought of as a particularly human problem, and therefore as something to do with the brain. The brain of chimpanzees is quite small compared to humans, and if we have similar principles involved in these two species, then clearly the brain itself is merely a contributing feature, rather than a stimulus for the occurrence of this pattern. And people have in the past speculated that it's something about culture. Maybe new kinds of tools that enable you to kill at a distance, or whatever it is. But again, chimpanzees are not using any kind of cunning tools in their violence. Therefore, the hypothesis that seems to me to be most reasonable is that it is something about the mind of the chimpanzee, and that humans have probably had the capacity for this sort of violence ever since our ancestors had brains the size of chimpanzee's; which was about five or six million years ago.

So I want to ask the question, why do we get these patterns of coalitional aggression, and suggest some ideas. And I'm doing this in the last, as it were, of Tim Brogan's famous four questions about how to approach "why" questions in biology. Which is why is it that natural selection favored the killing behavior in our species, as opposed to other species. And I'm going to give just a pathetically simple kind of answer. The key seems to me to be an imbalance of power when rivals meet with rivals from neighboring groups. So I have an imbalance of power hypothesis. The key that I suggest is that when one individual victim is attacked by at least three others, then those three can make the attack at very low cost to themselves. Because a couple holding a victim down allows a third to impose punitive damage. And this is in fact the case in all of the observations that have been seen; that it is at least three individuals who are involved.

So what is it about our species that means that we have a special tendency for parties when they meet to be imbalanced in their relative power? Well, the sort of special feature about chimpanzees is that they are fruit eaters. And compared, for instance, to the monkeys that live in the same forest, we have very big differences in the amount of fruit. These are meant to be frugiverous monkeys, if you look at the descriptions in the literature; but during periods when fruit varies in its abundance in the forest the chimpanzees insist on eating lots of fruits, and the monkeys are relatively uncommitted to fruit eating. And here we have the beginning of a very simple causal chain which takes us from the distribution of the food that these animals eat, up to their social patterns. And the key concept is scramble competition.

There are two kinds of competition we can recognize in social systems. One is scramble competition, the other is contest competition. Scramble competition is competition that occurs as a result of resources being lost merely as the result of the presence of another individual; regardless even of whether you actually meet. It might be by day or by night in the same environment. Whereas contest competition is fighting over stuff. And I'm suggesting that scramble competition is important. Here is the way that chimpanzees have a day. They wake up in the morning, they travel, and then they eat at a tree. And they may often deplete that tree of fruit, and then travel, and go onwards, and so on. And this turns out to be very different from the way that most primates do it. Because most primates forage as they go between their little fruit patches. The result here is that if you're in a large party, and you come to trees that don't have enough fruit, then at the end of the day they don't have enough calories, so in order to eat their two and a half thousand calories for the day, they have to visit more patches. And this increased travel represents a cost in terms of time or energy.

So that cost, it turned into efforts to minimize group size during periods when food is scarce. And here with a measure that really doesn't do justice at all to the chimp's perception of the environment. Simply the percentage of trees that have ripe fruit out of, I might say, four thousand trees is back breaking stuff. We find, and everybody does, a clear relationship between party size, and the amount of fruit in the environment. There are times when there is too little fruit, and they are forced into relatively small parties. And the idea is that there is just spatial heterogeneity between neighboring territories. So that sometimes one territory has got lots of fruit, and sometimes the other neighbor has not very much. And one is forced to travel in small parties, and the other can't afford to be in larger parties. And then the one with larger parties can go and commit mayhem on the one that's small.

Are chimps capable of assessing the significance of their own party size in relation to others? Behavioral observation certainly suggests so. And here a graduate student, Michael Wilson, is conducting playback experiments where we give the call of a chimpanzee, a stranger chimpanzee, near the edge of a community range, and we have various control systems with eagle calls, and so on. And then just after giving this call, you pack up the speakers, and run away. And the people who were sitting with the chimps don't know where the call was given from, and later it turns out the chimps are incredibly good at localizing it. We play from between two and three hundred meters away, and the chimps have sometimes gone and sat right exactly where the speakers had been earlier placed. What we find is that chimps are very good at working out how to behave in relation to their own numbers. So if you play a solitary call, then by the time you get up to three adult males in the group, then they will always approach. And a number of behavioral assays like that show them to be very sensitive to the relative balance of their own numbers, and the opponents.

So with that sort of information, here is the sort of half of the cascade that suggests to me that there has been selection for a violent brain. That we have intermittently intense scramble competition leading to a foraging system of fission and fusion; in which parties are sometimes large when they can afford to be, and sometimes small. And we see lots of details that support this, in the sense that you get things like individuals visiting the edge of the territory normally are in large parties. The small parties tend to be in the middle of the range.

You get occasional power imbalances as a result of this, which are to some extent predictable. And that means that careful raiding, in a sense of being able to assess your power in relation to the neighbors, can be done cheaply. If it is the case that three or more individuals can, at very low cost to themselves, attack. And so far we have no cases of the aggressors being wounded during the course of imposing this appalling punishment. So that one could in theory have selection for violence that is careful in the sense of being careful to assess the relative balance of power.

That doesn't say why there should be and the suggestion I want to make here is that there should be a selection in favor of killing members of the opposition if in fact you have contests between coalitions that will ultimately lead to variation in reproductive success, because it pays to reduce the military power of the neighbors.

And just towards that, I want to start with an observation that this is the typical way that you see a mother chimpanzee. And that is to say, alone with her offspring. So we think of these as very sociable animals. And that sociability is based on the males. The males are very gregarious. It's very unusual that we have, in chimpanzees, a species in which the males are more gregarious than females. Here are two samples showing how the percentage of all of the individuals in an age sex class within the community are represented as party size increases. As you get up to about fifteen or sixteen adults in the party, you find that a hundred percent of the males in the community are represented. That is not true for the females. Only twenty percent of them. So that represents the fact that females are very much less gregarious than males.

In order to understand why that should be, I resort to thinking about the cost of grouping. It is a regular feature in primates that eat fruit that as the size of a group increases, the distance they have to travel every day goes further. Much as I was showing with the scramble competition argument. It's just more intense in species that have fission and fusion. And you can use observations like this to calculate the percentage increase in the day range, according to the group size. And this will vary between species according to how fast or slow they are, and various other phenomena. In the case of chimpanzees, it turns out that females that have offspring are substantially slower, 2.1 feet per second here – than adult males at 2.8. And this is because the babies are an energetic burden, as well as being a timing burden. Sometimes they want to move, and the babies are still playing. So I assume they're traveling at their optimal speed. And we then have data looking at how day range changes with group size for both males and females. And if we plot them on the same equivalent scale, what we find is that for females, as a result of their slower speed, they have an increase in day range that is steeper and faster than the males do. And the optimal size, if they are to maintain their original total time spent traveling, is therefore different for these two sexes. And it turns out that the predicted amount of group size, in order to have the same amount of time spent traveling, is exactly what is observed. Namely two for the females, and six for the males.

And that sort of observation suggests that these differences in locomotor velocity are responsible for putting them on different cost of grouping curves that mean that they need to be at different party size.

And here are the females that don't have babies in two separate years to test it. So the presence of a baby appears to be particularly responsible for a loss of gregariousness in these females. That kind of thinking supports a benefits side of the cascade here, where scramble competition is weaker for males than for mothers. As a result, males are able to spend time with each other in a way that the mothers are not. Merely being able to spend time together enables males to form coalitions with each other. And we can see the equivalent of this when we bring females into captivity, and they're able to spend as much time as they like with each other, and they do form alliances with each other which they use aggressively against other individuals.

So the argument is that in the wild, males are regularly able to do that. And they have been able to do so so predictably that coalitions of males have been able to defend a territory, and it has then paid for males to reside in that territory, and to defend it, with their kin, against the neighbors. Once you have a system of male philopatry where males born into the group remain in the group, and defend it against the neighbors, then you have male rivals being replaced very slowly. In fact, if you kill a member of the neighboring group, he's only going to be replaced by adolescents maturing into new adults; and that is going to be slow. Hence it pays to kill, and so we can move to a violent brain.

So the logic of this hypothesis is that when you take the opportunity for individuals to go out, and raid the neighbors, on average it is the case that there are low costs, and that in the end there will be benefits, because you will have times when there are going to be contests with the neighbors when you can take advantage of their reduced power. But in any specific case, while the costs will be low, the benefits will be, for that day, unpredictable. It's not that they're going out, and seeking an opportunity to kill because there is a particular fruit tree on the border, or anything like that.

So natural selection integrates the costs and benefits over time, and as it were puts them into the brain. And that perspective seems to me to be fairly effective in working with humans. You have a number of similarities with humans. Male gangs go out, and seek victims. There is very little difficulty in persuading men to kill members of neighboring groups. And to say that rivalry alone justifies the violence seems to me to be not far from the way that anthropologists describe the reasons that people give for killing in raiding situations. They talk about revenge, they talk about witchcraft, they talk about long term historical relationships. But it's not that easy to come up with really good reasons for why are you going to go out, and knock off the neighbors.

Well, a nod to how we can sort of further think about this. It is ninety-nine point nine percent certain, I think, that this is the correct clatogram for the hominoid evolutionary relationship. So here we have orangutans, we have gorillas. And now we have humans spinning off before chimpanzees, and binobos. What I've been talking about is this species, chimpanzees. We have another species, binobo, which does not show this behavior, and shows no indication of showing this behavior. So that any kind of theory must explain what it is that's different about chimpanzees, and binobos in order to solve the problem of where this comes from. So we have two kinds of possibilities. One possibility is that we have convergence between humans and chimpanzees to the extent that these patterns are similar, and the other is that there is a pattern of common ancestry, which is actually what I think happened.

The binobo is a remarkable species which has the same body weight, and degree of sexual dimorphism as some chimpanzee populations; but is much more lightly built, and shows much less tendency towards violence. And it, critically, lives in larger groups. So then a big question becomes why does it live in larger groups? And I have some fairly specific hypotheses about that, which come back to a reduction in scramble competition for this species as a result of its curious biogeographical position on the left side of the Zaire River where it lives in an area which has a slightly different ecology from that occupied by chimpanzees and gorillas on the north side. I'm just going to leave that as a nod, and we can discuss it in question time if we want to, in order to move on to the larger picture.

One of the questions that this kind of analysis I think prompts sort of really severely for people like me is what did our ancestors do in their environments as far as stable groups, or fission – fusion groups? And so a key question is was there a pattern of scramble competition for early hominids similar to that that we see in chimpanzees? And that's a reasonably open question, but you can tell which way I think.

Well, that's all I want to say about lethal raiding. Suggesting this horrible concept that you have a long-term selection in favor of males using the opportunity to assess coalitional strength, and knock off the neighbors when they can. But however, I want to emphasize that there are lots of differences between chimpanzee and human patterns of violence as well. Some of these are mentioned here. I want to draw attention to this one – which is one that tends to amuse academics – that there is a theory of military incompetence. Who thinks that we need a theory for it?

Military incompetence occurs in the context, particularly, of battles. So here is the distinction between raids and battles. We've been talking so far about raids, which appear to be enormously the most important form of intergroup aggression in human evolution. And a raid is a sure thing. The victim is surprised. A battle, on the other hand, is not a sure thing to either side. And the classic kind of battle is one in which you have two opponents, both of them wanting to engage. Now, animal aggression theory says that you shouldn't have battles; because when two animals have the opportunity to assess each other, then the eventual loser should be able to assess that he's likely to lose, and therefore give up without a fight. Because if you're going to lose, it's better to lose without a fight than it is to lose with a fight. So the fact that battles very often happen is peculiar.

Now, battles do happen in monkeys a little bit. But they are relatively uncommon. And they are not highly escalated, and very rarely lead to anything lethal. I mean virtually never. Battles are known, of course, in tribal societies, but again, not very common. A little bit of lethality. They are much more common once we get into agricultural peoples.

So here's the distinction I want to draw between raids and battles. Raids involve surprise attacks. The weaker opponent is an unwilling participant. You have a very high death rate. It's known in chimpanzees and in humans. It involves accurate assessment by the stronger, and no assessment by the weaker, and no opportunity for this thing that I want to now talk about, which is self-deception. Whereas with battles we have escalated conflicts. You can give up at any time, but you very often just keep going to the point where there may be some kind of appalling attrition. The weaker opponent is a willing participant. Of course you do get battles where both sides' backs are against the wall, but I'm talking about the cases where both in fact are willing. You get a lower death rate. This is really a human characteristic; lethal battles. And in retrospect, the stronger individual had an accurate assessment, but the weaker one had an inaccurate assessment. And I'm going to suggest that they have self-deception.

Now this is an incredibly important thing. And it turns out that people think of this happening very often in the initiation of wars as well. That both sides think they're going to win. And you would think that natural selection would have put into our brain some fairly good mechanisms for working out whether or not you're going to win or lose, and try and avoid losing. And try to avoid getting into a battle that you are going to lose.

So what do the psychologists tell us military incompetence consists of? Well, the classic thing is that people are excessively confident about their own prospects. They systematically underestimate the enemy, and overestimate their own ability. And there are all sorts of group processes that go to sustaining this illusion, ignoring of intelligence reports being among them.

Now these are examples of self-deceptive biases that occur in all sorts of ways. People have positive illusions, they also have negative illusions. Men more often have positive illusions; women more often have negative illusions, and so on. People vary in the illusions, in the context, and so on, over life. Some examples are overrating a personal, or team prowess, exaggeration of your own contributions to joint tasks, and past successes. You know, this is the sort of thing where you ask a series of couples how much each of them does the housework, and it always adds up to about a hundred and twenty percent. The median good driver, according to everyone's personal estimation, turns out to be about eighty percent. We never get the fifty percent point. And everyone thinks that they deserve better in a dispute than they actually get, and so on.

So why do these things happen? There are two kinds of possibilities. One is that there is just an assessment failure, which doesn't make a lot of sense, if you think in terms of natural selection. And the other is that these are adaptive biases. And there's reasonable evidence that those sorts of things apply. Because in general, people are pretty good at making assessments. And you can argue that very often the information about an individual's competitive ability is there, and it's just only revealed when it is really needed. And depressive realism is one example of this, where people actually become more realistic when they become depressed. Until then, they have an excessive perspective on their own abilities.

So why should fighters have positive illusions? The theory, until about fifteen years ago, was that generals just tended to be individually stupid people. And it is doubtless true, occasionally it happens, but for systems analysts it doesn't seem very exciting. A sort of slightly more sophisticated version of that was that in a military hierarchy, what you tend to have is promotion of people with characteristics that are good at being promoted within the military, but are bad at taking the kinds of decisions, and making the sorts of assessments that are needed in a military engagement. But there are various reasons why this doesn't look at all a good idea. And that is that it's not just the military at all. This kind of tendency for exaggerating your own ability, and minimizing the ability of the opposition occurs way outside military contexts; such as politicians who are making decisions about the military, or in sports, and so on.

So another possibility is non-adaptive self-deception, where there are just biases, or stupidities that happen to be lodged in the species. But it doesn't seem, in general, as though the biases that one sees are fixed. They instead tend to vary with context in ways that suggest that they may be being used in appropriate contexts. And I want to just refer to the idea that what we should be thinking about is effective bluff; bluff which may have a variety of positive effects.

Now there are two suggestions here. One is the performance enhancement hypothesis, and the other is an opponent deception idea. The performance enhancement hypothesis is the one that would have helped the New England Patriots win today. They gave up with forty seconds to go, and need the ball, and didn't try and move it, and then the Miami team got the ball immediately in overtime, and scored. And the comments on the radio as I was coming up here this evening all agreed that this was a failure of nerve on the part of the New England Patriots. That they failed to think of themselves as champions; they allowed negative thinking to come in. Championship thinking is something that all sports psychologists know about. I don't think it's been talked about very much as a phenomenon that enables you to succeed, particularly in contests. But it seems very important that if in fact you allow yourself to start thinking about the possibility that you might lose, then that very process might contribute to your losing. And natural selection may have favored brain systems that therefore hide those possibilities in contexts in which you need to have everything going in order to be able to win; particularly in an arms race with an opponent that is suffering the same kinds of procedure.

Here is another example of the way that positive illusions might be favored. And this is an idea about bluffing which resonates with the animal aggression literature, but takes it a bit further than it has been taken before. That is that in an even contest between two opponents who are relatively evenly balanced, and in which the opponents are favored to give up if they perceive themselves to be of lower quality than the other; then it pays each to bluff as long as the costs of probing are not too high. And in a situation in which bluffing is favored by each individual, then you may get to the point where in order to bluff as effectively as possible, you have to deceive yourself in order to be able to avoid revealing any kind of hesitation, uncertainty about your own ability to win. So this is a minimization of behavioral leakage concept. Either way the idea is this: that as individuals deceive themselves into thinking that they are better than they really are, then each individual misreads the opponent, and the zone of negotiation in which each can give up, from an adaptive perspective, is reduced. And the probability of them both fighting is therefore enhanced.

So I think now of positive illusions as being rather equivalent to the teeth of male baboons. In which, from the point of view of the baboon species as a whole, it's a terrible idea that males have these things, because they go around cutting other baboons. But from the point of view of the individual male baboon, natural selection or sexual selection favors them having long teeth because it means they're more effective fighters against others. So positive illusions in contests are argued, by this idea, to be the equivalent. Very bad for the species as a whole, because it visits mayhem on our species. It means we have all sorts of military encounters, battles, in which people are prepared to fight harder, and more intensely than they should be. Because they're misreading a situation. But from the point of view of each opponent, it pays.

So that's an argument for thinking about why in wars we always tend to think that our own side has the moral virtue. We always have the moral virtue on our side, because we are misreading our own competitive ability compared to the other, and give ourselves positive illusions about our own qualities.

This is very speculative, but a study has been published specifically drawing attention to the fact that as you increase the testosterone in, in this case, individual women, therefore adrenal testosterone, you get an increased tendency for self-deception. And in our chimpanzees, we find that higher-ranking males have more testosterone. And I just love the idea that one of the effects of testosterone psychologically is to modulate the relationship between your perception of your ability, and your real ability. I think this would help explain why it is that testosterone doesn't have predictable effects on aggression, but if you get into situations in which it matters whether or not you perceive yourself correctly, or can read the world correctly, then enhanced difference between reality and your perception may lead to an increase in probability of aggression.

So, and therefore thinking that there are two kinds of ways in which natural selection has favored a sort of immoral behavior in human males. One is that in the raiding situation, sovereign groups of males like to go, and beat up on the neighbors. And in a battle situation, we deceive ourselves into thinking that our own moral stance and abilities are greater than the neighbor, and therefore we have incompetent forms of aggression.

So the third part is an idea about why it is that we have our very peculiar type of mating system, in which, in an idealized way, males and females are cooperative. And this may be at the level of a romantic painting, or it may be at the level of males supposedly going out, and hunting, and bringing meat back, and females gathering plant food; and the male and female cooperating nicely in the rearing of their Leave it to Beaver family in the (Pleistocene). This scenario is one that has not been well worked out, let's put it that way, in biological anthropology. And one can sort of, you know, make all sorts of accusations against existing ideas. But nevertheless, there is a strong tendency to think that it goes back a fairly long way.

The general idea is that from an ape-like ancestor, an Australopithecine, which was an ape, we start the human career when hunting comes in. And that the provision of meat does all sorts of things that changes us from being a sort of a roughly chimpanzee-like system, to the modern human system. And this slide is meant to reflect the fact that arrows can sort of go in all sorts of different directions. But once you have meat eating, all sorts of sort of funny things happen. And people can sort of argue it in every different direction. And to cut a long story short, there are a lot of problems with this idea. And we can take those up later.

So there's my sort of quick and dirty way of summarizing the hunting hypothesis, and now I want to give an alternative. I think what's interesting about this idea is that it is possible to generate a new idea in a slightly different way from the way people looked at it before. And that is by looking for a signal.

Food and sex are the two things that drive social systems ultimately, right? And human sex is very strange in primate perspective. And one of the ways in which it's very strange is that females are sexually attractive on a year-round basis, during most of their adult life, to adult males. And to some extent they are sexually receptive for that period as well. And this is compared to a chimpanzee; more than an order of magnitude difference. If you take the number of mating days between births, number of potential mating days, then for a chimpanzee it is thirty to forty, and for a human it is many hundred.

We have, it turns out, a fairly good, well-documented little theory for what effect the number of mating days between births has on the anatomy of a primate species. And it comes from this. Where here we have the operational sex ratio in a species; which is a measure of how intensely males are competing for mating opportunities. So you divide the number of males, divided by the number of females. You take a count of the inter-birth interval, and then this variable is very important. So that as you have more males per female, then mating intensity, the competition intensity should increase. If you have more mating days between births, then there are more mating opportunities, and should be less competition among the males.

So what you find is that the anatomical characteristic that's so striking is dimorphism in body mass. Male mass divided by female mass is well predicted by the operational sex ratio. Well, well in terms of anything we ever do with primates is an R squared of point four six. So a substantial amount of the variance is accounted for.

Well, here we have the pattern of male mass divided by female mass in human evolution. Where we're trucking along here, and then something new happens here. So you say to yourself, at what time in human evolution did sexual dimorphism in body mass drop? Can we recognize it? And the answer is, you can recognize it very easily. It happened right here, right at the beginning of the genus Homo. So if primates predict the way that humans are operating, what this means is that right at that time, we get a change in the mating system. It should be that we have short cycles, and few of them, up to the time that we have this drop in the degree of sexual dimorphism in body mass, at which time we come to this current system of long cycles, and lots of them. An extension of female sexual attractiveness and receptivity to the human pattern. And we see no change. The modern human pattern of between fifteen and twenty percent of dimorphism is absolutely retained over the great majority of the last two million years, up to about one point eight million years ago.

Well, that was sex. Now I want to talk about food. And I want to ask the question: when did cooking evolve? Because Lady Strauss said that humans are defined as the animal that cooks their food. And if you're an ecologist, and you think about the distribution of food before it is eaten, there is nothing that happens more dramatically in human evolution than at some point cooking happening. And we don't know when cooking evolved, so this is stuff you can play with, right? So I'm playing with it. And I'm saying to myself at what point is it reasonable to think that cooking should have evolved? So again I'm looking for the signal. I looked for the signal of sexual dimorphism. I'm looking for the signal of cooking. And here we have a timeline, and there are various signals that all happened right around the time of the emergence of Homo Erectus.

Now cooking just has a bunch, a ton of effects on food. And as we know from feeding cooked food to animals, they relish the effects. And we've tried to see what happens to an animal when you add meat to its diet, compared to when you start cooking a proportion of its diet. And without biasing the results too much in my favor, it looks as though you've got to get a tremendous amount of meat in the diet to even come close to what the effect of cooking the plants, in terms of increasing the calorific availability of an Australopithecine diet.

So it seems reasonable to me to think that whenever cooking happened, it should have enormously increased the availability of nutrients. And you say to yourself at what time in human evolution do you get this sort of response? And I already sort of flagged it a little bit, but I just want to show you how dramatic it is. But females, who are the sex on whom ecological pressures operate, truck along at just over thirty kilograms per females. Then here's where we separated from chimp-like ancestors; about five, or six million years ago. Here are these Australopithecines. And then bingo, right at the time Homo Erectus evolves, we get the modern human pattern of over fifty kilograms, and it stays right there.

So I know I'm thinking a wholly new idea about the evolution of the modern human system. And it relates to the idea that cooking should have happened much earlier than anyone else has allowed. I would like to point out that the archaeological evidence from eighty percent of archaeologists would say this is nonsense. Because the really conservative archaeologists say you can't push cooking back more than twenty, or thirty thousand years. Some people push it back down to three or four hundred thousand years. There are a couple of people who have found reddened patches of oxidized soil in East Africa, and say it happened with early Homo Erectus. So you can pay your money, and take your choice. But it seems to me it ought to have happened, and it ought to have had huge effects. And nothing else should have had as big effects. And whenever it did happen, it should have had those effects, and there's only one time it could have happened. So you get the idea. Cooking makes energy soft and dense. It could have had a ton of effects through reduced chewing, and a smaller gut, and improving juvenile nutrition, and making more time and energy available. So now I'm thinking that males are hunting not because they are needed as contributors to the female, but because they go off on a kind of luxurious, stupid activity. And, you know, that's a cavalier comment to make. But it is in fact the case that nowadays the way people are looking at hunting and hunter gatherer sites is it's very difficult to understand why people hunt; because the actual calorific returns from hunting are much lower than they are from plants, and the whole thing becomes a big mystery.

Then I think you have another causal chain, which is that cooking creates a high value packet. Because for the first time now, instead of animals picking things off bushes, and putting them into their mouths, now you've got things being gathered, and put in one place, and that creates something that you can steal. And once you've got something you can steal, then you need social arrangements to account for that. And the matter is made worse by the fact that if you're cooking, it sits there actually improving in value for ten minutes, twenty minutes, half an hour, however long the fire is being applied. Which means it's very easy for dominant individuals to sit there watching, and say five minutes before the food is ready thank you very much. I'll take care of that now. Which is what baboons do to each other. There are baboons that are digging for foods. A dominant individual will watch a subordinate digging for a food that is going to be ten minutes deep. And then moves in after eight minutes, and says let me do the rest. Thanks very much. He gets ten minutes worth of food for two minutes worth of effort. So that's the idea. Produce a scrounger system, in biological terms, in which of course the dominants become the scroungers, and the subordinates become the producers. So in other words, males rob the females. Everybody initially, as it were, collects their food, and puts a fire to it, and then within days, weeks, months, some bright scroungers emerge, and start scrounging. And females have got to do something about it, because they're losing their food to dominants. And so they need guards, and they compete for guards. And at this point, the way they compete is by taking advantage of the in-built bias of males to be attracted to sexual features. And they start extending their sex appeal, both through their body, and through their physiology. And this leads to an increase in the number of mating opportunities, because females are now mating all the time in order to compete with each other for the best male guards. And now you have a system that explains why we have less sexual dimorphism in humans at that time. I told you it was going to be speculative at this point.

So there you have it. Individual little males and females sitting by their fires. Then the males move in, and then the females come back. You can argue it other ways, but at least it has the merit of making us think about the impact of something that surely should have been massively important in human evolution. I don't understand if it only happened three hundred thousand years ago, or twenty-five thousand years ago, why we shouldn't see more of a mark of it in the evolutionary record. So I think of it as being just enormously important, and capable of explaining the emergence of the whole human system at around one point eight, one point nine million years ago. And the harnessing of energy from fire would therefore be a special human characteristic.

And just to tie that back to the earlier parts of the talk about violence, you know, I have this sort of strange sense now that humans were born in fire. And then you look at pictures of atom bombs, and you wonder if we're going to be dying in fire as well.

Q: To go back to the chimpanzee raiding. Can you elaborate on the benefits, and what the benefits of a raid are? So this individual from the other territory is dead now. That territory is weakened. But exactly when do the benefits come back to this focal territory? And who, in fact, is it that benefits from this raid?

Mr. Wrangham: The question is in the chimpanzee lethal raiding, what is it, or when is it that the benefits accrue to the aggressors, and what exactly are they? And who are the ones that benefit? These are all good questions that we don't know, and we can just sketch out. The oldest two studies of chimpanzees, the one in Gombi by Jane Goodall, and the one in Mahali by Toshi Zalinichida have both had entire communities wiped out. In both cases, the observers believe that it was a result of this pattern. In one case, there was no obvious benefit to the raiders, because what happened was that the, as it were, empty space was taken up by the further neighbor, who turned out to be an even bigger community that moved to occupy the vacant zone.

In the second case, the females of the victim community were all taken up, and became members of the new community. Their babies were all first killed, and then they had babies again fairly quickly, and then are reproducing in the aggressor community. So, you know, that's an empirical answer, but what do we expect? It seems to me that the benefits may be quite variable. I could imagine that in some circumstances what's going to happen is that the aggressor community may be able to expand its land, and all of the individuals within it may get an increased size of home range. In another case it may be just that the males benefit, and they're able to take over the females of the neighboring range as happened in that one case in Mahali. And it doesn't seem to be necessarily the case that anybody will benefit. And just as in animals, primates are programmed, I would say, to be as high ranking as possible. But they don't know whether or not high rank is going to pay off. And the empirical data are when you look at the relationship between individual dominance, rank, and reproductive success in primates, in fifty percent of studies, individuals who are higher ranking have higher success reproductively. In fifty percent, though, they don't. In those cases it's neutral. So I come back to the idea that what's happening is that the brain is integrating the average results. But it's going to take, you know, a lot of time before we know how often those benefits actually accrue.

Q: You started talking about violence occurring in chimps, then have ended up talking about sex. Is there a comparison, or sexual explanation that could be done there?

Mr. Wrangham: The question is I slid from violence to sex, and once I started talking about sex, and I was really starting talking about a system that is more similar in humans to binobos than it is in chimpanzees. And so how do we bring binobos into the picture? So it becomes quite an elaborate story.

If we, first of all, look at the pattern of mating competition in binobos, we see something a little bit human-like in the sense that ovulation among all females is concealed, and females do have an extended period of mating receptivity, which almost approaches the human pattern. We also see the same pattern of sexual dimorphism in body mass as between male and female chimpanzees. Which appears to undermine what I've been saying. On the other hand, everything else about sexual dimorphism in the binobos is totally unlike the chimpanzee pattern. They have the same sized heads, and the degree of dimorphism in behavior is enormously reduced. And the degree of dimorphism in robusticity is enormously reduced. And the degree of dimorphism in the teeth is essentially nonexistent. So that in terms of their arming for aggression, it looks as though body mass differences just don't work for binobos. But here's the big difference, as far as sex goes.

A binobo female will have sex with any male who invites her to. With humans, of course, we have concealed ovulation, but at the same time, we have a pattern of private, and individual male directed sex. Which suggests that the female is using her sexual activity towards improving or establishing, or maintaining, or getting something out of a particular social relationship. And that I think is the sort of key difference between binobos and humans in the relationship between sex and violence.

Q: Are you saying that binobo females never refuse sex? Or that every male has a chance to have sex with a female?

Mr. Wrangham: Obviously one should never say never. But they are extremely willing to have sex. So as close as anything I would say that a binobo doesn't refuse sex. Now, you know, she may do so occasionally. But to all intents and purposes, binobo females act, I would say, bored but willing.

Q: Do you think that these raids that you describe, that they're really a cruel play act. Meaning that there is a keen, you know, kinship gene that needs to be expressed. And one has to remind everyone in that group that I expect you to fall on the grenade for me, if you will. So could it be really that it is a play act? A very cruel play act. Because it doesn't happen, as you've said, when there's one or two. There needs to be a certain some critical mass. Is that possible? And also second part of it, do they ever show remorse after the violent act?

Mr. Wrangham: I didn't fully get the question, but I think the question is essentially what I think of as the collective action problem. Under what conditions, or how do we explain the fact that the males are supporting each other in a potentially dangerous activity? And you also asked: do they ever show remorse? They don't show remorse, they show intense enthusiasm and excitement for the kill. It is horrible, bloodthirsty. And this is so even when they know the individual that they are killing. And in some of these circumstances they have done so. One of these cases was a group that fissioned into two, and the smaller of the amoeboid fissions was killed, and so the aggressors were killing individuals that they had sat, and groomed with on long, sunny days, and bounced on their knees, and so on. They still had this tremendous lust for a kill. And after the kill they leave, and thump the buttresses of trees, and scream with excited joy as they come back into their ranges. Excuse the anthropomorphism.

It was a long answer, but I never actually got to the collective action problem. And I don't know quite how to think about the collective action problem. But it seems to me that it's a bit like the way that Robert Frank conceives of individuals as coming into economic interactions prepared to get involved. Prepared by natural selection in the sense that the individual may not know that it's good for them to get into this interaction, but natural selection knows that it is, because individuals who have in fact gone in in the past have benefited reproductively. The observation is that you don't get any hint of the presence of a problem. You don't get any hint that individuals are actually monitoring each other to say well, if you go forward, I'll go forward. Once you've got them, three or four of them, ready to go, then they go, and just support each other totally. It just seems to be in the mind at that point. No assessment.

Mr. Wrangham: When and how did religion play into this? Well, religion is often seen as something which unites people socially, culturally. And it can unite people to do cooperative things. And cooperative things can be cooperative merely within the group, or they can be competitive against other groups. So religion just fits very neatly into the idea that individuals become primed, particularly males, to form alliances with each other that they use at the expense of the neighbors. Religion is seen as a way to just enhance that system. But of course the best religions try and stop that.

Mr. Wrangham: What do I say about the phenomenon of primates counting? Primates have been known to be able to count for some time, and this latest science report seems to take it a little further in ways I don't fully understand. I mean, it's not surprising that primates should be able to count, of course, because they're constantly doing it when they are choosing whether to go to that branch, or that branch to feed more efficiently. So they're very good, I think, at assessing. And, you know, one of the striking features about the line of thinking I've been thinking about is that primates are obviously very good at assessing relative numbers, and so on, in certain circumstances. And yet in other circumstances, as I was drawing attention to, they don't. Now your first part of your question was what is the status of suggesting that testosterone has effects on self-deception for chimpanzees and humans in relation to lots of other animals? Because testosterone probably occurs in plants. I don't know, but it occurs in animals, certainly. And yes, I think that's a good question, and I don't know how to think about that. I can sort of think of various kinds of possibility. And one kind of possibility seems to me to be quite interesting. Although I'm certainly not going to argue too strongly for it until I sort of know more about it. But that is that in order to be able to deceive yourself, maybe you have to have a level of self-awareness. And so the extent to which it is possible for positive illusions to muddy your assessment may be related to your consciousness. And I thought this was sort of, you know, an interesting kind of way to think about it. Because then in theory if you could assay individuals' difference in their own ability, and their perception of their ability, then maybe that's an indirect way of getting at the extent to which consciousness occurs. So that's a kind of sort of, you know, appealing way to think about it. But I'm sort of dubious about it in a way. But maybe I'll just sort of leave it like that at the moment.

Q: Most of the modern organizations spend a lot of money on team building. And it's perceived, you know, if you are part of a team, you are more productive. I have been involved in a lot of those activities, and I find it's problematic. What can be learned from primate behavior about team building and group building?

Mr. Wrangham: Teams are said to be very difficult to organize to be effective. Rarely do teams have benefits commensurate with the costs. So how can primate behavior be used to think about that? So, you know, I'm naturally drawn to thinking about the fact that the kinds of teams that would work would be teams of a type that have been used in our evolutionary past. And the enormous majority of cooperation that we see is in the context of contest. So that applies to alliances among females, and alliances among males. And maybe a bit less to alliances between males and females, but even there, to some extent. And let's see, I'm wracking my brains, I suppose there are other ways in which you see cooperation, such as in defense against predators. But it's not elaborate.

You also see a little bit in hunting. There aren't many primates that hunt. But I'm of the view now that hunting should be seen as a sort of metaphor for hunting members of your own species. I'm very struck by the fact that binobos don't hunt members of their own species, and they don't hunt monkeys either, even though they like to eat meat. They're in the presence of monkeys in their habitat. They sometimes even catch them, and keep them as pets. But they don't go coalitionally hunting for them. So in general, I resonate with the idea that it would be very difficult to build teams, except when you're building them to fight someone else.

Mr. Wrangham: So the suggestion is that in these analyses of battles we shouldn't be thinking of winning and losing so much as playing the game, by which you mean probing the opposition to the point where you can then still win a cheap victory?

Mr. Wrangham: So the idea is that they're kind of an intermediate strategy of instead of going all out in a battle, or giving up without a fight, you should be able to probe a little bit, and establish how good you are, and then maybe reduce the time to the next battle, or get some merit like that. And I'm sure that's right. I mean, there are lots of complexities in the interaction between opponents that are not being taken account of here.

And of course I was making very sort of sweeping remarks about the nature of battles. But at the same time, I'll defend this perspective, in the sense that, as I understand it, there is no theory at the moment that accounts for the frequency of battles. And, you know, people are very confident about the fact that there is this military incompetence. So it seems to me at least a place to start to sort of think about very simple outcomes of either you do fight or you don't. And then yes, absolutely, some of this may be probing, but at least that gives a framework to go on.

Emergence

Yaneer Bar-Yam
NECSI
Philip Anderson
Princeton University
Stephen Kosslyn
Harvard University
Norman I. Badler
University of Pennsylvania
David Sloan Wilson
SUNY Binghamton

Mr. Yaneer Bar-Yam: It is a pleasure and an honor to welcome you here. I am the President of the New England Complex Systems Institute, and the Chairman of this conference. Let me use this opportunity to not only introduce this session, but really talk about the entire conference.

The form that my remarks are going to take are shaped by three interactions over the last few days; two with students, and one with a colleague. A couple of days ago I spoke with a student that told me that she was here last year at this conference; and it was very interesting, but she wasn't coming again this year. Why? Because it wasn't relevant to what she was doing. So I asked her, what are you doing? She said I'm using cellular automata to simulate ecosystems. I said well, what questions are you asking? She said well, there are a lot of people using cellular automata to simulate ecosystems. I asked her several times what questions she was asking, and I never really got a direct answer.

I thought that for the students who are here, I would spend a few moments to emphasize this point about the questions that we are asking, and the importance of the questions that are distinguishing what we are doing here from what many people are doing in other places. Of course, in the few minutes that I have, I can only talk at a very high level of abstraction; I hope that it will help guide some of the thinking of some of the students who are here. Let me go over the different sessions of the conference, and review briefly what the questions are.

– The first session is on emergence. Emergence is an effort to understand how the parts work together to make collective behaviors. I can echo Herb Simon who distinguished between strong emergence, and weak emergence. Strong emergence was the emergence of behaviors that are not found in the behaviors of the parts. Whereas weak emergence, as he defined it, was behaviors that could be understood through the behavior of the parts, but they only had meaning in the context of the behavior of the whole, and not if you were looking at an individual part. As he said, the first is not part of our scientific dialogue; but the second, where we can understand the collective behaviors, is in fact what we are after. And of course we are interested in understanding typology of emergence. How, when, and why emergent behaviors arise.

– The second topic turns out to be central to our understanding of complex systems in general. Whether we are dealing with them on a practical basis, or whether we are dealing with them on a philosophical basis - that is the issue of description. On a practical basis, in order to think about a complex system, we have to describe it. If I may go again to Herb Simon's book, he says It seems to be an easy problem to have someone program a computer to figure out how to get from position A to position B on a map. Or I leave it as an exercise to the student to show that in fact this is the case. Today, of course, we have computers that are able to find positions, and give directions; but the issue of the representation is crucial here. If you take an aerial photograph, and hand it to a computer, it's known to be an extremely difficult problem. But a representation, in terms of coordinates, and locations of streets is quite feasible. What kinds of descriptions are useful under what circumstances is really central to our understanding. Beyond that, if we're thinking about complex systems, those complex systems are themselves representing the environment in which they are found, and the nature of that representation we don't yet understand. Whether they're biological systems, whether it's in the mind or in the genome, all of them are representations of the environment that we don't understand.

– Self-organization returns to the issue of how complex systems arise. The dynamical process by which they arise, whether in physical or biological systems.

– The issue of networks is a little bit more specific. We have an understanding that there is an abstraction called a network from a complex system, but we don't yet have a good typology of networks. There are influence networks, there are transportation networks, there are communication networks, there are networks that involve computation. We don't yet have a very good understanding of how to classify them, for purposes of treating them.

– Time series is the problem of predicting the very largest scale behavior of a system. How, when, and why we can ensure prediction and/or when you cannot make a prediction is something that we would like to understand. And linking the time series, of course, to the underlying structure and behavior of the system is also essential.

– Agents and agent based modeling is a more practical issue. People are now trying to create models of certain collective behaviors in social economic systems, and how they work in reality, or don't work in reality. This is something that is going to be probed in that session.

– Finally, human organizations is of course the system that we often know the most about, but understand the least. And I will be mentioning a little bit more about this session in a few moments.

There are many other sessions at this conference, but I want to mention two topics in particular, because they're quite unusual, even in the context of a complex systems conference. There is a session on design which should be central to complex systems studies, because the principles of understanding how design works are of course directly related to the principles of biological, or social, or any other system design. In the context of engineering design, these principles are being studied. The other topic is medical systems. There are two sessions. One on medical complexity, and the other one on medical management. Both of them have not yet been adequately represented in the context of complex systems studies. In recent years, the medical management issue has become of substantial concern to many people in society, and the issues that arise there, due to the great changes that have happened in that system, can be well served by concepts that have been developed in complex systems. I pulled these two topics out, because they are unusual, but the other sessions are extremely exciting as well.

There was a second conversation that I had with a student on the way here. This was a student who had come last year to the conference, but had not attended Herb Simon's banquet talk. I asked him why. He said well, who is Herb Simon? Now some of you may be chuckling, because you know who Herb Simon is, and some of you may not because you don't. But it became clear to me, and it has been clear to me that we don't know who are the important people in other fields, and even who are the important people in our own fields sometimes, because we are coming from very different directions.

To emphasize this point a little bit, I would like to introduce to you the speakers of the last session of this meeting - the session on human organizations. In that session, speaking will be Michael Hammer, who is the father of reengineering the organization, and the author of the book by that name. He has been listed as one of the twenty-five most influential people in this country, by either Time or Newsweek. In that session is Peter Senge, who is the father of the learning organization. He is considered, in management circles, I am told, an equal to Michael Hammer. Henry Mintzberg is in that session. He's the author of the classic and very standard reference on human organizational structure, and its relationship to behavior. It was authored in the seventies, and he has, of course, done very important things since then. Larry Prusack is the head of the knowledge management of IBM consulting, and Bill McKelvey is formerly the Dean of the Management School at UCLA.

It may be hard to understand this, but most of these people have never met each other. We are bringing them together, and to have them be present in the same room at the same time is an outstanding opportunity to learn a tremendous amount.

With that, I will mention the third interaction that I had, which was with a colleague, who after the session yesterday, said to me, you know, something strange happened to me today. It occurred to me why all of these people are talking here. And this understanding that we have, this bringing into focus of the synergy, and the unity of the study of complex systems is something that we are after.

I would like to introduce Phil Anderson, our first speaker. Phil is a Nobel Laureate in Physics. The areas where he originally concentrated much of his attention was in disordered materials, from which he then generalized concepts to thinking also about patterns, and memory, and other problems that are of relevance to complex systems in general. He has more recently been teaching the entire physics community about superconductivity, which is a very active area of research in the last fifteen years. I asked him what kind of introduction he would like, he said as short as possible. So with that, I will give you Phil.

Mr. Phil Anderson: It's a great pleasure to be here. I've heard a great deal about the New England initiative in complex systems, and now I finally see it in action. What I'm going to do today is give you actually another introduction to this session; a kind of a sermon in praise of emergence as a general way of looking at things. This is a summary of the various ideas I'm going to try to discuss here. I'll take the usual point of view about what's been going on in the past century in science; namely that we've reduced the world to very simple ideas, rather than the other way round. I will look at the same material, but turn over my point of view, and look at it as starting from the bottom, which is simple, and arriving at the real world, which is complex, and suggest that that actually is a more general, and more useful way of thinking about it. That emergence and reduction, in many ways, are the same thing; just looked at from the opposite point of view. Then I'm going to talk about some specific things about emergence. In particular, one of the fascinating things about it is that the whole process of measurement in physics is itself an example of emergence, and a very complicated one; and it confuses people, as we all well know. It confused Bohr, it confused Einstein; and therefore it's quite proper that it confuses the rest of us. So I want to talk about why it confuses us, not making it clear, necessarily. I'm willing to say emergence is even more serious than that. In very many cases, the very concepts and categories that we place things in are emergent. They weren't intrinsic to the underlying theory, they emerge as we look at the consequences of the theory. They're not necessarily unique as well. The whole conceptual structure in which we think is very often itself emergent; not explicable directly from any kind of underlying laws.

And then finally, in a postscript, I'll talk a little bit about the opposite of reductionism, or what I call constructivism, and whether the strong reductionist point of view that you can explain everything using the underlying laws is possible at all. There I'll just be talking anecdotes, and examples, and suggesting that in general it is impossible, unless you can think laterally - and I don't think anyone, in particular any mechanical process, or any Turing machine is very good at thinking laterally. We'll just leave this outline up, and start talking on the various topics.

In modern science, what you see is very seldom, in fact never, what is actually there. At least that's the point of view of the modern scientist. The object or fact you're looking at is explained by something very far from direct human perception. This fact intrigues a lot of laymen. They find it very interesting and glamorous. But a lot of others, particularly people who are inclined to the humanistic side of life, are turned off by it. And this may well be one of the reasons why there's a very general unease about the whole of science. The fact that everything is analyzed in terms of indirect causes. This kind of thing may have started with Copernicus when he started the heavenly bodies moving independently of themselves. Maybe it started even earlier. Way back when astronomers gave up the idea that there was a god for each heavenly body, pushing it along around its orbit. It certainly became a serious problem when Galileo and Newton destroyed the intuitive Aristotelian notions of force and motion. Anyone's who's taught elementary physics knows that that's the point at which you lose half your audience: the point at which you stop thinking the thing which moves must be continually pushed in order to make it move, and try to abandon the various intuitive notions that go - those codified by Aristotle.

In the twentieth century, this process has pretty much accelerated. The immutable atom has become a composite structure which you can tear apart and analyze, in the first instance, into its nucleus, and it's electrons. Then the subparticles themselves become composite particles. The electron, fortunately, isn't yet composite, but the nucleon is a composite of - as we all have been being taught recently - three quarks, and an indefinite number of gluons. Lord knows how much more, and how much farther we will go in analyzing ordinary matter.

But even below the particles themselves is a vacuum. The vacuum itself is now endowed with all kinds of properties -complicated properties - and it has great complexity. Also we have a myth as to our origins. If you think about this myth, it's much more complicated than that we all were issued from the mouth of a great fog, or some similar tribal mythology. It's very spectacular, very bizarre. More so than anything dreamed up by some primitive tribe.

This origin myth itself contains two examples of emergence, and of broken symmetry. There was a phase transition, we're told, in the course of the Big Bang which broke up the identity of the weak and the electromagnetic interactions. And there is the emergence of nucleons and electrons - protons, neutrons, and electrons - from the primeval plasma of quarks and gluons.

Even in biology, we have reduced the kind of general facts - species, and metabolism, and all that - to a complex, and increasingly abstruse inner chemistry, and a terribly simple chemistry. But very, very far from the standard facts of biological behavior.

So what we find is that physicists, and scientists in general, love to do two things. One of them is to take things apart. To analyze them into simpler and simpler components. And the second, I'm afraid, is that scientists love to mystify. They love to say well, it isn't really this, it's much more - much simpler - not simpler, but much more remote from ordinary experience. A solid body isn't a solid body. It's mostly empty space, and it's made up of electrons, and nucleons. The nucleons are - etc., etc., etc. So what you see is not what's there. And as I said, this may well be part of the reason why the general public distrusts us.

The problem is that the physicist has taken upon himself the role of the shaman, or the mullah. Everything comes from a first cause, the first equation, and only he can investigate this with his very expensive equipment, and his very abstruse theories. The arrogance this attitude fosters has to be experienced to be believed. Such books as Stephen Hawking's A Brief History of Time covers the whole of science in six very brief chapters - ordinary science - and spends the rest of the book, the majority of the book, discussing the first picosecond of time, when he seems to feel all that matters happened. Everything after that was just chemistry. Much better, but still dismissive of other sciences is Steve Hornberg's book, The Dreams of a Final Theory, where he repeats his claims of many years ago that you can classify all of science by following things he calls little white arrows. Which lead downwards back to the fundamental problem. The fundamental problem is: what are the elementary particles made of? One of the elementary particle physicists even had the temerity to write a book called The God Particle.

The Achilles Heel of this approach, if it isn't obvious, comes from the repeated reference to a hypothetical perfect computer. Which could, in principle, follow out all the consequences of the basic laws, making the basic laws the only input to this computer that matters. If there is an enormously simple set of basic laws, which the elementary particle physicists are going to find for us, then we just start with those basic laws, and compute, and everything that happens is a consequence. The Achilles Heel is the repeated reference to a hypothetical perfect computer. The fundamental reason for this, is that you couldn't do it all. The basic fact I express with kind of a slogan. The slogan is that combinatorial numbers are bigger than cosmic numbers. Orders and orders of magnitude bigger. A small number, to some exponent - and it doesn't need to be a very big exponent - is already bigger than the size pf the universe. An exponent equal to the number of different characteristics that you can distinguish in a human being or an animal is already bigger than the total number of particles in the cosmos. In order to make this computer work, you'd have to make it capable of tracing down this enormous number of consequences.

A second problem with this perfect computer is the problem of what is the purpose? What's the question? Yaneer made a very good point. The important things in science are not the answers but the questions. Not the procedures, but are you asking the right question. The question you want to know, in a very general term, is why is something happening? But suppose you have some phenomenon? I'll take an example from my own experience on, high temperature (high Tc) superconductivity. The Superconductivity of yttrium barium copper oxide.

Supposing you had this perfect computer. You start with the properties of the atoms, and it computes, and it computes, and it says aha, yttrium - this combination of yttrium barium copper and oxygen atoms is a superconductor, and it reproduces all the phenomenology. Well, we've spent fifteen years, or twelve years, and something close to a billion dollars tracing out all these consequences of superconductivity, and yttrium barium copper oxide. We still don't know why it happens. But after we did this computation, we wouldn't know why it happened either, because all the computer would have done would have been to reproduce the operation of those fundamental laws. We don't know which gimmick it is in those fundamental laws that makes it happen. So we wouldn't have any answer to why, we'd only have the answer to whether it happens. But we know it happens. We can look at it, and see that it happens.

It was, in the first instance, as a cry of anguish at this attitude of my elementary particle physicist colleagues that I wrote an article called "More is Different," nearly thirty years ago. Actually it was more than thirty years ago. I was, perhaps at that time, still a reductionist. Still willing to mystify, and maybe I merely wanted a piece of the action. I don't know. But I did start looking at the problem from the opposite point of view.

The reductionist view starts from the top, and searches for simpler, and simpler first causes. I looked for the same material facts, and accepted that all these little white arrows probably existed. That you could reduce everything to simply loss. But asked whether this computer was really possible, or necessary. What one sees from the opposite point of view is more and more complex consequences arising from simpler and simpler causes. This is also the historical order in which things happened, if we actually believe the Big Bang theory.

So what you see in this enormous triumph of reductionism in the twentieth century is also a triumph of emergence. What we've shown is that the entire world is the consequence of the process of emergence. The development of more complicated consequences at the macroscopic scale from simpler consequences - simpler rules at the microscopic scale.

The message, then, is in the first instance that the world in which we live is the consequence not of some incredibly simple but hidden God equation, or some God particle. The idea is that we have a God principle, which is emergence at every level. I first became aware of this in a very self-centered way as the principle by which my own field of science arose from the underlying facts about particles, and interactions. But it was then, as I raised my perspective, that I began to see how general the phenomenon of emergence is.

This is one of the main points I want to make here. That even in the reductionist, quantum physicist's point of view, he's caught in the emergence trap at a very fundamental level. In the most recently fashionable type of theory, for instance, string theory - or the generalizations of string theory which the particle physicists are playing about now - they're talking about these simple, elementary equations being valid in various spaces of almost indefinitely large dimensionalities. Even with the distinction between particles, and fields, between bosons and fermions erased by a new symmetry. In fact, the most recent theories have even higher symmetries than that.

Now, if you look at it from that point of view, even the question of which of these various dimensions become the actual space and time that we look at is an arbitrary matter. You know, the original theory is symmetric among those dimensions, and it picks four of them, three space, and one time dimension, and those four dimensions are what we've got. So space and time themselves, from this point of view, are an emergent phenomenon. What could be more triumphally emergent than that?

In fact, when one begins to analyze the actual measurement process, that is, to discuss the famous difficulties and paradox that have dogged quantum for so many years, since the days of Einstein and Bohr; we realize that the problem again is one of emergence. The apparatus with which we try to measure the consequences of the simple theory is already at an emergent level, and is not obeying directly the simple laws of the simple level. It already has emergent properties, which have to be understood because we can understand how the measurement takes place. The quantities that we measure in the quantum system of interest are variables which are defined not by that system, but by the macroscopic apparatus with which we do the measurement. The quantities we're measuring are things like spin orientation, magnetic fields, electric fields, position orientation, and so on. We can't imagine a way to set up a coordinate system - for orientation or position, for instance - without having a rigid body to refer it to. And yet the underlying quantum equations of motion don't tell you what a rigid body is. And a rigid body is a derived property. It's an emergent property that happens when you put enough atoms together into a large thing which then self-organizes into a crystal. And only then, when you have it self-organized and rigid, can you measure the positions of the fundamental particles from which it's made up.

The physical way in which a rigid body operates is very interesting. It exhibits no dynamics in kind of the canonical experiment that Bohr or Einstein were talking about. You say there's some system of slits. Nobody asks how does quantum mechanics arrive at a system of slits? It just says they're slits. Well, these slits are assumed simply to act as boundary conditions. That means that their quantum state is assumed not to change as this measurement process takes place. In fact, if it changes, the measurement is spoiled, and that is the essence of something called the Debye-Waller factor in X ray crystallography. So there's this very fascinating process of what is called a zero-phonon transition in which the rigid body takes up a quantum variable, and whatever momentum, whatever angular momentum, and so on is necessary to make the measurement possible, without changing its quantum state. The fact that it can do this is itself a very complicated consequence of quantum mechanics.

This question of measurable quantities, and measurement, and measurable quantities arising spontaneously is made very explicit in the example of superfluidity, or superconductivity, because a superconductor is only a superconductor below some low temperature. We never saw them before coming on liquefied helium, because they're only low temperature objects.

The fundamental fact about a superconductor, or a superfluid is that it has certain macroscopic variables, things like position, momentum, temperature, and so on. And it has certain macroscopic variables which are not the classical macroscopic variables we're used to. It has macroscopic thermodynamic variables that correspond to order parameter, or to the expectation value of a quantum field. It has an order parameter that's a thing like an electric field, or a magnetic field, but it's a particle field. So it has an expectation value of the particle field. And that's a macroscopic variable. But before you lower the temperature before this critical temperature, there is no such macroscopic field, there's only fluctuating microscopic fields that you can't measure. But now we have a macroscopic variable-like position, we can compare that macroscopic variable for different superconductors. So we can set up a coordinate system for quantum fields if we have a superconductor, but we can't set up such a coordinate system if we don't have it. So here's a coordinate system, a whole macroscopic dynamics, and a whole classical, if you like, dynamics that's appearing out of nowhere that we have if we prepare the system, and don't have if we don't prepare the system. Once we have it we can do measurements. We can carry out the conventional processes of macroscopic physics.

This kind of tears us from our bearing, because we can't really see, in principle, the difference between this, and space, and time. Space and time may have arisen in the same way. Before there were macroscopic bodies, there wasn't any space and time, so we don't really know how to think about the original parts of the Big Bang, because there was no identifiable space or time. There was no way we could have measured space and time.

Some very fundamental things are emergent processes. Then, of course, as you get to more complicated systems, you have the emergence of all kinds of things. Life, for example. After worrying about whether space or time is an emergent property, life is relatively simple and straightforward. But the concepts and the structure by which we think about things are also emergent. It's a truism to say no one could have predicted the genetic code. The genetic code that controls all of biology is of course a frozen accident. But it's more, it's even a deeper question to ask: is the existence of a code a frozen accident too? Was it necessary that there be a genetic code? Is that the only principle on which life could have developed? We don't really know. If we played the tape over again, in Steve Gould's favorite phrase, would the consequence have been a genetic code, even though we all accept that probably the genetic code is a frozen accident.

In human affairs there are all kinds of examples of emergent phenomenon. We can look at different civilizations, different cultures, and see whether these emergent properties are more or less inevitable. Whether the equivalent of the code is general, or whether it's not. It isn't true that it never happens. All kinds of cultures have invented money. They've used different substrates for it. But they've all invented money, and in some cases this has been wampum, some cases stones, in some cases it's been gold, some cases paper. Agriculture was reinvented by different cultures quite independently, but with a totally different substrate. You can have an agriculture based on rice, or an agriculture based on potatoes, or an agriculture based on wheat, and different domesticated animals.

So there is this interesting factor of emergent phenomena which seem inevitable in certain systems, i.e., the human system. But other emergent phenomena may or may not be inevitable at all. It is a very deep question intelligent beings would have emerged from a different replaying of the tape. Just a wild guess, I would guess yes. Steve Gould would certainly guess no.

I think that's almost enough. The last idea I want to leave you with is the anecdotal evidence as to whether you really can predict what's going to happen. Even in something as simple as physics where you know all the equations, and you have only a very simple job; the fact is it almost never happens, but oddly enough it does occasionally happen. The minute Landau, for instance, heard that there was such a thing as a neutron, he is reported reliably to have said oh wonderful, what a wonderful thing to make a star out of. So the neutron star seems to have been obvious. It wasn't obvious to everyone else.

On the other hand, superconductivity as a phenomenon existed for thirty-six years before anyone figured out the quantum mechanical explanation for it. That quantum mechanical explanation looks trivially simple, at least to a trained quantum physicist, once you know it; but it took thirty-seven years for some of the brightest people in the world to invent it. On the other hand there's where we have all kinds of examples such as the quantum hall effect. We had the theory of it, and yet when this experimental phenomenon was discovered, it took a couple of years for anyone to come up with the theory. And another couple of years after that for the rest of us to believe the theory. So it's certain the standard predictor or this great Cartesian computer that will explain everything didn't work in that case.

The problem seems to be one about lateral thinking. I don't know what lateral thinking is, but I know a Turing machine can't do it. So we're left with the apparatus we have. Thank you.

M: Perhaps not a question, but I'm wondering when you talked about reductionism, and the basis of reductionism seems to be that we accept that causality works from the bottom up. When we explain the properties of a more complicated system in terms of the interactions of its components, what would you call it? I'm not sure emergence would be the right word, if you explain causality working from the top down. So you explain the behavior of a part by the fact that it has to belong to a particular higher level system.

Mr. Anderson: I don't know a name for it. It's the fundamental assumption of modern science, and it's worked very well, that it always works. Is that perhaps holism?

M: I have a question mark in my ??.

Mr. Anderson: Yes. The Gaia hypothesis is very close to that in biology. Namely that all of biology is in some way organized so as to keep itself going as a whole. There are holistic approaches like that in various things. I think in biology, the Gaia hypothesis has very definite status of some sort. I take no position on it. But there is this very fundamental assumption that we're all making, which is that the causation does happen from the bottom up.

M: I'm just struck by the use of dichotomies all the time. I wonder if you would comment. The fact that everything is always emergence versus reductionism. Top down versus bottom up. It seems to me maybe science just loves dichotomy. I wondered if you think in the next century we can kind of embrace ambiguity, for example, rather than dichotomy.

Mr. Anderson: I was trying to say that in fact it isn't a dichotomy in any real sense. The goal of the reductionist isn't - that's an impossibility. That was the point I was trying to make. He can't do what he's ostensibly doing. On the other hand, he is - that was an idea that I wasn't going into. He is doing something useful. It's just that what he's doing that's useful is not what he thinks he's doing that's useful.

Mr. Bar-Yam: It's very exciting for people who come to this community from physics that the concepts of emergence are going back into physics, and having a strong influence on the thinking there.

Our next speaker is Steve Kosslyn, who is at Harvard. His talk will be about understanding various levels of structure, and abstraction, and processes in vision. The work on biological neural networks that happened earlier in the century had a tremendous influence on inspiring work in the field of complex systems. It's very often important to go back to a field that you borrowed from, and learn what else they have learned there, and gain some additional insights and concepts. One of the things that is important in the context of what I talked about before, in terms of our own community, and the formation of our own community, is that we are also building higher levels of abstraction on the more directly observed concepts in a specific system-related study. Our field is in a sense creating new levels of abstraction in science just like the visual system creates abstractions. It's interesting that in recent years it has become recognized that those higher levels of abstraction are in fact affecting the lower level representations. Similarly, we are presumably having impact on the other fields, from which we are taking the information.

So I will turn over the microphone to Steve. He has broken his leg recently.

Mr. Steve Kosslyn: Thank you, Yaneer. And now for something completely different, as they say. Struck by this notion of dichotomies, and whether it's useful to think that way, particularly in a context of emergence and reductionism; because what I'm going to present is sort of an example of both at the same time. I'm going to be speaking about visual mental imagery, and how it arises from the brain.

First let me see if I can get you a sense of what I'm going to be talking about. Could you try to count the number of windows in your living room? Do you know how many there are? Go ahead. How many are there in your living room.

M: Five.

Mr. Kosslyn: Lovely. His eyes moved to the side, and they sort of jerked a bit. If you try to do this, you'll probably have the experience of visualizing the room, and kind of scanning along, counting. That's what I study: how you see with the mind's eye. The reductionist piece is I want to know how the brain does it. The emergence piece is I don't think the brain evolved to visualize. Rather it's like Steve Gould's idea of exaptation, where the nose, for example, evolved to warm air, direct sense, that kind of thing. Once you've got a nose, you can now use it to hold up glasses. What I want to claim is that something similar is going on with mental imagery. That the machinery that's in there for vision can be usurped in a different context, and used in the service of visual thinking. Some of the aspects of the way we think are therefore molded by the way the visual system happened to be. The talk has two basic parts.

The first part, I want to give you some sense for the phenomenology of visual mental imagery. And some appreciation for what I think of as a miraculous fact, which is that our introspections, our conscious awareness of certain aspects of mental imagery reflect information processing. For some reason, we're aware of some aspects of information processing in our brain. That's the first part of the talk. I'll give you phenomenology, and some behavioral sorts of data.

The second part is we then move into the reductionist piece where we try to have some sense of where this comes from. At the very end I may say one or two words actually relevant to the kinds of concerns you ordinarily have. So visual mental imagery, as I use the term, refers to representations that produce the experience of seeing in the absence of immediate appropriate sensory input. For example, if I asked you what shape are a German Shepherd dogs ears. I hate to make victims out of people who happen to sit in the front row, so maybe I'll pick on someone further back. Do you know what shape a German Shepherd dog's ears are?

M: Pointy.

Mr. Kosslyn: Thank you. Alright, indulge me for a moment. Do you know what shape a German Shepherd dog's ears are?

M: Pointy.

Mr. Kosslyn: One more time, please. What shape are the dog's ears?

M: Pointy.

Mr. Kosslyn: Lovely, lovely. Alright. The first time he answered, his eyes moved to the side, probably to minimize visual interference, because you visualized the dog's head, and kind of looked at the ears. Now, if you had not had a doubt there at the end about the veracity of your answer, and were not wondering why I was probing repeatedly, you would have simply remembered your verbal response, and no longer needed to consult the image. There would be a qualitative distinction between those two kinds of mental representations that would have been evident. Namely, the first time through you visualized it, the second time through, I don't know what, but the third time through you wondered why I was asking you, so you checked again that third time. Eye movements are a giveaway.

Alright, try this one. Which is darker green? A Christmas tree, or a frozen pea? Does anybody know? Do people have the experience of visualizing color in this case? (Anything to do with) texture. Which is bumpier, an avocado or a golf ball? So forth. So there's a sense in which it feels like you can represent the object - but it's not there. It's in your head. That's what I'm talking about with visual mental imagery. Let's start with three properties of imagery that are evident to introspection. So introspection, looking within; which you were just doing if you were doing the demos. There's a lot of demos in this talk, but you have to supply them for yourself. You'll get much more out of it if you actually try them, I think. The first part concerns the relation between introspection, and information processing. The first aspect I want to discuss is the notion that distance is somehow embodied in a visual mental image. If you tried to count the number of windows in your living room, you probably had the sense that you were scanning along, as if there was something that was extended in space in the representation. That's a common introspection. How can we document that that introspection maps into information processing? Try this one.

Imagine a horse - close your eyes, please. Imagine a horse seen from the side. Mentally focus on the place where its tail meets its body, as if you were staring right there. Can you do that? Now tell me if its ears protrude above the top of its skull. Yes or no? People commonly report they have the experience of sort of shifting their mental attention along the horse's back up to the head. Now try it - focus in the center of its body rather than its tail. You ready? Tell me what shape it's ears are. Again you scan up. Now focus on his head proper. Again tell me whether his ears protrude above the top of his skull or not. What people commonly report is the further they had to scan, the more of the horse they sort of went through, the more distance it felt like they went over. Do you have that introspection?

Here's what we're going to do. The experiment uses response time as a kind of tape measure. It's analogous to a cloud chamber where you don't actually see the cosmic ray, but you see its signature. We use response time as kind of a behavioral signature of certain mental events. In this experiment, people first memorize this map. It was drawn so there are seven distinct locations. (I'll use my crutch as a pointer.) There's little Xs indicating exactly where they are. They're set up so that every inner pair distance is distinct. Twenty-one distinct distances. People learn to draw this map. Actually, they learn to draw where the Xs are, and they just label. They get very good at this. Later they close their eyes, and they hear the name of one location, for example, hut. They're asked to mentally focus on that location, to wait, then they hear the name of another location. It might be tree, it might be bench. Their job is to "look for" - their eyes are closed - the second location on the map. If they find it, they push one button. If they can't find it, they push another. We have them start from every location, and scan to every other one equally often. And of course there's an equal number of cases where they can't find it. Here are the response times when they could in fact find the second location. So we have graphed on the X axis distance that they had to scan, their eyes closed; the Y axis is time. What you see is a rather systematic relation. The further they have to scan, the longer it takes. Another version of this was done some years later using much a better technique. The subjects see this ring-like thing. Three of the squares are randomly filled in. It goes off. For fifteen milliseconds a little arrow appears some place within the center. Their job is to say if the arrow had been superimposed in the ring, would it have pointed to a filled square or not? The trick is that there were different distances. Three distances from the arrow to the square pointed to. Here's what you find. The further they have to scan, the longer it takes. Although we never mention the word "mental imagery", there's no opportunity for demand characteristics, or other possible contamination, such as the subject trying to figure out what it was we wanted.

M: Do their eyes move at the same time also?

Mr. Kosslyn: Yes, the eyes move; but they don't move in a smooth pursuit, they just jerk, which is a distance not proportional to the distance taken. It's like a saccade, it is a saccade, actually. This is a curious thing. We did it with Air Force pilots, as well as control subjects. You can see the slope is the same. This is an aside. Try this. Imagine the upper case letter N, like Nancy. If you tilt that ninety degrees clockwise, is it another letter? Yes, or no. What letter is it?

M: Z.

Mr. Kosslyn: That's a mental rotation task, which Roger Shepard actually developed. It turns out these pilots are super. There's virtually no slope as a function of the amount of rotation they have to rotate objects. So typically the further you have to rotate, the longer it takes. The pilots show no effect of that. Our idea was that because they're oriented in three-dimensional space so often, they get quite good at reassigning coordinate systems. Of course, we don't know whether they were like that before they became pilots or not, so that's an interesting question.

The first set of experiments showed that the further you have to scan across an object when your eyes are closed when you're visualizing it, the longer it takes. Which suggests that the introspection for something extended in space, really reflects properties of the representation.

Second question was are there limits on spatial extent? This is a very difficult task. Try this one, if you can. Close your eyes. Imagine an elephant off in the distance facing to the side. So you're staring at its flank, right at the center of it. Now walk up towards it in your mental image, so it seems to loom larger. You're getting closer and closer. It's looming larger and larger. What I want you to do is to pay attention, keeping fixated at the center, to how far away you are at the point where the edges start to blur, and it seems to start to overflow. Can you do that? It's difficult. About twenty percent of people can't do this one.

Now try the same thing with a rabbit. Imagine a rabbit off in the distance. You're staring at the center of its side. You're moving towards it so it seems to loom larger in your image, getting closer and closer, to the point where the edges start to blur you're so close. As long as you fixate in one place. How many people could do this task? Could you raise your hands if you could do it?

M: How fast are you moving?

Mr. Kosslyn: It's up to you. Here's what I want to know. How many people who could do it felt they could get closer to the rabbit before it seemed to blur than to the elephant? Could you raise your hands? Oh, what a lovely group you are. How many felt they could get closer to the elephant than they could to the rabbit? So you can see by sign tests, or a Chi square, that it's highly significant. Those of you who got closer to the elephant, I'd love to know why. Here is the idea. That if there are limits; fixed visual angle, as it were, the larger the object is, the further away it would seem. So we did this experiment, again, probably about twenty of them. There are three labs now that have done these. The one I'll show you the results from, which took cardboard that we cut out - a black cardboard, construction paper - rectangles that we vary orthogonally the height and the width. We show the rectangle to somebody, ask them to study it, close their eyes. Visualize it as if it were mounted on a wall at the far end of a hallway. Do the same task we just did. Imagine you're walking towards it until the edges start to blur. Stop. Then we gave them a tripod, and said position it on the floor the distance you would have been if it had actually been there the way it appeared in your mental image. What we've got here is the distance on the floor, and this is the size. It turns out the length of the diagonal is the single best predictor of the distances. Not the area, not the height, not the width. It's the single greatest extent. What you see is the larger it is, the further away it seemed when it overflowed, and the angle is remarkably consistent. It's about twenty degrees of visual angle, in case you ever wondered about how big an object is when it seems to overflow a mental image.

Last bit on this part of the talk. Try this. Close your eyes, please. Imagine holding your arm out at full arm's length, and resting on one fingertip is a butterfly. What color is the butterfly's head? Now, try the same thing, but now imagine in the palm of your hand the butterfly is resting. And the palm is about eight inches from your eyes. Close your eyes. And again, what color is its head? Now, the interesting introspection is that many people report in the first case, when it's held at arm's length, that they have to sort of zoom in to see the head. The notion is that it's as if you've got sort of a grain. If an object is too small in a mental image, its details are obscured. So here's an experiment. This is actually from my Ph.D. dissertation, 1974. I've done the same work all this time. 1974, it's been one line. Subjects are shown four sizes that are represented by boxes that are actually different colors. So this might have been green, yellow, red, and blue. Subjects are told when you hear green, say, prior to the name of an animal, visualize it so it's as large as you can visualize it without it overflowing. The other sizes are going to be scaled down relative to that. And of course each color is used equally often for different subjects at each size, so it's all counterbalanced up. The task was they visualize an animal. It would be like, 'green bear'. Hold the image. It would be at one size. Then they'd hear something like 'curved front claws'. They would have to say yes or no using their image. Looking at their image to see if it's true. Here's what you find: the response time is a function of size, the same way the boxes were arranged. Smallest, up to largest. What you see is when it's small, particularly when they do find it, you get a big effect. It's much slower to find the properties.

It not only works with objects as a whole; we looked at properties of objects. For example, for a mouse, we'd ask people to visualize a mouse, and tell us whether it had wings. The answer is no. Is it large, had a back, or whiskers? The trick is there are two kinds of properties. Here are large properties, here are small properties. But the trick is the small ones were more highly associated with the animal. So whiskers for a mouse, or feet for a duck. Whereas the large properties were less associated.

Two things to notice. One is association strength is what's crucial when you're not using imagery, as opposed to size. The other is by the third time I asked you what shape a German Shepherd dog's ears are, you've coded it verbally, and that's a much more efficient system, because it's sort of a lookup. As opposed to having to form this image, and do this mental scanning business. It turns out five year olds, by the way, look like this, whether or not you tell them to use imagery. They're faster with the larger things, than the smaller. It's true whether or not it's more highly associated for them.

The idea is that you've got two different kinds of representations. One is some kind of abstracts, or list-like thing, which is ordered in terms of how distinctive or important or associated a part is for the object. If you ask people tell me properties of mice: you'll get whiskers out much more often than back, for example. But that representation is useful in different contexts or circumstances than this kind of representation. The beauty of this kind of representation is there's tons of implicit information in it. You probably never thought about the shape of a German Shepherd's ears until I asked you about them, but that information was embedded in your visual memory. You could reconstruct the image, and kind of recycle it. You can dig out information that was only implicit, making it explicit, which now can be used in language, and various other contexts in ways this stuff really can't, at least initially.

F: What about the factor of color? Doesn't that compound the situation, making it a little different for different people.

Mr. Kosslyn: It might. There are big individual differences in this stuff. This tradition was started at the beginning of scientific psychology, the late nineteenth century, based on individual differences. Because that was the most striking thing that Galton, and people like that noticed; the gigantic individual differences. You're probably right about color interacting with shape, insofar as the color gives you distinctive information. It will probably mitigate the effects of grain if there's a distinctive color. The butterfly example might not have been the best one, but it's the only one I thought of on the spot.

Let me just summarize what I've done so far. It's the end of the first part of the talk. Introspectively it feels as if when you visualize there are at least three kinds of things going on. One, the spatial extent of distance actually embodied in the representation. Your introspection suggests it, and the data suggests the introspection for some reason is reflecting information processing. Two, the spatial extent is not unbounded. There are limits to spatial extent. If an object gets too big, it seems to overflow. The angle is constant for different objects. And finally, there's a grain. If something is too small, parts are obscured. The second part of the talk: why would this be? The working hypothesis is that there are shared systems in visual perception, and visual imagery. There's plenty of behavioral data that suggests this is true. If you have a visual task at the same time you try to do visual mental imagery, you get interference. There are various perceptual phenomena which are evident in mental imagery. For example, there's something called the oblique effect. If I gave you a set of lines that are oblique, forty-five degrees, you have lower resolution in perception than if they're vertical or horizontal. You find exactly the same thing in mental imagery; there's also brain scanning results.

The next part of the talk is moving into the brain, and I rely primarily on results from positron emission topography. Here is how this works. The subject is peacefully laying down with their head resting right here within this camera. I work in collaboration with Nat Alpert at MGH, and the crew there; the radiology department. The way we do it is you breathe a very small amount of CO2 where the oxygen is O15. There's a half-life of about two minutes. So there's about a five step reaction in the lungs where this oxygen gets converted into water, goes into the blood. The more different parts of your brain are working, the more blood is taken up. The more blood that's taken up, the more this radioactive isotope goes with it, and you get fission, with a positron released. About two millimeters is the shortest distance before it slows down enough, and has a great enough probability of hitting an electron to cause annihilation. So that's the resolution limit, about two millimeters. You get annihilation, you get positrons going in opposite directions along the line. There are these tubes with little crystals on the end of them. What they detect is near simultaneous arrival of positrons coming from the head. The slight differences in arrival time are used to compute where along that line the annihilation took place. Over a period of about forty seconds, we get a three dimensional image of the brain. Actually of radiation density in the brain, which reflects blood flow, which reflects metabolism. We do this while a subject is doing some task. There's a little tray with a Macintosh computer on it. They're doing something while we record. We use pairs of tasks. The brain is never off, so you always have to compare one state with another state. Here's probably the best evidence that I've got for the idea that imagery and perception share common mechanisms. What we have here are areas that were activated both by imagery and perception. This is the left hemisphere, this is the right hemisphere of the brain, this is the lateral surface, this is the medial surface; so this is the middle part. Each of these little triangles indicates where there was a significant maximum of activation, as these score three point one. These were areas that were activated in common by imagery and perception. Areas that were activated either just by perception, that's the triangles, or imagery, that's the circles. What you get is about two-thirds of the same areas are used in common between imagery and perception. Imagery with a very different task than perception. So there's good evidence that common mechanisms are used in the two. Which is interesting, because look at this monkey brain.

This is from a monkey brain now. This is the first visual area, area seventeen, also called V1 in the monkey brain. This is a slide from "To Tell it All," about 1984. They train a monkey to stare at this pattern. There are lights staggered along here that are flashing. The animal is injected with something called 2-deoxyglucose, which is a radioactively tagged kind of sugar. The more different neurons are working, and the more the sugar is taken up. It gets lodged in the neurons; it doesn't get broken down. It's radioactive. After about ten minutes time the animal is put to sleep and it never wakes up. Its sacrificed, and its brain is taken out. What you see here is the first visual area. And these dark lines indicate where there were neurons that took up more of the radioactivity. The striking thing is, you can see the right visual field has been projected onto the cortex - this is from the left hemisphere - so that the relative topography is preserved. Isn't that incredible? They're like a picture in the head. The cortex is using space to represent space in these areas. There are about fifteen areas like this in the monkey brain.

The reason I find this interesting is these topographically organized visual areas embody distance. That's clear. They have limited extent. That is, they only evolve to take in information from the eyes. The eyes only subtend a certain visual angle. You don't see behind your head. And they have a grain. There's something called spatial summation. They record from a neuron in these areas. There's something called receptive field. The receptive field is a part of space, but if you put a stimulus the neuron likes, it will drive that neuron. If you put two stimuli within the receptive field, they get summated, so it sort of averages them. It blurs over distinctions within the receptive field, so that's a kind of grain.

Looking at this, the thought was perhaps visual mental images arise from activation of these areas. This is the first visual area I just showed you. It's the first cortical area that gets input from the eyes. The thought is that you have information stored in memory that can go all the way back to the first visual area.

Here is an early slide of the connectivity among visual areas from Van Essen's lab. Here there are a couple of important things. Every area below this line is topographically organized. This is V1. This is also called eighteen, V2. They have lots of different names. The second critical thing the anterior IT, which is inferior temporal lobe, PIT, post-inferior temporal lobe, these areas are under the temple, and they're crucially involved in storing visual memories. Now the format, the way they're stored is not at all picture-like. There's a columnar organization here, a population code; at least in the monkey.

Last bit. Notice these lines. There are no arrowheads. The reason there are no arrowheads is every area that projects an afferent connection to another area received information from that other area via efferent connection. Those backwards connections are of comparable size to the forward ones. So there's an enormous amount of information flowing backwards in the visual system. I'm going to say more about the specifics of the connectivity at the end of the talk, but for now, this will do.

The basic idea was you've got stored information in some compressed code. Sometimes I use an analogy of magnetic patterns on a VCR tape. In order to see, we reconstruct the implicit information, like what shape the German Shepherd's ears are, you need to use these back connections to reconstruct the spatial pattern in some of these areas. One other little thing. As you go higher up this hierarchy, receptive fields get larger and larger. So the resolution is getting smaller as you go up. So if you want really high res, the claim is, you want to reconstruct the spatial pattern in one of these really early areas.

Here is in humans now. This is what I tell my undergraduates is a deep Mohawk. It's a midsagittal slice. This is area seventeen, V1 in a human. The important thing is it's medial. The topography is hard to see because the cortex is folded in funny ways. The crucial thing is at the very back where the fovea projects. The small high resolution piece right in the middle of your eye. And as something gets larger, it projects increasingly forward, anterior along the cortex. There's a demonstration of that by Peter Fox, and his colleagues out of Wash. U. in the early days of PET scanning. They show people altering a checkerboard right on the fovea; very small. Or it spares the fovea, and stimulates further out. Or spares an even larger region, and stimulates further out. What would you expect to happen, in terms of activation, as you go from here, to here, to here? What should happen along the cortex? How should the activation move? So the fovea projects there. What should happen as it gets bigger? It should move forward. Of course that's just what they found.

Here was our first experiment. We wanted to know if the first visual area, which is topographically organized in humans, is also used in mental imagery. So here is the first experiment. We had people in two conditions close their eyes. Try this. Close your eyes, please. Visualize an upper case letter A, as small as you can see it, and still have it be visible. Got that? Does it have any curved lines? Does it have an enclosed area. We did tasks like that. The idea was when it's a tiny image, it should stimulate where along the cortex? Really far back. The other condition was we had them visualize it as large as they could without it overflowing. The idea was that the part of the cortex that was stimulated a lot by the little image here would be barely stimulated at all. But other parts of the cortex, that represent the parafoveal, the larger portions would be stimulated more by this image than the other one. What we did is exactly the same task. You visualize letters, the only difference is the size. We simply compare blood flow. These are sixteen people. What's more activated by the small images than the large? Brighter colors mean more activation.

Here's activation in area seventeen. Very posterior, more activated by the small than the large. We look at this here, large versus small. It moves forward, and there's actually a bit more activation. Here's the same data seen from the side. Here's the small versus large. You can see it's very posterior. You see it moves up, and it's a big larger.

When we first did these experiments, published in 1993, no one was convinced. People would think if your mental images can affect the first visual area, it's going to distort reality. It goes all the way back. That's the first area that's registering what's out there in the brain. People said well, maybe these are just different areas, in fact. They're not really part of the cortex, and one is being activated by the large because it's harder, or something. Other people said well, you've compared two imagery conditions, so who knows about relative values of imagery? There were actually quite a few complaints, so we did a series of other experiments. This one was published in Nature, so I suppose it's the best of them.

Here's the deal. We did three imagery conditions now - so we established a continuum - rather than just two. The subjects before each one study a set of pictures of objects. After they study them, they're taught two of these probes. For example, right higher. They're told later in the imagery task you're going to be hearing the names of objects you just studied. For example, guitar. Hold that image, and then you'll hear something like right, higher. And you have to use that drawing to decide the right side was higher than the left side. Here it's not true. The left side was higher. To get the answer right, you have to use those particular drawings. They don't know in advance what we're going to be asking them. There are three imagery conditions. The only way they differ is in terms of the size at which they form the images. So prior to scanning, they study a box at sometimes a quarter degree visual angle, four degrees, or sixteen degrees. They're told in this condition you're going to be asked to visualize those objects until they fill the box, and get really big, medium, or small.

The stimuli are always auditory, their eyes are closed. They hear something like guitar, right side higher. In fact, there were four versions of those stimuli. Before they actually know anything about imagery, they come in, and they just hear the stimuli from one set, one version. They just respond alternating. That's going to be a baseline. We have identical stimuli, auditory stimuli, coming in. There are twelve people. We rotate so that each auditory stimulus occurs equally often at each size as the baseline. In addition, we used a resting baseline.

The reason we did that is there was an even more unfortunate complaint about our original imagery experiments than alternative explanations. Some people couldn't get the result in France, and in Sweden; although other people did get the result. So we're trying to figure out why not. And what we noticed was all the labs that did not get activation of the first visual area during imagery used a resting baseline. They asked people to close, relax. In fact one - the Swedish lab - tells you to visualize blackness. So we thought, hmm, if you're comparing imagery to a condition where they're asked to visualize blackness, maybe you're just subtracting out the imagery because of the baseline.

So we had five conditions. Three imagery, a listening baseline - which we always do with identical stimuli; only the task is different - and a resting baseline. Here's what you get. Here is the imagery data - small, medium, and large - compared to a listening baseline. So there's identical stimuli, identical numbers of responses. What you find is in the right hemisphere these dots indicate a small mental image. You can see it's about as posterior as you can get and still be in the brain. The square is the medium sized. And you can actually see that it's a larger area that's subtended by the large mental images. This is probably the outer envelope, and maybe this is internal details, is my current theory. It actually worked pretty well.

Now look at this same imagery data when you compare it to the resting baseline. You get nothing at all. Why not? Here is relative blood flow in a set of areas that are activated in imagery. Just pay attention to this for now. This is what you get in the listening baseline. This is what you get in the resting baseline. So for some reason, during rest, you get enormous activation of the first visual area, which was subtracted out.

We initially thought that the debate had been resolved. That it was just a question of the baseline. Then this lab in France, in Cannes, re-did their experiments using a listening baseline, and they still didn't get activation in V1. I looked at their results, and noticed they were using spatial tasks. Such as imagine walking through your hometown. Turning left, and turning right. And at the end they ask you where you are on a map. You don't need high resolution. So the idea was maybe what's going on is you need high resolution in an image to force the activation all the way back down the system. Here's what we did to test that idea.

We had people study gradings, four of them. These are relatively low resolution. So they would study these. They were told this is quadrant A, this is B, C, and D. The task later was that here is something like AC tilt. So they were asked whether the second-named quadrant, C, had greater tilt, clockwise, relative to the first name. Here the answer would be yes. It might have been number of bars, length, width, or various other properties. Again, they didn't know when they were studying them. This is one condition. Another condition, we have higher spatial frequency gradings, so you need high resolution to resolve these. That was the idea. Here's what you get. Again, there's a listening baseline. Here's the low resolution. You get good activation in area seventeen; not a surprise, because there are sharp edges. The third harmonic's going to be high spatial frequency. Here's the high resolution. Again, you get good activation in area seventeen. Your eyes are closed. Here is probably the more important comparison. This is high resolution minus low resolution. And in fact we found a part of area seventeen that does get activated more by high resolution.

Now look at the other comparison. This was cool. What's activated more by low resolution? This area is the one that other people got when they don't get area seventeen. And you also notice you get inferior temporal cortex, and also fusiform, which these areas have much larger receptive fields. So low resolution, you're getting more activation in the areas that have larger receptive fields. High resolution, more in areas with small receptive fields.

I spent last year in France working at this lab that wasn't able to get the result we got. We designed what I thought was a very good imagery experiment that requires high resolution, and we still didn't get it.

M: It's the wine.

Mr. Kosslyn: It could be the wine. But I noticed something interesting. Before I get to that, the first question was is the activation we've got in area seventeen functional? Is it actually indicating any work is being done, or is it just sort of along for the ride? Here is the TMS results. This is transcranial magnetic stimulation. You put a coil right on someone's skull, and you put through it either a burst, or a series of bursts of a lot of electricity, which creates a big magnetic field. And if it's big enough, it actually spasms the neurons directly under it, and you can create essentially reversible lesions. You need

FDA approval before each of these experiments. People in the lab do them on each other. This is in collaboration at the Beth Israel Hospital.

What we did is we used the low resolution gradients experiment I just showed you that we knew we got activation in area seventeen. Put the coil right over the back of the head. It's fairly precisely delivered, it turns out. Maybe a centimeter, a centimeter and a half, right medial cortex. So here's two conditions. When we tilt the coil so the magnetic field is missing area seventeen, that's called sham or real. We did it both in imagery and perception. You see it's actually even a little bigger. Also look at the accuracy rates. The way we did this is before the experiment, we gave them a series of tickles back there, and then test immediately afterwards. We're looking at a carryover effect that lasts maybe ten minutes or so. You've sort of spasmed this area. The accuracy drops way down, compared to the sham for imagery. That suggests it really is functional.

Here's another suggestion that it is functional. If we ask people to close their eyes, visualize letters in a scanner, and then answer questions like whether they have any curved lines or not. If we look at area seventeen, here's what we get. Each dot is a person's brain. Well, actually it's the blood flow - it represents the blood flow in area seventeen of one person's brain. V1, seventeen; same thing. The brains are all normalized to a mean value of fifty milliliters per minute per hundred grams. So the first thing you notice is most people have more than mean value in area seventeen where they're doing this imagery task. These are response times while they're actually in a scanner. The four people who are slowest also have the least blood flow. There's no difference in accuracy rates. Everybody's doing the task. It suggests that there is some functional connection between blood flow in this area, and performance. It's not just epiphenomenal, or along for the ride. This, in conjunction with the TMS results I think is fairly compelling.

So what's going on? I don't know what's going on. But when I was in this lab in France, we were testing these subjects. I noticed there was something strikingly different from how we do it at Mass. General. This is probably the only part of the talk that's relevant to your interests, maybe. It's pitch black. It was absolutely pitch black. You couldn't move in the room, because you'd bump into something. Not only was it pitch black, but they had a black tent around the PET machine, around the camera. So there was no light at all coming in there. So I started thinking about this. Could it be that the diffuse light that comes through is crucial? Maybe. The important piece of anatomy, for me, is that the afferent connections, the ones that are flowing into the visual system, are very precisely targeted. The efferent connections going backwards are a mess. They're very diffuse, they're not at all point for point reciprocal. It's misleading to say they're reciprocal. That's the term in the field, but it's misleading. That makes you think they're coming back where they came from. They're not.

The thought was, maybe what's going on is not like playing a stored tape, and painting on a CRT an image in V1. Rather, maybe what's going on is it's more like putting an attractor to a network. Where you have to have some random activation in V1, and there's some biasing among the connections; some visual learning that went on there when you first saw the thing. You're dropping a catalyst, essentially, into a saturated solution. That's what these backward connections are doing. If that's true, then your imagery ought to be worse at pitch black than with slight diffused light coming through. A very counterintuitive prediction.

I came back from France, and immediately did an experiment; the simplest one to do. We had people do the experiment with the letters, where they closed their eyes, they heard the names of letters, and then something like curved lines, and they had to answer whether or not it had it. There were three conditions. Either their eyes were closed, and it was pitch black with a blindfold on; their eyes were closed, but slight diffused light was coming through; or their eyes were open. Just pay attention to this for a moment. This is where they had to make a symmetry judgment. So is the letter A vertically symmetrical or not? Yes or no? Amazingly, you get a significant difference in response time where they're faster if there's diffused light than if it's pitch black. Not only that, they're more accurate. There are error rates. There's barely any difference between actually having your eyes open, and having them closed with light coming through. And in contrast we have a spatial judgment. We asked is there a straight line on any border? Which is a different kind of task, which we already had reason to believe, from studies that were done in Europe, doesn't necessarily require high resolution or shape; there's a parietal system that will do it. It goes the other way. This I'm less sure how to interpret. But what's striking is that when you clearly need to use the shape, having diffuse light coming in is actually better than pitch black. So what we're doing now in France is turning the lights on. Their eyes are still closed, and we'll see if it works.

Let me summarize the talk. There were two main parts. I started off with phenomenology and information processing. I pointed out that introspectively it seems as though there are three aspects of images. Distance is embodied, but it's got a limited spatial extent, and there's a grain. Each of those introspections predicted behavioral data, response times, or judgments. We then asked the question can we explain these data in terms of the brain? There's the reductionist piece. We noticed that there are these areas in the brain that have those properties. They embody distance, they have limited spatial extent, and they have a grain. So the question was: are those areas playing a role in visual mental imagery? We showed that not only are they activated when you visualize with your eyes closed, but the spatial properties of what you visualize, that is the size, affect the areas very much the way you find in visual perception. Moreover, if we temporarily zap one of these areas, as it were, not totally disabling it, but impairing it, that affects the times in this task. Moreover, the amount of blood flow, the less blood flow predicts the response times; and there seems to be a functional relation. But the issue was: why is it that some labs are finding this result, and not others? It's not just the baseline, although it might contribute. It's not just resolution, although it is the case the higher resolution you need, the further down you get activation. It may be something having to do with the way the system actually works. That you need to reconstruct this pattern by taking advantage of activations already there, and amplifying, organizing, and augmenting it, okay, using attractors to form the image.

So that's the hypothesis. Stay tuned. We'll know the answer probably within the next six months. Thanks very much for your time.

F: Your talk was fascinating. So it is actually the nature of the reciprocal connections to allow imagery?

Mr. Kosslyn: The reciprocal connections... that's the nose. I mean, they weren't there for imagery. But what they were there for, I suspect, is cooperative computation. Where you're using stored information downstream to augment what's coming in, filling in gaps, organizing, and so forth. I think as soon as you could do that, you could then do it in the absence of the input. So as soon as an organism has these reciprocal connections, and the ability to use top down processing, I think it gets imagery for free.

M: And where would that be?

Mr. Kosslyn: What kind of animal?

M: Yes.

Mr. Kosslyn: Good question. I don't know. I'd have to look. I've never actually thought about that. Certainly mammals. At least all the ones that I can think of. Probably not insects. Somewhere between the two

M: What about birds?

Mr. Kosslyn: That's an excellent question. Does anybody know if birds have reciprocal connections? I don't know if anybody's looked. Excellent. Thank you.

Actually we did some PET studies of dyslexics also. A curious thing. Behaviorally, when they form images, the overall time is about the same. If you visualize the upper case letter F, you're not aware of this, but you're actually forming it a segment at a time, in roughly the order in which people typically make an F. For dyslexics, this is not true. They don't put the segments up in the same order. So even though the end result is the same, and they're equally accurate, they do it in a different way. Now why that is, I don't know. But as I recall, rotation was about the same in them, actually. Their brains look different. I'll tell you that. We just had them read words aloud over four blocks of trials in the scanner. What we find with normal people is there's typical areas that get activated, and by the end, those are almost all off, and new ones are on. It's not just with practice you get less activity, you actually get qualitative changes in strategy. The same was true of the dyslexics, but they were completely different areas. They start off with much more frontal activity. A huge amount of frontal activity. It was much more of a conscious strategy, reading, for them. That all got turned off by the end, but other areas - I have no idea what they're doing there - were turned on. It's just being written up now, actually. Well, thanks again.

Mr. Bar-Yam: Norm Badler is from the University of Pennsylvania. He is supervising a large group of people that are building models of human physiology and motion. He, very early on, recognized that we had both the capability, and the importance of creating such representations.

Mr. Norm Badler: It's a pleasure to be here. And I think that Monty Python had it right when they said "and now for something completely different". Though I think by the end it may actually link in a little better. Coming, as I am, from the computer science side, I guess if I had to use one word which would characterize what I do, I would say I'm a modeler. I'm not necessarily an experimentalist, nor a theorist. I look at the world, in this case at humans, and try to understand what I can enough to build models of them for a computer presentation.

And not knowing exactly what to expect in this group, I thought I'd even take one step backwards first, and just make sure we all are on the same page regarding what a virtual environment is. A lot of people play games, and they see 3D, so it's not a big deal; but let me just start with saying a virtual environment is a 3D world that we model within a computer system, and it's fairly obvious that we need geometry for shape and form. In order to get a virtual human to interact in that world, we also need to provide at least two other pieces of information. We have to actually make it believe that the geometry is substantive, so it doesn't necessarily pass through walls, and bump into things. And also needs - we basically avoid the perception problem by assuming there are sufficient labels on things that if we refer to them, our virtual human can understand what we're talking about. So just a little bit of base understanding there; all of which can be argued in the light of a real human system. But that would take us too far afield.

So what are virtual humans, then? They're simply computer stuff - models, 3D shapes - that are embedded in these virtual environments. And there are generally two different purposes to which such virtual people may be built. The first is actually the way we came at this problem, which is that we wanted human surrogates. We wanted people that were structured, and built well enough - and I don't want to define that really - well enough where they could be embedded in a three dimensional environment, and used as a substitute for people. These occur in engineering environments, and whatnot.

The work that we did on ergonomic humans left the University of Pennsylvania a couple years ago in a company called Transom Technologies. Just last month, Transom Technologies was merged into a public company, EAI. I guess that means I now have the only public NASDAQ virtual human. I don't want to advertise for the company, but I want to give you an idea of what they do. This company actually sells the engineering community. For example, just this month Ford Motor Company bought a site license to this as a way of evaluating their products for various users. Operators of machinery, as well as drivers of vehicles. A minute level of detail, down to maintenance activities and the like, can be done here.

The second way we can use virtual humans is representing ourselves, or other constructed participants in a virtual environment where the actual performance of some specific task, or how well we fit in the environment is not particularly important. This is the area that we've moved into more in the research environment. I'll try to explain why the title of this talk was from instructions to action in that context.

Let me give you a little bit more of the background on virtual humans. First of all, humans are extremely complex. There's no way I, or any of my colleagues would ever get up here, and say yes, we've managed to build a virtual human. I'll try to pull this into some kind of taxonomy, very, very briefly, to give you an idea of at least five dimensions along which we can create virtual humans of varying degrees of complexity. One is in appearance. We can go all the way from simple cartoons into three dimensional geometry. Then we can get into biomechanics, and even some physiology.

In function, things at the low end, basically cartoons, are just there to entertain. We gradually want to put in more and more of the human functionality. We get the skeleton in, then start to build in physiological properties like fatigue. Maybe injury, or the acquisition of skills. Which is a funny transition, but if you're skilled you try to avoid injury. And then you get into other things, including cognitive models, and then actually working in teams. The idea of these dimensions is simply to give you some idea of increasing complexity as you go toward the right.

In the time dimension, it's how long does it take someone to create and animate a virtual human? If you go to the movies and you see special effects, it's all done off-line. Lots of creativity, lots of time, and lots of money. As you move towards the right, you end up with things that run more and more rapidly, that have to make decisions, that interact among themselves, and that ultimately appear to be coordinated in real time. Not all of these things are possible today, but they give us some targets for the kind of veracity we want from these virtual humans.

In the autonomy domain, if we start with an artist's drawing, or some scripted animation, there's no autonomy at all. As we move to the right, we get some reactive kinds of animation, which sometimes are called artificial life. There's good stuff in artificial life, but it sort of reaches a point, and then it can't go very much further, because autonomy includes making decisions, and actually communicating amongst agents. And ultimately taking initiative, and actually figuring out what you want to do.

On the individuality dimension, it's easy to make generic characters like Jack. He represents anyone. He can be sized differently. He can be given different color shirts. But he's basically a generic character. Also, artists are good at making hand-crafted characters, and they impart an individuality that way. But there are many things that we need to add in order to get to the point where ultimately we can create a specific individual. So the state of the art is not such that we can figure out how to create you, or me - our appearance easily may be recreated, but not the way we react, and the way we think. That remains an unachieved goal so far.

Given those five dimensions, let me just say that there isn't any one solution to a particular problem. The application area has different demands in these dimensions. The background information here is no one model suffices for anything. If you're in the special effects world, for example, appearance is very important, but function is not. You can take a lot of time to make things, and they don't need to be very autonomous. But they probably are somewhat discriminated as individuals. Say you want it to look like a character that's already an actor in the scene, for example. If you go to something else like tutoring systems which have to interact live with a student (sometimes they're called pedagogical systems) appearance is sort of important, functionality might not be so important. But they should react in real time with the student, and be more autonomous, in the sense of being able to make some decisions about the study plan, or helping the student, and so forth. So one can make these distinctions, and they just help to show that any kind of system that I show you is not meant to be a panacea for all possible applications, and vice versa.

Now, the way that we typically control virtual humans, in fact the way that that commercial software I showed you, and many other systems work is the good old point and click paradigm. You have some kind of menu of things, or you might be able to grab its hand, and move it around. That's okay. Almost a considerable industry now is directly sensed motion. You put on sensors, and gloves, and various kinds of devices, and your movements can control the virtual human's. That's all pretty well understood. What I want to focus on here is a much more human problem of how do we get these virtual humans to understand language commands? How can we just tell them what to do? The reasons for this is that in our normal, everyday lives we don't go around wearing bodysuits, and we don't go around pointing and clicking at people. We tell people what to do by giving them instructions. We give them manuals, we give them verbal instructions. Sometimes we show them, so there is a physical aspect, but very often language is a way of communicating intention and behavior. That's what I'd like to focus on here. How do we actually convert from language instructions into the behavior in a virtual human?

We've given this a name. "Avatars" are embodiments of people directly controlled by someone. In order to decouple the spatial characteristic of actually sensing the motion, or actually manipulating them onscreen we simply call them smart avatars. So a smart avatar is a virtual human that understands instructions.

Now, to take this big problem, and try to find a mapping from language at one end, which is very abstract, very conceptual, and action in that virtual human at the other; who is geometric, who has motion, and maybe some physics underneath it. How do we do that? In my own experience, I started in this whole area going the other way. I started out as an AI computer vision person who wanted to look at people with computers, and describe what they did. Back when I was doing my Ph.D., computers were much slower, and vision was very difficult to do. I decided that synthesizing a human, and making the image processing simpler would be the way to go. Well, I've been synthesizing people for the intervening twenty-some years. But I've always believed, as far back as my Ph.D. work, that the number of levels between action and language couldn't be too deep. That may be an ontological argument we can get into. I couldn't believe there were eighty levels, or something like that. All our experiments and systems over the years have led to something which at least works for us, which is approximately a four, or maybe five level architecture. I'd like to give you the sense of that here. The first level, or the lowest level - because I'm a bottom up kind of person - is a set of motion generators which actually move the body around. The next level on that, which may be surprising, but actually is one level that seems to be almost universally arrived at from multiple perspectives now, is an abstract computational machine level that I'll call Parallel Transition Networks. Then a layer on top of that called Parameterized Actions. Those of you who have a natural language background will actually see this as almost an alternative) name for something called case frames. And finally, the Natural Language Instructions, or text itself.

I'm not saying that these are the only four levels. I'm not saying that you absolutely need these. But I'm going to try to justify why and how it works out well this way for us. Because after all, if we can get things to animate from these levels, we have at least achieved a demonstration that the complexity is sufficient.

At the first level is the guy you saw. He's got a number of routines that have been written that allow him to gesture like I'm doing while I talk. He can reach for things. He can grasp things. He can walk, he can turn, he can climb, he can do some posture transitions. He has a very interesting visual attention mechanism, which is built from cognitive science principles, but that's another talk entirely. He has a number of things that he can do with objects. He can take motion which has been created some other way, and play it back, sometimes in context. For those of you who are big on subsumption kinds of architectures, I don't ignore that; it's just there are some reactive behaviors which are embedded inside here. For example, walking can subsume a reactive layer which includes collision avoidance. I'll just lump it in this layer for now.

Level one, the Parallel Transitions Networks, is a virtual machine. Our computers that we use are all sequential, usually uniprocessor machines, while people are inherently parallel. I'm doing this, I'm walking, I'm talking at the same time. If we want to have multiple people interacting, it's easier to think about these things happening in parallel; whether or not the underlying computation truly is or not. PaT-Nets are simply a conceptual structure with a programming language attached that let's us deal with the parallelism, and a very simple computational model, which is that when I'm doing something, or when he's doing something, a process is a node. And when that process ends, or something else happens to cause the process to change, there's an instantaneous transition out of it into another process state. We can also do some message passing, and synchronization on this.

This idea of finite state machines is certainly not new computationally. Having parallel ones is not exceptionally brilliant, but there are now at least five or six major animation efforts that have arrived at the same computational model independently. That doesn't mean that it's the right model, but independently voting is a pretty good way of validating, at least, the veracity of that particular model.

We've been using PaT-Nets in our lab since about 1992, and a series of works came out of this, which I'm going to show you short snippets of. The first one that we did in conjunction with Justine Cassell, who's now at the MIT Media Lab, was two whole virtual humans - agents that had a conversation with each other. All aspects of the conversation were mediated through parallel transition nets. They controlled the gestures, they controlled the turn-taking, they even controlled the dialogue, the eye gaze, the face motions, and all that. I'm going to show you just a little bit of that. Ignatius is a bank teller, and George needs money. They hold this conversation in order to get this task accomplished. In order to get a whole human, a large number of things need to be coordinated. Remember this is five years old. They're having an actually rather strange conversation.

First of all, their hand gestures are based on what they're saying. There's intonation in the speech generation, and there's some beat gestures that go along with the accents. There's head nods, which have to do with turn-taking. The faces and lips are animated according to the speech. The eyes move, again, for the conversation. Even though they look a little silly, and they're a little stiff, the idea of putting all the pieces together was what was important. We then applied this to games. We did hide and seek. We had four people playing hide and seek all day long on the computer. Then we applied it to looking inside the body. A very large component of our center is doing biomedical modeling research - hearts, and lungs, and all that. Before that, we tried for a trauma simulator, or a trauma management system that we did for Rick Satava at DARPA called Medisim.

This was Medisim, and we used PaT-Nets to model the physiological response of someone who was injured. There, for example, a soldier gets a gunshot wound in the field. This guy comes along who is the medic, or his buddy, and has to treat him. He's not going to do virtual surgery on the battlefield, he simply wants to stabilize him enough to get him to help, so this goes very quickly. It's sped up about five times; it really should be fifteen minutes, but it's going to only take three.

What you're seeing, the student wouldn't typically see. It's the central nervous system response. The casualty lying on the ground can do certain things by command, the medic can do certain things by command. Your job would be to control the medic to do the right things to stabilize the patient. You don't necessarily know what's wrong. You can do things like give commands; in this case, to expose the wound. Now you can see he's breathing, which is a good sign; but he's turning blue, and you notice his physiology is going down. You don't know that. You just see this. He's got a chest wound. The student would typically be telling the medic what to do. In this case, the instructions are optimal. But if you do the wrong things, he will expire. It doesn't seem to be a bleeding problem, so the dressing is on. His breathing starts to get labored. There are various tests. There's some capillary refill tests that are being applied. One of the diagnostic things for this particular injury is that the neck veins are distended. Now it turns out this guy has a tension numothorax, which - I'm not a medical person, but basically means the lung is collapsed, and it's pushing on the main vein that leaves the heart. Even though the circulatory system isn't impaired, basically the blood flow to the brain is stopped. The solution is to relieve the pressure on the lungs, so the blood can flow again. That's done through something called the needle aspiration.

Now the medic is doing the needle aspiration. In goes the needle, and just in time, at minute twelve. By minute fourteen it's too late. You see as soon as the needle goes in, his vital signs start to return, his color comes back. It's basically the pattern that's sort of running this simple physiology under there. It was done in conjunction with a trauma surgeon. His eyes open up. Yes, we saved him. That's great.

One more experiment then. We had a visitor from Japan, Tsukasa Noma. He did a Jack Presenter. Basically this looked like a weatherman who narrated, and talked, and pointed to things on a screen. And after that we built a system called JackMOO.

Now the trouble with these things. Dealing with the PaT-Net level - remember this is our second level, or level number one. What's missing in that level? These things work, but we had to code everything by hand. So much for emergent behavior, right? If you code it up, what's new? We realized every time we tried to build a level to control this animation, that we were always building an artificial level. That artificial level might be good for computers, or computer scientists, but it wasn't the way people were conceptualizing the situation. We realized that in order to understand how people were executing instructions, we really truly needed to understand what the connection between language and animation really was. In recent work with other people, including Mark Steedman, and Aravind Joshi at the University of Pennsylvania, we've been developing a parameterized action representation. This mediates between what you get out of language, and what you need to drive the animation. Basically language comes in at the top, and PaT-Nets come out at the bottom. We don't have to go all the way down into the low level motion control. It's just got a bunch of stuff in there which is not crucial for this discussion. The advantage of taking this point of view is we build off of the enormous natural language community. We're saying look, our agent has got to have the power of a full natural language parser at its front end. If it has that power, it should be able to do some interesting things.

We're using this XTAG tree adjoining grammar, which has thirty- seven thousand terms in it or so. So it's got a good coverage. It knows a lot of English grammar. It probably could read the Wall Street Journal, but there aren't too many instructions in there. The corpus of instructions that we're looking at is actually quite different than ones that most parsers are tested on. So we're actually building a translator between this XTAG parser, and our PAR representation, which I flew past you very quickly. The idea is to handle very complex goals presented linguistically. Things that have referring expressions, things that set conditions to happen in the future, or questions that are just about things that are going on in the virtual world.

We're also expecting to use language as a way of priming the agent's abilities. There's this question of if I build Jack to do something, how do I tell him what to do in the future. It's basically an education process; and a lot of education is of course show me, and try it, but some of it certainly comes through language. We're looking at language as a means for priming, or giving hypnotic suggestions, if you like, to an agent so that they can be executed when the time is appropriate.

Now some of the experiments in this. We've been looking at greetings that are contextually dependent. The way two people greet each other is not stereotyped. We don't always do it the same way. It depends on the culture, and the gender, and the age, and all that. The way we go to a place is highly contextual. The way I sit in a chair is going to be different than the way I go to bed. There are lots more activities involved in that. It may involve opening doors, and navigating obstacles, and all sorts of thing. I may set up relationships. For example, if we get that far in the video we'll see two individuals who set up a follow me relationship, so that one follows the other, but doesn't exactly mirror the same movement. We can also create autonomous agents that just do their things.

What you'll see briefly: actions are triggered by simple commands. At the low level, there are virtual human movements done from synthesis, some from motion capture, some from an attention process. The autonomous waiter is a process simulation that doesn't reflect any real person at all. When we tell these guys what to do, if there are certain preparatory actions that have to be done, they do them. When we tell the guy go to bed he doesn't just fly into bed, he has to get up from his chair, he has to walk to a ladder, climb the ladder, go to the bed, and sit down on it. All these things have to be part of the model of executing those commands. In this particular one, there's no face, or conversational signals at all.

What's happening here are these avatars are controlled by text that's being taught. Bob walks in. Bob types greet Norm. It's a little slow, and the motions are not blended real well, but. Bob has got to greet - hi I'm Bob. Sorry I'm late. That's canned speech. Now Norm that's sitting down has said okay, I will greet Bob. He could ignore him, right? So he stands up. It's a little awkward here, but he has to get out. This isn't just a chat room. They've actually got to get their hands in spatial proximity, and they do their little handshaking thing. Now, when I'm done with the greeting, my avatar goes back to drinking there. Sitting and drinking. The autonomous waiter is over on the left. Whenever anybody who sits down has an empty glass, he comes over, and fills it. Bob says greet Sarah. Sarah is Asian, so he bows to her, and she stands up, and bows back to him. He has a different greeting, depending on the culture. I'm Sarah. So this is elaborate greeting rituals here. David's over in the red shirt on the left side of the table. All these things require a lot of moving around, and repositioning. David says greet Bob, so he has to do the stand up, and shake hands routine. The point is, the actions have to be embedded somewhere. They don't come in the instructions. An important question that we need to ask is how much real planning goes on, and how much stereotype planning can be embedded. We believe a lot of stereotype planning can be done.

Now he's greeted everyone. Bob's going to go take the remaining seat. I think he's going to be told to sit down. Sit on chair. Sit on chair. There's only one chair open, so he gets over there. Anybody who's a glutton for punishment, we can watch the rest of any of these at lunch time.

But the trick here is to get all this stuff - there are five people that are interacting. They're actually all doing something, even if they're just sitting. Now the waiter should come over, and he says - go to bed. Now there's going to be two things happening. David says he wants to go to bed. But meanwhile, the waiter is triggered because someone sat down, and has an empty glass. So this is where you need these parallel things going on. You can see Sarah's drinking. Making funny noises. The pouring motion is actually motion-capture, so that looks pretty good. The climbing is actually done from a dynamic simulation. Sarah, would you like to go for a walk? Yes, okay. Norm says to Sarah would you like to go for a walk? Now there's a little negotiation here. Sarah has to agree. Sarah says sure, I would like to. Norm, who's on the right, will get up and leave through the door. Norm types leave. The starts of actions are done by typing the commands, and then they run PaT-Nets underneath. And so Sarah, though she's agreed, won't actually leave until she says Sarah typed follow Norm. Follow Norm. Now she'll follow. So I'm smart enough to leave the door open, and Sarah will be smart enough to close the door, and basically I've told you the rest of the story.

I had to use the word emergent somewhere, so I thought - oh, it's hidden. I thought one thing I'd say is we are actually seeing the emergence, which is probably the wrong sense of the word, of a virtual human industry now. Not only from games and special effects, and not only from the ergonomic models; there's animation products. What we're finding in this as we look at this as an industry is that the modeling, that is, the geometry is relatively easy. The control is the hard part. And I don't want to talk about standardization.

The conclusion is that even in that simple mucilage, the individuals execute similar instructions in very different ways. They're context-dependent. You can't just associate a chunk of behavior with a chunk of language. The context sensitivity is crucially environmentally dependent.

The second point is, even though we can use other sorts of interfaces, and we still do so, we're trying to push very much on a focused look at what we can do with just language, and how we can connect language into action to extend the usability and the applications of these virtual beings. We actually invented a new word called an actionary which - it's unlike a dictionary, in that a dictionary defines terms within the same conceptual framework - the actionary takes linguistic things, and transforms it into action; context-sensitive action. We're in the first stages of building a very small, but structurally sound actionary at this time. Thank you very much for your attention.

I left the blocks world because there weren't very many interesting actions in a blocks world. There were very interesting structures there. When you get into human actions, and you start to look at the verbs that people use to describe the way they interact with the world, it's certainly richer than the Winograd world of referring to blocks. The language part of referring to spatial descriptions is also very different than the language referring to actions. One of the things our natural language people have been very good at helping us discover is that the world of language in terms that is, let's say, dynamic or kinematic is much richer than that of static or states-based descriptions. There are differences in the linguistics. There are lots of termination conditions, and manner expressions, and so forth, that are very important in this world that aren't there.

A little bit on the gesture Jack guys. We've been doing faces since 1991, and other people certainly have been doing faces. The state of the art in face animation now is really pretty phenomenal. You can download stuff on the Net that's pretty amazing. The trick is to try to actually get the facial expression to reflect the internal state of the person; and that's the hard problem. Even new work just reported two weeks ago on facial expressions that relate to something called performatives, which is just essentially the reason for making the communication change the facial expression. We, and many other groups are really looking at the deeper problems now. I don't think they're any harder. If anything, the face is a more tractable object than the rest of the body. What's different in the face is so much is driven by the internal state, we don't really know what that internal state really looks like. We don't know how it's manifest, necessarily.

F: What about the use of Medisim?

Mr. Badler: The medical simulation was designed as a training system. Our software was actually embedded, and then removed from a system down at the Information Sciences Institute. It's a system called Steve, which was just published in the SIGART bulletin, which is a pedagogical system to teach people actions. There's a question of whether you really need a full up body to be a surrogate teacher. I think the answer is you probably do, but then you've got to get to this problem of modeling a specific individual. Not to replace teachers, but to let them get some sleep. The dial a teacher late at night who looks and acts like the teacher you see during the day and helps with homework is an interesting idea. Whether it's needed, the pedagogical researchers amongst us will have to inform us.

M: I have one question. Can you say a few words about the internal structure of the system, of the physiology that gives rise to the motion?

Mr. Badler: Sure. He's got a full skeleton, including shoulders that work properly. He doesn't have muscles, so all the activations are done by computations that rotate joints, or move them appropriately. We certainly helped pioneer a technique called inverse kinematics, which allows him to reach for things. Specifying end affectors, and then having the body joint attached, and move in a reasonable way, not necessarily realistic. Things like walking is generated by a walking synthesizer that understands curve paths, and the like. All these pieces are just built up in sort of a bottom up fashion. When you put them together, and control them at the upper levels, then you hopefully get a behavior which may not look great, but at least it looks like many things are happening at once.

M: You skipped over standards involved...

Mr. Badler: There are basically four or so levels that are looked at. The MPEG-4, which in all likelihood will be an international standard is the most interesting. The compression part is actually not interesting to me, but it contains a standard body, and a standard face, and parameters for not only animating them, but changing their size. That's very likely to come down in a television set near you in a few years. There are some nontrivial noises to take our PARs - this parameterized action representation, and consider that to be "a standard" for going between language, and the other levels. It's not clear how much we want to push that. But the industry believes that certain standards - humans are basically structured the same - is good for economic benefit amongst the various players.

Mr. Bar-Yam: Biology has been the source of some of the most reductionist thinking applied to complex systems in science.

It seems, however, that there is some movement in the direction of more carefully understanding how collective behaviors arise in biological systems. David Sloan Wilson, who is from SUNY Binghamton, is going to tell us about understanding how collective behaviors arise. In particular, through evolutionary processes.

Mr. David Sloan Wilson: And now for something completely different. I want to begin by talking about two ways of looking at the world that exist around the world, and probably throughout human history.

The first view sees higher level units, such as groups, and ecosystems, as having purpose and order in their own right. Joseph Butler said in 1726, "It is as manifest that we were made for society, and to promote the happiness of it, as that we were intended to take care of our own life, and health, and private good." This view often finds religious expression. There's a wonderful quote from the Hutterites, which is a communal Christian sect, which says that "True love means growth for the whole organism; whose members are all interdependent, and serve each other. That is the outward form of the inner working of the spirit, the organism of the body governed by Christ. We see the same thing among the bees, who all work with equal zeal gathering honey."

The other view sees purpose and order only at the individual level, and tends to see all higher level units as mere consequences of individual action. (Margaret Thatcher really said that, by the way. I checked it with a researcher at the House of Commons.) This view, of course, has been very influential in academia. This is a quote by the late Donald Campbell, a social psychologist, who says that, "methodological individualism dominates our neighboring field of economics, much of sociology, and all of psychology's excursions into organizational theory." This is the dogma that all human social group processes are to be explained by laws of individual behavior. That groups in social organizations have no ontological reality. That, where used, references to organizations, etc., are but convenient summaries of individual behavior.

The purpose of my talk is to show you that this ancient controversy over individuals versus groups can be placed on a scientific foundation. There's limited justification for the group level view in both nature and human nature. In other words, the concept of groups as organisms can be, but is not always, legitimate. And finally, that natural selection and complex systems together are required to explain higher level adaptive organization.

These are very large topics, so the best I can do is to provide you with a whirlwind tour. For more details, I'd like to suggest that you attend the multilevel selection symposium, organized by Sean Rice at this conference, which is being held on Tuesday evening. And also to read my book with Eliot Sober, which was published by Harvard University Press last spring.

How do we place this ancient controversy of individuals versus groups on a scientific foundation? We need to begin with Darwin's theory of natural selection. What Darwin showed was that in a population of organisms, three ingredients are required for the process of natural selection. First, these individuals have to vary in some of their measurable traits. We call that phenotypic variation. Second, this variation has to be heritable; there has to be a resemblance between parents and offspring. And finally, there has to be fitness consequences. These traits that we measure have to make a difference, in terms of survival and reproduction. When these conditions are met, then individuals will acquire the phenotypic traits that enable them to survive and reproduce in their environments. We call these traits adaptations.

Traits such as the long neck of the giraffe, and the hard shell of the turtle, and the cryptic coloration of the moth are all easy to appreciate as adaptations which cause the individuals who have them to survive and reproduce better than those that don't. But the concept of adaptation becomes more subtle when it comes to social behaviors, and other traits whose sphere of influence extends beyond the individuals that cause them. The reason for that is that natural selection is based on relative fitness. It doesn't actually matter how well you survive or reproduce, it only matters that you survive and reproduce better than everyone else in the population.

Imagine that I do something that increases my fitness, and increases the fitness of everyone else in the population exactly to the same degree. That trait will be neutral, with respect to natural selection, because it has not caused any differences in survival and reproduction. Or imagine that I do something which benefits you at the expense of myself. Well, that trait will go extinct, as surely as bad eyes, or faulty teeth; because natural selection is all about having more offspring than anyone else. Or conversely, let's say that I do something spiteful which decreases my fitness, but decreases the fitness of everyone else even more. Such a trait would evolve by natural selection. And so these disturbing consequences of adaptation can be summarized here.

These are the problem with relative fitness. Natural selection within a population is totally insensitive to the welfare of the population as a whole. It is sensitive only to the size of the slice, and not the size of the pie. Many behaviors that increase relative fitness can be bad for the common welfare, and it's even possible for a population to evolve itself to extinction, such as the overexploitation of resources. These problems might sound familiar to you, because they're similar to the social dilemmas that abound in human life. Just imagine how many of our own problems, large and small, are caused by personal gain at the core. This will help you to see the degree of the problem, not just in human life, but in the entire organic world.

It might seem that our scientific theory of individuals and groups has given everything to individuals. This is exactly the conclusion that countless people have drawn from evolutionary theory. However, it is not the correct conclusion; and it was Darwin himself who saw the key to bringing groups back into the picture.

Darwin was familiar with this problem of relative fitness, and he proposed the following solution for the human moral virtues. These are the traits expressed by humans - we wish they would express them more - that seem to benefit others at the expense of self. So Darwin said, "It must not be forgotten that although a high standard of morality gives but a slight or no advantage to each individual man and his children over the other men of the same tribe; yet that an increase in the number of well-endowed men, and the advancement in the standard of morality will certainly give an immense advantage to one tribe over another. A tribe including many members who, from possessing at a high degree the spirit of patriotism, fidelity, obedience, courage, and sympathy; who are always ready to aid one another, and to sacrifice themselves for the common good would be victorious over most other tribes, and this would be natural selection. At all times, and throughout the world, tribes have supplanted other tribes. And as morality is one important element in their success, the standard of morality, and the number of well-endowed men will thus everywhere tend to rise, and increase."

What Darwin is saying here is actually very simple. He's saying that it's true that altruists have a lower fitness than non- altruists within single groups. But it's equally true that groups of altruists are more fit than non-altruists. So that if we could imagine a process of natural selection operating at the group level, groups that differ phenotypically, and that variation is heritable with fitness consequences; then we can explain the evolution of altruism as a group level adaptation.

It's easy to verify Darwin's intuition with a mathematical model; here's a minimal model that has two groups, which vary in the frequency of altruists. They're represented by the black slice of that pie diagram. We start with two initial groups. One group has twenty percent altruists, the other has eighty percent altruists. In this model, the altruists contribute - they do something which causes somebody else in their group to have five additional offspring, and as a result they have one less offspring.

What we find are two things. First of all, of course the altruists are selected against within each group. The size of the black slice is declining within each group. In the left- hand group, the frequency has declined from .2 to .184. In the right group, it's declined from .8 to .787. Natural selection is operating against the altruists within each group. That's the bad news. The good news that the group with altruists is making many more offspring than the group without altruists; because every altruists, after all, is contributing a net increase of four offspring on the part of the group. When you include both levels of natural selection together, you find that the between-group advantage of altruism can outweigh the within- group disadvantage, and so the actual total global change in the frequency of altruists has gone up from point five to point five one six, despite the fact that the local frequency has gone down within each group.

We can generalize this to what is known as multi-level selection theory. What this does is envision nature as a nested hierarchy of units. An individual is envisioned as a population of genes and other sub-components; a group is a population of individuals; and a metapopulation is a population of groups. Natural selection is envisioned as operating at any level of this biological hierarchy. The necessary ingredients of phenotypic variation, heritability, and fitness consequences can exist at any one of these levels. That means that any one of these units can evolve into an organismic unit. There's nothing privileged about the individual in this scheme.

On the other hand, where adaptation actually evolves depends on the balance between levels of selection; where the natural selection actually occurs in this hierarchical arrangement. So it might be, as we usually envision individuals, that they're exquisitely adapted units that have evolved to maximize their relative fitness within groups. However, when we envision an individual as a population of genes, and we imagine evolution taking place among those genes within the individual organism, then the genes become the individual - the organisms, and the individual just becomes a population of individuals, often dysfunctional. A bag of quarreling genes, basically, without function of its own. We can think of a group as a dysfunctional collection of individuals striving to maximize their self- interest. Or, if natural selection takes place at the group level, then the group becomes the organism, and the individual becomes the organ.

Now we have a theory of natural selection which does not give everything to individuals; and yet, on the other hand, we also cannot be axiomatic. We can't say in any grandiose terms that groups can be thought of as organismic units, or ecosystems can be thought of as organismic units. It depends on conditions; and those conditions can be quite stringent.

There is a statement that I wish I could burn into the heads of everyone that thinks seriously about evolution: adaptive organization, at any level of the biological hierarchy, requires a corresponding process of natural selection at that level. That comes next. If you want to wax poetic about groups as organismic units, the burden of proof is on you to demonstrate a process of natural selection operating at that level.

Let me give you an example of altruism in nature, just to show you that we're not talking about theory. This is the lifecycle of a parasite, a trematode parasite called Dicrocoelium dendriticum. It spends the adult stage of its lifecycle in the liver and bile ducts of cows. Then the adults lay eggs, which pass out through the feces, and those eggs are ingested by snails. Within the snail, the eggs hatch into an asexual stage, which passes two generations, and then the population of parasites now exits the snail in a little mucous capsule, and this mucous capsule is then eaten by ants. The average ant eats about fifty parasites, which enter the ant by boring through the stomach wall of the ant. At this point, one of the parasites migrates to the brain of the ant, the subesophageal ganglion, and becomes what is known as the brain worm. The brain worm changes the behavior of the ant. Instead of going in and out of its nest foraging, it climbs the tips of the grass blades, and just stays up there, which of course facilitates being eaten by a cow. This is one of many fascinating examples of parasites that have evolved to manipulate the behavior of their host in order to complete their lifecycle. What makes this example interesting from our standpoint is that the brain worm, the individual that actually causes that to happen, does not survive the passage to the cow. It has sacrificed its life so that the other members of its groups will enter the next stage of the life cycle. If ever there was an example, a biological example, comparable to the soldier throwing himself on the grenade, this is it.

An evolutionary biologist who wants to know how such an altruistic behavior can evolve, the way we usually proceed is to imagine a parasite population which does not behave in this way. We imagine a world with parasites, but none of them do this thing. This is reasonable, because cows eat plenty of ants that are just going about their business. They don't have to climb to the tips of grass blades.

Into this world without brain worm types, we introduce a single mutant brain worm type in the egg stage. Now follow it along. It gets eaten by a snail, along with other eggs. For simplicity we'll say that the snail eats five eggs; the brain worm mutant, and four other eggs. They hatch, and they grow asexually to a population in which the brain worm type is now at a frequency of one over five; twenty percent. Then that leaves the snail, and gets eaten by an ant. We'll say the ants eat just one capsule; fifty parasites. Now what we have at this stage of the lifecycle - the parasites are divided into a multi-group population. Every group has an infected ant. The vast majority of infected ants are infected just with the non-brain worm type. Now we have this special group which consists of the brain worm type at a frequency of twenty percent. So that's ten brain worm types, and forty non-brain worm types. At last the behavior manifests itself. One of these brain worm types becomes the brain worm.

It's the nature of this particular trait that only one individual gets to be the altruist. And so this makes the altruism a little less extreme than it did before, because the extreme sacrifice of one individual is diluted by the non- sacrifice of the other individuals. Nevertheless, I think it's easy to see or to appreciate that this is an altruistic behavior. If you want to think about it in human terms, imagine that you're in a group of fifty individuals, of which a dangerous activity has to be performed by a single person. Ten of you decide to draw straws to see who's going to perform this dangerous act; and the other forty say we think it's great that you ten are going to draw straws. Go for it. It's clear that the forty who are not drawing straws are freeloaders, comparing to the ten who are at least taking the risk, even though one individual is the only one who's going to actually do the dangerous deed.

In natural selection terms we can see that the frequency of the altruists within that group is going down. Ten minus one equals nine. However, it's also true against that within-group disadvantage we also have the between-group advantage of altruism. That group is more likely to end up in a stomach of a cow than the other groups. And it turns out that a very modest group-level advantage is sufficient to counterbalance the within-group disadvantage of altruism; allowing this mutation to spread from a very low frequency.

So here we have a living example of what Darwin was talking about, in which a trait is selectively disadvantageous within groups, but nevertheless evolves because of the advantage it has at the group level.

Let's turn this example around, and imagine a population that consists only of brain worm types, and introduce a mutant non- brain worm. We go around the lifecycle, and now what we have is a whole bunch of groups consisting entirely of brain worms, and then one special group which consists of forty brain worm types, and ten mutant freeloaders. Mutant cheaters. There is still a selective disadvantage against the altruism within that one group, and there is still genetic variation between the groups; but there is no phenotypic variation between the groups, because it only takes one individual to become a brain worm, and all of these groups have at least one brain worm type.

So now we have a situation where within-group selection against altruism, is not opposed by between-group selection, and we expect the cheater to invade at a low frequency. What we expect here is a polymorphism in which the equilibrium is reached where the advantage of exploiting the altruists within groups is exactly counterbalanced by the disadvantage of being in groups without any altruists. Many multilevel selection models predict polymorphism. These models are quite relevant to human social interactions; including the non-human species.

Now so far I have been talking about multilevel selection, but I have not been talking about complex interactions. And in fact most evolutionary models of social behavior do everything in their power not to consider complex interaction. Often behaviors are assumed to be coded directly by genes. The properties of groups are assumed to be very simple functions of their individual members. For example, we might be interested in population regulation. The idea that if a group becomes too large, it overexploits its resources. We might build a model in which individuals differ in their personal fecundity. Some individuals practice voluntary birth control; others reproduce prolifically, and the size of the group is merely the sum of the individual fecundities. That would be a standard kind of group selection model.

These assumptions are made not because they're realistic, but to simplify the math. I believe that when we replace these simple interactions with complex interactions, then it will have profound consequences for the process of multilevel selection. The best way to see this is to describe one of the first laboratory experiments that was done on group selection by Michael Wade at the University of Chicago in the 1970s.

Artificial selection experiments have been done for centuries to mold properties of individual organisms. That's how we've derived our domesticated plants and animals. What Michael Wade realized is that you can use artificial selection procedures in the laboratory to select groups in addition to individuals; at least potentially. His experiment was simplicity itself. He worked with flour beetles, so he had a common stock population of flour beetles. He'd simply make up forty-eight glass vials with flour, and into each vial he'd put in sixteen beetles at random from this common stock population. He then waited thirty-seven days, and he measured something about these groups. So he measured the size of the group. Then in one experiment, he selected for large group size. He took the groups with the largest individuals, and he used those as the parents to initiate a new set of groups; and then the experiment proceeded. In another treatment he took the smallest groups, and used those as the parents to found a new set of groups.

Imagine that we're doing this experiment, and ask yourself the question: how much variation in group size would you expect among these groups? If group size is just an additive sum of the individual fecundities, the answer is you would not expect much variation. Variation among groups would be dictated purely by sampling error. If you initiate the groups with sixteen individuals, you shouldn't expect that much variation among groups. What Wade observed was lots and lots of variation in group size. The reason is that group size is not caused just by individual fecundities. In fact, individual fecundity didn't vary much at all. Instead, there was a long list of individual level traits which interacted with each other in a complex fashion to produce what you measured after thirty-seven days, such as group size.

One example of this interaction is that flour beetles are cannibalistic. Larvae cannibalize eggs. So if you imagine all the eggs are laid at one time, synchronously, when all the larvae hatch, they have nothing to cannibalize. They'll all grow up as a cohort, and they'll turn into a large group. Now imagine that for any reason the eggs are laid asynchronously. Some larvae hatch before the other eggs. They turn around, and eat the other eggs; and that develops as a small group.

So that's one example of a complex interaction in which a small difference in synchrony of development can translate into a large difference of the trait that you're actually measuring: group size. This can be generalized to a statement about complex interactions. Basically, complex interactions are influencing one of the fundamental ingredients of natural selection: variation among units. With simple interactions, variation among groups is based on sampling error, and therefore declines with initial group size. If groups are initiated by small numbers of individuals, then you expect a lot of variation. Perhaps group selection can work in that situation. But as soon as groups are initiated by many individuals, then if the interactions are simple, you would expect very little variation among groups. Therefore, group selection would become an unimportant force. The theory of kin selection is founded on this principle.

With complex interactions, even small initial differences between groups can be magnified by the interactions into large phenotypic differences. This of course we know as the butterfly effect, or sensitive dependence on initial conditions. So here's one profound implication of complexity on the process of natural selection. It has to do with phenotypic variation among units. But we need more ingredients.

We need to know is this phenotypic variation heritable? If you take the groups that are largest, and you break those into a next generation of groups, are they also going to grow into large groups? If so, then by definition, the trait that you're measuring is heritable. There's a correlation between the offspring units, and the parent units. The only way to do this is of course to do the experiment. Wade's experiment, after nine generations, produced a ninefold difference in the trait of group size. The low line produced on the average of twenty individuals after thirty-seven days, and the high line produced a hundred and seventy-eight individuals.

To summarize, what we've shown here is that artificial selection procedures can be used to mold the phenotypic properties of groups in just the same way that we're accustomed to molding the phenotypic properties of individuals; and that complex interactions played an important part in that process.

If we can mold the properties of single species groups, how about ecosystems? This is an experiment that we're performing in our laboratory. My collaborator here is William Swenson, who is at this conference also. What these are are high tech flowerpots. They're like petri dishes, so they have covers that prevent microbes from coming in. And so imagine that you filled these flowerpots with soil, and then you sterilized them. So you have sterilized soil. And into each pot you put six grams of unsterilized soil from a natural source. Then you plant arabidopsis seeds. Arabidopsis is a very fast-growing plant that's used a lot in laboratory research. You let these plants grow for thirty-five days, and then you measure their biomass. Now if this were an individual-level selection experiment, we'd collect the seeds from the largest and the smallest plants, and we'd use that to grow a new generation of plants. But this is an ecosystem selection experiment. We throw away the plants, and we pick the soil communities that are under the largest and smallest plants; and we use that as an inoculum to colonize a new generation of ecosystems.

I think you can see that the spirit of this experiment is exactly the same as a standard individual-level selection experiment, or the group selection experiments. Except now we're selecting ecosystems rather than single species social groups. These might seem like small ecosystems, but in biological terms they're just the opposite; they're enormous ecosystems. Six grams of soil contains literally millions and millions of creatures. Microbes, fungus, nematodes, algae. We don't know what's in there; that's how complex it is. It would take many lifetimes to figure out what actually is in there reductionistically.

We're doing three of these experiments. The first is the one that I just described. The second is just like it, except that the inoculum size is .06 grams of soil, rather than 6 grams of soil. So that's a hundredfold difference in the size of the ecosystem that we're passing from generation to generation. So a sampling error is important in creating variation among units. Then we should certainly expect more variation in the second experiment than in the first.

Finally, in a third ecosystem experiment, we're selecting aquatic ecosystems for high and low pH. In this case, we simply take some pond water, we split them up into test tubes. After a period of time now - which in this case is three days, this is a much faster experiment - we measure pH, and then we use the high and low pH units to colonize a new set of test tubes. So we select for that trait.

What are the results of these experiments? First of all, we get lots and lots of phenotypic variation among units. It does not matter that we initiate these units with jillions of creatures, and that we make the physical environment as similar as we can. Sensitive dependence still takes over, and causes these little ecosystems to take off on different trajectories, which is reflected, among other things, in the phenotypic trait that we're measuring. So we get phenotypic variation. The next question is: is it heritable? If you take an ecosystem which grew a large plant, and then break it up, does its offspring also grow large plants? The answer is very interesting.

What we see here in each case - this is the 6 grams soil experiment, the .06 gram, and the pH experiment. There's an interesting pattern of initial divergence, and then a collapse, and then in two cases a more permanent divergence. I think that this can be explained quite nicely in terms of complex interactions.

Imagine that we begin this experiment, and all these little ecosystems take off on their separate trajectories, which has consequences for the trait that we're measuring. Very few of these units will have actually come to any kind of a local equilibrium. When we form new units from the old units, there's no guarantee that they're going to have the properties of the parent, because the offspring units are going to take off on trajectories of their own. At some point, however, while selection is searching this very complex parameter space, legions will be found which are locally stable - which have the properties that we're selecting, and in addition are locally stable, and maintain their structure as they get passed from generation to generation. The final divergence here has lasted for about thirteen generations in the pH experiment, and for about eight generations and counting in the six gram soil experiment.

Let me just show you what this looks like. This is a soil ecosystem which has been selected for high plant biomass, and this is one that's been selected for low plant biomass. These are two ecosystems, now very different in their properties, that are descended from a single ancestral lump of soil, basically. An ancestral ecosystem had the potential to have these very different properties, which you can bring out merely by the process of selection.

The pH experiment differs by a factor of one and a half to two pH units. That's measured on a logarithmic scale. This means that the ancestral aquatic ecosystems differ by a factor of as much as a hundred in hydrogen ion concentration. All derived from a single ancestral cup of pond water.

These experiments have practical implications, in addition to their scientific interest. Charles Goodnight will give a talk on Tuesday night, and review the group selection experiments, including one on chickens, which has increased the productivity of egg production by a hundred and sixty percent in six generations. This one selected strain of chickens will save the poultry industry millions of dollars if it becomes widely used. I think you can see how we can use this procedure to evolve designer ecosystems that are designed to solve practical problems such as soil remediation, enhancing crop growth, inhibiting weed growth, and so on.

The title of my talk was "Altruism in Nature and Human Nature." Now I need to talk about multilevel selection in our own species. Let's return to this wonderful quote from the Hutterites, which compares their communities to single organisms, and social insect colonies. This kind of comparison has been made throughout the ages, and for good reason. Human social groups, large and small, seem to be organized adaptively, often with much differentiation of labor, in a way that invites comparison to the other adaptive units of nature. Nevertheless, most evolutionary biologists have rejected this comparison, because the genetic structure of human social groups is very different than the other adaptive units. Individuals, the cells of individuals are genetically identical. Honeybee workers are very related to each other, and reproduce through a single Queen. In contrast, the members of human social groups are often weakly, or even completely unrelated to each other, and they're reproductive competitors. How can such a different kind of a society become organismic in the same way that the first two have?

This argument, at the bottom of it, is based on the same assumption of simple interactions, as I described earlier. It's based on kin selection theory. Kin selection theory assumes, just to simplify the math, that behaviors are coded directly by genes. What this does is it makes a very tight linkage between behavioral variation, and genetic variation. The only way to get a group that's behaviorally uniform is to have that group genetically uniform. This is manifestly false for human groups.

Here are three very typical quotes from the anthropological literature of various cultures around the world. "The basic axiom of everyone is that all its existence and welfare are completely in the hands of the gods, in particular the clan gods, who shower their beneficence not on one, but on all. Any attempt to monopolize the gifts of the gods must inevitably incur just punishment by the common benefactors of the clan." For the Apaches it was said, "Within the camp, the norms requiring the sharing of food were so pronounced that the entire community could be considered a single production-distribution- consumption unit." And for the Navaho, "The actions of the members of the society were oriented to a normative order to accepted values and beliefs, and to the correctness of certain sanctions for inducing conformity."

These groups are behaviorally uniform not because they're genetically uniform, but because they share a similar system of values, and enforce those values. They have a moral system which creates a kind of uniformity which has very little to do with genetic variation. When we generalize from this we find that if we evaluate human social groups in terms of the fundamental ingredients of natural selection: phenotypic variation; heritability, in terms of a correlation between offspring and parent units; and fitness consequences, we find that the comparison between human social groups, social insect colonies, and individual organisms is not so farfetched after all.

Let me conclude by just showing you how it is possible for a human social group to be like an organism. This returns us to the Hutterites. The Hutterites practice extreme altruism. They give all their belongings to the church. They try to stamp out a sense of self-will by wonderful sayings such as "a grape is nothing until it is pressed with other grapes to make wine", and "a grain of wheat is nothing until it is ground with other grains to make bread". In Europe, before they migrated to North America, Hutterites willingly chose torture and death rather than renouncing their faith. In Northern Canada, where they reside now, their altruism works very well for them. They thrive in a harsh climate that most of us wouldn't want to live in, and their birth rate is higher than almost any other culture known. Of course, this is what we should expect, because nothing works better than a group of altruists.

On the other hand, an evolutionary biologist just has to ask that question; the cheater question. What would happen if we took a Hutterite, and mutated him. Turned him into a cheater, put him in there, just the way we did with the brain worm? The answer is, that cheater would not fare very well, because the Hutterites are protected. Their moral system is virtually cheater-proof. That's why they can afford to be altruistic, because they are protected against cheaters.

The point of this is to say that altruism is not unconditional. If someone, a member of the group, behaves in a non-altruistic fashion, they are quickly excluded from the group. This is how you can maintain a group of pure altruists.

The prescriptions and the rules for cheating behavior are actually quite explicit. Despite the fact that they're described in religious terms from 1650. They have a wonderful game theoretic feel to them. In case of minor transgression, this discipline consists of brotherly admonition. If a brother or sister obstinately resists brotherly correction and helpful advice, then even these relatively small things have to be brought openly before the church. If he persists in his stubbornness, and refuses to listen even to the church, then there is only one answer to this situation; and that is to cut him off and exclude him. To rejoin the community there is repentance, he is received back with great joy in a meeting of the whole church. They unanimously intercede for him, and his sins need never be thought of again, but are forgiven, and removed forever. This is a selection of behavior, it's not a selection of individuals, which manages to get a group of individuals behaviorally pure despite the fact that they are genetically, of course, diverse.

My favorite example of a Hutterite anti-cheater adaptation occurs when their colonies fission; just as bee colonies fission after they get too large, Hutterite colonies fission. What they do is they find a new site, they put up buildings, and now it's time for some to go, and others to stay. It's always better to stay than it is to leave. What is to prevent some sneaky Hutterite from sort of rigging things so that he's the one that stays, and some other sucker goes? The Hutterite solution to this problem is to divide the community into two groups, and to drop a list for each one that's balanced, with respect to skills, and age, and so on. On the night before the split, everyone packs their bags, and one of these lists is chosen at random. They go, and then everyone else unpacks their bags, and stays. I hope you can see the eerie resemblance between this social convention, and the genetic rules of meiosis, which ensure the fair passage of gametes between generations. The comparison between a human social group, and an individual organism can indeed be much more precise than we've thought.

I will finish with this slide. This slide shows two kinds of complexity. The left-hand slide I would say is complexity without a purpose. It can be studied without any reference to natural selection at all. The right-hand slide is complexity with a purpose. It's a diagram of a nerve cell which is designed to maintain itself, and transmit electrical signals in just the right way to serve the purposes of the organism of which it's a part. This kind of complexity is often explained in terms of natural selection, often without reference to complexity.

Numerous attempts have been made to explain adaptive complexity purely on the basis of complex interactions, without reference to natural selection. It's my personal opinion that these efforts are doomed to failure. Natural selection is required to evolve adaptive complexity. On the other hand, natural selection by itself without complex interactions might also be doomed to failure.

One point of my talk has been to show how natural selection and complex interactions together can be studied, rather than apart. The other point of my talk, of course, is to show that adaptive complexity can extend above the level of the individual to higher levels of organization. Thank you very much.

Description and Modeling

Steve Lansing
University of Arizona
John Sterman
MIT Sloan School of Management
David Harel
Weizmann Institute of Science
Richard Palmer
Duke University
Gerald Sussman
MIT

Mr. Steve Lansing: I meant to say a word or two of introduction. What I thought I'd do is just share with you an historical footnote that occurred to me. Our subject is description and modeling. Obviously, exploring what those tools do is one of the main reasons that we're here, and we're looking for patterns - the right level of abstraction to characterize those behaviors that we see in our experiments in different fields.

It occurred to me that this parallels some of the discussions that happened about a century ago with statistics. Not only in the sense that it brought natural and social scientists together to explore new tools, but also because it set in motion some very interesting philosophical speculations, not only by philosophers like Nietzsche, and Charles Andrews Pierce who wrote about the effects of statistics. Nietzsche didn't like the idea that there could be laws in history. But also James Clerk Maxwell also wrote about what the effects of this basically central paradox are. This is my point. A central paradox that emerges through the use of statistical tools, and also the tools that we're using; which is the question of indeterminacy at the level of individual behavior generating deterministic patterns at the higher level, at the emergent level. That proved to be a very fruitful area of speculation.

I'm going to quote you just a little line from James Clerk Maxwell's book on molecules in which he quotes Alfred Lord Tennyson talking about, "Flaming atom streams fly on to clash together again, and make another, and another frame of things forever." He quotes this to say, here is this - the clash of atoms generates and endless stream of frames; in effect, emergent properties. What Maxwell asks is whether the indeterminacy at the level of those clashes is merely measurement error, or intrinsic to the interactions. Of course that proved to be a very fruitful question to ask, and in some ways was antecedent to the development of quantum mechanics.

Perhaps in the same way, I'm suggesting to you a way of thinking about what we're up to this afternoon as we compare our different experimental areas where we're looking at emergent properties. It seems to me that in exploring that tool in these different areas we're looking at the possibility of emergent properties that are at a level of abstraction that will appear only as we explore them individually. In that way then we may in a sense be posing some of the same questions that were posed with the invention of statistics.

Our first speaker this afternoon is John Sterman, who directs the systems dynamics lab at MIT, which is the premier systems dynamics program on this planet; and we look forward to his remarks.

Mr. John Sterman: Thank you. When Yaneer asked me to speak, I said I'd be delighted; but then I saw the title of this session: modeling and description. I had no idea what it meant, so I decided I would describe a model that I built for you today as a way of illustrating what I think system dynamics is, and what it can be used for. I'll only be, of course, describing one tiny slice of what system dynamics is, because it's used today in a vast array of disciplines and domains, from physiological models to helping companies re-engineer their supply chains. Since I sit in a management school, I'm going to talk about a business application.

The issue that I've been interested in for some time in this aspect of my research is why it is that so many firms have had such great difficulty successfully benefiting from what you could broadly call improvement programs. Things like TQM, or BPR. If you're not into the jargon, that's Total Quality Management, and Business Process Reengineering. There are about fifty other TLAs, that's a Three-Letter Acronym, which describe the names of particular programs that you as a company can hire consultants to train you in; purporting that it will improve your customer satisfaction, raise the quality of your products, shorten your cycle times, and of course improve your bottom line. What's interesting about this is that the vast majority of companies that have attempted to benefit from these programs have utterly failed. There certainly are some fraudulent, or untested and bad programs out there. But there's so many clear, unequivocal successes where companies have made major improvements in all aspects of their operations using these tools that you can't say that they just don't work.

Why is it that most of these companies have failed to do this? We call this the improvement paradox. What's even more interesting: among the companies that have used them successfully, at least initially, there are often unanticipated side effects that can hurt the company in some cases, even after they've been successful in improving quality.

I'll just give you a couple of quick examples which motivated our research. We documented particular companies. In one case they improved the quality of their operations so much that they effectively doubled the production capacity with the same labor and equipment, and yet the company didn't benefit. The stock price actually fell. In another case, improving manufacturing in a very large company, in fact one of the big three auto companies, made it harder for them to improve product development. Where you would normally think that any organizational learning would help those who come later to do better based on the experiences of those who came earlier. In another company, having employees who are active users of the product, and incredibly committed to the use of this product, it actually made it harder to improve. Whereas all the literature and organization theory says if your employees are also your customers, you're going to have more commitment, more enthusiasm, and it should be easier to motivate them to generate improvement.

This is a very pervasive problem, and it's one with significant dollars at stake, because companies have spent, over the past decade or so since the Japanese quality revolution, many, many, many hundreds of millions of dollars to try to improve. If they could, there would be many, many hundreds of billions of dollars of benefit.

Let me tell you the story of an actual, real life company that we investigated. There is a paper available on this. The company is Analog Devices. Analog Devices is down here in Route 128. It's an MIT-founded company. They are extremely successful. As the name might indicate, they make A to D, and D to A converters, digital signal processor chips, and other specialty integrated circuits which you would find in your cell phone, your CD player, and so on. They're very, very successful, founded in the mid-60s. Through the mid-1980s they had experienced compound annual growth rate in sales of around twenty-seven percent per year. That's extremely good, and on average they were highly profitable. At the time this work began, they were, and still are, a New York Stock Exchange listed company with sales at that time of about five hundred million dollars. They were the big player in their niche of the semiconductor industry.

Mid-1980s, their founder, and then CEO Ray Stata, decided that although they were very, very successful, they weren't doing well enough. He believed that the problem was management, not technology. They were a clear technology leader in their segment. The issue was that he felt that they had let technology carry them along, and they needed to improve the quality of their management. In particular, the quality of their learning in a very dynamic industry. Some of the evidence for that is shown right here. On time delivery was running at about seventy percent, which is very poor in the semiconductor industry. They had five hundred parts per million defective parts that they were shipping to their customers. The average yield of their wafer fabs was a pretty poor twenty-six percent on average, and the cycle time in the factory was fifteen weeks. This was a pretty long time.

So looking around he said it's not the technology that's the problem. We have to figure out how to learn better. They implemented a total quality management program. They did everything right. They sent all their top people to Japan. They brought in the best folks available to train them. They implemented incentives that were aligned with their quality improvement goals; and, in fact, it worked great. In three years, this was where they stood. They had increased on time delivery to ninety-six percent. They had cut defects by a factor of ten. They had doubled the yield of their wafer fab, and they had cut the cycle time almost in half. These are spectacular results. I don't know if you have been involved in any quality or reengineering projects, but any winner of the Baldridge Award, or any winner of the Deming prize in Japan would be delighted to have this much improvement across such a wide range of metrics in just three years.

So here's the question: what should happen to their financial results? Well, by merely posing the question, you immediately know they didn't do what intuition tells you they should have done. If you ask this question of folks after reading a case on Analog, everybody will tell you well, if you can make such dramatic improvements in just three years, and do it better and faster than any of your principal competitors, you must benefit. You're going to see higher market share. You should see higher return on investment, return on equity, more profit, and a higher stock price. After the stock market crash in 1987, and their shares were at 18.75 with healthy earnings per share, and a pretty good return on equity. By 1990, after all of these successful improvements, the share price was 6.25, and they were losing money.

This just shouldn't happen. So why did it happen? This is not a rhetorical question. What are your thoughts?

M: What were their competitors doing?

Mr. Sterman: Indeed, if their competitors did this even better, and faster, and sooner, then they would have fallen behind. That's not true. They did it first, they did it better. The competitors, of course, were trying to do TQM, and other things as well, but they did it later. Analog was first and best. Other thoughts?

M: The economy went the other way?

Mr. Sterman: Yes, a recession. And indeed there was a recession in the semiconductor industry in 1990 as well as the economy as a whole, and it did hurt them. What's interesting is they did worse during that period than their competitors did who hadn't done this. So yes, the recession knocks everybody down, but it shouldn't knock you down more, it should knock you down less. That doesn't work either.

M: What were the costs of this?

Mr. Sterman: They were pretty trivial. That's another good theory is it just costs so much to implement TQM that you lose. But that turns out not to be right either.

M: What was market share?

M: Technology changed.

Mr. Sterman: Okay. I'm hearing lots of ideas now, and let me move on. I think now you're starting to get into questions that are more getting at the -

M: Isn't it true that quality has little to do with profitability?

Mr. Sterman: Well, that's the cynical view, right? But you can point to a lot of companies -

M: Microsoft, for example.

Mr. Sterman: Well, okay. Yes. A lot of people take away from experiences like this the idea that TQM just doesn't work. But that's not right either; it clearly works for a lot of companies. But here's an example where it didn't. And it worked, right? I mean, if TQM didn't work, then you wouldn't have seen this. The quality clearly improved extremely dramatically. The disconnect comes that it doesn't translate to the bottom line. It's actually hurting it. So what's going on?

M: Were the products still in demand? Did they keep up with the technology or spend all the time with management?

Mr. Sterman: You must be an engineer. Let me try to tell you the story that we've evolved to address this question before I run out of time. Along the way you should shout out your questions. We went in, and developed a formal model of this company and its experience here. One of the things you notice is you have to then model how fast could a particular process improve if you were to do TQM, or reengineering, or whatever TLA program you're using? If you were to do it right, what's the maximum rate of improvement you might experience? It's not a constant of nature, but it's also not arbitrary. You can actually model this. If you just ask this following question: how fast can a process improve? It's going to depend on how much you can learn every time you iterate around the so-called Deming cycle or PDCA cycle. This is an FLA, I guess. This is jargon for the Plan, Do, Check At cycle. That's particular jargon for TQM, but the point is that learning proceeds by experimentation. Begin with a hypothesis, design an experiment, implement it, see what happened, and then try and evaluate the results. If it made an improvement, you try to implement and routinize that innovation in your operations. If it didn't work, you go back, and iterate again until you come up with ideas that will actually work. Improving the yield of your process, for example.

How fast can you improve the process? It depends on how much you learn every iteration, every experimental cycle, and also, what is the cycle time for experimentation? If you can do a lot of experimental cycles in a month, you're going to learn faster than if you can only do a few. This can be characterized by the improvement half-life, which is what you'd obviously expect it to be. How long does it take to cut the defect rate of any metric, any process, in half?

In the simplest possible model, which has been proposed in literature, the rate of decline of defects for the process you're interested in - it could be on time or late delivery, it could be yield of your wafer fab, factory cycle time, whatever it is - is just linear in the gap between the defect rate, and some theoretical minimum. This would be an empirical parameter. You get pure exponential decay; not terrifically interesting.

One thing that's interesting is to characterize how the half- life depends on the process. A colleague of ours named Art Schneiderman - who was at that time the Vice President for Quality at Analog - he had done an extensive amount of empirical work on this. It's simplified by this graph which we call Schneiderman's Space. This is the projection onto these two dimensions of Schneiderman's Space. Basically it says that the half-life is longer if the complexity of the process is greater. It's a very simple idea that's not appreciated by management. Technical complexity is one dimension of the complexity of the process. More interesting is organizational complexity.

Technical complexity is the straightforward notion that a simple lathe, like you might have in your home workshop, is going to be easier to understand, and therefore to improve, than a hundred million dollar, numerically controlled stamping machine that Ford has to stamp out body panels for the Ford Taurus.

Organizational complexity is much more interesting. What that says is that the improvement half-life will be longer if there's more organizational boundaries that have to be crossed in order to assemble the team of people that need to be working together to do the improvement activity. It's going to be a longer half-life if the activity is cross-functional. You have to have people from different backgrounds - most obviously marketing and engineering - working together. If you have to go outside the boundary of the firm, and bring in your suppliers and your tooling vendors, and your customers, and your bankers. If they need to come together to redesign your product development process, for example, complexity is much greater, and the half- life will be much greater. Up in this quadrant where you have processes like product development partnering with your customers, and integrating the supply chain, and other general and administrative functions, you're going to have much slower intrinsic rates of improvement than what's happening at the factory floor level with particular machines, even if you can do your TQM process correctly.

Here's an example. This is real data from Analog showing the process defects generated in the wafer fab. You get about a year and a half half-life for defect reduction. If you were to show the same thing for product development across a wide range of firms, you'd find that the half-lives are more on the order of two to three, or even more years.

I'm going to skip over some things that Analog did to align the incentives of its managers with this new focus. But one of the things they did was they began to evaluate the division presidents on the basis of how much they were improving, and not just what their short-term financial results were. So every quarter when they'd have the division Presidents retreat, and they'd take all these people off site, and show them, and normally they would just go over the financials, and people would get rewarded if they were making money in their division. Now Art Schneiderman would put up graphs like this. These are the names of the divisions, and this is the company average. Art would calculate your improvement half-life based on your data.

If you're this division President here, you're sitting there licking your lips, and imagining the new corner office you're going to have at company headquarters after you get promoted. If you're this division President here, you're wondering who you can send your resume to. This was highly motivating. There's a very important incentive story here that I'm going to gloss. They would do this on a regular basis. When you look more deeply at why didn't this work, it's not because TQM didn't work. TQM worked great. Then something else happened as a consequence of their success that caused the difficulty.

We developed a system dynamic simulation model of this. This is an overview of the model. I won't go into any detail on this, except to point out that the model has an extremely broad boundary. Which is to say there's almost no exogenous inputs to this model. Almost everything that you care about here, including the stock price on Wall Street, the behavior of the customers, the behavior of the competitors, the behavior of the labor force; they are endogenous. In fact, there's only three exogenous variables in this model. One is economic growth, one is the dividend yield on the S and P 500, which is used to value Analog's expected earnings, and give you the stock price. The third is a metric that represents the general rate of diffusion of quality concepts throughout American industry. And that was happening largely because of the population of Deming, and Duran, and Japanese management techniques, and so on. Those are the only exogenous variables. Everything else is one hundred percent endogenously generated within the model, and it's a very tightly coupled system.

This is not like a lot of economic models that you may have seen where everybody has perfect rationality. We view that as a clearly counterfactual axiom in economics. Instead we do field work to try to discover what are the decision rules that the individual agents running individual functions, such as the TQM effort, or new product development, or the manufacturing operation. What are those people doing? How do they make their decisions? We try to represent their cognitive limits, and other aspects of their bounded rationality. This is not like a traditional economic model where everybody is perfectly rational, and solving out equations that you can't even solve yourself.

The model is relatively large. Probably on the order of a three or four hundredth order system. 'm obviously not going to describe the equations in detail. We did do extensive field work to develop this model. Interviews all up and down the organization. They gave us an enormous amount of archival data. We also looked at all the public records, as well as at many proprietary databases that were available. In that way we grounded all of the assumptions, or as many of them as we could, in first hand field investigation; which I think is critical in this kind of model building effort. We also looked at how well the model reproduced history.

Here's the summary of the statistical fit. The ratio of market value to cash flow. The model tracks it reasonably well. The model is not perfect, but it captures the trends here reasonably well. Here's the stock price. As the stock price goes from that peak in '87 down to this low point at six and a half here, the model tracks that pretty well, and remember quite endogenously. As that's happening, the market value of the company is falling. The free cash flow continued to be pretty good. So what that meant was that the company became very vulnerable to takeovers.

In fact, right around this point here, '91, the ratio of market value to cash flow was basically six years. That means that you and I could have put our Visa cards together, and bought a fifty-one percent controlling interest in this company. That's a pretty dangerous situation for a company to be in, and it plays a role in the story still to come.

If you look at some of the operational metrics, here you see the manufacturing cycle time, then the yield in the wafer fab, defect rate, and on time delivery. The model's picking it up pretty well. Not perfectly, but quite well. I don't want to go back on that, but rather tell you why what happens happens.

I described this half-life process for improvement. But in that basic naive half-life model, improvement happens autonomously. You start a TQM program, and then you expect things to just improve with some characteristic half-life. That's not how it works. TQM is not like a machine tool. You cannot go out and buy a perfectly functioning turnkey six sigma improvement program. You have to grow it from the grassroots up. That means you have to persuade employees to commit their time and energy to learning the tools of improvement, and then applying them to any real situation. Since that takes time away from their real work, that's a risky thing for employees to do. Since many employees are very cynical about the value of these TLAs, it's doubly difficult to enlist their effort. We would joke that all the employees, when it comes to this, are from Missouri. That is to say, you've got to show them that it works before they're going to commit their effort. Here is one of the basic positive feedback loops that's critical in getting any improvement program off the ground.

There is some initial management improvement to commitment. The CEO comes back from Japan, and says we are going to do this, because it's working for everybody else over there, and they're kicking our butts. So that sets up a new push. All the employees are being pushed to participate. Well, that's not going to do it. You're going to get a few people to do a few things, but not much is going to happen. There's going to be eventually some improvement in quality and productivity if they can do it right. It's the observation that this appears to work, that's going to enlist the vast majority. That closes a positive loop kind of diffusion process where more people committed generate more improvement efforts, which lead to more improvement, which lead to still more commitment. That positive loop is like any other diffusion or epidemiological contagion process, but at the organizational level. But it's not that simple.

You have to have a few other enabling conditions in order for commitment to rise. You have to have support. You can't say everybody has to do quality, and then not provide any training, or any support resources. A lot of companies fail because the support resources that they provide become inadequate when there's a big burst in activity. That creates a balancing or a negative feedback loop which can choke off commitment because people say I couldn't get any training, so I'm clueless, and don't know what to do. Or I called up the guru who's supposed to help me, and they never returned my phone call. That can choke off an effort. That didn't happen in this case. They had plenty of resources, at least initially.

The other input here is perceived job security. You have to have some confidence that if you participate in this improvement effort, you will not be downsized as a result

I'm not going to go into the model formulation details, but here's one little chunk of the model just to give you a feel for what the equation looks like. The rate of defect reduction is, as before, proportional to the gap between the current defect level, and the theoretical minimum. But it also depends on the commitment of the work force to improvement. In addition, there's an entropic term here that says improvements erode over time, because people turn over in the organization, processes change, products change, and so on.

Commitment itself is another state variable, which is evolving in a diffusion model framework. C^* here represents management's goal for commitment. Everybody is going to participate in quality. We're all going to go to the training. That's what C^* is. That produces this management push term: theta times the gap between the management-mandated commitment level, and the actual commitment level. The interesting part is the second term, which is the word of mouth term. This is the learning from observing everybody else's results. Now you can see the positive feedback loop here, because commitment here drives its own adoption, its own growth. Of course it's limited eventually. Commitment can't be greater than a hundred percent of everybody working on this all the time. W represents the strength of this effect, and its direction. W can be negative. If people had lousy support, were worried about losing their jobs, and didn't observe results, then W, which is the sum of these nonlinear effects of results, support, and job security could be a negative net evaluation, and people would say TQM, that's the worst thing ever. Stay away from it. Don't get involved. And commitment would die away.

On the other hand, if results are very positive, there's adequate support, and people aren't afraid of losing their jobs, word of mouth will be positive, and this will grow in a positive feedback fashion. That's a simple way to get at one of the key formulations in the model. In the Analog case, what does that formulation do?

Here's the simulation of the model. From '87, when the TQM program starts, through the end of '89. The black curve is the level of commitment to quality in the manufacturing organization, and the dotted line is the level of commitment in the product development organization. For the first year, everybody's on board the TQM bandwagon. It looks like it's doing great. Manufacturing, it really takes off, and there were dozens, and dozens of quality improvement teams operating throughout the different manufacturing facilities; but in product development, it faded away. That turns out to be important, and we have to ask why that occurred. One consequence of this is that you're going to make dramatic improvements in manufacturing. This is simulation, but in fact the data would look like this if you could plot them numerically. We got this from interviews. But yes, it did in fact do that, and I'll show you some quotes in a little bit. So imagine that you could in fact now dramatically improve manufacturing. Let me give you a little thought experiment. I snap my fingers, and when everybody comes back from lunch we've magically doubled the productivity of every process in the firm. Has profitability changed? Not one cent. Because we still have the same payroll, we still have the same capital plant and equipment. Nothing has changed except everybody can finish their work in half the time.

In order to get the quality improvement that's generated by one of these programs to the bottom line, you have to do one of two things. You have to either sell more, or reduce your expenses. Well, so if quality and productivity rise as a result of all of this working nicely, one thing you can do, given the demand, is downsize. That then allows you to meet the same demand and get the same revenue with lower expenses. That gets your productivity improvement to the bottom line. And that loop is very popular. That's the Chainsaw Al strategy.

What then makes that a bad idea? Well, there's a few other side effects. In particular, if you lay people off, it's highly disruptive. You take established work teams, and work groups, and quality improvement teams, and you break them up. You have to reorganize. Management's attention is diverted away from improvement, and onto reorganization and severance. Even if you don't downsize, the mere existence of excess capacity - which everybody on the factory floor understands is there, because they don't have to work as hard to get their work done anymore - will cause the fear of layoffs, which will feed back to reduced commitment. Nobody wants to work themselves out of a job. So if that belief spreads, it will choke off commitment, quality improvement stops, and you don't get any of the benefits. So there's two negative feedback loops that can thwart your program.

You can solve this problem if you can grow demand as fast as capacity is growing as a result of your improvement program. Let me show you a little field work that we did that illustrates this effect here. Here's a boss speaking to a subordinate, and he says, "Well, Pendleton, as of noon today your services will no longer be required. Meanwhile, keep up the good work." That's a cartoon from the New Yorker, but our interviews were very much along those lines. People are not stupid about this. They understand that downsizing is often a code word for layoffs to come. If we're going to succeed in avoiding those negative feedbacks that can choke off improvement, we have to grow demand.

Now let's look at how demand is modeled. Demand you can think of as the product of the market share you have in each of the markets in which you compete, times the total demand of those segments. Total demand depends on the economy, and some other things we'll see in a minute, and market share should be benefiting from your quality and productivity growth. Right? I'm now delivering on time, much better. I have lower defects that I'm shipping out to my customers. I should benefit by having a larger market share, and this reinforcing loop should then allow me to grow demand at least as fast as capacity, and avoid this problem. Why might that not happen? Well, it's going to take some time because my customers are going to have to design in my chips to their products, and design out my competitors products. That will take some time. There is a delay there.

Analog was already the market share leader in many of the market segments that they played in. What would happen is, for example: Hewlett Packard, one of their big customers, would say, it's just great that your quality's improved so much. We're really grateful. We think it's terrific, and it's about time. But you know, you're already our biggest supplier in these chips, and we can't afford to let you become our sole source, so we aren't going to give you very much more business. That would limit it. By the way, that has clear strategic implications. That means a big company's going to find it harder to avoid excess capacity than a small one that has small market share.

How about what the competitors are doing? Indeed in our model, as in the real world, the competitors are not exogenous, and they don't stand still. They also are trying to improve their quality and productivity, and they were stimulated in their efforts by Analog's success. They all watch each other very closely. The different agents in the model, if you like to think about it that way, they're watching each other. And when the competitors observe that Analog has invested in TQM, and it's working, then they say we better do this too, and we better do it faster than we otherwise would have. They then catch up. That creates a negative feedback that causes the customers to benefit a great deal, because now Hewlett Packard, and all the other customers are getting much better products from all their potential suppliers. The relative competitive advantage of the different firms doesn't change very much.

There's not a lot of leverage to grow demand through the market share channel. Very little of the TQM tools - none of them are privately appropriable. You could hire a million consultants who would train you how to do this. Everybody could imitate what you were doing. Now they could do it well or poorly, but there's nothing you could do to make this privately appropriable.

M: The big winners are the TQM consultants.

Mr. Sterman: Right. That's in fact correct. But, back to the story of our firm.

The goal is to grow industry demand. So of course you'd like to have lower interest rates, and loose monetary policy; but that's low leverage. What else can you do to grow industry demand? You've got to get new products out the door to market. The engine of growth for any high tech firm like this is new product generation. Let's look at that. If I can improve my product development process, then I can get more new products out the door with the same number of development engineers, and the same R and D budget. Those new products, given our strong technology lead, should open up entirely new markets. For example, I make my chips smaller, with lower power consumption, less heat generation. Now they can go in a whole new generation of handheld consumer electronic devices such as cell phones, opening up a huge new market.

M: Or you can make an alliance with some software manufacturers so that your stuff becomes obsolete in a couple of years.

Mr. Sterman: That's happening very fast anyway. But yes, you can partner with either downstream hardware manufacturers like Nokia, or Ericsson, or with software people so that there's software for your DSPs, your digital signal processor chips, that make them more attractive than other people's DSPs that don't have good software, and so on. That's part of what I mean by product development. Product development here encompasses not just the engineering task, but also the customer needs assessment process, and the partnering with technology suppliers. It includes the technology supply chain, not just the engineering of new products. Of course, Analog's TQM program involved improvement across the board, not just in the factories. They set up metrics and quality improvement teams for every organization, and every function, including product development, including sales and marketing, including their supplier and distribution arrangements. But look at this positive feedback loop. It says we get faster product development, more products out the door, more industry demand. Not excess capacity, but actually insufficient capacity. That means that we're doing better. Everybody sees how wonderful the quality improvement effort is. We get more commitment, more results. Everything spirals up. If that loop dominates in that fashion, you're going to be in great shape. There's only one problem. The delays in this loop are very much longer than the delays in the manufacturing improvement loop. Why? It is organizationally and technically complex. Improving product development is just very much slower, even if you do it right, than improving manufacturing. Then of course there's delays in getting the new products actually developed. The product development cycle is much longer than the manufacturing time.

Skip a little details here. We actually got from them a complete database of all their product histories. Here you see the age of the product versus the growth rate of sales for that product. There's a clear effect where the younger the product portfolio, the faster the sales growth will be. It's a very strong effect. When you put that in the model, if you can stimulate product development, you're going to do very, very well.

In addition to the long half-life for improving product development, there was another important feedback effect that choked off improvement in product development. Remember I showed you that commitment to improvement in the PD area went up and then died away. How come? They had so much improvement activity going on that they could not support all the improvement teams. There's only one CEO to go around, and slap people on the back, and tell them what a great job they're doing, for example. Management attention has to be parceled out. If you're the Vice President for Quality, where do you spend your time helping people? You should, in some sense, be helping the people who need the most help: product development, where it's tougher. But that's not what happened, because you're under pressure to demonstrate that this tool will work. Where are you going to go to show that this tool will work? You go to where the half-lives are short, because that's where you get the fast results. That results in this feedback system here. And you can think of this as a kind of path-dependent situation where the amount of support that's going to operations, as opposed to product development, will increase as they need support - because you answer your phone, whoever calls you, and then you go help them - and with results. The better they do, and the more quality improvement activity that's happening in operations, the more support they get. That helps them do even better, and helps you - encourages you to spend even more of your support effort helping them, at the expense of product development. You're systematically helping those who need it least, and starving those who need it most.

Here's a couple of quotes from the field work that I think are interesting to illustrate how this worked out. "Some people saw the value of TQM right away, but the majority needed to see results before committing themselves. A few resisted, and could never be persuaded. We needed to demonstrate that TQM worked for Analog, to bring the skeptical majority on board. We spend a lot of our time on projects we thought would succeed early on. Every single one of those was in an operations area, and added to capacity.

Many engineers didn't think TQM could work, and thought it interfered with their autonomy. The requests for help we got came mainly from the operations side. There's a cultural issue here that biases the allocation of resources. The CEO was a little more blunt. He said well, there's some closeted cynicism about TQM in the company. Among the engineers it isn't even closeted. They think it's crap. He didn't actually say crap, but that's how we sanitized it. In the simulation, it was a four letter word.

In the simulation, what happens is as TQM really takes off, they have to start deciding who's going to get their help, because there's only so many hours in the day. So here you see the adequacy of support resources falls away from a hundred percent. It's not terrible. They're doing better than a lot of companies, but they had to decide. Those positive feedback loops I just showed you meant that progressively more and more of the support went to manufacturing, and less and less to product development. The result is that the product development time in Analog barely improved at all, compared to what it might have done. That reinforcing feedback loop - the only hope you have to save you from excess capacity - is not working. The result is you've got a huge amount of excess capacity, and now everybody knows that there's more workers than are needed.

You don't have to lay them off. You could just tough it out, and say yes, we've got more workers than we need now, but we'll hang onto them because it's the right thing to do, and wait for the demand loop to kick in. They couldn't do that, and it's because of one more unanticipated side effect that was extremely interesting. This is data now. This is not simulation.

In 1985, this is their cost structure in dollars per chip. The average chip sold for sixteen bucks, cost of goods sold, 7.60, gross profit 8.70, indirect costs - that's primarily GSNA and R and D, 6.35; leaving you operating profit per chip of 2.36. A markup ratio - that's the average selling price divided by cost of goods sold, of two hundred fourteen percent. What does TQM do to your cost structure? It lowers your direct costs really fast, because that's where manufacturing is driving your costs. It doesn't lower your indirect costs very fast, because that's where R and D and product development are. In 1989, direct costs had fallen by around sixteen percent, and prices had fallen by about the same percentage. Why? Because they used cost plus markup pricing. They have thousands of items in their catalog, and they can't do a zero-based pricing analysis for each one. It's just like a department store. They take their unit-direct costs, times the markup ratio, and that's their price. So they lowered their prices about as much as their costs went down. Indirects didn't go down nearly as much. The result is operating profit falls by nearly fifty percent. The markup ratio didn't change at all. At this point, Wall Street gets very, very anxious, and they say things like this, "We caution our readers that this company has a history of frequent earnings disappointments. Sales may be difficult to predict, but it's hard to understand why Analog cannot cope with this problem by adjusting expenses accordingly." and so on.

The result of this is all of the sudden top management becomes worried about this takeover threat because the stock price has been beaten down. To demonstrate to Wall Street that they're serious about cost control, what do you do? They laid off twelve percent of their work force. The first layoff in the entire history of the company. The immediate effect of that of course is to destroy all commitment to TQM. In the simulations what happens is perceived job security plummets - and eventually it recovers. But you have perceived job security plummet, and so the second half of the simulation that I showed you before shows commitment to improvement in manufacturing tumbles. Basically TQM effort dies for several years. In fact, that's exactly what happened. We have lots of interview data to show it.

M: Where were the layoffs? Were the layoffs in manufacturing?

Mr. Sterman: Primarily, yes. You know, the engineers didn't get laid off. Here's a little quote. This is from a TQM manager in one of the divisions. "A lot of employees in a particular plant were working their asses off for TQM, and their reward was their operation was moved to the Philippines in search of lower labor costs. TQM was another path to a layoff." That then closes the loop.

What you've got here is a story of unanticipated side effects. Where management's mental models were not dynamic enough, or sophisticated enough to understand that improvement per se isn't enough. TQM you could think of as some horrible mutation that, although it benefited the fitness of certain operations locally within the ecosystem of the company, was not consistent with everything else, and caused enormous problems. We ran our model forward. We did this work around '92, and we ran our model forward to see what would happen in the next few years. It suggested that TQM would rebound in a more balanced fashion between the two different functions: engineering, and manufacturing. In fact, that did happen. Here's what the model predicted about the stock price. Now this simulation was done around here, and it predicted that the stock price would rebound greatly. The model is basically flashing a huge "buy" signal. That's exactly what happened. Today the company has got sales of over a billion dollars. It is doing as well as you could do in a context of world depression in the semiconductor industry. They're still the industry leader in their niche, they're doing extremely nicely. But my view is that if Mike Milken had not been in jail during this period in 1990, '91, they would have been taken over by a hostile raider, and this whole thing would have been gone.

We've done some policy analysis and some other things here that I can maybe allude to if there's any time for discussion. What I've tried to illustrate is a couple of things. First, with system dynamics we're often attempting to look at real life problems with major consequences for the clients. This was not a consultant study, this was research; but it obviously has important policy implications for the firm. Secondly, we're developing behavioral disequilibrium models where the agents had bounded rationality. We're not assuming that people are economic creatures in the textbook, axiomatic sense, because that's just counterfactual. In order to validate and calibrate your models, you have to do extensive field work using all the available data sources. Not just the numerical data, but also interviews, observation, participating in the operations of the company, as many different data sources as you can to triangulate on the truth as best you can find it. It's a lot of fun. Well, thanks.

M: What happens if you run the model with the focus reversed, so that product development gets more and more of the effort, and you do it earlier to try to balance the rate of improvement?

Mr. Sterman: That's a very interesting issue. It turns out the answer is contingent on competitor behavior. If the competitors are kind of following you, and so you don't do big TQM improvements in your manufacturing, and so they do theirs also later, then this works great. You get new products out the door in a way that balances demand and capacity better. That avoids the excess capacity layoff problem. However, if your competitors don't understand this, and so they go down the wrong road, and generate huge improvements in their manufacturing operations, then in the short run you become less competitive. The long delay in getting new products out the door means that in the short run you have not improved manufacturing as much as they have. You're now going to lose market share to your competitors. Lower market share - and you don't just catch up later, because the whole system is driven by positive feedback. Because now with lower market share you have lower revenues; that means your R and D budget is lower. Now, although you might be improving product development, you have fewer R and D dollars to work with. So you get less out. You get fewer new products out, and you can actually spiral down into bankruptcy. It's contingent on the game that's played between you and the competitors. In this industry, the competitors would not have waited. They would have gone ahead anyway. So that turns out not to be a great policy. Go ahead.

F: I have two questions. One, wasn't there a short term advantage to the firm, in terms of a delay before the competitors actually caught on? And second, did you run the model with the price of the chip unchanged? In other words, if they hadn't lowered the price of the chips, what would have been the effect?

Mr. Sterman: The answer to the first part of the question, which is was there a short term benefit: there is, but it's very small. That's contingent on the particular historical situation that Analog already had the dominant market share. The improvement in quality that they had short term didn't translate into a big bump in market share. If they were a much smaller company, I think that would have made a difference. We did run the policy that you suggest, which is to compensate - to understand how TQM unbalances the organizational routine about market pricing. And all you have to do is increase your target margin - markup ratio by five percent. Prices still fall, they just don't fall quite as fast; and this is the result. It makes a very substantial difference. Now you can't avoid some harm in the recession. But it turns out that the drop in market value to cash flow ratio here is sufficiently moderated that the threat of takeover is greatly reduced, and depending on how much they can hang tough in the executive suite, they could have avoided the layoff altogether.

M: It's a very complicated model, doesn't it contain the economic variable of price.

Mr. Sterman: Oh sure it does. Price is fully endogenous in the model.

M: In the end, isn't it coming from outside?

Mr. Sterman: First of all, technically in the model, the price of Analog's products is a completely endogenous variable. Rather than assuming that price is set, either by some market equilibration process, or by some economically optimal decision rule, we modeled it the way they actually do it. The way they actually do it is by cost plus markup. The markup ratio evolves very slowly. Their field work to support that is incontrovertible. Furthermore, a lot of firms do that. Boeing doesn't do it because they're selling a very small number of product lines, each of which costs a lot of money. So it's worth their while to price in a more sophisticated way. But a company selling thousands of different SKUs, each one worth about sixteen bucks, it's not worth their while. This is in fact how they set price.

I do think it's true that management did not understand the connection between TQM, and pricing. In their mental model, TQM is over here. It's about the factory, it's about operations, it's about learning, it's about improvement. It has absolutely no links, in their mental model, to the pricing decision process. Pricing is viewed as a flexible routine that they have that will accommodate changes in costs and market conditions. They needed to see that connection, because the markup ratio had evolved to be at a pretty good level. I don't want to say optimal, but pretty good level over many years; during which the ratio of direct and indirect costs was stable. The minute that ratio changed, in a rapid disequilibrium fashion, that routine was no longer helpful. It created these unanticipated side effects.

Mr. David Harel: I'd like to thank the organizers for inviting me to this conference. It's really amazing. I've never been invited to a conference in which I don't know the session chairman, and there's only two people here whom I've ever met before. The sad aspect is I have to leave about half an hour after the talk is over, and I'm really disappointed. This is a conference I might just come back to, even if I'm not invited. The title in the program is Statecharts. I'll be talking a little bit about statecharts, but mostly about the motivation and the general ideas that led to these developments. By the way, maybe I'm one of the most hard-core computer scientists here, yet these are going to be handwritten transparencies. Which just goes to show you that the shoemaker walks around barefoot. I did want to say one thing about the notion of complex systems. I'm not quite sure how long the notion of a complex system, as practiced by maybe eighty percent of the participants here, coming from physics, and economics, has been around. We've been using the notion of complex system for almost twenty years in a different way. I think the division between these two kinds is important to remark upon.

We talk about systems that have to be built. They're going to be complex whether you like it or not. You cannot make a simple nuclear submarine. You cannot build a simple F16, or a simple communication network for AT&T. So it's not a question of analyzing a natural system, and figuring out how it evolves, or how it works, or what its structure is, which is the other side of the coin; but it's a question of starting from nothing, and building something that's really very complicated. This is what a system development process should be like. Everything goes nicely and smoothly. It's on budget, it's on time. Complies with some standards. In this case some military 2167 standard, and so on. Of course the most important thing, the customer accepts it, and everyone is nice and happy. There are no people in this picture. The reason is obvious. Everything is so nice and smooth that they're all beyond the horizon down there.

The real world looks a lot more like this; with loops, and problems, and pitfalls. This picture has no people in it either, but for a completely different reason. They're all down here in this place that you might term the specification gridlock. Let me show you in a zoom in what this looks like down there. You get people sitting around scratching their heads trying to understand requirements documents or specification documents. I'm not going to be too particular about the distinction there. To be a little more responsible, there's a system that has to be built, and there's a group of people, and they want the system. They essentially know what the system should be able to do. Then you have the people who have to build the software, the hardware, and produce that system. At the very least there's a problem of communication between those groups. The problem is much, much deeper than just communication. I'm going to stress again and again the notion of behavior of the system, and I'm going to draw a very superficial analogy between the behavior of the system, and the engine of an automobile.

Maybe the most important thing I can convey to such a varied audience here is much less the details of the approach, or the approaches that had been taken. Our approach is only part of what is needed here. Maybe some of the personal social aspects of how these things happen are of interest. I'm really a theoretician. Most of the time I do complexity theory, as in P and NP and decidability, and automata theory. I got dragged into this in 1983 doing some consulting on a fighter aircraft program for the Israel aircraft industries that was later scrapped. I hope not because of this, but because of the fact that our country's simply too small to build a fighter aircraft. It did fly, though. A couple of prototypes were built and flew. It was called the La Vie. I was kind of amazed at what was going on there.

They needed help in describing the behavior of this complex system. The talk by Norman Badler this morning was particularly interesting. But I should point out that in terms of control, the state space in his systems are extremely simple compared to these. What makes his work very interesting I think is the translation from natural language to the state space, and to the actions; and of course the graphics, and the entire effect. But a few agents with several actions that are going on. Orders of magnitude more simple than an F16, or a nuclear plant, or even a BMW - the 700 series of BMWs are going to have sixty-eight microprocessors in them, all of them talking to each other in real time.

When I got to the Israel aircraft industries, there was a team of experts working on the avionics. Avionics is everything that's in the aircraft, except for the aircraft itself. The communications, the radar, the electronic warfare, and so on. There was a radar expert, and a hardware expert, a software expert, an electronic warfare expert, and so on. I wasn't an expert in anything that they had to build. You would grab the radar expert by the collar, ay tell me about the radar, and he would give you the algorithm that the radar uses to measure the distance to a target. Or the hardware person will tell you about the wiring of the SCSI and then the chip placements. The communication people will tell you how they format the information that runs along the fiber along the length of the airplane.

I had one big advantage over these people, and one small advantage. The big advantage, I didn't understand anything about radars, and about electronic warfare, and about hardware. Or even about that kind of software that they needed. I had a completely blank approach. I wasn't preconceived into terminology, and to certain modes of thinking. The other slight advantage was I was brought up as a mathematician, and it was a little harder to fool me with these two thousand page documents that they would bring out written in structured natural language like eight point, twelve point, one point six, eight point, twelve point, one point seven. I would consult there a day a week.

One day, after learning some of the buzzwords, I would ask these deep and profound questions, such as what happens when I press this button in the cockpit? They'd say oh, in volume one, page seven hundred and eighty-one, clause eighteen point six, point one point ten, it says if you press this button, such and such a thing happened. Then I'd say but what happens if an infrared missile is locked on a ground target? They'd say oh, that's a different story. Volume two, page seven hundred and sixty-four says in that case something else happens.

To me this was astonishing, because it seemed, from a naive point of view, that the first thing these people should have done was to figure out exactly what they wanted the system to do under all possible circumstances. Lay this desire out in a rigorous and clear fashion. These are the two cue words. You want things to be formal and rigorous, not just diagrams with arrows, but things that have precise mathematical meanings. You want them to be clear so that people can understand them. Which goes across diagrams that just have a lot of spaghetti lines running all over them. Of course it goes against just having a document with two thousand pages that you can read from here until China, and not really understand what was going on.

Incidentally, the third or fourth question I would ask them, they weren't sure of the answer. They would have to go and ask the experts. The fifth or sixth question, they had to call the Air Force, which was the customer, and say well, what indeed does happen in this particular combination of cases? The seventh or eighth question, no one knew the answer. This again was an astonishment, because these things are built, and they fly sometimes. Of course, who makes these decisions? Ultimately a lot of these decisions are made by some petty programmer who comes along, and writes some piece of code that's linked together with another piece of code. Things happen which were not anticipated. It's no joke that the first or second prototype of the F16 used to flip over on its back every time it crossed the equator. This is not a joke. They figured out a couple of months later that there was a minus sign missing on one of the flight control algorithms.

I'm not making this problem up, and I didn't invent the problem, and the problem is not just a problem of the Israel aircraft industry. This is, apart from getting up on the table, and jumping up and down I don't know how to say this more dramatically. The heart of developing complex systems - complex engineering hardware/software systems - is the ability to describe behavior clearly and precisely, and to do things with it to which I'll get in a moment.

This is taken from a very large document describing the operation of a chemical plant. A particular company that I won't mention the name. I spent a couple of hours trying to find wherever in this document a particular piece of behavioral information that was of interest to me was written. I found it appearing in three places. Earlier on in a clause that has to do with security it says, "If the system sends a signal hot, then send a message to the operator." A lot later as it closes, it talks about temperatures. It says, "If the system sends a signal hot, and T is greater than sixty degrees, then send a message to the operator."

Finally, towards the end of the document, there's a clustering of crucial critical behavioral issues that someone had summarized. Listen to this one carefully. It says as follows. "When the temperature is maximum, the system should display a message on the screen, unless no operator is on the site, except when T is less than sixty degrees." I got my Ph.D. in logic, and I have never been able to figure out whether this is equivalent to the others. It doesn't really matter; even if it is equivalent, you understand what's going on here. These were written by different people at different times. Now put yourself in the position of the manager of the software or the software programmer himself or herself who has to make - and this took me two hours to find these three places. And they have to write the software that controls these things.

Many years ago, the horseless carriage was a good thing. If someone would come along and say that there was an engineless car out in the offing, that wouldn't be a very good thing. By the way, the title says, "Myths, Facts, and Challenges." Because this talk is only about forty minutes long, and I have transparencies for about a three hour tutorial, I was contemplating whether I should leave out the facts, and just talk about the myths, or leave out the challenges, and talk about the facts. And one option was not to tell you which is which. But anyway.

Two myths here are, first of all, that engineless cars are useful. That's obviously a myth. Another myth in this area is people say I don't do real time systems, and therefore I don't have a problem. As I'll say in a couple of minutes, real time- ness, the fact that the system has to respond in real time makes a system an order of magnitude more complicated. That is not the crucial issue. If you take a simple reactive system, like a teller machine that you will use, or a VCR. Let's say your teller machine does exactly what it does normally, but it has to respond to everything say three hours after. For example, you press a button, three hours later it does what it's supposed to do. It doesn't make the description of the behavior of that system a lot easier. You still have to decide what causes what, what things happen in parallel, what can trigger a nasty result, and so on. The real time-ness just makes things even more complicated.

The facts are that houses, and bridges, and things like that are there to be, but software, and systems, and cars, are there to do things; the key issue is behavior, the key notion is reactivity. This notion first appeared in a joint paper with Pnueli, who is, by the way, the 1996 winner of the Turing Award, which is our Nobel Prize, for his work on program verification. I hope I'll have some time to say some words about verification also. The term is actually due to him. The point here is that in contrast to systems that are transformational, they take inputs. They take all their inputs, they do their thing, they produce all their outputs, and they go to sleep again until the inputs come back again. The role of a system which is dominantly reactive in nature is to sit out there and react. If you kick it, it does something. If temperature goes above a certain level, a valve opens. If you press a button, something happens. This is the key issue. Whether or not the system is concurrent or sequential is important, but not as important as reactivity. Whether or not it's real time is important, but not as important. Whether or not it's distributed, and you have to put some parts here, and some parts there that have to communicate is critical; but the most critical thing is the reactivity of the system.

Maybe thirty years ago there were a lot of computerized systems that were nonreactive, you would go to your bank, and put in a request to buy some stock, or you put the check in, and it would collect all these. Friday afternoon it would all be typed in, and it would run a batch calculation. Everyone's' accounts would get updated. Today even an MIS, a management information system is highly reactive. Your check goes through the machine, things start happening all over the place. Even money transaction systems are highly reactive, but obviously embedded real time systems like command and control systems, telecommunication, avionics, aerospace, automotive, and so on; and interactive software, for that matter, too. Reactivity is a fact of life. As we say in Hebrew, with the food comes the appetite. As our machines get smarter, and larger, and smaller, and more powerful, so our desire to have more complicated systems grows, and we have to be able to catch up in our methods for describing behavior.

The grand challenge in this whole story is to devise a framework for developing complex reactive systems consisting of a means for describing and analyzing the structure of the system. You have to figure out what the system is made of, even if it's software. Software has structure, it can be structured by functions, by objects, by tasks, by procedures. But driven by, propelled by behavior. There's lots of things you can do with a motor car without an engine. They're not very interesting. Well, maybe most of them are not very interesting... Anyway, the engine is the most important thing.

Two words of caution here. If you're thinking about how to address this problem. First of all, it's a language design problem. You can't sit in your ivory tower, and come up with a language or an approach, and stuff it down the engineers' throats. They have to like it. There are examples in our field of formalisms that are equivalent, mathematically equivalent, polynomial time equivalent, equivalent in size, in everything, in terms of the power of expression. One of them they like, one of them they don't like. The ultimate judges are the engineers. The other thing, of course, is it's not just a language design issue, it's a mathematical issue too. These languages are not just pictures, or not just programming languages with nice key words. They have to have underlying, rigorous semantics that have to be good enough to build the kind of tools that we mentioned.

Here is another myth. CASE is computer aided software engineering, or computer aided systems engineering. In the late seventies, early 1980s, there was a proliferation of CASE tools. The myth at the time was that maybe these were the solution. The sad fact was that that was the era of engineless cars. Most CASE tools were at best nice graphic editors. They draw bubble diagrams, and block diagrams, and structure diagrams, and state diagrams with nice arrows, and stuff like that. They couldn't do much more than the nice graphics, and documentation, and project management following the project, and the dates, and the pieces. Some structural analysis of the structure, and pieces that were missing, and arrows that were dangling, and stuff like that.

What was missing was the engine. To take it to the extreme, if I were to come here and propose a new programming language, and tell you about its terrific features, and it has call by value, recursive procedures linked to object events, and stuff like that. Would convince you this is a lovely language, and you would say that's neat; and do you have a tool to support it? I would say sure. What can I do with your tool? I say you can sit in front of the screen, and type in your programs, you can look at them. There's documentation abilities. Give me the list of all integer variables starting with a "G". You can do automatic indentation. You can figure out whether the types, and parameters, and stuff like that all match, and everything. Then you say okay, that sounds neat. Can I now sit down, and run some programs? I say no, no, no. No compiler, no interpreter. You can't run your programs; you can write them, you can look at them, you can analyze them, stuff like that. That was really what was going on in the early 1980s.

To be a little more detailed about what this challenge, this behavioral challenge breaks up into, it has three parts. One is the engine itself. You need a good (again, good is clear and precise, not one on the expense of the other) language for modeling pure reactive behavior. Forget about the structure. You want a language that will enable you to say if this happens, this occurs in parallel to that, and this causes the other thing to happen thirty milliseconds before a third thing happens. If in parallel something arrives from here, then something else happens, and so on.

Another challenge, which some people gloss over - once you've decided that you like, state machines, or Petri nets, or VDM, or Z, or whatever you choose for the engine itself, you can't just say oh, I have a car. I need an engine. Let me call someone, buy an engine, and stick it in there. There are a lot of things that have to be connected up. When you need a language for modeling, submarines, and F16s, the structure is important, the project management is important. The connection with the behavior is crucial, and you need a language set in which all these things are linked up rigorously and clearly together. The third thing, which is the ultimate goal of this whole story, is to build tools that are really useful. Tools that are really useful do for you a lot more than just enable you to draw pictures, and look at your system, which is important when you're doing a live system. Even drawing the pictures, and thinking about it helps; but we ultimately want a lot more than that. Let me try in a few words to say what more we like.

You want convenient modeling of the system, but more importantly you want to be able to execute the models, and simulate them. Model execution - this is the most important thing. Just as you want to be able to run a program using a compiler or an interpreter, you want to be able to run your models. This might seem a little strange, if you've never seen anything like this, that you would be able to draw pictures and diagrams. Diagrams are not there just to serve to illustrate an idea, or to illustrate connections between things; as diagrams are used in many lectures, and in many textbooks. The diagrams are used to run the system, and they can be executed because their semantics is sufficiently precise. You can use them to analyze dynamic behavior, and to figure out whether things will happen, and so on.

Finally, if the model is sufficiently precise that it can be executed and analyzed, then it should be sufficiently precise to be able to lend itself to code synthesis. By this I don't mean to build a skeleton of code that then the program is sitting in, and fill in the important parts, but what you really want is to rise one fundamental level above high level programming languages. Once upon a time we programmed with zeroes and ones, then we programmed in machine language, in assembly language. These were compiled down into zeroes and ones. Then we started programming in high level programming language, which we're kind of creeping up in levels of abstraction. It's essentially the same idea, and these are compiled down into assembly language. Now we're talking about modeling with what I'll be calling visual formalisms, which you would like to compile down into a high level programming language, and free the programmers from having to program. Or to put it in a more constructive way, to shift the effort to the thinking, and the modeling, and the analysis on a much higher level than writing the code itself. You want to be able to synthesize code from the model automatically.

To say a few words about what are statecharts, and where they come into the story, the question to ask here is if you want a visual language for modeling the behavior of a system, what's wrong with a good old state transition diagram? We all know what finite state machines are, automata, if you want. We all know, these are states, and these would be the events that causes transition. This is a start state. This is a condition. If you're in this state, event A occurs, you go over here on condition that condition C is true at the time. This is fine. It's visual, it's clear, and it's formal. There are forty or fifty years worth of mathematical understanding of automata theory. Everything is fine here, except that for anything larger than a five or ten state coffee machine, these on their own are completely useless. They're useless for a variety of reasons. I don't think I'm going to get into them. Basically, if you take even a simple system like a VCR, and it might have two hundred and fifty or three hundred different states, and you're asked to figure out what those states are. Put them out on a piece of paper, and connect up all the possible transitions, you get a big mess. A very big mess; even on a very small system. I don't have to drag you through a mental leap to a much more complex system than that.

So what are statecharts? The notion of states of the system and events was very important. Unfortunately, the meager formalism of simply figuring out the entire state space, spreading it out flatly, and connecting it was far from being workable. In state charts, the idea is to add three or four very elementary - it's just the right way of partitioning a state space into its pieces, that act in parallel where there's flexibility between the hierarchy and the concurrency. It's not as though you have a bunch of communicating finite state machines, but whenever you enter this vector of states, you (at the moment) can expand or shrink, depending on the way it's structured. There's a notion of broadcast, history, communication, and such. What I will do, though, is just to show you what they look like. I'm taking a risk here, because the last time I showed the next two transparencies, someone said excuse me Sir, is this part of the solution, or part of the problem? Being hand-drawn on a small transparency, they look a little cluttered. The point I'm trying to make is these aren't just pretty pictures. These are mathematical objects that have precise semantics. The semantics can be, and are embodied into tools within.

This is part of a statechart of a mission computer controller for an airplane. Each of these purple things is a state. This is a normal state of maintenance. It has maybe some more stuff inside. This is a much more important state, which is the operations state. It's broken up by these three dashed lines into three parallel or orthogonal components. What that means is if you're in this state, and this event occurs, you enter this guy over here - which, by the way, is also the default. You don't just find yourself here or there. You're here, and here, and here simultaneously, unless something happens to take you out of here. In which case you just drop everything, and you leave. And there are ways of not dropping everything. And there are conditions that you can set here. To give you a feeling that things are much more complicated than this, I just want briefly to zoom in. This is the general mode of the aircraft: air/ground, and navigate. I just want to show you how much deeper the behavior is inside of these modes in here. So here is air/air and air/ground, which are another couple of levels of description. What you see for example here is as you go deeper inside, you might expand in the number of things that are happening simultaneously, and then you shrink out when you leave things. I don't have any numbers here, but as I said earlier, the proof of the pudding is definitely in the eating!

The raw engine language is fifteen years old now, and I'm very happy with the way it's proliferating. We made some changes to the language in response to feedback. I think it's fair to say that this is fast becoming the main language used to describe behavioral specification in a variety of situations. I just heard the other day that in a space probe to Mars the little land vehicle was programmed using statecharts, and other things like that. By the way, this audience might be interested in knowing that we have a project going with our life sciences faculty, in which they're trying to model parts of the immune system using statecharts. I'm not quite sure how successful that will be.

Friday I was down in Otis Elevators, and discovered very interesting things about elevators; just as a side remark. I don't know how many of you know this, but most of the major elevators have little modems in them, and the headquarters of Otis can call the elevator in the World Trade Center, and talk to it. It can get some information and feedback. That's kind of frightening.

I gave three challenges for describing behavior. One was the engine itself. The statecharts language is somehow the response to that. You have to connect this up with the structure of the system, and you have to build the tools. The two main approaches that are taken have been taken about ten years apart by methodologists in software and systems engineering to describing the structure of a system. One is called structural analysis. It really is a function of decomposition. If you're talking about your ATM, the structure is: balance, deposit, communication with the bank machine. These are the functions that the machine can carry out. You can subdivide those, so you decompose the functions up to a level with which you're happy. You identify the flow of information that can flow between these functions. This is called the functional decomposition. This is still the structure. It doesn't give you any answers to questions about what happens when you press this button. That is done using statecharts that specify the behavior of each of these functions. The thing I said earlier about having to link the languages together is a nontrivial issue of linking the data flow on the structural analysis, and the functional decomposition with the states, and the behavior. This is nontrivial, but it has been done. The other paradigm is object orientation where you describe your system in terms of objects, and the things that they can do for you. The interface of the objects, the request you can make of an object, and the deliverables that those objects make. Again, the structure of the objects itself can be very complicated. It can be hierarchical itself. But again, it's just objects. This is a very fashionable, and also very useful paradigm. But you can't just say I'd like to think of objects, and aggregation inheritance, and all these nice things about a large system without the engine. You can also not say I'd like this, and I'd like statecharts inside. There's a lot of work to be done that has to be done in making the statecharts, or things like statecharts. I'm not going to say that statecharts are the only way of doing this. Some industries use Petri nets, and SDL diagrams. There are other means for describing behavior. Essentially, the responsibility of linking the languages together exists even if it's not statecharts that you use, but something else.

In the structured analysis realm, I'm going to skip a lot of stuff here. The language for describing a functional decomposition is called activity charts. These are merged together conceptually, and technically, and mathematically with the statecharts. This has led to a tool, built by a company that I've been involved with for many years. A tool called Statemate. Recent versions are called Statemate Magnum. In fact, there's a book that just came out last week from McGraw-Hill that you can read about this stuff. This is in the structured analysis realm. In the object realm, things have different names, and maybe I won't be able to get to that.

The tools themselves do lots of things for you. The Statemate tool enables you not only to see the diagrams, as I said a little cynically earlier. You can run the diagrams, you can run the system. The diagrams get animated automatically. You can run the systems from the diagrams, or from a mockup display of the system itself, or maybe the system itself physically. If you can link it up to the software, you can actually run the system from these panels, the diagrams animate. The nice thing is, if something happens that you don't like, you say okay, pause. You click an icon, you take this out, or you lift it up, you move it over there. Maybe change the name of an event, or something. You recompile down. A few seconds later, your system is running again. All this is done before you've written a word of code, and definitely before you've built any real hardware.

I could go on talking about the kinds of abilities that these tools can do for you. You can go home on a Friday afternoon, essentially, and instruct your tool, please fly this airplane for me for three hours; and conjure up the events at random. Not completely at random. You don't want it to pull the ejection handle the same rate it moves the throttle, but you give it probability distributions, and so on. This allows for certain things. Certain very bad things - you don't want a bomb to be dropped if the radar is not locked on a target, and you want the right things in trace files, and so on. These things can all be done.

I do want to say a word about visuality, and a word about formality. When I call something a visual formalism, I don't mean an iconic language. You turn on your PC, and you see little icons.

These are visual, but not diagrammatic. This is not a visual formalism. If you're out there to design a visual language, my recommendation would be first and foremost do things topological. I'm not a brain researcher, and I'm not a visual imagery researcher. I've heard a lot of interesting things here. My intuition and experience is that people understand topology very well. Things that are connected, things that are disconnected, things that are inside each other, things that intersect each other. Topology is a more fundamental branch of mathematics than geometry. More things are invariant on the topological properties than geometry. Then go to geometry. Size, shape, color, location, and stuff like that. Then you can also go to iconic.

So you might have noticed this from the diagrams I showed you very briefly, that the notion of insideness is very important, the notion of intersecting is important. What is connected to what is very important. Then you can go to different color codes, and different shapes, and sizes, and so on.

The notion of visuality is supposed to hint at intuition and clarity. The notion of formality is supposed to hint at formality, and mathematical rigor. There is a myth in the software engineering world that formal specification is a formal-looking specification. Some kind of feeling that anything that has a lot of Greek letters in it is formal.

I've seen Greek letters put into a formalism which is very informal, and very ill-defined. Things like statecharts have no Greek letters, and very few textual and algebraic constructs in it. It's completely rigorous, and completely full. The fact is artificial language is formal if it has a rigorously defined syntax and a rigorously defined semantic domain. You have to say what you want your things to mean. A program for you might mean the input/output relationship. Mine might also take into consideration the values of the variables throughout the program. That's a different semantic domain. You have to figure out what you want your things to mean. Then of course you have to have a rigorously defined semantic function, which can be transparent to the user. The user doesn't have to see the equations that underlie these kinds of languages, it can be embodied in a compiler; but it has to be there, and it has to be rigorous. The sad fact is that working out the semantics for visual diagrammatic languages for describing dynamic behavior (as opposed to describing structure, which is what most visual languages do) is complicated. And I'm not talking about simple languages for simple systems. If you want something that works for a nuclear plant, or for a communications network, then things can be very difficult. This should be contrasted with sometimes things that people sometimes publish. I call these the M and M style semantics.

There's a terrible book by James Martin and Carmen McClure called "Diagramming Techniques For System Development," or something. It's full of hundreds of different visual languages, none of which has any semantics. The only thing they say are things like green circles mean this, and yellow arrows denote transitions. The minute you try to put two things together, and figure out what happens if you take a green arrow, and you put it over a green arrow; there's no explanation, there's no intuition, and there's no attempt to work things out the way they really should. This is doodling, and we're not into doodling here.

I just want to mention that in the object-oriented world, the statecharts are linked up with things called object models that describe the structure of the model in terms of objects, and aggregation inheritance.

I should mention that the area of object orientation is a lot harder. The reason is in the object-oriented world, the structure itself is dynamic. Objects get created, and get destroyed, and have multiplicities, as opposed to functions. Your ATM can compute a balance, or it can do a deposit, but it doesn't have a hundred deposits, and the deposits don't disappear, and get created. If you're talking about objects, and you're doing a radar system which has targets, you want to describe the behavior of a target, which is a member of the class of targets. As the system starts running there are lots of targets. They get created, they get destroyed, they interact with each other. The whole idea of language design for this kind of application is a lot more complicated.

I do want to mention one research topic that I'm very excited about, and interested in. Just one word on two of the challenges here. One is formal verification of systems such as these with objects or without objects. Formal verification, or rigorous verification is, of course, the idea of telling a computerized system what you think the system is doing, and have it verified for you rigorously. Just as no one is going to leave this lecture going to look for triangles whose angles don't sum up to a hundred and eighty degrees - planar triangles, that is - because there aren't any such triangles. Although no one has ever seen any out of all the triangles in the world - if you prove that your program does what it's supposed to do, you don't have to go looking for counter examples. Now this sounds very, very naive. And in fact the myth that was true for many years was that program verification is a useless academic exercise. The facts are different. A couple of interesting facts.

First of all, there's no possibility of a general fully automatic verification. This is an undecidable problem. The ability to have a tool such that you give it your program or your system description and you give it what you think it's doing, and it says yes, this is correct, or no it's not. This is in general unsolvable.

Almost paradoxically, if your system is correct. If indeed your elevators will never open the doors in between floors, if that's the case, and your model has that property, you can prove that. There is a mathematical framework in which this can be proved. The problem is finding those proofs automatically, which in general is impossible.

The work done on program verification has had profound impact in the last five years. Particularly in hardware companies, where rigorous verification is being used to verify the behavior of chips. The challenge here - and I think the time is right for that challenge - is to lift the techniques up to high level systems with hardware and with software together modeled using high level visual formulas that get translated into code.

The final thing I want to tell you about is one of the research projects that I'm very excited about. That has to do with what I'll be calling the grand duality, or the grand dichotomy of system behavior. Let's say this is the structure of your system. It might be objects, it might be something else. It's a lot more complicated than this. There's hierarchies inside this, also there's many more - and multiplicities, as I say, and dynamic things being created.

One of the social issues in system design is engineers find it very easy to talk about scenarios. They say ah, how do you shoot the missile? Well, you press this button, and then the system sends a message to the infrared detector, and it does the following. Then you click here, and it does that, and this does that, and that does the other. Then the missile fires. How do you communicate with the ground station? Oh, this you do as follows. They tell these stories. The stories involve a lot of the objects, or a lot of the pieces of the system. So essentially you're doing scenarios. These are described using something called message sequence charts that I won't get into. What you're telling is the red story. The red story does something here, sends a message here, does something over there, and so on. That's the end of that story. Here is the green story, and then there's a black story, and so on. This is something that's done relatively easily early on in the process of describing the behavior of a system. Now why can't you just take these, and turn these into code? Well, that's the problem.

The problem is that what you want in the end is not the red and green and black stories. What you have to come up with in the end - because this structure is not just done there for convenience. This is what the system will look like physically. If it's software, then physically its pieces of software appropriately partitioned. What the system will look like at the end is of course like this. Each piece of your structure will have its appropriate pieces of behavioral specification.

For example, a statechart. If this event comes in, you do that. You send a message over there. Go to sleep. If this event comes in, you move into that state, and you do this. You have no idea what part of the story this little piece of the behavioral thing is coming from. It might come from five different stories, it might be relevant to ten different things. It might trigger things over here that this guy doesn't know of directly at all. But this is what the system will look like. Yet, we need to figure out, first of all, consistency with the stories.

What is more important - and this I think is the major open issue in this field, is essentially going from here to here automatically. This is something that we have ideas for, but it's far from being solved. You want to be able to take your engineers - and when I say scenarios I don't only mean positive scenarios. Negative scenarios too. This happens, and then that happens, then you do this. This must not occur, under any circumstance. This is a no no scenario. You have yes yes scenarios. And you want these stories, or these counter stories, or these branching stories. Sometimes they have one, sometimes they have ten, but essentially you're telling stories of how the system will behave that cut across the boundaries of the complex pieces of the system.

You want to synthesize out of these, for example, statecharts; but not just from ordinary scenarios, but from more complicated ones that have existential and universal quantification. Stories a lot more complicated than I'm showing here. But what you really want to be able to do is to synthesize the first cut out of the behavior of a system which is efficient and compact from the stories that engineers tell you early on.

The final grand fact of this whole saga - and I know I've been very brief, and very superficial - is that there are a lot more things in this business that we don't know how to do than things that we do know how to do. A lot of humility is in place. The road here was very exciting, but it's long and hard, and there's still lots to be done.

Why objects are more difficult I think I've explained very briefly. I can give more explanation. The comparison to a linked list is not appropriate to this story, because a linked list is a data structure. I'm not talking about data structures, or data manipulation, or data retrieval. It's two different animals. Object orientation - not in databases. Object orientation in reactive systems is a lot more complicated than functional decomposition. Mainly because the structure itself is dynamic, and growing, and shrinking, and has multiplicities, and has cues of events that are not present in functional decomposition. I make no claims about whether dealing with object-oriented databases is more difficult than dealing with fixed structure databases. That's not the subject of this. Now what happens with the statecharts? Statecharts have to be parameterized. They have to recognize one another's parameters. What happens if you send an event to the class? Who gets to listen to it? Who gets to hear about it? If you aggregate another object from these, do they all get aggregated? Does one of them get aggregated? What about the indices? When you put things together I can ask fifty language decision questions. You can't just answer them one by one. Every answer you give to one influences the five hundred other questions that arise. It just turns out that this is a much more complicated language design issue, and tool construction issue than in the non object-oriented case.

Mr. Lansing: Richard Palmer is a physicist at Duke University, and also at the Santa Fe Institute. He'll be talking on a physicist's view of dynamical systems.

Mr. Richard Palmer: Something like that.

Mr. Lansing: Something like that. Alright.

Mr. Palmer: Thank you for inviting me. Some fascinating talks today. I've really enjoyed this meeting. I was asked - what was I asked to talk about? This is a modeling - description of modeling session. The original title had something about statistical methods. So I translated that into statistical mechanics ideas, which was the nearest thing there that I do to statistics. But it wasn't quite following what I was asked. But in another way I am going to follow what I was asked. That is, Yaneer asked us all to first sort of outline what we do - the big ideas - and then give examples. That actually is my plan for the afternoon. This is mainly directed to nonphysicists. The physicists should either leave, or be bored.

I want to start off by a quick introduction for nonphysicists as to what statistical mechanics is, and my views on it. Some of the failings of statistical mechanics as applied to complex systems. Then I'll go over some of the more positive things that I think come out of that; I call it "Attitudes of a Physicist". We turned it around instead of "Physicist with an Attitude". Then I'll talk about a number of examples. I'll try to do a little bit of each of these examples to try and drive home some of the points I'm making in the first part.

This talk is actually very different in attitude, or my attitude, I guess, than the previous two. I see complex systems very differently. I'm coming from a scientific, physicist's point of view trying to see something simple beneath all the complication of the world. That doesn't apply if you're going to look at something in a complex engineering problem, which I can't say much about. This is more a scientific attitude of trying to understand simplicity beneath the complexity. That will become clear as time goes on.

First, the quick introduction for those that don't know statistical mechanics. I think of statistical mechanics, or statistical physics at two levels. There's a formal part of it - standard methods, and so on, that don't work too well in a lot of complex systems - but then there's a much larger set of ideas that come out of that. A rule book or an attitude that I think do apply.

First, let me talk about the narrow part. What statistical mechanics really does, in my view, is to try to compute the behavior of usually a large system - lots of agents, lots of particles - by in some sense averaging out all the irrelevant stuff (throwing away all the details that don't matter). That's my overall view of what statistical mechanics does. One way of doing that comes from the so-called ergodic hypothesis. That, loosely speaking, says that a physical system in equilibrium will visit every possible state it could, given long enough. Ergodic means wandering. If you want to look at a long-time average of what a system will do, you just average out all the possible things it might do. That's the sort of basic principle that underlies a lot of the success of statistical mechanics from Boltzmann, and Gibbs, and onwards. But which breaks down, as we'll see, for complex systems.

An example would be the gas in this room. You don't need to know where each and every little oxygen and nitrogen molecule is, and which way it's traveling, and all those details. You could imagine writing down and running a gigantic computer simulation, because you really wanted to accurately model the gas in this room. That's crazy, because all that really matters for most of the properties are just the temperature, and the density, and a few other properties. It wouldn't matter if the molecules in this room were hexagonal, or some sort of weird shape. It doesn't matter. There's a lot of detail there you can throw away. There's just a few things that do matter.

There's the example of a magnet with little atomic spins here, or little atomic level magnets that are below a certain temperature tend to line up. If you have a magnetic field all you need to do is know what the magnetic field is, and what the temperature is. A few macroscopic, large-scale things like that. Then you can average over all the detail things that the individual spins might be doing.

The more technical part of statistical mechanics at that level is really a way of averaging over all the irrelevant stuff while keeping the relevant stuff. The key is deciding which is which.

Let me just say a few more things about statistical mechanics. There's lots of clever ways of doing all this. There's the standard textbook methods that tell you how to calculate these averages. In fact, what you usually do is calculate something called a partition function, and divide everything else from that. I'll show you the first part of a text on statistical mechanics by Richard Feynman. This is page one of his statistical mechanics text where he says, "The key principle of statistical mechanics is as follows. If a system in equilibrium can be in one of N states, then the probability of the system having a certain energy EN is 1 over QE to the minus EN over KT." There's a Boltzmann factor there. Where this Q (which I usually call Z) is just the sum of all the possible states of an exponential of the energy with a KT. T is the temperature and Q the partition function. Then he goes on to say, "This fundamental law is a summit of statistical mechanics. The entire subject is either the slide down from the summit, as the principle is applied to various cases, or the climb up to where the fundamental laws divide, and the concepts of thermal equilibrium, and temperature, T, are clarified." He goes on from there, but that's the essence that we learn in school (at least in physics school) of what statistical mechanics is about. He mentions later on that one of the issues is what is equilibrium. Of course, most complex systems are not in equilibrium, so that's part of my point. He mentions a phrase I like, "If all the fast things have happened, and all the slow things not, then a system is said to be in thermal equilibrium." The obvious question about that: what's fast, and what's slow.

As was said on Feynman's first page, there are lots of clever techniques. There's a hundred years or so of developing clever techniques from actually doing this. It usually relies on having some fairly well-defined Hamiltonian, or energy function. But you can adapt that idea to other areas such as fitness in biology, or cost function in computer science, or error functions, and so on. You can do the same sort of techniques, to some extent, if you have all the other requirements. Not just physical systems, but all sorts of other things. There were still a lot of people who believe that all you need to do ultimately is calculate this partition function, and divide everything from it, and you've understood any system of interest. I call them statistical mechanics fundamentalists. A couple more points. Some people say well, is this just probability and statistics? Is it just doing the probability? Is it something any mathematician would do? Yes and no, but principally no, because you need to add probability and statistics, an awful lot of techniques, an attitude I'll come back to, approximations, various methods, paradigms, and so on. Many, many years of accumulated techniques; often not mathematically justified. It goes a lot beyond just doing the obvious probability things. Statistical mechanics is more than that.

Another question I'm not going to get into now. How could you justify this procedure? I just pulled that procedure, or Feynman pulled that procedure out of a hat, and he said half the story is how you climb up to that summit. There are various ways to do that. You can pretend to derive it from mechanics. Pretend is the operative word there, because it usually doesn't work, except in the very simplest systems. You have to make lots of assumptions. It can be done at various levels about ergodicity, or about metric transitivity

The other way of doing it, which I prefer, is to know all this about a system, and then make your best guess as to what will happen. That's something that can be quantified in terms of minimum information, or maximum entropy. You come up with all the standard statistical mechanical formulas in all the relevant cases. Coming to it from saying let's be honest, we're just guessing. We're taking into account everything we know, and beyond that we want to be as least biased as possible. That is, put as little extra information into the problem as possible. Then you come out with answers.

So there are two ways of trying to justify statistical mechanics in its formal sense.

That was the introduction for the nonphysicists. I see the physicists over here smiling like crazy. Yes?

M: I liked the Feynman quotes.

Mr. Palmer: You liked the Feynman quotes. Okay, good. Let me now mention some of the failings of statistical mechanics, particularly in the narrow sense that I'm using it at present, when applied to complex systems. I could talk for a long time on these things, but I'm going to keep it fairly brief. Some of these will come up again in a more positive framework.

Firstly, most complex systems that we're interested in are nonequilibrium systems. They are driven systems. Alfred Hubler talked yesterday about the importance of looking at turning up the flux or the energy flow, or matter flow, information flow through a system. If you turn it up high enough, it becomes complex. That's his definition of a complex system. I think it's one of the important factors of what makes a system complex.

There are some statistical mechanical methods that can deal with nonequilibrium, but they're a lot more complicated. Applying equilibrium methods doesn't usually get you very far, except to see what are the stable states that a system might get to if left alone to become an equilibrium. But if you're thinking about a living system or something, the equilibrium state represents death, for example. It doesn't represent a living system. One of the projects I've been working on is modeling markets, like stock markets. They're computational models. And we find in these models that markets can do one of two things. There actually seems to be a phase transition between the two.

They can do something very like real markets in which there's lots of volatility, and there's all sorts of correlations, and the agents do technical trading, and learn, and change, and so on. Or they can do something that looks exactly like what the economic theorists say that markets should do. You probably heard of random walk markets and efficient markets hypothesis. Which is that there's not much trading at all, and there's really no information in the price that you can use to predict anything. There's very few correlations, very little trading, and it's sort of a much simpler system, actually, with much less volatility. Essentially that's the equilibrium case. Our model, interestingly enough, can either do the equilibrium thing, or can do the real thing the real markets seem to do. There's a transition between the two. It's a whole long story about what causes it to go to one regime or another.

Another thing that's a problem with statistical mechanics itself is what's often called "broken ergodicity". That the ergodicity that's needed - that is, a system does everything it might do, given long enough - just doesn't work. A lot of physical systems get stuck in some substate, or subspace of the space that they might possibly explore. That's sort of obvious. My attitude is this is not in the least bit surprising. What is surprising is that some systems do explore almost everything. Ergodicity is surprising, not the lack of it. If our computer chips were ergodic we'd be hopeless. That would mean the internal state of the chip would just depend on the external voltages on the wires that went into the chip. But having internal state is absolutely crucial to something. Almost any system of interest these days, having some internal state that depends on a past history, and what you did to it makes a system interesting.

Phil Anderson was the inventor of this idea of ergodicity. In terms of what Phil was talking about this morning, it's this broken ergodicity, or in certain cases broken symmetry, that really is responsible, to a large part, for emergence of new patterns, new orders, new structures, new universes. From the plasma of the Big Bang, how different particles condensed out, and so on. In each case it's that they're not doing everything they could do, they break the symmetry, and the system does fewer things. These choices get locked in. Frozen historical accidents, if you like. If you want to tell the history of the universe, you have focally to look at broken ergodicity, and broken symmetry, and so on, as new orders, and patterns, and so on develop.

Now a problem that really relates to the sort of physicist's attitude that I'm going to express in just a minute or two. A physicist tends to have this attitude of universality, such that if they take the simplest case of something, that will represent the typical case. I'll come back in more detail to that. It's not at all clear that that works for a lot of the complex systems that we look at. It's a question mark. If it does work that there's a large class of things that are more or less equivalent, then you're well excused, indeed well advised, to look at the simplest case of something. That tends to be a physicist's attitude. It's my attitude, when I can get away with it.

Another problem with statistical mechanics is that the agents tend to do a lot more than little atoms. They tend to do things like planning their own trajectories. If they're economic agents they have foresight. They're thinking about the future. They're choosing their strategy. That makes things a whole lot more complicated. They may be making internal models of the world. In a certain sense, the whole world may be influencing the parts. It's often useful to think in terms of a whole influencing a part. Downward causation. At one level it's all just atoms, and molecules, and so on. In my example of the stock market, our agents are trying to get rich, individually, as it's told to them in the computer. What they do to get rich is buy and sell stock. Why do they buy and sell stock? The price of the stock is varying, and they're trying to model when it's likely to go up, and when it's not. If they're good at this, if they have good internal strategies - which are based on various linear predictors, and genetic algorithms, and such things - if they're good at that, they make money, and if they're not, they don't. Why is the stock price going up and down? That's because the agents are buying and selling stock. The whole thing is circular. The agents buy and sell stock, which makes the price go up and down. The agents are trying to predict what the price is doing, and make money on it.

The agents are responding to what I would call the whole in this sense. The price of the stock is an emergent property of all the activity of the agents. The price emerges from the agents, but then that price in turn influences the agents. It's sensible to think in terms of the whole. In this case the price of the stock influencing the parts, the agents.

A final conclusion I'll come back to strongly. It's not a problem so much as it's the best you can do. If you try to apply statistical mechanics methods to complex systems, you usually find that you can't really treat specific instances. A lot of complex systems have a lot of specification in them. Physicists would say there's a lot of parameters or information in the Hamiltonian. There's a lot of specification. There's a lot of particular instance. You're typically not going to be able to treat a particular instance. What you can often do is to treat some sort of average over instances. You look at trends, or average trends, or averages or distributions of things, not particular instances.

That was part two of the talk; the failures. Part three was the attitudes of a physicist, and then we'll get to examples. This is taking statistical mechanics from a broader frame. Initially I started out thinking in the narrow sense of computing partition functions. This is the broader frame of what someone like I think about, or the way we approach science. If you're trained as a statistical mechanic, you typically have a lot of these attitudes. They contrast fairly strongly with the attitudes of some scientists in other fields. I think it's worth making them explicit.

First, idealization. What we try to do, typically, is leave out as many details as possible when we're trying to understand. This is trying to do science. This is not trying to design a jet plane or something, in which all the parts have got to be there. We are trying to understand why nature is the way it is in a certain regime. We try to leave out as many details as possible. Dyson, I think it was, said, "A model is done when nothing else can be taken out." Not when nothing else can be put in, but when nothing else can be taken out while having the essential essence of the phenomenon we're trying to understand. The more simple a model is, the more you tend to get insight. This is true in the physics world. The question, of course, is how true this is in other worlds. I'm not trying to say that this physicists' attitude is the one and only attitude, and the right way to go in all cases. Of course its not. I'm a great believer in diversity of approaches. I'm trying to explain my approach, and the typical approach of many physicists. I think it's a useful one in many cases.

More specifically, in lots of situations, some variables are relevant, some are irrelevant. This turns out to have a technical meaning. As I said earlier, in statistical mechanics the problem is deciding which is which. I talked already about gas molecules versus gas laws, and magnets, and so on. Technically they're important parameters in a problem. They're called order parameters if you're a physicist. The magnetization is an order parameter in the magnet. The interesting thing is that in complex systems, new order parameters can condense out through time as systems self- organize.

There's a very technical part of physics that deals with these things about relevant and irrelevant parameters. I'm not going to go through in detail. Renormalization group is a way of exploring parameter space. If you imagine you have some model, and it has lots of parameters in it - not just numbers that you could twist and turn, and adjust with a potentiometer, but other switches with which you could turn on this effect, or that effect. You could imagine a wide range of possible models being phrased in some general model with lots of parameters. The physicist has this image. As this idea of renormalization group lets you map one setting of the parameters onto another. We think of that in terms of parameter space. In the space of possible parameters there's a mapping from this point here to that point there, so that the systems with those two sets of parameters will have sort of equivalent behavior. I have to say sort of. Some things will be equivalent. Important things will be the same. Some less important details may be different. Particular numbers of things will be the same. It's like you're in the same general category of behavior. You haven't gone through some bifurcation to some totally different behavior.

Here's an imaginary picture of parameter space. This blue region here is supposed to represent some vast space of possible parameter setting. Including lots and lots of different models of things. These green lines I've drawn, renormalization group trajectories. I talked in terms of mapping from one group to another; but there's a way of making sort of a continuous line.

The point here is that if I follow this line here, all along this line you're going to find an equivalence, in a certain sense, of the outcome of the model. It's not going to do something drastically different. Important numbers like power law exponents probably won't change along that line. Some other details may, but it's the same qualitative behavior, anyhow. This is true in a lot of systems in statistical physics. It's been particularly used for studying phase transitions, but it's a fairly general idea.

If you follow these trajectories in some vast space, what can happen? You can go through the usual list. It turns out you can't get cycles and things. You basically get limit points, or fixed points, as they're called. These trajectories flow into certain fixed points, and for the different ones, then they must be separated by what are called basin boundaries. You have these basins of attraction, and basin boundaries between them. In these different basins of attraction, there's different behavior going on. Within a given basin, which is called a universality class, in a certain sense there's very similar, qualitatively similar behavior. Of course if you try and cross this with your parameters, then you see some bifurcation or phase change to a different sort of behavior.

This is deep in the heart and mind of most physicists, certainly coming from statistical mechanics, that there are these universality classes. Once you have this picture, if it's true, then of course you don't care about exactly where you are within this class; you choose the simplest example of a model within that class. That justifies, in a certain sense, what I was saying initially about simplify, simplify, simplify. Idealize as much as you can. Throw out all the details, while keeping in the same universality class.

The question, which I raised earlier, is whether this sort of picture is at all applicable to the sort of complex systems that we study. I don't have an answer; I want to raise the question. Someone mentioned earlier that raising questions was the important point. Of course it could be the equivalent parameter space breaks up like this into millions of different regions, so you can't move anywhere without getting into a different region of behavior. Every bit counts, if you like.

In which case this whole picture, and the physicist's attitude, is useless. Tell all physicists to go home, and you've got to study the particular system. But physicists at least hope that this isn't true

One last set of attitudes. Forests and trees; as in not seeing the forest for the trees. It often may be the case that the simplicity that physicists love to look for, the generality, the regularity, may lie on a higher level than the particular system you're looking at. I'll exemplify it in a minute. It may even be at an aggregate level, or a distribution. It may not be seen in an individual case. Some regularity may only be seen where you average over a lot of cases, or take a distribution of cases. That leads you to do averages, look at distributions, look at the way things scale as you change the parameters. That's not the same as looking at a particular system.

Let me mention Per Bak's example of a sand forecaster as an example of this. This is always controversial. Per Bak, probably most of you know. A statistical physicist, he co- invented (with Tang and Wiesenfeld) this self-organized criticality idea. In particular, the sandpile model is a very simple model of that. The image, of course, is a pile of sand. You're dribbling grains of sand down onto this. Avalanches occur at certain times. It builds itself up to a so-called critical state in which you get large avalanches, small avalanches, medium avalanches, and so on, with the distribution of sizes, it's a power law. Has no intrinsic scale. Okay.

Per Bak in his book, rather modestly called How Nature Works tells a story of a sand forecaster who's living in a sandpile and is on the news every night in the sand world saying what's been happening. Here's a story he tells one day. "Yesterday morning at 7 a.m. a grain of sand landed on site A with coordinates such and such. This caused a toppling to site B at such and such. Since the grain of sand resting at B was already near the limit of stability, this caused further topplings to sites C, D and E. We have carefully monitored all subsequent topplings, which can easily be explained and understood from the known laws of sand dynamics, as expressed in simple equations. Clearly, we could have prevented this massive catastrophe by removing a grain of sand at the initial triggering site." Everything is understood. Now from the point of view of people living in this world, that's not so stupid, actually. You may indeed want to prevent avalanches in your neighborhood. That's the "not in my backyard" attitude.

There are several morals of this story. One is being able to give an historical account of something like this after the fact is not equivalent to being able to predict it beforehand. More importantly, this detailed history level of what happened in a particular instance, a particular case, a particular time, totally misses the higher level regularity - this whole power law distribution of the sizes of avalanches. If they move this grain of sand, there's going to be another avalanche somewhere or other. It's arguable whether it would be helpful to this population of people living in the sandpile to know the distribution, but at least it is to us, looking at the outside. The interesting thing about this sandpile is not the particularities of one avalanche, but the distribution of avalanche sizes, and the fact that the system self-organizes to get to a particular case. Per asked the question of the sand forecaster. You can describe these individual avalanches, but why couldn't all avalanches be small ones? The sand forecaster has no understanding of how to answer that question, while living in the sandpile as such.

Enough of the generalities and attitudes. All of what I've been saying justifies - at least in systems where the statistical physics sort of approach works - looking for simple, generic models. Even if they're not, in some cases, immediately testable, realistic, comprehensive, or specific; the idea is looking at toy models and trying to get some general understanding out of them. Often this is at the level of averages, or aggregates, or trends, or scaling distributions. Another way to say this, coming back to my point about simple and simplifying, is we should not be looking at the kitchen sink approach where it's best to put everything in but the kitchen sink. Instead we need the KISS approach. This is an Army phrase: Keep It Simple, Stupid. Keep it as simple as possible. Peter Schuster says it nicely that our models should be a caricature of nature. Emphasize certain features that we want, but downplay others, and have just a few details.

That was the general overview. What I can do now is talk briefly about some examples.

The examples I was going to look at are here. Statistical mechanics, genetic algorithms, combinatorial optimization, and models of extinction. This could be at least another four hours of talk.

On the first one, statistical mechanics of genetic algorithms. I'm going to assume that you all know what genetic algorithms are. You try and solve an optimization problem using a Darwinian sort of approach. There's been a lot of theory on that trying to understand how these genetic algorithms, or for that matter, real genetics, work. There's been some computer scientists and mathematicians who have set up the whole process as a giant Markov process, which it is. In the computer there's new strings in the genetic algorithm generated from old ones. Unfortunately there's not much you can do with that. You include every detail. Every bit counts. You don't learn very much. You learn that it should have an attractor, or something like that. It should converge eventually. I was actually thinking about this some years ago. But Jonathan Shupiero, and Adam Prugel Bennette came up with a much nicer approach in which they left out a lot of the detail. This is very much in the sense of the statistical mechanics I was just talking about.

What they was to say well, in a genetic algorithm you have a bunch of these strings; members of the population with fitnesses. Call them F1 through FP, P is the population size. Represent that by some sort of histogram. A fitness histogram, or a fitness distribution. Then ask how this fitness distribution changes as you execute the genetic algorithm. So this fitness distribution changes as you do selection, and mutation, and crossover, and you keep on going around from one generation to another. They looked in great detail at how this happened. They've thrown away a lot of the detail just to look at the fitness distribution. Just knowing the distribution of fitnesses, or the fitnesses of two strings that are going to do a crossover does not tell you everything you need to know to figure out what the fitness of the progeny will be from a crossover. But they try to estimate these things for particular cases. When they can't make a detailed estimate they use a maximum entropy approach to make their best guess. They may have to introduce one or more order parameters. But basically they have a lot of success in filling out almost all the details, and just looking at the distribution of fitnesses, and how it evolves through time. They don't even look at the distributions, to be honest; they just look at the first few moments or cumulants of the distribution. They get recursion relations for those.

They work detailed mathematics for particular problems. They're pretty successful in that. They can predict, for example, how you should adjust parameters. How long the thing should take to converge, and so on. Here are the lessons.

I was going on and on earlier about a hundred years of tools; and they use a lot of these. They use a bunch of clever techniques that have been evolved in the field of statistical mechanics to do this.

Knowing this fitness distribution doesn't answer all the questions, but it does give some sort of general understanding, it does aid in choosing parameters in your genetic algorithm, and so on. You can actually estimate some other quantities. Here's the sort of same old message again. Approximation, idealization, abstraction, averaging, and so on work in this case.

The next talk was on combinatorial optimization problems, such as the traveling salesman problem. Many of these were attacked by simulated annealing methods. That's not what I'm talking about now, though. The general case is that we have some problem of a given size, and we have a particular instance. We have configurations of that instance.

Think of a traveling salesperson. N is the number of cities, I is a particular placement of cities, X is a particular tour through those cities, and that has some associated cost. What we want to do is to find the tour that minimizes that cost. I have a few graphs here for lots of different examples of the traveling salesperson. There is the standard thing. Trying to minimize the tour around the cities.

So what do we actually want? (this is different, depending on who you talk to). Ideally, if you're really a salesperson who's got to visit all these cities, you want to find the shortest route. Curiously enough, when I was eighteen, between high school and university I did exactly this problem. I worked for a local building supplies company on their computer, among other things; and their problem was how to route their trucks around England from their suppliers to their local distributors in each town. The Marley Tile Company, for anybody who knows any British companies. Indeed, a big part of this was actually the traveling salesperson problem of minimizing the total route length, with lots and lots of constraints. That certain towns close early on Wednesday afternoons, and such things; and certain drivers are drunk between certain hours. All sorts of constraints. This was all coded in. So if you're in that business, you really do want to know the tour, in this case, at least roughly. You may not want the absolute minimum, but what minimizes the cost function for a particular instance. You might want to know how many there are too. But if you can't do that, you might like to know what the minimum tour length is. These later ones are not useful if you're the Marley Tile Company, but they're useful to a physicist or to a scientist trying to understand the problem.

You might want to find the value of the minimum, or you might want to know the average length for some ensemble of instances. This is actually the sort of thing that statistical mechanics can do. Now rather than take a particular set of cities, take cities averagely placed in some area, and then try and calculate what the minimum is. You can actually do that. If you can't quite do that, you can do the very last one, the limit of that, as you take an infinite number of cities. A bunch of people, including Philip, and David Sharington, were involved in calculating this using quite sophisticated statistical mechanics methods. That's the sort of thing that you can calculate. Here the moral is, as I've said earlier, you can't do particular instances, but you can sometimes do averages.

The next talk was going to be on genetic regulatory nets. These last two are getting away more from the sort of formal stat mech, but still keeping these sort of ideas of the statistical mechanic. This is Stu Kauffman's story.

In the genetic system, a lot of genes turn each other on and off by one mechanism or another. Here's a diagram from Albert's, the cell biology book. Here is a particular gene. All these other things with colors and so on here, are things that can control whether it's on or off. Whether it's producing protein or not. This is very common here. This whole length of DNA folds up. Some of these things, if they're there, it's on, if they're not, it's off. There's a very complicated logic to this. Most of the effort in this area is spent in finding out the details of particular cases like this.

You can codify a particular case, in terms of truth tables of. Of course what most people have been doing has been trying to find the appropriate truth tables, and understand why or how particular systems work. Here's some of the roots you might follow in this. The first root is what I was just talking about. Figure out the wiring diagram in a particular case. What Stu Kauffman did, quite a few years ago now, was to say let's back off from this particular instance, just as we did in the traveling salesperson problem, and explore general features of switching networks - of things that turn each other on and off with logic gates. What sort of general features can we find? So he studied what he called random Boolean nets. Switching networks, basically, but with random connections, and random truth tables, and on on.

The third thing you might do is look at this experimentally. To see what's turning on and off, and try and reverse engineer the switching network that's doing it. That's actually very important right now. Future technologies can rely on this sort of thing. It's called postgenomics. A lot of medicine in the future is going to rely, I think, on interfering with these switching networks in subtle ways. Add that chemical for a second, then that, and then that, in some special sequence so you kick it into a new state. There's a lot of medicine, and so on, going on with this thing now. But I want to talk for a second about this middle case, the random Boolean nets.

What Stu did was to look at the random case. The important result was that to get the desired properties to match with something that could conceivably be appropriate for a sale. Going through sale cycles, and having a bunch of different sale types. He needed two input nets. So these are switching networks with two inputs per logic gate. That was the only case that worked, until he started biasing his truth tables. Then he found he could get the same thing with three to five inputs. But in any case, for these nets that had these appropriate biological properties, these things go through cycles. You put them in some state, they go to another, to another, to another, and then come back. Stu interpreted, rightly or wrongly, that this is the lifecycle of a cell, and different cycles corresponding to different cell types. We know we have lots of cells in our body - liver cells, skin cells, toenail cells - that have the same DNA, but do different things. The general assumption is they have different things turn on and off. They're different states or cycles of these switching networks.

What Stu found was that in the cases that were biologically relevant, the way that the number of cell cycles, and indeed the length of these cell cycles, scaled with the number of switching elements, number of logic gates, number of genes, was the square root. This is the sort of thing that you're never going to see if you look at one particular cell, or one particular genetic network; this sort of regulatory of the whole. He found it by stepping back from the particular problem to the general case. Then of course he made a plot of the DNA per cell, which roughly corresponds to the number of genes, and plotted against that the number of cell types, which he'd interpret as different cell cycles, over animals ranging from tiny little bacteria up to cells, there's a nice straight line on a log log plot. Indeed it turns out it has slope very close to a half. That is, the number of cell types does scale as the square root of the amount of DNA.

You can believe or not believe his story, but it's a very pretty one, and it's a nice example of how we can step back, and use the sort of statistical mechanics approach; go to the general case, and maybe learn something from it. Not about a particular case, but about the general case.

In conclusion, in genetic algorithms, we figured out we need to pick the right variables, and throw away most of the detail, just looking at those distributions. With combinatorial optimization, we could only compute averages, but the statistical mechanics approach does work for that sort of thing. With genetic regulatory nets we see highly idealized models, but that nevertheless give some sort of trends in how the scaling goes.

If I had time, I would have talked about extinction models for the extinction record of the dinosaurs, and all the little extinctions in between. It turns out that if you plot the number of extinctions per geological period, and you follow distribution, you get a power law for that. People have been trying quite hard to explain that over the last few years, including with some very simple, highly idealized models again. They do seem to give distributions, and some of the right scaling. In actual fact, though many of these models, including the famous Bach-Sneppen model, do not fit the data. You can actually eliminate them because they give the wrong exponents. You can prove that their possible exponents, the power laws that they give are outside the bounds of the data.

That was my last point. That even though some of these statistical mechanics models are very idealized. Some people say they could fit anything. In fact, they can't. Even some of these models that are actually testable are indeed falsifiable. That's another whole story, but I need to stop there.

Mr. Lansing: Here now is Gerald Sussman, from the artificial intelligence lab at MIT.

Mr. Gerald Sussman: I am an engineer, and now we will talk about engineering. This is not about averages, or about explanations of things that happened once. The whole idea is this: if you think about the progress of technology in the last very short time, the exponetial pace, it seems like it won't be long before your personal computer will be the size of a grain of sand, and you'll be able to make them by the barrelful. If you make them by the barrelful, then what do you do with them? What we're worrying about here is a particular effort which we call amorphous computing, which is trying to understand the methods for obtaining coherent behavior from ridiculously large numbers of objects. Objects that are full universal computers that talk to nearby neighbors. They can't talk to very far away neighbors, they're limited by physics. There are so many parts that they don't have any names. There's just billions of them, if you want to get into that mode. Our job is to figure out how to instruct them to do work for us. As an engineering goal, that's quite different from observing what might happen.

I'm going to talk to you about the particular development of engineering principles and languages that could be applicable to such a myriad of particles that do things, such that we can get them to do particular jobs for us. And indeed, the reason why this is important now, is that we've got the confluence of microtechnology of the silicon type, and molecular biology; These are both ways by which we can manufacture things that are precise and specific. Even if they're bacteria, by golly, they're just part of the stuff out of which an engineer builds things. We're looking for a basis for a new kernel technology. Why might this be interesting? You might want to think about the possibility of buying concrete by the megaflop. Imagine if you had stuff - material - that came, as part of its ordinary engineering properties, with computational power and actuation in it. We had a graduate student long ago who made an active bridge. It sat there and made itself stronger by applying little restoring forces to prevent buckling of the compressive members. That was cheap. You could imagine making much better engineering designs by replacing strength and precision with intelligence.

The particular project I'm talking about today is the invention of programming paradigms and languages. That's the critical thing. You can't do engineering with a synthetic problem without having a language to express what you want to do and what you want to build. We need language for describing what we're doing, for controlling amorphous computing agents. We're going to do a few prototypes here as well.

What's an amorphous computer? Here's an amorphous computer simulation.

What you see here are dots. Each one of those dots in this picture symbolizes an individual little computing agent. Perhaps this computing agent can talk to a few neighbors. We visualize these by making colors that say some particular part of the state of this machine is interesting to us. So green might mean that this particular machine feels unhappy. It's at least trying to tell us something. We, as programmers, are going to decide what colors they take on. This is just for simulation purposes right now. The critical thing is every one of these machines has the same program. It's structurally identical to every other one, except some of them may not work. When you buy a gallon of 8086s, ten percent of them are going to be bad, if they're from a good manufacturer. That's what you're seeing there. This is not cellular automata. They're not laid out in any pattern, in any regular array, nor are they synchronous. Okay? That's very important. This is a very different situation. There's not much you can do. You don't know what's North, South, East, and West. The model is that there are computing elements sprinkled on a surface, in a volume. There are too many to individually program, or name. Each talks to a few neighbors, but not reliably; and it's not synchronous, nor regularly arranged. It sounds like you can do almost nothing with this. But you can buy it by the gallon.

How are we going to program it? One of the things that an engineer does is makes up stories. The stories may be false. It's very important to have mythologies by which you decide how to do work. So we're not going to talk about real biological mechanisms, although we do have ideas that come from biology, that may be complete nonsense if we look at them from a biological point of view. The cartoon caricatures of biology are sometimes useful for organizing this intellectual endeavor.This was done by one of our graduate students, Veronica Nadpaul awhile ago. Supposing I wanted to make something that was a bifurcating tube of some sort. One thing I could do is imagine having a weak membrane that bounds a pressure vessel, and as the pressure increases and the bulge expands, the little part gets weaker where it's expanding. And then that bifurcates. And this automatically happens. Unfortunately, that's not amorphous computing. That's an amorphous computer simulation, but amorphous computing doesn't do that because I want to know what thing I'm going to get. In other words, what I got there was one of a large number of possibilities. I couldn't predict, a priori, what it was going to be like, and real engineering depends upon me knowing what I'm going to get. I have to be able to predict, and therefore design, the structures that are being built. I'm going to tell you about the efforts going into how we can capture, control, and design this stuff.

Supposing, for example, you wanted to make something with a precisely specified geometry. Here's a nice example of something with a precisely specified geometry. It's got all these carefully constructed interfaces. The shapes of these devices are carefully controlled. They have to be within the right tolerances to do the right thing, such as bend over.

How do you make things like that? One particular idea is there is sort of a local SIMD paradigm that was started by Ron Weiss last year. These computing elements include some binary markers, and they tell you so many details. Each computing element's program has many independent rules that manipulate these binary markers. They're triggered when a message is received.

If a rule is applicable, i.e., if a certain Boolean combination of markers is satisfied when a rule is applied, it may set markers, and may send further messages. The messages may have hop counts to determine how far they will diffuse, and the markers may have lifetimes after which they expire. Now there is a description at one level of detail of the language. It's not hard to program such a thing. Start out with a polarized material. All these particles have exactly the same program in them, but some of them have been initialized to have a different initial state. All these yellow ones are the same initial state, the green ones the same initial state. This is a very unstable device.

What this thing does is it produces a sequence of waves which slowly expands and builds what we would hope to be something like an alternating structure - a structure of segments. Interestingly, that structure does not depend on the details of which neighbors can talk to each other.

The program for doing that doesn't look very complicated. There are these descriptions, the set of messages, and this is the test for the various markers, and then how certain markers get changed, and how a new message is sent. There are little rules that look like this.

You can make a more complicated system that also causes growth to happen. I won't tell you any details about that, except for the fact that languages have to have the property that they are extractable. Think of it this way. You build things out of resistors, and capacitors, and inductors, and transistors, however, at some point you want to make amplifiers, or maybe TTL. Then you build things out of the TTL. You have to have layers of detail with which you can describe things.

Here is an example at the segment level of what's happening. This is a timeline, and this is the segment number of the thing that's being constructed. We've abstracted what you just saw into two rules. A C over an R turns into an R over a C. One of these things turns into one of these things, and this is how the stuff propagates.

Now, whenever you start building things, you need some sort of primitive. What are the things you can do? I've given you some high level ideas, but let's talk now about the simple things that you could possibly do.

One example of a simple thing, is wave propagation - that simulates the diffusion of material in a biological organism, for example. Somebody puts out some goodie, and then that goodie spreads, and then maybe it counts down as it goes. Meaning the density or the concentration increases as you go out. Then there's things you can do with that, and then we can do things like making coordinate systems, refining coordinate systems, activation, inhibition. Eric Rauch has been worrying about conservation and exchange processes. Veronica and Daniel have also worked on communication network construction.

Here's a nice little piece. If you start out in the middle, and this guy puts out some signal, then there's a count in the signal, and the count's down, you would get a wave. If you were looking at some sort of parity for the colors of how far away it is; you get a wave. If you have two such waves, one coming from this end, and one coming from that end, you end up with a slightly nonorthogonal coordinate system, which is reasonably good around here, but you know you can do better. If you use Laplace's equation, you can turn nonorthogonal coordinates into orthogonal ones. LaPlace's equation can be approximated by everybody asking his neighbors what's the value, and then says, I'll make myself the average of those. If they do that fast enough, you get some sort of smoothing, and orthogonality out of it. There has been some work done by Kevin Lynn on how local coordinate systems can be combined to make a manifold so you can talk about global properties of a structure.

There is also other sort of primitives in our bag of tricks. There is the famous reaction diffusion, which is not very good for talking about big structures. They have very complicated phenomena going on in them where I have to control carefully how things are connected together. But they're pretty good for making little things like patches of dots or stripes. If you need some of those, those are probably the right thing.Veronica and Daniel have also been working on building communication networks in these things, which are sort of like cellular networks, but they have to self-organize.

Now you've seen the things that you can put together to make bigger structures. You have to produce the TTL data book of amorphous computing tricks. I suppose you need that sort of thing. By combining these ideas, there is a way of getting detailed topological control of what's going on in such a system. I'm going to show you some work from Daniel Coore.

Here we're going to have a caricature of biological differentiation allowing us to impose structure on an amorphous network such that it builds a bunch of C-Mos inverters. That's a tall order. This is not a good way to make C-Mos inverters, however, that's not the intention. The intention is that electrical circuits are specific, well-defined objects that we can tell whether or not we've got one, because we know what it should be. It's a proof of the attempt to do something deliberately, rather than as a matter of accident that you get it.

Here's the basic idea. What Daniel did was he organized the process in terms of a botanical metaphor, where there are growing points. They are a way by which stuff passes along. What's the stuff? The stuff is activation of some sort. Remember, all of these machines are programmed identically. There has to be some initial condition that starts it going. The program has to, for example, have a little operating system underneath that allows the diffusion to happen. That's automatic. Then there's going to be a specific control program that makes structure. This control program that makes structure is basically turning this parallel process into a serial one. The way to do that is by saying even though it's happening in parallel all over the place, I'm going to say what happens in a particular place that's growing. The design rules are going to be implemented, in this case, by aversions and attractions.

We start out with some polarized material with an initial position. This has got two power rails that were just put there, but they can be grown now. The poly contact sprouts a growing point that bifurcates and grows toward the pheromone secreted by these two guys. It's sort of a strange way to think about it. Then when it gets close to them it bends over, because it doesn't really want to get too close. It doesn't like the smell when it gets too close. Then after it dies out, a part over here sprouts a growing point that goes this way, and this way. Another one sprouts going down and up. When these hit the power rails, they produce contacts, and the middle ones get attracted toward each other. They hit, and they produce a contact. A polycontact starting a new inverter, which starts growing; and this process repeats, growing the next inverter.

Here's an overview of a nice thing that you can do with an amorphous computer. If you let this run for a long time on a simulation, you can make yourself a long string of inverters. It turns out any particular circuit you like.

Growing point language, that Daniel will tell you about, is capable of doing lots of things besides making circuits. You can draw various types of geometric diagrams. Also, you can do Euclidean constructions to some accuracy, depending upon the density of these particles.

I've so far told you about caricatures of biological processes. I'm now going to get everybody angry by telling you about real biological processes. One of the things we're doing is attempting to take charge of real biological processes so they could do work for us. For example, one of the things I'd like to do is make bacteria that will make particular machines for me. Or they might just do something very simple, like make a special material. I've always wanted to have something like wood, but which was molecularly perfect. Wood is great stuff, but it has knots, and it's not really perfect. Wouldn't you like to have stuff that was molecularly perfect? We might want a material made out of murine, with appropriate crystals put in the right places. Wouldn't it be very nice if we could just program the bacteria - get a vat, put some sugar in, and out comes this stuff. That's the kind of thing I hope that the future has in store.

I'm going to tell you about the efforts to take control of biological cells, mostly led by Tom Knight, with various graduate students, and staff here. The idea of this is the opposite of what we just talked about. What we heard was there was someone who was interested in understanding, for example, what happens when you have a random distribution of logic gates. Or, for example, what logic gates are involved in a particular cell. Instead of that, I'm interested in the opposite - the synthetic question, the engineering question - compiling logic circuits into pieces of DNA that encode genetic regulatory networks where logic signals are represented as concentrations of DNA binding proteins. The action of the regulatory network implements the desired logic function. We want to run Microsoft Windows on an E. coli. It's a little hard. You have to have a big E. coli. And it would be about as reliable.

The basic idea is very simple. If you want an inverter, you have concentrations of DNA binding proteins. Here you might have a DNA as a fragment that encodes a gene that makes Z. Here's some RNA polymerase that's going to pop down here, and attempt to cook out some Z. Now of course it goes through MRNA translation as well, but I'm not going to talk about that. If I happen to have a repressor sitting on this site, then it prevents that from happening.

We did a simulation first of tetramer DNA binding proteins, because the tetramers have the right nonlinearity, if you look at the kinetics, to produce an inverter curve that's robust and that doesn't depend upon a lot of the parameters. We did some tetramer examples, and we saw how you could make simple logic circuits. Here's an example of a simple one. There is actually some people cooking stuff up on the ninth floor of Tech Square trying to make real things like this. That's George Holmsey, Ron Weiss, Dick Papadakis, and Tom Knight. I stay away from that part of the lab. I have very strong aversion to biological materials. I like electrical stuff. Electrical stuff has the property that it doesn't come out, and get you. It's very hard to hurt yourself with physics; but biology and chemistry is much more serious.

Anyway, this is the mechanism Tom Knight and I worked out. If we have a gene, BG, that produces material B, and there is a DNA binding protein A, N being four, it produces a combination BGA, and that's not active. Similarly, if B is another such binding protein that makes an inactive one, and if these disassociate, because all reactions are reversible in biology, just like in every other part of chemistry, and if I have RNA polymerase here that takes BG clear, and makes kinetic constant K5, it makes the material B. A and B break down at some appropriate rates, and this produces an extremely nice inverter - one that I would actually want to build something out of. It turns out that there are some materials that are hard to produce, but I think there's at least one or two proteins that we know of that actually act that way.

A simulation produced by Ron Weiss recently of a ring oscillator using the materials from the Lambda phage system - the Lambda phage repressor that represses CI at two sites - using that, he managed to make a very good simulation using the actual numbers from Mr. Ptashne's book, and a ring oscillator that actually oscillates correctly. I'm sure this is the world's slowest logic. Twenty minutes per cycle, or something like that. But that's alright. If the job that this bacteria does for us is not computing, because that's pretty useless - I'd much rather compute with electronics. But if the job is to make my electronics for me, then I don't care that I just add sugar to this vat, and a month later I pour it off and I've got a Cray 1 deposited. I'd be happy with that. If we use these as second order construction machines, that's fine. And they have a lot of error correction capability, such that if this guy has a bad circuit, he's going to get eaten.

Now that requires cooperation among the various bacteria, so there has to be some worries about doing the communications. How would you get this bacterium to talk to that one? There's a little programming system that's being built now by Ron Weiss called BioSPICE which attempts to say, supposing I have cells in particular geometric places? And I happen to have particular concentrations of particular proteins in those places; and they are allowed to diffuse according to reasonable patterns of diffusion. Can I actually do something with that? That's, by the way, the first step in using the amorphous computing programming ideas, and getting that into the biological application. With this he managed to implement in a four by four grid, two cells. And do this particular nice little discussion, and they have a little latch and some inverters. One cell can send out a message, and the other one will latch it.

They're working on a compiler now, which is a compiler for the turning the logic circuits into actual plasmids, which have real known DNA segments in them. That's beginning to have a library.

It's time for me to summarize a bit. What are we really after? The biological part of this is that we're interested in a new manufacturing technology. Living systems are special. They can self-assemble, they're self-powered, they clean up after themselves, they provide an interface to the chemical world, and you can also pour them out - that's nice.The idea of making little machines that are precise is a problem, though. This problem will occur in the next fifteen years as we make a transition to smaller and smaller devices. Since they're made by a statistical process, wherein you use ion implantation and things like that, the number of impurity atoms that appear in a particular place goes down because the total number of impurity atoms you've got in there is limited due to the solubility of the impurity material in this underlying silicon - since that's a limited value. Ultimately, that means that if you're using statistical methods to make things like transistors, then the variation from transistor to transistor becomes unbelievably bad when you make the transistor small enough. The only ways we know to get around this problem is to put the atoms in particular places.

There may be hundreds of different ways of doing that. We don't know how the technology will break. One possibility is by convincing bacteria to do it for us. They're pretty good at things like that. As I said, if things are made out of one molecule, then they are perfect. I don't care how they're laid out, so long as they're topologically correct. I don't particularly care if the network looks beautiful or horrible. There have been lots of proposals for making things like gates by crossing over pieces of conjugated polymer, and bonding them in the right way. There's a US patent by Mark Reed from Yale on a method of connecting these things to a gold pad so that you can test them. He actually had some designs in his patent for a gate.

Now, nobody really knows how to make these things in large quantities such that they connect together in a particular pattern. If you could train bacteria to make them for you, and put them down in a particular pattern, and lay it out like cells in our body make bone cells, then maybe we can make very dense, perfect electronics. Perfect meaning each object is identical to every other one. That's a long shot, but it's always worth working on things that have the property that one out of five of them is a winner, as far as I can see. In any case, there are still lots of applications for other precision chemical construction, outside of that; and we of course need to do some of that work.

Let me just finish with what the challenge is. I started with this, and I will end with this. The challenge is to reliably obtain desired behavior by engineering the cooperation of many parts without assuming any precision interconnect, or precision geometrical arrangement of the parts, and without any time precision - it must be asynchronous. And to invent the computational substrate that can support this kind of engineering.

One of the problems is assembling such structures. Eric Winfrey has got some really great ideas about self-assembly of biological things. It may very well be the case that by using heterogeneous materials, for example, fragments of DNA that have particular sticky ends in the right ways, one could manufacture a pattern which was in fact the right thing for a little piece of such a system.

M: You talk about engineering style, but there have been engineers who use something that your talk lacked - evolution or fitness improvement. Why can't you just breed the bacteria? You know, programs emerging or evolving out of competition with others?

Mr. Sussman: Just to repeat the question. This has to do with why it is not necessarily the right thing to do to try to evolve things complicated, rather than try to understand them, and build them up; which is what you're really suggesting. I think the real answer is that that sometimes works, and if it works, I'll use it. I'm not opposed to using whatever falls into my lap, but I want to have more control over the situation than just that. I want to make sure that in some situations I can actually say what I'm going to build. Certainly I want it reliable, and precise. And it may not be that I'm talking only about biology. If I'm building, I want to be able to build things that may not be doable by some automatic process that you feed it some sugar, and it produces lots of copies of itself, to some extent. I really care about the fact that we do capture the control of the situation well enough to be really good engineers, where we can specify beforehand what is the problem, and what is the result? And how the result connects to that problem.

Funding

David Litster
MIT
James Anderson
NIH (NIGMS)
Mike McCloskey
NSF

Mr. David Litster: The topic is something that none of you have much interest in. It's funding. My name is Dave Litster. I'm a physicist from MIT. I've currently fallen in with bad company, so I have an administrative position there as well as Vice President for Research. I've known Yaneer since he was in short pants, and he called me up, thinking that I might know something about research funding, and invited me to come up, and chair this session. So I'm happy to do it, and maybe like you I'll learn a few things.

I prepared about five minutes worth of remarks to get things started off. We have Jim Anderson from one of the National Institutes of Health. Medical Science is it?

M: General Medical Science.

Mr. Litster: General Medical Science. And Mike McCloskey from NSF. I thought I would say something from the university perspective. The only university perspective I know is MIT, but I think it's very similar to what's going on in your universities; and I'd be interested to see if it works out that way.

MIT is maybe not completely typical of research universities. Just to illustrate that, this is a plot I like to show where you see total research funding at the institution, and on the Y axis you see how much of it comes from industry. That's the top twenty research universities. Top being measured in terms of dollars, not quality of the research program. And you can see that MIT is rather an anomaly there, along with Kent State. I don't know what that means, but it just means that maybe we're not typical, although I think there's a lot of interesting things going on.

So one of the big things that's been happening is that the science and technology support, 6.1, 6.2 money from the National Science Foundation. This has got some 6.3 thrown in it as well. Ever since the end of Star Wars, the Department of Defense money has been in decline. Some of you may say: so what? At MIT we don't say so what, because it turns out we get a lot of our research support from the Department of Defense. We get more defense money than most schools do. It actually is important for lots of people, and I don't think Congress realizes this. This just shows where the basic research in engineering gets supported. In almost all disciplines, the Department of Defense is an extremely significant sponsor. I think when Congress likes to whack away at basic research in the defense budget, they don't realize that there's a potential for doing a different kind of damage than they expect. But there are real trends going on in national research support. If we leave out the defense support from the federal government, and look at the other federal money, you can see what's going on here.

These black bars are the National Institutes of Health, and the gray bars are everything else which is not DOD. And my guess is that that trend is going to continue. It's having lots of impact. It's influencing the kind of research that people are thinking. Most engineering departments are now thinking they should be doing research in support of human health somewhere down the road, instead of research in support of better weapons somewhere down the road. And I think it means something else that maybe we'll hear some discussion of tonight. If this goes on - and Harold Varmus is among the others who have pointed this out - the National Institutes of Health are going to have to get into the business of supporting research of a kind that they have not traditionally supported in the past. Because you can't do the kind of human health, medical, and biological science research that the NIH traditionally supports without basic research in mathematics, physics, chemistry, complex systems, various engineering disciplines, and so on. And as these two lines continue to meet, and possibly the black ones even go above the gray ones, the NIH is going to have to think about what it does.

What else should I say? Well, MIT doesn't have a medical school, so we haven't benefited from this increase as much as many places that do have a medical school, if you look at different numbers.

M: Does MIT research include Lincoln Labs?

Mr. Litster: No. I've taken out Lincoln Labs, and I've taken out major subcontracts, because we have a few labs that are just pass-throughs. We do a lot of work in plasma fusion, but that gets passed off to people. But I think one of the things that's happened at MIT in response to the changing funding climate is that the amount of industrial sponsorship is going up. That has a lot of complications, which I will not go into, because they're probably not terribly interesting to you.

If you look at our budgets - and this is probably true for your institutions as well, if you're at a university - our research funding has not been growing particularly fast over the years, compared with the rest of our budgets for instruction.

That has consequences. We have a whole series of infrastructure costs. I will call them operating the libraries, heating the buildings, keeping them clean, paying the salaries of administrators, stuff like that. Those costs are incurred for two reasons. Partly in support of the university's educational mission, and partly in support of the research activity; assuming you can separate research from education, which is tricky. And so if you look at those cost pools at MIT, and I suspect you'd see the same thing at your place, and see what fraction of them did the federal government pay, and what fraction is the university paying with the students' tuition money, as a function of time, there's been changes. If you're quick, you'll see those don't add up to a hundred percent; and that's because we have some research that's not supported by the federal government, but by industry, and so on. So I think that's been going on at every university in the country.

If you look at financial aid to undergraduates - and this is why tuitions rise faster than inflation; because we all play Robinhood. We take in some money, and use it to increase financial aid to subsidize the others. And once again, we got into this need-blind admissions basis back when the government paid a significant portion of the share. Now, it's mostly the universities that are paying it. I think the same thing is happening at your institution. Which shows that institutions are complex systems.

Finally I'll show two more little things that show developments. If you look at the funding history of a private research university like MIT, Stanford, Chicago, and you go back in time a ways, they got a ton of money from research, and some from gifts investment, and some from tuition. But if you look at where our money comes from now, MIT still gets more money from research than the other two sources, but you can see those are starting to approach each other - they are roughly equal in size. And if you look at a state supported university as a function of time, you'll see that maybe research is not quite so big as at a private university like MIT or Stanford. But again, instead of gifts and endowment income, they get money from the state. And again, the three sources are roughly equal. So in many ways, state institutions are becoming more and more like private institutions, or vice versa.

That's all I want to say about what seems to be going on in private universities.

Mr. James Anderson: I'd like to thank the organizers for the opportunity to talk about the NIH. I think that most of you are going to be kind of pleased with what I'm going to have to say. But first, I'd like to ask a question. How many people in the room are now, or have ever been supported by the NIH? Okay. So we've established one point. The NIH is already funding work in this area.

I've got the proverbial good news, and then the other kind. David has already told you, from a different perspective, what the good news is; and here it is. The NIH has a lot of money. A lot of money, at least by our standard of reference, not the Department of Defense. This year we will have fifteen billion dollars. As you can see from the pie chart up there, the lion's share of this money in extramural awards goes out to folks like you. It leaves the NIH, and it goes out there. That's the good news. There's a lot of money out there. The not so good news is you've got to get some of it. This is perhaps part of the problem.

I love hearing people tell me what the NIH does or does not do; such as the NIH is not interested in funding this, or the NIH prefers to fund this. The reason why this is an issue is it's the Institutes of Health. And there's a good reason why there aren't arrows and linkers between all these boxes up here. There are twenty-one institutes and centers that give research grants at the NIH. And they act more or less autonomously. They're chartered individually by Congress. Congress individually assesses how well they're doing, and their individual budgets are set. What one needs to do is navigate through this group of independent funding agencies to find out where you might find support. That's your task.

Some of these institutes are pretty well defined. The National Eye Institute. I think we could say that, without knowing anything more, you would be able to tell me what their mission was. I doubt that anybody in this room would be able to tell me what this Institute does; the Institute of General Medical Sciences. Which is okay, because that's where I'm from. This one has very, very, very fuzzy borders. The others, despite their title, have very fuzzy borders too; and that seems to be the trend. But in order to get support from the NIH, you have to understand this system. That's what I hope to, at least in part, help you to do.

Just to reiterate. Each of these institutes has a mission, and it has its own interpretation of what that mission is. It has its interpretation of what's important to it. It has different administrative procedures. And you cannot generalize, except on the highest level of organization. You can't generalize about what the NIH may or may not do.

I can't report to you what the NIH is doing in the area of complexity, but I certainly can report to you what my Institute is doing. What you need to do is figure out a way to get this kind of information yourself.

I'm going to lead you through several ways of doing that by using as an illustration my own institute. Each of the Institutes now has a Web site. They are more or less useful, depending on the Institute. Ours is a particularly informative one. But somewhere on the Web page, the Institute is going to tell you about itself - what its mission is and what kind of research, broadly defined, it supports. It's going to tell you who its members are and it's going to give phone numbers. Almost nothing can beat calling people up, and talking to them. However, there is an official way that the NIH communicates to the general public its changing interests and changing policies. That is the NIH Guide to Grants and Contracts. Everybody ought to know about this. You can access it directly on the Web. What it is is a formal way of announcing what is of special interest to any one Institute at the time. And the – even though you may be aware of the overall mission, generally, this is going to give you the most specific information about how to go about finding out what the Institutes are interested in; how you should go about getting a grant.

This funding information takes the form of program announcements, and these are just called PAs in the lexicon. Also, requests for proposals. Requests for proposals are for contracts, because the NIH not only awards grants, it also awards contracts. And there's another mechanism called the request for applications. And I want to talk about the RFPs – I mean the RFAs, and the PAs specifically. A contract is, as you might imagine, it's the government's way of setting out specifications for some service, or some good it wants to have at a particular point in time. And most of you probably wouldn't be involved in applying for those, I think. Maybe that's not the case.

M: (What) fraction of your money goes out in the form of contracts?

Mr. Anderson: Oh, the great lion's share goes out in the form of grants, not contracts. I think the last time I checked was ten percent or less. So the lion's share of money is going out in grants.

And there are two ways that grants are solicited, as I said. The first is the RFA. And this might seem like a detail that you don't really need to know, but you really do. The request for applications is a way that the Institute will take an area of science, an area of biomedical science that it's very particularly interested in. It's not as prescribed as a contract, but there's one particular area they really want to get applications in; and this RFA has several characteristics. The first is that it's usually put out by only one Institute. In this case, this example I just pulled from a recent release, two Institutes, or an Institute and a Center are cooperating on this. Frequently, but not always, a Letter of Intent is required, so that the Institute can communicate with you whether or not what you have in mind is what they have in mind, and save everybody some trouble. There is only one receipt date for an RFA. They're only going to accept applications one time for this limited topic, which happens to be cryopreservation. There's only going to be one award date. And this RFA will also tell you what mechanism it's going to use, and it's usually only one mechanism. This happens to be a cooperative agreement. Okay.

So the RFA is a one time solicitation for the purpose of that competition. And the other thing about the RFA is they want to tell you how much money they have available to give out. And in this case they intend to commit two million dollars for the first year, and they'll tell you how much the individual Institutes are going to contribute. And they expect that up to four awards will be made within an award period of three years. So they really – they circumscribe what their commitment is. And it's only a one-timer.

M: They expect you to let it melt after three years?

Mr. Anderson: Well, after three years you can come in for an ordinary RO1, at that point. But it all depends on the purposes.

M: What was that word? Normal R1?

Mr. Anderson: Oh, RO1. I'll talk about that in a second, because that's an individual grant mechanism. But this is how most of the Institutes, but not ours – that's again one of these differences you have to be aware of. This is how most of the Institutes focus attention on one very circumscribed area of research, which likely as not has been mandated to them by Congress. Because these Institutes individually have their own pressure groups from the community, and they have pressure from Congress that they are – in fact, Congress will tell them you are going to spend five million dollars for – you know, put down your favorite disease of the month. And then this is probably how it will show up as an RFA. Let's see.

Then this is another little fillip. The RFAs generally are evaluated by a review group which is convened by the Institute. So the Institute that put out this RFA, actually runs the review. This is not true for other grant applications, which are reviewed by an entirely different part of the NIH. So those are the main things about the RFA. As I said, our Institute – yes?

M: You mentioned pressure groups. In this case however Congress actually mandates it?

Mr. Anderson: Not that particular one, absolutely not.

M: So it sort of looks like there may be three or four labs out there that do this kind of work.

Mr. Anderson: Often, that is the case.

M: So how does one know where the money is coming from?

Mr. Anderson: You call up the person whose name will be at the bottom of that RFA. You didn't see the actual RFA, you just saw some excerpt (of) the language. But just as the program announcements that I put out here, there are always contact names at the bottom, where you call up, and you kind of get a feel, a lay of the land. Now that was perhaps a little bit more of an unusual RFA, because the area was so specific. Often they're broader than that. Like proposals to assess the spread of drug-resistant bacteria, or something like that. Which is much broader, and would get a far greater response.

M: Do you have advisory committees which help you choose some of these things?

Mr. Anderson: Yes. That is also true. The RFAs are – when an individual Institute is going to put out a solicitation of one form or another, this usually is vetted, or approved by an advisory council. Every Institute has an advisory council which serves that purpose, as well as to make sure that the review process is operating normally. No grant can be paid unless it has been approved by council. And that's a detail, but it is important.

The program announcement is different. The program announcement takes a much broader area of science; in this case structural biology of membrane proteins. And you'll notice in this case quite a few of the Institutes have signed on to it; indicating that this is an area of interest for them too. And here several different grant mechanisms are part of this package.

In this case, people can also come in under this general area of interest with small business innovation research applications as well. So it tends to be broader; the program announcement. And here again, the applicants are urged to contact the staff, but they don't have to. There's no letter of intent. You're always in good shape if you call somebody up, but you don't have to. You don't have to do that; you can just send it in. Okay. They are to be submitted, and will be accepted at the standard application deadlines. The same deadlines which are for all the applications which are not solicited by a program announcement.

So there's not that much unusual about the way that these applications are going to be received, and about how they're going to be reviewed. In this case, it's going to be an appropriate review group, in accordance with the standard NIH review procedures; which means a separate unit of the NIH is going to review it. Applications will compete for available funds with all other approved applications. There is no set- aside. There is no number like five million dollars is going to be spent. We're going to make so many awards. We can make all the awards we want, or none of them; depending on the quality of the applications that are received.

The program announcement is stating that this Institute is really interested in this particular area, and wants to see applications, and will make awards to those that score well. And the other advantage is that if they don't score all that well, and the Institute is really, really interested, then with council's help, we can in fact fund them anyway. I should mention that. That's important about the NIH.

The only authority for making a grant is the head of the Institute. People think that the Division of Research Grants, (what now is) – the reviewers are funding. No they don't. No. The only person who can award a grant is the head of the Institute. Not even Varmus can make a grant. Yes?

M: In this case, this was underwritten or sponsored by five ??. Is that correct?

Mr. Anderson: Yes.

M: (And) they reached a consensus. And how is this distributed out of their (allocations as to funding)?

Mr. Anderson: Well, by signing on, they have stated that they will consider these applications, assuming they score well, for funding. You know, that they will fund the good applications. Their individual councils may or may not – that's not necessarily the case – may approve this. But generally speaking, the director (of) the Institute has to pass that we will participate.

M: This is not a grant, it's five grants that are all related to each other.

Mr. Anderson: No. No. The grant that comes in the door at the NIH is going to fit some particular mission. And if it's one of these five, then it will be assigned to one of those five Institutes for consideration for funding. They don't have to fund it, but it all depends on the mission. Now this is such a broad area that you can well imagine that if you were looking at the structure of proteins that are related to, let's say, heart, heart muscle proteins. That that will wind up in the Institute of Heart and Lung. And General Medical Sciences probably will fund a lot of ones that are based on model systems. It's just simply that applications that come in will be distributed to these five Institutes who have pledged that they are really interested in this area. That's what it means. Yes?

M: Is it best then to target a specific one of the Institutes (when you're sending a grant in)?

Mr. Anderson: We don't usually advise targeting, because – I mean, what should drive it is the science. This is my good idea. I want to see whether it's going to fit into your program. Rather, what's your program? And I'm going to try to figure out something that will somehow shoehorn into it. In fact, probably the truth is in between there. But that's when you talk to the person – the persons that are at the bottom of these announcements, and say well, this is my idea. Or you could even discuss, you know, the research project itself. And is this something that really fits what you say your interest is? And is it likely to go to this Institute? And the director, like myself, might say not only that, but I'm going to put a little flag in the system that will make sure it gets to me. So these are the kinds of things you can work out with the staffers at the NIH. They may make another suggestion and say that really is probably more interest to NINDS, and here's the name of somebody to talk to over there. That's how you work the system; by talking to people within it. Because, despite the fact that there weren't those lines between the Institutes, people like myself try to keep contacts with a lot of different people that we know, we share interest in, and we trade things back and forth.

Okay. That's mainly what I was going to say about the NIH itself. I'm going to talk about now program announcements that NIGMS has issued within the last year, which should be of particular interest to this audience.

We had several workshops last year in which we addressed this issue of quantitative approaches to complex biological problems. So we announced five related initiatives with the purpose of promoting quantitative interdisciplinary approaches to problems of significance. Particularly those that involve interactive behavior of many components. So that's pretty much what this conference is about. And these initiatives fall into three groupings.

The first is just simply the support of interdisciplinary research. That is, we're interested in making grants, research project grants. Now this language I want to explain a little. We have the objective of attracting investigators trained in mathematically based disciplines to the study of biomedical problems. It became pretty apparent during the course of these workshops that we held from – that were attended by experts in the cutting edge biological problems, that there was too much data, and too few people chasing it. And the people who are extremely good at generating this data are not necessarily the people that are really good at analyzing it. So what was emphasized was the fact that we need to bring in people into an area where conventionally they have not played the game to help us out. That doesn't mean that if you yourself happen to be interdisciplinary all in your own way you can't apply for a grant. You certainly can. Again, the science – the quality of the science will dictate.

The second broad area is the development of workshops and other vehicles to train biomedical scientists in new approaches. That is, people who have gone up to the molecular biology ladder, for example, or the cell biology ladder who are expert in that field, and now want to use techniques that they had not been trained in. And reciprocally to acquaint people who may be physicists, engineers with the scope of the biological problems that are particularly amenable to their expertise. And finally, the promotion of interdisciplinary training for scientists at the pre- and post-doctoral levels. That we do already, but we're really emphasizing it now, hoping to get more folks at the earlier part of their careers, to get a jump-start.

So these are the names of the five initiatives. There's supplements for existing grants for the study. There are what I just described. The genetic architecture of complex phenotypes fellowships and short courses.

This probably doesn't apply to anybody here, but I'll just mention it quickly in passing. People who already have an NIGMS grant can get additional money to bring somebody on. So suppose they have a very well-developed problem in some area of cell biology, and they get the idea hmm, I really need to bring a physicist in here, because I'm starting to work with forces within the cell spindle apparatus. I don't know how to handle this kind of problem. Well, this will provide a supplement, so one can bring in somebody for some period of time to help expand the scope of that project. And if any of you know anybody with an NIGMS grant, and think this would apply; well, great for you. Look them up. Okay.

And this is the meat and potatoes of our initiatives. And that's where we're really looking for brand new grants. Okay. You'll notice that the Institute of Mental Health also signed on this program announcement with us. And they have had a long- standing interest in using the more quantitative techniques to analyze these problems, so it's no surprise. Okay, so the purpose. Supporting research projects that develop quantitative approaches to describe, analyze and predict the behavior of complex systems; especially those requiring the integration of potentially large amounts of molecular, biochemical, cell biological, and physiological data, dah, dah, dah, dah, boilerplate. These projects are expected to require the participation of individuals with diverse expertise, and to be collaborative and cross-disciplinary, dah, dah, dah, dah, dah, dah. And then we just lay out the areas of our interests, both NIGMS, and NIMH. So we're serious about it. We're looking for these kinds of proposals, and we're going to fund them. Any questions? Yes?

M: (Inaudible question)

Mr. Anderson: In the case of NIGMS, you need have no concern about that whatever. Because what we do is we fund research into fundamental principles, really, which at some point will be applicable to an actual human health problem. But with the exception of a small part of our portfolio, our research is mainly in model systems like Drosophila, like yeast, like bacteria; which don't have any direct, you know, one for one relationship to any disease at all. In my portfolio, one third of my grants are in plants; arabidopsis. We fund over a broad scale. This is an interesting one, which may not have any appeal here. I thought I would mention it. Genetic architecture of complex phenotypes. And you see this was signed onto by a lot of Institutes.

Let me explain genetic architecture. In the case of complex systems, we're talking about problems where you have an enormous amount of data, and you're trying to make sense of it. In the case of complex phenotypes like obesity, like heart disease, like diabetes, you may have precious little data to go on. And one is trying to find out, let's say, for a particular disease, let's say, how many genes are involved. Not even what they are, but how many genes might be involved in the manifestation of this disease, and how do these genes interact with their environment? What's the gene interaction environment? In other words, analyses that are at a very formal level that may or may not get down to the actual mechanistic level. But the objective is primarily to use statistical techniques to find out, to at least look at the outline of this architecture.

Let's see. So here, "Expect to expand our understanding of the roles of genetic and environmental variation, their interactions in causing phenotypic variation in populations. Increase the quantity and quality of population-based data, the development of mathematical and statistical tools for analyzing measured genotype data." Dah, dah, dah, dah, dah, dah. These are all on the announcements that I have along the front. Here are the short courses.

So, scientists who are studying these hard problems need to have backgrounds in both biology, and in the analysis and interpretation of data. (It's important) biologists may also require instruction in the language and applications of mathematics and statistics in order to collaborate with mathematicians. Scientists with mathematical skills who wish to apply may also require instruction, etc. So we're inviting people to put on short courses, conferences, whatever, in order to try to bridge this gap, and begin to get – they begin to more widely disseminate both information about, and also how to actually do some of the analyses which you're familiar with.

This is really a re-announcement. We've had individual fellowships, of course, from day one. But here we're emphasizing the fact that people can use fellowships in order to – at the post-doctoral level is what I'm talking about – in order to acquire some additional training that might be useful, both in the quantitative disciplines; and again, for people who have a quantitative background, to do a fellowship in a biomedical lab, where they can begin to be introduced to biological or biomedical problems. So we're just emphasizing. It's a program we already have, but we want you to know that you're eligible. Let's see. I think we're running to the end. I'm not going to mention that one. Yes. I think actually I'll stop there. That's good enough for the time being.

So I'm delighted to answer your questions. I'm sure you must have some, because this is – I hope. Yes?

M: I didn't see very much discussion of the more theoretical side. How about computational models?

Mr. Anderson: Oh yes. We're looking for computational models as part of it. The best thing for you to do is to read the program announcement. That was just an overall objectives; but the program announcement really starts using examples, and laying out, you know, in more detail the kinds of things that we're looking for. And if these program announcements disappear, as they are almost going to do, I've got postcards on the end which has our Web site, and all these announcements can be found right on our Web site, so you can make reference to them. And as I say, every announcement has the names of people you can call up, and get most of your other questions answered as well. Yes?

M: (Inaudible question) (Pause)

Mr. Anderson: So you're asking for an outcomes analysis on the research that the NIH supports. Well, it is certainly true that you can find out with the abstracts, and the names, and dollars, everything that the NIH – all the grants that the NIH supports. So if you're interested in a particular area, and you wanted to find out does the NIH already support work in that area, you could find that out.

Now I think the gist of your question was can people who are specifically looking to determine whether the NIH is getting the maximum bang for their buck go into this more transparent source of data, and actually start doing those analyses? Is that the gist of your question?

M: (Inaudible response)

Mr. Anderson: I'm not – well, I'm aware of people's speculation about that. I mean, one emergent behavior that people are concerned about is that the NIH is very conservative. That what has happened with the competition for grant money is that it's become harder and harder to get a truly innovative proposal through the NIH. So that, I think, might be one example of what you're talking about. And it's the way the system is configured that causes this behavior. That's – there's a lot of speculation about that. And I think there's – personally, I'm just giving my personal view, I think that there is some truth to that particular one. But I also think that it's a lot less significant than most people might think. I mean, I've seen a lot of extremely innovative things come through, and it's usually just the quality, how well it was written, how good the investigator really explained the significance of what he or she was doing, and convinced the reviewers. The reviewers are people just like you and me. I mean, they can see good science when it's really presented that way, and good science can be really obscured by bad writing, you know, that sort of thing. Anyway. I'm not sure I completely answered your question, but – yes?

M: If you analyze by past applicants ?? (successful applications) – (Pause)

Mr. Anderson: Oh yes. Yes I can.

M: Something that we can learn.

Mr. Anderson: Grantsmanship. We're talking about fundamental grantsmanship. Well, one of the first characteristics of a really successful grant is that it's not an incremental improvement on something that's already being done by forty other people extremely well. That it really does have to have some significance that is generalizable. I think that's one hallmark just simply of the kind of science that's proposed. That if you're looking at some little tiny niche problem or question, and you, the investigator, have not related that to the broader context of what's going on in biological research, that you're going to be behind the eight ball. You can't assume that the reviewers are going to automatically know what that significance is. So I would say that's – number one is you have to explain to reviewers why this is important, what you're doing. And, let's see, another general principle is – well, I'll restate that. Don't assume the reviewers will give you – will know anything, really. And you have to lay it out. And you have to lay out exactly why you're going to do the experiments, and the order in which you're going to do them. And why if you get this kind of data out of the experiment, you're going to have to change to some – you might have to change your approach. And these are some of the approaches that you might have to use as an alternative. Running out of time? Okay. I can talk to you folks after it's over.

Mr. Litster: Yes. We can have sort of a general question and answer session afterwards. I think we will now hear from the NSF. (Tape change) – 1999 is about equal to the NSF total research and related budget. It's a little smaller. But that's not to say the NSF itself didn't do pretty well this year. What, you've got 8.6 percent or something for research and related. So there are opportunities with the NSF too, and Mike will tell us about some of them.

Mr. Mike McCloskey: Okay. Well, it's getting near the end of what's been a long day. So I'll try to be reasonably brief here. Fortunately, the NSF is quite a bit simpler than the NIH, and I suspect you're more familiar with it; most of you, anyway. Complex systems research is supported by the regular, disciplinary basic science programs in most, if not all, of the NSF science directorates. In CISE, computer and information sciences and engineering; in BIO, the biological sciences; in SBE, the social, behavioral, and economic sciences, and so forth, and so on. Those sort of things are things that you probably are fairly well aware of already, and I'm not going to dwell on them. What I am going to talk about is one specific program that cuts across the science directorates at NSF, and is designed to support research just of the sort this community does. This is KDI, or knowledge in distributed intelligence. Some of you I know are already familiar with it. What I want to try to do is give you a sense of what these cross-cutting programs like KDI, how they work at NSF, and also tell you specifically about this program. Where it's been, and where it's going in the future. As well as giving you some insights about what does make a successful proposal to a program like this.

Now right now the NSF is emphasizing three basic themes that cut across most, if not all the sciences supported by the Foundation. One of them is educating for the future, having to do with just exactly what it sounds like. Effective education, developing a work force prepared to meet the challenges of blah, blah, blah. Another is life in Earth's environment, which is also very broad. These are at fairly early stages of development. The one that's most fully operational at the Foundation is KDI, knowledge in distributed intelligence.

Now each of these themes involves a number of interrelated activities across the Foundation, many of which are not particularly visible to the research community. But at least in the case of KDI, a major aspect of this theme is a special proposal competition for research on knowledge in distributed intelligence. And this is something that has a dedicated budget, and evaluates proposals separately from proposals submitted to the regular programs; and this is what I'm going to talk about.

The basic idea of KDI is that the information revolution has created opportunities for all sorts of advances in the way we do science, as well as in the way that we work and live. So some of the things emphasized here are exploiting the faster processing speeds, as well as the storage and networking capabilities, to provide rapid and effective access to information; generally for dealing with very large data sets. Data mining, and things like that; or for collaborating across space and time. Developing co-laboratories of scientists that will share data, tools for analyzing, and even instrumentation, in geographically and temporally separated groups. Studying complex systems.

One basic notion is that the increase in brute computer power has rendered a lot of problems that were just a few years ago computationally intractable susceptible to attack. And so one aspect of KDI emphasizes pushing forward the boundaries of what we can do computationally in studying complex systems like the atmosphere, or like the brain, or you can supply many additional examples. And also here, KDI encompasses research aimed at advancing our understanding of learning and intelligence in both natural and artificial systems.

These priority areas are summarized under three headings within KDI, three focal areas. Knowledge networking: which is basically handling massive data sets, and communication, and so forth, and so on. Learning and intelligence systems: which is learning and intelligence in natural and artificial systems. And new computational challenges: which is developing new computational methods, and attacking problems that were computationally intractable until recently.

Now, the way KDI works, and the way many if not all of the cross cutting programs at NSF work, is a little different than the way the regular programs work. Now, KDI encompasses all of the science directorates at NSF. All of them are involved in the program. I won't try to read them. You can do that just as well as I can. And what that means is that these directorates are contributing part of their budgets to funding research in the program. And it also means that they are contributing expertise to managing of the initiative. Most of the programs at NSF have one, or at most two program directors who are responsible for the programs. These program directors have quite a bit of autonomy. They do make decisions on what's funded, they do manage the review process, subject to signing off by higher authorities.

The programs that cut across these different directorates are managed differently. KDI, for example, is managed by a working group that consists of program officers from across the foundation. So the representatives from all of the science directorates here, and the working group collectively arranges the review process, makes the funding decisions, and in the earliest stage of something prepares the solicitation, prepares the program announcement.

My position in KDI is that I have a been a program officer within the social, behavioral, and economic sciences. I'm a cognitive scientist. In fact, I'm just temporary at NSF. I'm what's called a rotator, like many of the staff there. At the end of this year I go back to my real job at Johns Hopkins. So my role has been to chair the working group that manages the KDI competition, or manages the KDI program. And this way of managing these things has some implications for the way you write proposals; which I'll talk to you about a little bit later.

Let me tell you what transpired with KDI in the last fiscal year, in FY 1998. Early in the year, the beginning of February, a proposal solicitation was released asking for proposals on these topics that I've already discussed. Specifically in the knowledge networking, learning and intelligence systems, and new computational challenges areas. We had a fifty million dollar budget for this. The response from the research community was overwhelming. We received almost seven hundred proposals. We received six hundred ninety-seven proposals. And for us internally, that was a fairly massive effort to deal with those. So they were peer-reviewed by eighteen multidisciplinary panels, involving a total of over two hundred panelists. So all of these proposals were panel-reviewed by three or four outside reviewers. And these panels were all put together, as the slide says, to be multidisciplinary. To involve scientists from several different disciplines.

Now, that's the good news, from our standpoint. We had a good sized budget; a very large budget for NSF. We got a tremendous response to it. The bad news, from the research community's viewpoint, is that of the six hundred ninety-seven proposals, we were able to fund only forty. That's a funding rate of less than six percent. As you might imagine, we're extremely happy with this forty that we funded. They are really outstanding. However, we're also extremely unhappy in that we have at least an equal number of very good proposals that we're not able to fund. And in fact, we could easily have spent a hundred fifty million dollars on these proposals, and still have been very happy with what we funded. So again, this is something that has implications for you as you think about this, and other programs. Just a point of information. The budgets for the proposals we funded were in the range of three hundred thousand dollars to two point four million. That's total over three years. Those are fairly large grants. In fact, they're quite large grants for NSF; at least at the lower end. They're not particularly large by NIH standards.

Now, let me give you some sense of what sorts of things were supported in the KDI competition from FY '98. You've heard multiscale mentioned a lot. And this was a basic emphasis within the new computational challenges part. The notion that many difficult but important scientific problems involve multiple scales. And so some of the projects that we ended up supporting was one on multiscale modeling of cardiac electrophysiology. Electrical activity within the heart muscle from the levels below that of the individual cell to the heart as a whole. And how the lower levels interact to create the patterns at the higher levels. Multiscale modeling of defects in solids. For example, metal failure; as when the wing of your airplane falls off. Again, modeling the defects at the molecular level on up to the macro levels. Multiscale modeling of fluid flow.

Global optimization methods in a variety of domains. For example, structural biology, 3D modeling of virus structure, or protein folding, seismic oil exploration, pattern recognition for geologic structures predictive of oil, and so forth. Agent- based distributed simulation.

And my next slide is cleverly headed "More Examples of Topics Supported by KDI". Here we begin to get more into what's called learning and intelligence systems; but here again there are complex systems issues. So there are several grants that in some way involve motor control in natural and artificial systems. For example, the motor control involved in an insect being able to fly, or a person being able to walk. Or the various robotic applications that grow out of things like that. Statistical learning. How humans and other creatures pick up statistical regularities in the environment, and exploit them. And in particular, what limitations, what constraints there are on the sorts of regularities that can be detected, and how they can be used. One of the applications in one of these projects is in language learning, and how regularities in what sorts of sounds follow what other sounds. Or what sorts of sounds occur in particular places in words, and so forth. How babies learning language exploit those sorts of things. Human and machine speech processing. One project we funded here tries to look at both visual and auditory information, and speech recognition, and speech production. One eventual aim of this is to produce artificial speech generation systems that will have visual cues from mouth, you know, simulated mouth movements, and so forth. And we can go on and on here. And I will go on for one more slide.

Computational challenges in astrophysics and cosmology. There are a lot of very difficult computational problems here, as well as the need to deal with massive data sets from sky surveys, and the need to mine those data sets for interesting patterns. Decision making in stochastic systems, so forth, and so on. There are also educational applications that are of interest here. This initiative includes the education and human resources directorate at NSF. In particular the use of information technologies for science and math education. So you can see there's a very broad range of things that fall within the domain of KDI.

This is what we did in FY '98. Now, in FY '99, which started the first of this month, the plans are still somewhat tentative, but this is what it will probably look like. First, we are going to have another competition. The budget will probably be about the same. The procedure will be a little bit different; motivated by the very large volume of proposals we got this time. We're concerned about the amount of work that so many people went to when we were only able to fund a small number of proposals. So the plan at this point is to have preliminary proposals. Brief, say five page proposals due probably in February, although that may have to get pushed back, depending on when the program announcement is released. These will be peer reviewed, but the form of the feedback will be that the proposal will be either encouraged or discouraged about submitting a full proposal. So you'll be told we think this proposal could be competitive; you're encouraged to submit. Or you might be told we don't think this is likely to be developed into a competitive proposal, so we don't encourage you to submit. However, proposers who want to submit a full proposal, in spite of getting discouraging feedback at the pre-proposal phase, will be allowed to do so. We anticipate having full proposals some time in May; and then with awards ultimately being made around the end of September.

Now, something else that's important to bear in mind is that these special initiatives, these cross cutting programs at NSF are viewed as temporary. They're viewed as ways of jump- starting an area. Ways of encouraging the research community to attack some issues, or to approach some issues in a particular way that the Foundation thinks are ripe for highly productive work. They are not meant to become institutionalized. And in fact, there's a very strong bias in the Foundation against institutionalizing special programs like that. So KDI, as a special proposal competition, will most likely not exist beyond the current fiscal year. So that what we anticipate, although this is tentative, for fiscal 2000 and beyond is that there will no longer be a separate KDI proposal competition. The notion is that KDI research will then be integrated into the regular research programs. That through this special program for a few years we hope to have changed the way, to some degree at least, the sciences are done, and hence the types of proposals that will be coming in to the regular programs.

Mr. Anderson: The money will get pulled back into the appropriate divisions? Or is that figured out yet?

Mr. McCloskey: The – well, there are various ways to answer that question, of course. It's hard to track what happens to a particular dollar, as you very well know. Yes, in principle, the Foundation will be spending about the same amount of money on KDI-related things, but it will just be distributed to the directorates. In practice, your guess is as good as mine. Does that mean that most of the basic research programs will be getting substantial increments in their budgets? Probably not, will be my guess.

So a lesson from this is that if you think you have a project that's particularly appropriate for KDI, the time to act is this current fiscal year, this coming calendar year.

M: What I was thinking of is suppose you happen to be one of the forty winners.

Mr. McCloskey: Yes. What do you do? That's right. If you are one of the lucky recipients of an award, either in fiscal '98, or fiscal '99, sure, you hope that the three year grant you get is not going to be the end of it. We would hope that you'll be developing a very exciting project, and will want to continue it. The foundation's position on this is that you then should be able to go to one or more of the regular programs for continued funding.

One thing NSF does reasonably well is to jointly review proposals across existing programs. We expect that as these KDI projects come back in for renewal, there will probably be the sorts of proposals that will be reviewed by two, or three, or even more programs. And that doesn't always work perfectly, but I think it works reasonably well at NSF. Again, we'll have to see what happens. One danger of the approach that NSF takes is that in trying to jumpstart an area, and then hope it will continue on its own, and moving on to something else, there is a danger that one develops or initiates a lot of activity, and then isn't there to support it later on. I can say that as someone who is temporarily at NSF, and without fear of very much retribution. Okay.

I want to say a little bit to you based on my experience with KDI both this year, and with the first part of it that also ran last year, which was learning and intelligence systems. What makes projects successful and unsuccessful? So if you're thinking about submitting, maybe to help you decide whether to or not, and also how to develop your project.

Well, it should go without saying that the science needs to be excellent in all parts of the project. And as Jim said, things that are merely small, incremental steps are not likely to excite a lot of interest. Particularly with a program like this where the aim is to fund groundbreaking research. The aim is to fund things that are truly new. However, we do see a lot of proposals where the investigators, although proposing a multidisciplinary project, are really most interested in part of it; and that part may be strong, but the rest of it is not strong. Those sorts of proposals die a very ugly death in the panels, where there will be reviewers from the sciences that were not emphasized, as well as reviewers from the sciences that were.

KDI is a basic research program, so the proposals need to have a strong focus on fundamental research questions. Basic research. Proposals that don't fare well are those that are, say, aimed at developing a specific application, a knowledge network for – choose your large database project. Now, the program does support projects that develop applications. But only when the aim is also to gain some more general lessons. For example, how to set up a co-laboratory in general, or so forth.

Another place that's often trouble is that the parts of the project need to be well integrated. Now, as multidisciplinary projects, they tend to have multiple parts to them. There will be this research thrust, and this other line of research, and so forth. If those are just, as often happens with proposals like this, really very separate projects that have sort of been cobbled together, and some verbiage at the beginning has been written to give a veneer of integration to it, those also don't do well. There does need to be some true commonality across the different lines of research.

KDI specifically emphasizes multidisciplinary research. And multidisciplinary is intended in quite a strong sense. The project should actually be making advances in at least two or more scientific disciplines. In other words, a project is not multidisciplinary just because it happens to involve a biologist as well as a team of computer scientists. Okay? Or, as often happens the other way around, a team of biologists with a computer science researcher as well. If the project, say is using well-worn computer science to address some questions in biology, that might be a fine proposal for the biology directorate programs, but it will not do very well in KDI. A project that involves biology and computer science needs to be addressing basic questions, and advancing the science in both of them.

The next is again something that should be obvious, but often isn't there in the proposals. Make sure that your team has all the necessary expertise. Ask yourself, you know, what is needed to do this project, and make sure the team has it. We see many, many proposals where a team in one or a couple disciplines believes they can solve problems that at least if you ask the scientists in other disciplines that are involved, desperately need scientists from other areas as well.

And this one requires a little bit of explanation. Relevant to two or more NSF directorates. This is related, but not identical to the requirement to be multidisciplinary. For example, the directorate that I'm in, social, behavioral and economic sciences encompasses many, many different disciplines; from cognitive science, to political science, to economics, to anthropology, and so forth. You could certainly have multidisciplinary projects that were entirely within those sciences.

The successful proposals in KDI tended to be those that were relevant to two or more of the science directorates at the foundation. And the reason for that is quite simple. And that is that multiple directorates are putting money into the program, and the more directorates you can interest in your project, the more directorates may be potentially willing to commit money to it. So that the successful projects were, say, things that were jointly funded by CISE, BIO, and perhaps engineering, or by some combination of directorates. So look at the directorate structure, and think about how your project relates to it. Okay.

Those are characteristics of the project. Briefly let's look at some of the form, as well as the substance. What makes a successful proposal, independent of the content? One thing to keep in mind with programs like this is that the reviewers will probably be more diverse than you would be accustomed to for a proposal targeted to a disciplinary program. You will probably have one, maybe if you're very lucky two reviewers who are fairly familiar with the research area. But unlike, say, in a disciplinary program, where there's a concentration of expertise in that discipline, and you're likely to get somebody who knows a whole lot about it who is an expert on your topic, there may be no one on the panel who has extensive expertise in your area. Plus, you're likely to be reviewed also by scientists from other disciplines who are less familiar with it in general. So you need to take special care to make the merit of the project, and its importance clear to scientists in a broader range of disciplines. And this is not trivial. It should not be done by dumbing it down, because then you don't give enough information to the scientists who do know something about it.

The research plans should be well specified. We often saw proposals that seemed like they might be good ideas, but since they had such broad scope, and since the length of the proposal is limited, they managed not to really tell us very much about specifically what would be done; and reviewers really hate that. So you need to try to give some sense of what specifically you'll be doing, to the extent that's possible within the page limits. You need to make clear what roles the different people on the research team will play; showing that they do supply the expertise, and that it's going to be applied to the problem in a good way. These are often complicated research arrangements involving collaborations across departments, across institutions, and so forth. You need to make it clear how the project will be managed. Because often, we suspect, projects are put together fairly quickly, and put together with no real serious consideration of how they would actually be done, should the money come their way. And finally, you know, they may involve multiple institutions. Many of the successful ones did. Many of them didn't; many were single institution. And the program does encourage partnerships with industry, with government labs, and so forth. Although there are some rules about what sorts of things we can provide money for, in the case of for-profit corporations, and entities like that. So you would do well to check with the NSF personnel if you're thinking of a project like that.

Okay. I do not have handouts with me. NSF is doing almost everything electronically now. So the KDI proposal solicitation was not last year, and will not this year be released in paper form. It's available on the Web for perusing or downloading. This is the KDI web page. All of it, except the /KDI is the basic NSF Web site where you can find all sorts of useful information about NSF. On the KDI Web page right now, there is the proposal solicitation from fiscal '98. The one that's already over with. Plus there is a link to a list of the KDI awards that were made where you can see who got them. You can read the abstracts, you can see the amounts, and so forth. I strongly suggest looking carefully at that if you're thinking about the program. And, of course, feel free to get in touch with the people involved. There will be contact information in the new solicitation when it's released; hopefully very soon.

What's happening right now is that senior management at NSF are in the process of appointing a new working group to work on KDI in FY '99. Taking a bit too much time for it, I think, because the proposal solicitation needs to be released almost immediately. So watch the Web site over the next month. Hope to see the solicitation there with the deadlines, and other information. Okay. I'm going to stop there, and then we can have any discussion you might want to have.

Mr. Litster: Thanks Mike. I think both speakers were clear. I learned quite a bit from each of them. But I guess we can sort of open it up for general questions and discussion now.

M: I have a question for both of you. I don't know if this is the right way to ask the question. There seems to be several emphases in granting this money. One of them is that the key is the institution that it goes to. The second is the individual researchers that it goes to, third is the subject matter. Are these all equal? Or are you primarily emphasizing one or the other.

Mr. McCloskey: If you mean – the institution you mean MIT or wherever?

M: Right.

Mr. McCloskey: At NSF that is not an explicit part of the review. Obviously, the reputation of the researchers, which to some extent reflects where they're from, is taken into account by the reviewers. However, the awards that we made, they were made to small schools in the Midwest, as well as to the MITs and the UCLAs, and so forth. So that's not an explicit consideration. Certainly the subject matter is. Proposals. We did receive many proposals that were excellent science, but simply didn't fit with KDI, in that perhaps they weren't truly multidisciplinary, or they weren't really directed at the sorts of issues that NSF feels define the program.

M: I think I (could answer) very similarly. There is an explicit criterion to look at the environment in which the research is going to be done in order to be sure that in fact it can be done (with the) kind of resources that are available (at an institution). ??. Who else is going to use (the) ??. The subject matter, of course, ??. And ?? work that is described ??. It's hard ?? to imagine ?? (broad topic) ??.

M: Actually specifically to you, if I may – (Pause)

M: ?? around that. We have extensive involvement in structural biology. And that has its own complexities. We fund a lot of theorists in the area of protein folding; which probably (isn't quite what you need). In my portfolio there is modeling metabolism, and (other theoretical work).

M: I have a question about (the funding level). (Pause).

Mr. McCloskey: The rules for what NSF is willing to support in the way of salary are just the same for KDI as anything else. These budgets are fairly large, not because we're giving PIs more months of salary, but because these tend to involve many PIs. They tend to involve sometimes fairly substantial equipment purchases, and many postdocs, and graduate students, and so forth. So they're just larger scale projects. It's not that we're supporting different kinds of expenses than we are in the other types of grants. But yes, two months of summer support is about what you should expect as a maximum. If, say, someone is on soft money, then that's a different matter. But for people on regular academic appointments, two months salary is a pretty strict limit.

M: (Question not taped) (Pause)

Mr. McCloskey: That's a good question. It's certainly true that single investigator proposals did not fare very well. But I think not because they were small, but because they weren't truly multidisciplinary. It is possible for single investigator proposals to be multidisciplinary, but of course it's more likely when you have multiple investigators. Some of us on the working group did have some sense that larger projects may have had an advantage merely because of their size. That if you look at a good project that's small scale, and you look at a good project that's large scale, the large scale project is likely to look more impressive simply because it is larger scale. So one of the issues that at least we're going to encourage the working group for this next year to consider is whether there needs to be some taking into account of the size of the proposal. I haven't done any more formal analysis than that because I've just been buried in dealing with seven hundred proposals. But it may well be.

F: My question is to both of you, and I had started this subject with Jim. I'm just getting a Ph.D. now. Are there programs for new researchers? Or what kinds of recommendations would you have for someone like me that has a very clear proposal in mind that is not yet affiliated with an institution?

Mr. McCloskey: Okay. Well, a couple of things real quickly. Grants are made to institutions, and not to individuals ninety- nine point nine nine percent of the time. So that when you submit a proposal, it will need to come from an institution where you're affiliated. Presumably you will be shortly, I hope, and then you'll be able to submit proposals there. With respect to special programs, NSF has a program called Career, which is early career awards. And this can be a good program to apply for earlier in your career, but not necessarily. It is not a protected competition where young researchers are competing only with other young researchers, and therefore you might have a better chance than in the regular programs. At least the way most of the directorates handle the career program, those proposals compete with the other proposals in the regular disciplinary programs. So if you were to submit a proposal – I don't know what your field is, but some particular field – it would be a career proposal if you chose in that format, but it would go to the regular panel that evaluates all the proposals. And we can talk more later. There can be advantages and disadvantages to applying. There is not any special program that specifically tries to create a protected competition for young investigators.

M: The conference chairman wants to say something.

Mr. Bar-Yam: Yes. I think that there's a basic comment about the difference between complex systems, the study of complex systems, and disciplinary research that should be made. I have had the sense for a number of years that if you give our research to a couple of disciplines to evaluate, they each evaluate it by their own criteria, but complex systems is also a strategy of its own. The existence of universal issues in dealing with systems from different contexts is not the same as the questions that those fields are asking. The KDI program doesn't seem to address that. Neither, of course, does the NIH program, because the NIH program is really focused on applying mathematical techniques in the biological sciences. Which is a particular area of interest. But the NSF program explicitly doesn't do that, because it actually creates two or more barriers. If you had the ability to fund complex systems research in the context of a particular discipline you would do that. However, what one is asking here is: do you satisfy the criteria of not only one discipline, but also of another discipline? And that's not necessarily the solution to the problem of recognizing the existence of new approaches to the study of complex systems in general.

Mr. McCloskey: Agreed. I think you're right. But you do have to remember that KDI is not a program about complex systems, it's a program into which a lot of complex systems research can fit. Now a few of the projects that we funded did try to apply a particular type of methodology, or a particular perspective across several different scientific disciplines. I think one of the global optimization projects that I mentioned is trying to use the same set of techniques in biology, in geology, and in satellite communications. That may speak more to your notion that there is a common set of techniques, or a common set of tools, and so forth that could apply across disciplines. But that is the exception rather than the rule. It may be that it will be good to target research that does try to pull out those generalities, but KDI doesn't do that. You're right.

M: Let me clarify. Even though the program announcements are really emphasizing the biomedical aspect, and collaborations that we would hope to see happen, we also support the search that is not targeted. That is, let's say, more theoretical, more (tool) building. That is not directly targeted to anything. There is a small portfolio (in) NIGMS, for example, that has a lot of grants to develop new statistic methods. Not statistic methods for solving biological problems, but just new statistical methods, period. And I think this is something that you should inquire about.

I also want to make the point that anyone at any time can send in any proposal to the NIH; you don't have to wait for any particular announcement. That is more or less the meat and potatoes of the NIH. Most people don't phone up, and ask us: should I submit this proposal? They just do. And then through our internal tracking system it winds up in the hands of somebody who's willing to accept that. Meaning that if it scores well it's most likely going to be funded. And oftentimes people like myself don't even know we're interested in it until we see it.

Mr. McCloskey: A more general comment is that one thing that some scientific communities do much more effectively than others, I've seen through my experience at NSF, is to make their needs known to NSF. So there are various ways to influence the directions that NSF, or I'm sure NIH goes. And I would encourage this group or any other to take advantage of them. Those would include talking with the assistant directors of each of the directorates. Each of the directorates has a head. It's confusingly called an assistant director. They are assistant directors of the foundation. So, for example, Yuris Hartmanis is the Assistant Director for CISE. Actually, he's now about to leave. Make your voice heard in those offices. Also, each of the directorates has an advisory committee that meets a couple times a year to give advice to the directorate on what they should be doing. And serving on those committees is a good way. Even serving on the review panels is an excellent way to have some voice, because panel meetings always include some time for hearing from the panelists about what should be done. So if you don't like the way we're doing something, let us know.

Mr. Anderson: You might be interested in an NCI, National Cancer Institute announcement pretty recently. They're calling for white papers in a particular area related to cancer research. So this is an instance of this where the Institute is saying we may not know what we want to know. Write some papers, and tell us what we ought to be doing, or what we ought to be thinking about. And in fact, I've got a copy of that.

Mr. McCloskey: And I should have mentioned also, what NSF does in developing this sort of program is hold many workshops, usually, in different communities, or in different multidisciplinary, inter-disciplinary groups like this to hear from the community what the perceived needs are, and then uses that information in developing the final form of the program.

Mr. Litster: Would you say, Mike, that for someone who's far enough along in their career that they have tenure that spending time as a rotator would also be valuable experience?

Mr. McCloskey: As long as you don't end up chairing a working group on something like KDI. It was a very – it's been a very valuable experience, although it's been harder than I was led to believe to maintain my own research program while I was doing this. But sure, I would encourage you to participate any way that you can.

Mr. Litster: Mark Twain had a saying that describes that. He said a man who picks a tomcat up by the tail is gaining valuable experience which will stand him in good stead for the rest of his life. Well, shall we thank both of these gentlemen for –

Complex Understanding of Today's Global Economics

Dean LeBaron
deanlebaron@compuserve.com

We are puzzled by global economics. Forecasts are universally disparaged for this most dismal of sciences, so much so that many refuse to call it a science and, instead, attach the black art label, with heavy reliance on promotion. Yet we are accustomed to aggregate growth achievements, around the world, on the order of 2% or so. So something must be happening right...or so we thought until a few years ago.

Things started to unravel and unravel big time. The precepts which lead the Asian tigers to prominence selected them to have the first troubles, first with their financial structures and then with their economies. And there was no warning. Even now most of us still don't know what happened.

And it spread to Russia, threatened Latin America. Meanwhile Japan, the number two economy in the world, slid into a recession/depression from which it seems powerless to recover.

The US remained almost immune...until April when its 16-year equity bull market gave its most vigorous signal yet that all was not well, with the average stock declining since then by about 50%. And the US seemed to recognize, slowly, that it could not be "an island of prosperity in a sea of depression."

Old ideas are only abandoned when they have been proven faulty, and surely the old economic premises have given a resounding signal that they are no good. We can do no harm by accepting the challenge to use complexity to find new ones.

And so I propose to select the ten principal issues I believe will help us understand the global economic environment. For each of these, there will be the conventional, structural, linear analysis leading us to prescriptions for solutions. And I shall propose, for each, a complexity-grounded suggestion, often with a model, leading us to quite another choice.

And leave it to you to determine, in the end, if complexity has anything to offer in building models of this economy and policy proposals which can be studied in detail and, if convincing, be implemented.

1 Issue: Can the institutions established by the 1944 Bretton Woods convention, IMF and World Bank, determine who gets assistance or not?

Conventional: Tight monetary policy, balanced budgets, low inflation will attract capital and build up supply, often export goods. Seven volumes essentially the same. Sound financial policy will lead to economic growth.

Complex: Growth can come in many forms. Internal demand (China and Eastern Europe). Export (Japan and Tigers). Economies are not necessarily homogeneous. They can shift from one state to another as in a social system for agrarian (like the US adopting emergency agricultural subsidies this year) to ignoring Glass-Steagall. There is an idea on the part of developing countries that prescribed behavior—democracy, human rights (what is that?), environmental concerns—will lead to cheap, long-term money.

It is quite possible that the advice of the postwar period for development of war-ravaged areas was good for the early days of developing markets, coupled with large amounts of money when none other was available. But it may be that growth, at whatever cost, is more necessary. And postwar Japan, under General MacArthur, and Chile were hardly paragons of democratic virtue. The advice prescription from complexity is to adapt to the times and not have a rigid set of rules in seven notebooks.

2 Issue: Is long term less risky than short term?

Conventional: Regression to the mean. Wait long enough and economic balance returns. Economics of equilibrium. Long-term capital credit. Look for small historic anomalies, with confidence add leverage, and wait for market to rebalance itself. (OK with infinite amount of money, not with loans.) These were common tenets of today's modern capital theory and application.

Complexity: Time is discreet, not continuous (as in Bob Merton's book). Information comes in bursts to the market and all time is not equal on the x-axis of a two-axis plot. Backtesting tells you little if you are careful. Most financial people follow the dictum of statistics "if you torture the data long enough, it will confess to anything." Economics of increasing or decreasing returns. Momentum and expectations may create their own trends (Soros' reflexivity).

3 Diversification is a good...

Conventional: If risk (volatility) is equivalent to return and we know how to increase risk, all we have to do is mix assets that are uncorrelated to increase aggregate (portfolio) return and reduce risk. Alphas, betas and R-squared are part of the friendly analyst lexicon. Close to the means, these measures seem to be quite stationary over the time periods we have studied them.
Complexity: Past correlation tells us little about "stress" times. The factors that determine values are non-stationary and combine in different ways. Evolutionary history of markets is unlikely to combine in the same way each time. And when they part, the results may appear chaotic.

4 Globalization is a benefit to all participants

Conventional: Free economics lets each specialize in their best skills to the combined benefit of all. Money will flow to the highest and best return per unit of risk and continuously stay close to balance. Without barriers there are no massive dislocations except in the most exceptional exogenous events.

But if globalization, free flow of goods and money do not work, bring institutions to work for stimulus, bailouts usually at the financial sectors of the affected economy because that is where the trouble started. RTC, international currency agreements, Japanese banks.
Complexity: Doubtful policy makers can determine when to flow and when to curb. But it may be necessary to put barriers around a system in trouble to see it does not contaminate. (We may be this with Y2K systems next year...only hook to those that are Y2K compliant.)

Have to have the system "punish" itself by selecting its own victims.

5 Economic model construction

Conventional: Most models are simple, historic, static and have two variables. Most have face validity with end period dominance. Most consensus models are proven useless or wrong.

Models are developed by history and may be quite sophisticated but depend upon the assumption that the past is the prolog of the future and all info is captured by the model period and no simplifying assumptions are predetermining the outcomes. These models are generally static, contain few variables selected by face validity and are often unstable in different market environments. Investment stat is rather crude.
Complexity: Forward testing. Simulations (...Bob Monks on Russian corporate governance...Brightline). Understand how the assumptions selected by the modeler will determine outcomes...Heisenberg.

6 Central banks can regulate capital flows

Conventional: Belief in the US that Fed sets interest rates and, through monetary policy, can influence level of business and interest rates. Fed is all knowing. Fed chairman is most powerful person in US. Forex trading is about 1.5 trillion. Fed might trade 10-20 billion a day, hardly significant. A kiddy car on the Mercedes autobahn.
Complexity: Capital flows are so large they have to be set by the markets. Markets act best when they have good info...purpose of the Fed is disclosure to the agents not secrecy.

7 Bailouts

Conventional: Leave system alone until it is in trouble. Then rescue...Mexico, hedge funds, RTC.
Complexity: We tinker with the system to our peril. The system needs to have an automatic destructive cleansing. And the bailout gives markets a message of tempered risk. Attempts to have a positively skewed payoff pattern often results in unintended consequences.

8 Insider info

Conventional: Keep market fair by preventing insiders from trading during corporate blackout periods and make government officials use blind trusts.
Complexity: Encourage insiders to get their info priced into the market by trading...perhaps revealing their insider status but not the insider identity. Companies could use posted FAQ's for continuous offerings.

9 Importance of level of inflation

Conventional: Low single digit inflation is a good even though it may be a path on the way to deflation. Figures are believed although revisions are the norm.
Complexity: Adaptable to a range of alternatives, not a single point. And expect that a number reflects some plus and minus. Prepared for leaps and surprises as the norm. Measurement lags...hindsight based.

10 Risk free

Conventional: Interest on US treasuries is the risk-free rate in all calculations. It is easily measured and usually reported to four decimal places. At the moment its real rate seems rather high. Quality preference is highest per unit of government yield and has been increasing for four years.

Complexity: Perhaps risk-free is the riskiest. US is the biggest borrower in the world and is only perceived as credit worthy because others give us their money to hold. How stable is this perception?

(and for a "Baker's Ten")

11 Performance measurement...globally, of companies and managers

Conventional: All historical and all short-term. If you improve the quality of the numbers you improve the quality of the results.
Complexity: Of historic interest only. Forms a basis of discussion.

Despite the work and the money spent in studying global economics, we know very little. Largely we are studying old, outmoded precepts. When we incorporate the principles of complexity we have a chance, just a chance, to understand this adaptive world better.

We know that systems often adapt best at the edge of chaos but to do so they need better disclosure to the agents and minimal barriers to propagation.

It is our job to encourage policy makers to risk new approaches in improving fitness to the landscapes surrounding us...and of which we are essential and active agents. We can be applying our complexity work in biology, physics and other fields to policy prescriptions.

Agents in Action

William Fulkerson
Deere and Company
Rob Axtell
The Brookings Institution
Abhijit Deshmukh
University of Massachusetts, Amherst
Van Parunak
Environmental Research Institute of Michigan
Win Farrell
Coopers and Lybrand Consulting

Mr. William Fulkerson: I want to welcome you to this, our last day, and our session on agents. You may recall that there was an article in the Wall Street Journal by Tom Petzinger, who brought me to fame. "Meet Bill Fulkerson, Staff Analyst, a rumpled man who once taught at Central Missouri State University. He toiled anonymously in Deers Engineering Labs for nearly two decades, and has received precisely one promotion over those years ..." That's what management saw when they read the article. And then they saw this last part which says, "... his business card says 'Project Manager, Adaptive Manufacturing Systems"; but it was an invented position, he admits. A substitute for his official title, which remains Staff Analyst. "I haven't been promoted," he says, "but I'm having more fun." Well, in fact, I have been promoted one more time, and I'm still having fun.

Let me share with you some ideas that I often share with college people when I'm talking. I think it pertains to the way you should view agents today. These are three quotes. "I played first chair in a high school band until they gave me an instrument." Many of us are very excited about the potential that exists for agents for complex systems, and can imagine all the things that we might want to do with it. But other people out there, like Bach, who say, "All one has to do is hit the right keys at the right time, and the instrument plays itself"; the true masters. I encourage you to be today, and as you leave, a little bit more like Michelangelo, who said, "If people knew how hard I worked to achieve my mastery, it wouldn't be too wonderful after all."

That's part of the challenge that I'd like to lay out in front of you - to forget the euphoria, and to forget the joys of mastery, and to revel in some of the hard work that may well be ahead of you.

Well, some people ask why is Deer and Company working with complexity? The short answer is that we follow Dean LaBaron's metric. A lot of people think of the real world as a problem like this; you've got a really good idea, but there's a lot of people not following you, and not going the direction you want to go. You come out with some very fancy mathematical explanations. Here's your mathematics, David. And if you take an action, all of the sudden you're making a lot of money, and people are working with you. You don't have to throw your pointer to be effective. In fact, the real world is a lot more like this, with a lot of people with ideas going different directions making a lot of money one way or the other. The mathematical model, the Descartian system that was there is oversimplified. It could be that the landscape that you're dealing with is as complex as this Martian landscape that I stole off of the front page of the Scientific American not too long ago. Or it could be that even that's too idealized, and it's more like this fractal image that we see. Or we could fantasize, and say that the mental model that we're working with is one of these beautiful, self-organized fractals like this. But in fact, what probably is happening is that we've got one of these convoluted, colonic looking things that comes back on itself, that's really not very pretty at all. And so the challenge is how do you deal with worlds like that? The answer is to deal with agents. What one has to do then is be comfortable playing at the edge of chaos; taking several prototype solutions, and working with them, making no serious commitment to any one of them. We should be taking what we can learn from each one, putting them together, and continuing to build. That's the power of agency.

With that short explanation or introduction, I'd like to introduce our first speaker today. We'll be hearing from Rob Axtell, who's a Fellow in economic studies at Brookings. He received his Ph.D. from Carnegie Mellon University, and has his interest in agent modeling of social systems, global science change, and policy and environmental regulations. He has published with Josh Epstein the book Growing Artificial Societies: Social Science from the Bottom Up.

Mr. Rob Axtell: Thank you for the introduction, Bill, it's a pleasure to be here at the NECSI conference. There are at least two distinct uses of agent systems. Bill has alluded to that in his talk. I'll make it explicit here. In particular, the way I use agents, and the way it is described in my book Growing Artificial Societies, with Josh Epstein, is we use them in what economists call a positive way, in the sense that we're trying to describe existing social systems. We're trying to build models of social systems using agents; whether those be markets, or firms, or whatever. This is a somewhat different notion of using agents than in the agency ideas, primarily in distributed AI where we actually want to invoke agents as a solution technique for solving some industrial problem, or maybe having Internet auctions. So there are these two distinct uses. In the terminology of economics we would say that the positive approach differs from what's called the normative approach. The normative one is where you're actually going to try to build the system from the ground up. The thing I'll talk about today here is going to be a positive description of human social systems using agents. Essentially I want to describe an extension of the conventional methodology that's been employed to date. What I want to argue ultimately here is that we know a lot today about using agents to build any kind of multi-agent structure where there's kind of a flat landscape of agents. What we don't know very much about at all is how to build institutions, or organizations of agents.

Now there is a branch of work called computational organization theory. This is an agent-based approach to understanding organizations. Typically what's done here is an organizational infrastructure is built in software, and then populated with agents. What I want to describe today is more or less a research program, with only one working example - a program in which we will simply instantiate some kind of incentive structure, and fill it full of competitively behaving agents. In particular, the general notion of forming organizations and institutions is what I want to talk about. At the end of this talk I'm going to actually argue that this notion of using agents in this way is actually a very powerful technique in social science, which can change the way social scientists do their work.

Many important ideas from complexity, associated with Santa Fe and beyond, all typically involve decentralized multi-agent interactions of some kind. I would like to argue that the dominant methodology for studying all of these systems is in fact agents. When we talk about a physics model, we're not really talking about sentient actors in the physics model. I'd like to argue that the agent methodology is something which unifies all these different areas. When we want to build a model of the immune system, or a market, or a model of artificial life, or want to employ the GA, when you want to build a model of a sand pile. We're always talking about building models with a lot of heterogeneous entities that interact.

When my book came out, people said okay, you have a little creature that runs around on a landscape. This is the sugarscape model that's described in the Growing Artificial Societies book. People said, isn't that just a cellular automata model? The answer is it's hard to disentangle CAs from agents always. But is it because CAs are spatially affixed? The fact that you can have autonomous entities that interact according to any particular topology of interaction, that fact generalizes the notion of CA, and so in fact CA is a special case of agents. I'll try to make this more precise as we go. Agents are the general methodology.

What, then, is an agent-based model? Typically we have some population of things, let's call them agents. In the work I'm familiar with, as few as five, up to as many as tens of millions of agents. Each agent has some internal states and behaviors. Technically, for this reason, it turns out agents are very conveniently implemented as objects in software. Agents can be autonomous, or semi-autonomous. I mean, they have their own clock of action, or their schedule to act in a certain way. They typically interact with one another, and possibly with an environment. We have the notion of local versus social interactions. This may seem completely obvious, and not needing of mention, but in all of the received economic theory we have, agents don't interact with each other - agents interact with a price system or a price vector. Simply adding agents who interact with one another is already a novelty in economics. Surprising as that may seem, that is in fact the state of the art today.

Agents are often purposive. Typically they may have a utility function they want to maximize. To me this is an important notion about agent models - we typically don't preordain the aggregate structure. We don't make any aggregation assumptions in building these models, we simply look for what emerges from interactions of the agents; although emergence is a loaded term here usually. We simply will spin the model forward in time, and look for a macro structure to emerge. It turns out in all of economics we usually have to make very specific aggregation assumptions, which can foil the validity of the results. I just wrote in here that not only does the macro structure emerge, but subsequent generations of agents emerge from the ancestors.

There are many examples of this. The CAs are examples of agent- based computation. As I mentioned, when my book came out people said isn't it just the opposite? Isn't your case a specific case of CAs? I like to argue the opposite. All these population-based approaches to optimization are in fact multi- agent software systems - genetic algorithms, the theory of games - all of these.

I'm going to give a very qualitative topology here of different agent systems.In the notional class of all agent systems, there are ones which have local interactions, and non-local interactions. In particular, you can think of evolutionary systems as having both of these. Whether it be evolutionary games, or systems like GAs. In GA, of course, we have nominal interactions. We have a Grim Reaper come through and purge the entire population. In evolutionary game theory we have only local interactions. CA typically involve only interactions with neighbors, so these are in the local interactions box. Now the people in computer science work primarily with local interactions, but there's some work with global interactions. Orthodox game theory sits on the border. The power of the agent-based approach is that it lets us study whole classes of systems for which it's very hard to make purely mathematical progress. I've heard Murray Gell-Mann say at Santa Fe that someday we'll have a mathematics, hopefully, which can explain all of this, but as of today we don't really have that. In the meantime we're forced, out of necessity, to solve these systems by spinning the models forward in time.

The main point I want to make about this slide is that I believe that all these systems are in fact social systems. They're social systems in the sense that they involve social interactions. When individual agents interact with one another, it's a social interaction, in the sense that's used by economists, or sociologists. And in fact the main argument here is there are a vast number of ideas in social science that are relevant to analyzing multi-agent systems. In particular, things like the various kinds of equilibrium notions that can happen in these systems. We oftentimes motivate agents via utility maximization. That is exactly like what's done in economic theory.

An agent-based system can only be shown to be an optimal configuration not with respect to some kind of global optimality, but with respect to evaluating the welfare of individuals. It's kind of a multi-objective notion, so you're forced to deal with Pareto criteria. The use of markets is something common both in social science, as well as in agent- based computational systems

Nash equilibrium is a very specific game theoretic notion that is now ubiquitous in microeconomics. You may say well, what does that have to do with agents? In fact, you can find it in a whole host of agent-based models used in some of my work. This notion arises naturally. In some of Ky Nagle's work on traffic models he actually argues about the solutions being Nash equilibria or not.

Common knowledge is something which is a game theoretic notion. Originated in the sixties by Olman, but now studied by Halperen in the eighties and nineties in distributed AI.

Mechanism design is something which is a game theoretic notion for how we in social systems would design a mechanism. For example, by which a monopolist would not be able to extract more welfare from society than he or she should. Or mechanism design might be, for example, how can we design an auction such that the bidders will be forced to bid their true bids, and not have incentive to collude or something. Exactly those mechanisms are being used by Tom Sandholm at Washington University to design automated contracting protocols for Internet agents.

So there's a strong overlap between multi-agent systems and social systems, at least conceptually. But what happens if some orthodox notion from social science fails to arise, fails to emerge in multi-agent systems? In particular, what if there is a theory of general equilibrium, a theory of how markets work? What if it turned out - and it does - that it's impossible to get a multi-agent system to produce a point at which supply equals demand? Does that mean we simply have agents who are too simple? I believe it means that in fact the axiomatic foundations of economics are unrealistic, and make in fact the solution concept intractable. For example, in almost all of mathematical social science, we assume that agents have strictly convex preferences. They're expected utility maximizers. What that means is agents would be indifferent between a one in a million chance to win one million dollars, and a one in five chance to win five dollars. Almost everyone in this room would express a definite preference over one of those lotteries, but you should be indifferent if the expected-utility hypothesis were valid. Constant returns to scale is a constant assumption in economics. It's always known to be violated at the plant level. Perfect rationality agents - agents who can compute what all the other agents in the room are going to do. People who are not economists are amazed to hear the current state of economics, because it sounds so bizarre. Bob Lucas won the Nobel Prize two years ago for the theory that all of you have exactly rational expectations about how the future markets will play out price-wise. But you also know that everyone else in the room does too. For that reason, there's no expectational uncoordination, and so markets are perfectly efficient. Now it seems unreasonable that one person could have a valid model of how the markets are going to work. If we're all perfectly correlated, then all of us have exactly the same model. But perfect rationality is in fact the basis of all modern economics.

The notion of the core is a game theoretic notion. In essence it says that the only exchanges that are allowed in markets are ones which no agent could block. I'm only allowed to make a trade with Bill, if in fact Yaneer says it's okay. Now this seems completely unreasonable, but in fact this is the orthodox received notion of a solution concept for games in game theory. So it turns out, just as a technical matter, the number of possible coalitions of the type that Bill and me, vis a vis Yaneer, scale like the number of agents to the eighth power. The number of agents to the power of the number of agents. That's computationally intractable. But this solution concept remains the axiomatic foundation of almost all of game theory.

There are many unrealistic parts of social theory that are very hard to reproduce with agents. We're forced to use simpler representations, and simpler models than these axiomatic models.

In summary, multi-agent systems in general fail to yield idealized outcomes of economic theory. When Mike Wellman at Michigan tried to build agent markets on the Internet which would be wall raising in character - meaning they would actually solve economic problems the way economists say they would solve them - he found it very computationally difficult to make that happen. It turns out a new theorem now in economics, and a new theorem in computer science from a couple years ago, is that the difficulty of computing the Brower fixed point theorem, which is what all the theory of markets hinges on, is in fact an NP complete problem. It's going to be very hard for a so-called wall raising auctioneer to actually execute the calculations.

It seems when Wellman finds difficulty getting his auction box to perform effectively, it's tied to this conceptual difficulty which has been neglected up until recently. One response to this is we need to build agent-based models of real markets, which may be at variance with the received theory. I've written social systems are multi-agent systems, but I want to also argue the opposite; that multi-agents are social systems as well - it cuts both ways.

In particular, a whole class of these axiomatic assumptions which are problematical in social science, but which the multi- agent approach helps us alleviate, are the ones associated with aggregation. The Nobel Prize winner Ken Arrow wrote in the fifties about the difficulty of aggregating preferences. Meaning there are various kinds of paradoxes associated with the fact that if there are a suite of candidates for office, you might prefer one set of candidates, I might prefer a different one; and there's no way for us to rationalize our preferences for those candidates. Aggregation in general is an important problem.

I want to argue here is that there's an individualist bias that arises from this in all of social science, but in particular mathematical economics. It says that the way to understand an economy or a society is to imagine it as being simply one representative agent. We're going to model the whole thing that way. So for example, you often will hear loose talk about a market as having its own psychology. The market was happy today, the market was upset about this or that. This is the notion of the market as one single giant mind. Which of course may have some metaphorical use, but it would be more useful if it was a theorem that it behaved that way. A hundred years ago in Alfred Marshal's famous textbook on economics he described the idea of a representative firm. If you understand how this one representative firm works, you can understand the whole production side of the economy. This is now thrown into disrepute by multi-agent models of firms.

What I want to focus on now is that notion that macroequilibrium is the result of microequilibrium. Traditionally the methodological stance of game theory has been to notice that at the macro level of society we don't have large fluctuations, in general, in any variable. If you look at the macroeconomic data, it's very stable over time. If you look at the way people interact with each other, it's very stable over time. The case I'll deal with right now is the firm size distribution in the US. Now we all know that firms come and go. Firms die, and are born every day. But in fact the firm six distribution has been stable for the last hundred years. There are a few very large firms. There are progressively more greater numbers of smaller firms. This follows the Pareto distribution very closely.

This stability lasts over a hundred years and across countries too. Britain has exactly the same slope of the Pareto distribution for the firm size as we do. That's been stable for a hundred years. People view that kind of global equilibrium, or that kind of macroequilibrium as arising from microequilibrium. They believe it must be stable at the micro level for that to be true. So all of game theory starts out from what? Game theory says let's solve for the equilibrium of a two by two interaction, and that's going to inform us a lot about how the macroeconomy works. Just to give you a sense of that, I would like to argue mathematically that this is not a necessary feature. There's some very high dimensional microsystem, X, which is being transformed dynamically into XT plus one. It's just a mapping from RN to RN. But now let's build a reduced form model of this. Let's look at the macro picture of this. There's some aggregation function, A, which projects this high dimensional, very heterogeneous micro world into some macro state like GNP. This is a heterogeneous world of Dunkin Donuts on the corner, and you being at the conference today, and John Deere has a new tractor, etc. And this is the world of GNP.

Presumably there's some aggregate dynamics like this, G. What is the relation of these two systems? They must be related somehow, since they purport to be the description of the same system. The aggregate state, at time T plus one, is just going to be the aggregate of the microstate at time T plus one. It's also the case that you can write the dynamics in this way. That the aggregate state of T plus one is an aggregation of this microstate with the macrodynamics, G. Now the question is, are these the same? The general answer is no. Let's think about equilibrium now for a moment.

When you're at a fixed point here where X of T equals X of T plus one, a microequilibrium would certainly imply a macroequilibrium, since if this fixed point obtains, it must be the case that this thing is going to equal that thing. But notice that it's not generally true here. For example, it turns out any vector X, which is going to be in the null space of the aggregation operator, is going to vanish. You're going to have a fixed point in the aggregate dynamic. It's very possible to have macrostationerity and macroequilibrium, without having a microequilibrium. This seems to me the basic problem with orthodox game theory - it makes this scaling assumption. It says that if at the macro level we observe not large fluctuations, the micro must be in equilibrium.

Quickly I will go through a simple agent-based model of the firm, and essentially the punchline is it turns out it's going to be a non-equilibrium model, due to the way the agents are competing with one another. But it's going to yield exactly the empirical size distribution function. Here's how it goes: in the current neoclassical theory of the firm, every firm is a profit maximizer, there is a principal, a CEO at the top who has multiple agents in his employ, he actually writes contracts for each agent, and the whole thing is in equilibrium. A multi-agent system model of firm formation evolution is what I'm going to describe now. Here's the basic setup.

Imagine a group of N agents. Each one can provide one thing to production. We'll call it effort. Something between zero and one. The total effort in the firm is E. The total output is going to grow with increasing returns to scale, meaning that Bill and I working together can do more than we can as individuals. That's all this means. Agents receive equal shares of output, but this is not essential to the argument, or to the results. Agents in general like to work, they like income, but they also like to engage in leisure time, time that they don't work. That's all there is to it. Here's the behavioral rule for the agents. At time zero we're simply going to start them all out as individuals, singletons, working for themselves. Then we'll simply pull them at random, or pick one at random, and say would you like to work with one of your neighbors? Yes or no? Neighbors in some notional sense. And then, because of the increasing returns, in general they will want to work with neighbors. And so what will happen is firms will start growing. Now once firms start growing, we're going to continually select these agents for how much effort they want to provide to the firm. What happens is, when a firm gets quite large, the amount of effort that's rational to put in to maximize the utility function is actually quite small. Now why is that? It turns out it's just because in a very large firm, if you put in a little bit more effort, you get very little of the income, but you actually receive a lot of reward from leisure if you put in less effort. So once a firm gets large, it suffers from free riding, and the average effort level in a firm will start falling. Then we can give agents the option to leave a failing firm, and go out and start up a new firm. That's the basic setup of this thing.

Let's look at a realization of this thing. At time zero, the agents are arrayed from top to bottom in red dots. There are I think five hundred or a thousand of them. They're all working alone now. An agent who's colored red is the founder of the firm. What I'm going to do is simply pick one at random, and that agent will decide whether he wants to join with this agent or that agent. And if the answer is yes, then I actually will just move the agent from here over to here, and color it blue. Blue means you're one of the employees, the red guy is the founder. You're going to see over time an emergent, or the evolution of firms. I'll run it forward a few periods, and stop it. What you see now is a group of maybe a couple dozen agents who have formed a firm here. It's happening up here a little bit as well. There are lots of singletons in the population. You see it runs very fast. It turns out the maximum firm size here is only on the order of a hundred or something. You can dial it up to make it larger, if necessary. The main thing to look at on the statistical screen coming up here is that in the population as a whole, the average effort level is around .25. But in the largest firm the average effort level is .09. In these large firms people are not putting in very much effort, because they can still get high reward for their efforts, or high reward because of the increasing returns aspect of production. In that model there's constant flux at the micro level. There's constant growth and death of firms. People are constantly changing jobs. There is in fact a stationary distribution of firm sizes. It follows a power law, and the exponents can be estimated, but this is exactly like the real empirical data.

Now for those of you who know the work of Per Bak on power laws, in the background of what Perbach says is that it's very hard to get these kinds of fat-tailed distributions coming out of equilibrium systems. They're inherently nonequilibrium. This model of firm formation that I've just described is inherently a nonequilibrium model. It's an analytical result that all the Nash equilibria are unstable.

If you then were to imagine a social scientist approaching the theory of the firm from an equilibrium perspective, it would seem that there's no way an equilibrium model would ever produce this kind of result. This is one example of where the multi- agent in perspective has given us a sense in which the entire focus of game theory on equilibrium solutions, and the current theory of the firm on equilibrium contracts in particular, is problematical at best, and probably just misguided.

I gave this talk once, and some applied economist said gave me this data: of the five thousand largest firms in the US that existed in 1983, how many of them exist today? The answer is half. The other half don't even exist. They're not autonomous entities anymore. They've either been absorbed, or dissolved, or bought. The point is there's a vast turnover in the US firm size structure, and this seems completely incommensurate with equilibrium models. The multi-agent system perspective lets us study these nonequilibrium dynamics. For me this is the power of this approach.

Mr. Fulkerson: Our next speaker is Professor Abhijit Deshmukh, from the University of Massachusetts. Abhi got his Ph.D. from Purdue within the SUNY system, and then Florida State, and is now at the University of Massachusetts. He will talk about some of the perils of dealing with agents.

Mr. Abhijit Deshmukh: We are going to look at agent systems, and what issues are involved in actually using these systems. Finally, how do you guarantee what kind of performance you get out of these systems? Before I begin my talk I want to preface it by saying that I am a big proponent of agent-based systems. This is a contrary perspective that I'm trying to take now.

The first issue that I'm going to talk about is when to use agents. A lot of discussion that people have seen in literature is that agents are extremely good for systems, but are they good for everything? As I have noticed with some business applications, typically things tend to be used because of the hype. Agents are cool, and so I want agents. But why do you really want agents? What is it going to buy you? What is it going to give you in addition that you don't have in existing systems?

The second thing I'm going to talk about: agents are not solutions to all problems. In fact, I just heard that Microsoft came up with a solution to kill this agent. The point I'm trying to make is how is it that your system is going to benefit from from using the motions of agency.

That gets to my last point: agency allows you to solve problems in a distributed manner in a dynamic environment. That's the advantage, or the main goal or main focus of using agents. If you get to a point where you're trying to apply this religiously, and applying them to all sorts of problems that they don't fit into, it might then be detrimental to the system and to the problems.

Finally, what are they good for? Agents typically allow you to look at emergent phenomena that you probably would not see in a highly centralized system. What you're looking at is distributed intelligence that is either evolving or coping somehow with the environment. We have a model of ants here, because some of the work I've been doing with a biologist colleague of mine at Stony Brook has been looking at the behavior of ants from an individual perspective to look at the emerging behavior. Some of the things we have found are kind of counter to the conventional wisdom that Wilson has in his papers, and a lot of other social biologists have, with respect to how ants look for food. The decision metrics that people use is going to the highest value. If ants are faced with a choice, they choose the highest value. If you implement a society of ants with those rules, you do not see the match between the experimental, and the simulated observations. There are various explanations that we can draw, or various explanation we can give for those things, but the idea here is that social systems have evolved the rules that they use over the years. Probably bad rules have been thrown out, and the good rules still exist. The notion of looking at emerging systems using agents gives a good application domain for agents.

Now, if you really wanted to use agents, the question of functionality of an agent is important. What should an agent be? What is the domain that you want to define as an agent? Suppose you wanted to solve the problem of global warming. Do you model your entire planet as a single agent? Or at what point do you start breaking these agents? What are the tradeoffs involved if you have a lot of agents as to having a few agents that are very, very sophisticated. And these do have tradeoffs based on what you're trying to implement. Coming back to my definition of agents: are you really exploiting the concurrency and distributed nature of this approach? If you're not exploiting that, then what are you getting out of this exercise?

How do you put goals into agents? Here is an example. Are these really your true goals? You go to buy a car. You look at the utility of the car, of the engine in it, and the design of the car with respect to how good the car is in terms of function. Now studies have shown in a lot of automobile marketing literature that the metric of this added utility, of added features enhances the utility significantly. But this is a non-lineal utility enhancement, in the sense that if you go to buy a car that is clearly not giving you the functionality, the added utility doesn't make a difference. But if two cars have comparable functionality, then this plays a big part in the decision process. Now the question, coming back to designing agent-based systems, is how do you incorporate these utilities? How do you, first of all, elicit these utilities out of the systems you are trying to mimic, and how do you incorporate these real utilities into agents that can make these decisions, or that can represent you in the marketplace, or in the decision process?

Second is global and local goals. If you look at true distributed systems, you have a certain set of local goals that each agent is trying to satisfy. The example I have is from a shop floor situation where you have a bunch of parts, and you have machines, and somehow you have to meet the production goals. Now the metric that a local manager may have is to have highest utilization so as to justify the existence of the shop, and maybe even get more resources in the future. By doing that, that might in fact hurt the overall productivity, or the overall profits of the firm, because you might get delayed or because you're missing the due dates. How does the local and global goal conflict, or how does the interplay relate to the overall performance of the systems? That's an issue that a lot of implementation examples don't really look into. Finally, how do you look into ethical and moral issues? This is getting to a point where agents are now representing decision makers in issues that relate to global policy. How do you incorporate ethical and moral issues into the decision processes of agents? That's an issue that hasn't been looked at.

Coming to a situation where we have agents that are trying to make decisions for you, where the humans are essentially either out of the loop, or just serving as monitors, the question of ambiguity in agent communication comes into play. I've come back to a scale where each agent is communicating. If you had a design process, for example, that was being conducted by two agents, and one of the agents was really an electrical engineer, and the other one was a mechanical engineer. And they said well, the load is not good for me. If one communicates that the load is not good for me, what does it really mean to the other agent? What I'm getting at is the communication itself might be ambiguous if you don't have the context or shared intent that is conveyed between the agents. As you increase the number of agents in the system, it becomes extremely difficult to convey that to everybody.

One of the examples I have is simple Web searches. The word I typed in the

Web crawler was "quack". The simple word "quack". How do you convey the intent, or the context in which you are relating this word "quack" to a particular search agent so that you can get exactly the type of information that you want? I got all sorts of information from duck hunting to bad doctors. This is really an issue when you come to a large society of agents that is operating autonomously.

And finally, the issue of using standards. Not just standards that I've mentioned, the communication standards, but also some kind of ontological standards for different applications of agents. That hasn't happened yet. There are no de facto standards for agent ontologies, or agent communications that exist currently. But that's an issue that needs to be addressed, or it will be a pitfall, in terms of implementing these systems. This is especially so if you want these systems to interact with other agent-based systems that you haven't designed yourself.

The next one really comes into play when you look at what agents should learn, or how should they evolve through the system. Some of the work that we have been doing, in terms of learning, has shown that the maximum gains you get in the performance as a system happen with the first very small amount of learning that your agents perform. After that, the marginal advantage is fairly limited.

The idea here is at what point should the agents stop learning? At what point is the computational complexity of learning greater than the advantages your system gets out of it? That's an issue when you implement some of the mechanisms, and the orders that are associated with these mechanisms. Each one of these mechanisms that are listed have different types of computational requirements, and they produce different types of learning.

The last one is designing agents with too much, or too little intelligence. This notion has come from game theoretic approaches. When you have hyperrational agents, or you have very myopic agents that are essentially looking at one decision. Hyperrationality doesn't necessarily buy you a lot of goods in the long run. The question is what is the optimal level of intelligence that your agents should have? These are design issues for an agent-based system that I'm talking about.

The next one is really a system-level issue, or a set of system- level issues. If you have a set of really distributed agents, will they ever converge? I'm going to run an experiment and I need three volunteers to serve as agents. I'm going to show what I'm talking about in terms of convergence. We've got one. I need two more volunteers. Bill can do it. And... alright. These are my three agents. I have given them certain goals. Each of these agents has certain goals, and they have one variable that they can control. Now they affect each others' goals.

I have a blackboard over here. Each of these guys can see the values of the variables they control being set, but they're not going to know each others' goals. All you know are what values they are currently set. Okay? I'm going to start with some number. Some number that I'm picking. What it means is that agent one, the value that they're controlling is zero; agent two it's two, and agent three is two. Now what I'm going to ask agent one is to change their value so as to basically meet their goal. Your goal is back here, okay, with the other two guys' values?

M: Yes.

Mr. Deshmukh: Alright. Agent two, are you okay with your value?

M: No, I want to change mine to three.

Mr. Deshmukh: Agent three. Now that you know that it's zero and three.

M: So I want to change my value to one.

Mr. Deshmukh: Okay. Now agent one?

M: Well, I'd say one.

Mr. Deshmukh: Okay.

M: I'm happy.

Mr. Deshmukh: You're happy.

M: I'll change it to zero.

Mr. Deshmukh: Ah. Alright.

M: One and a half.

Mr. Deshmukh: Well, let's make it two.

M: I'm still happy.

Mr. Deshmukh: You're okay.

M: Yes.

M: I have to choose minus one now.

Mr. Deshmukh: Ooh. You have to choose minus one. Okay. Alright. Well, I thought this was going to converge in two or three iterations.

M: It would have if I had given you the right answer maybe.

Mr. Deshmukh: Keep that in mind, because I'm going to talk about rogue agents. The process I was doing was essentially a distributed search process. I was polling the space in a particular sequence. I went one, two, and three. I could have gone by saying that you all just start computing as fast as you can, and just keep giving me answers on what you see. Or, I could have said that I'll look at the deviations that we had. So for example, agent two changed from one to three, so that was the largest deviation. So I'll poll that agent more frequently than other agents that seemed to have converged. I can come up with processes which are extremely creative, if you will.

What you can see is that the system that I gave there was an extremely simple system. From a mathematical point of view there were three linear equations that I gave. With this system you can see any reasonable polling policy converges to a fixed solution that is acceptable to everyone. Now we are just talking about a single point solution here. We are not talking about the optimality, or a maximization function. Now what I'm going to show you is another example with another set of linear equations that essentially diverges from most of the polling policy. What I mean by divergence is you don't come to a consensus. The point I'm trying to make here is that the convergence of these distributed systems to a particular fixed solution, or to a point, not only depends on their own objective function, but the relationship of these objective functions, and how the agency process, or the negotiation process, or the option process, if you will, is set up in the system. That governs how well you're going to converge.

Those are the issues that you need to look at when you formulate an agent-based problem.

The second question is how good is that solution? Well, we don't know. We don't know what the optimality of that solution is.

Now the last two points relate to whether or not your agents really have to negotiate. The question is do agents always have to negotiate? First of all, what do we mean by negotiation? The way I'm defining negotiation is not a one shot deal, such that I ask you the price, you tell me the price, I agree, and we are all set. I look at negotiation in the form of you're trying to change either the space in which you're looking at your solution, or the actual number of dimensions in which you are looking at the solution. The example I have is one of the negotiation or auction mechanisms I use to buy an airline ticket. There are auction platforms now available for buying airline tickets online where you basically are quoting a set of prices, a vector of prices, and the auction mechanism, or the market mechanism determines whether or not that is going to be available. You have a certain number of dimensions, and the market determines, or the airline industries determine whether or not that's going to be feasible at a given point in time. The point I'm making here is that if the space is feasible, and you have a potential solution, you do not necessarily need to have negotiation. Negotiation is not a necessary component of having agent-based systems.

Next I will talk about more system-level issues. When there is no supervisory mechanism, how do you detect stalling and deadlocks. An example of deadlock is A1 waits for A2 waits for A3, and nothing happens. Now who detects that? How do you detect that something like that has happened in a system that is operational? They need to be monitored. By monitoring, are you getting away from the true essence of agency? What does that buy you?

Also is the question of the question of the theoretical limits for number of agents you may have in the system, or the theoretical limits for informational capacity of the channels that you have. Essentially if you exceed those, what you are looking at is an information gridlock when nothing else is happening, not because of the computational requirements, but because of no information channel capacity being available. That's a design issue from a system level perspective.

The next point I want to make is how do I design an optimal agent system? I know how to do that with operations research kind of problems; I know what kind of models I can use, given the condition of the variables. But for agents, there is no single model that incorporates all of the aspects of agents. You have different models dealing with different things. Right now, the tools that really are used, are essentially something that is a simulation environment - a Swarm kind of platform that we use often. There is no analytical descriptive model that exists currently that can tell you what is the best agent-based system for you? What is the best aggregation or disaggregation level that you're going to have? We're talking about nonequilibrium models when you have the system evolving, where suppose if you look at the coalitions that may form in your system where two companies might decide to bid on a project together, and submit it as a single entity. In that case, the structure that you may have had, in your analytical models, would not work at all. That's an issue when you look at designing these optimal agents.

The last thing is what if I have agents that either don't perform as they should have, or are intentionally negligent? How do you deal with these issues when you start looking at agent- based commerce? For example, my agent, by negligence, or by wrong algorithms used at wrong times, purchases some ten thousand gnomes for my house. What do we do if I have an agent representing me in the market for buying things, who is responsible for this activity? Obviously, if that's my agent, it's going to come out of my checkbook. The second issue is if you have fake information sites, or if you have misinformation being fed into the system. How does that affect your system, and performance of the overall goals?

The next one is basically you have agents that have no authority to place orders. For example, I have an agent that talks with Amazon.com on a site, and that has no authority to buy new books for me, but just goes on buying. How do you verify the authority of these agents? The last one is what if agents are corrupted by the network? Who is responsible for that in an electronic commerce environment? For example, you set up an order, and the order form gets corrupted. How do you deal with these issues when you're looking at extremely large agent-based commerce mechanisms?

To summarize: agents do have significant advantages for modeling complex adaptive systems. However, you have to make sure that it is the right system that you're applying the agents for. This is still a very young field. We need to know a lot from an academic point of view and an implementation perspective to actually make these things happen in the broader context. From the practitioner's point of view, I would say this is a good area, but beware until you actually see the results before you put your money in this thing. That's my conclusion out of all this. Thank you.

Mr. Fulkerson: Our third and last speaker is Van Parunak. He is currently a scientific fellow with the Center for Electronic Commerce at Ann Arbor. He got his Ph.D. at Harvard University in Near Eastern languages and civilization, and his Master's of Science degree from the University of Michigan.

Mr. Van Parunak: I want to describe to you some experience we had in a program called DASH - Dynamic Analysis of Supply Chains - in which we built a Swarm, agent-based model of the supply network in order to explore some of the dynamics that can be experienced in what is intrinsically a distributed system in a changing world - a complex adaptive system. My work in this is done in collaboration with Bob Savit and Rick Riola.

I'm going to begin by briefly outlining what a supply chain problem is and what the problem domain is that we're discussing. Then I will walk quickly through three different approaches to modeling this domain. First of all, I'll describe the agent- based model that we use. I'll show you how it was put together. And of the many interesting behaviors that we observed, I'm going to focus on one, which I call the inventory oscillator. Then I'm going to show you two different equation-based models of the inventory oscillator. One in a systems dynamics formalism, the other in an Excel spreadsheet. Then I'll show you an analytical model in which we can prove certain behaviors of this oscillator. Then I want to try to draw a lesson from what was gained by taking these three different modeling approaches to the same domain problem.

A supply network is a common phenomenon in modern manufacturing. You may or may not be aware that none of the Big Three automakers in this country manufactures its own seating systems. They are provided by companies like Lear-Sigler or Johnson Controls. And recursively, Johnson Controls does not make the seating systems that it sells. It puts the seat together, but it purchases components from a variety of soft goods suppliers, and hard goods suppliers. Now this supply chain right here, with about twelve suppliers to Johnson Controls, is one that we instrumented and monitored during a pilot deployment of some manufacturing technologies, so we know it quite well. But it's only a fragment of the entire seating supply chain for Johnson Controls. There are about a hundred and forty more companies in that system. You'll notice that the supply linkages between them go around in circles. There are many to many, not just one to one. Supply network is probably a better term for what we're looking at here, but supply chain is the commonly accepted jargon in the industry.

There are a number of business drivers that have led people to accept change like this. In the interest of time I'm not going to read those through with you, but just point out historically the trends that we see. Since the early 1970s, the amount of value in a vehicle added by the Big Three OEMs has been dropping consistently. Probably the largest drop here is from fifty percent value added by General Motors back in 1970 down to something less than forty-five percent in 1995. So there's increasing tendency to draw one's supply from outside of one's own company.

The supply chains have lots of problems associated with them. I've highlighted three that can be traced to the dynamical behavior of the supply network. The information that flows through a supply chain is often late, and often wrong. Supply chain dynamics impact both delivery and cost, and suppliers can't trust the forecast that they're given, and they often do not trust the forecasters; that is, the companies that are giving them forecasts. Now those three in particular, among the others that are listed there, can be directly related to dynamical behavior in a supply network. To get a handle on this, we constructed an agent-based model in the Swarm platform. It was a very simple model. It had a set of manufacturing entities. At the top, we had a consumer, who consumed goods; at the very bottom we had a supplier who supplied the raw materials; and we had two intermediate sites that transformed raw materials into finished goods. Information flowed down this list through mailers whose purpose in life was to delay and corrupt the information that was given to them. To model those phenomena in real life, each site had a production planning and inventory control algorithm embedded in it. In this case, a simple MRP algorithm, for those of you who are acquainted with such things, that transformed the orders that it was receiving from its customer into the orders that it would in turn pass on to its suppliers. As the sites produced the material, the material was shipped back up the chain. That was the simplest supply chain we could think of. When we constructed it, we said this is going to be real boring, and then we can start making it more complicated step by step until it does something interesting, and then we can do our research. We never got to the point of enhancing it. This simple linear chain kept us busy through the duration of our funding. It exhibited a number of interesting behaviors. Just with this very simple chain we saw something that people have often seen in supply chain studies, and that is the amplification of variation. That is, the top level consumers' orders have very small variation. By the time you get to the bottom level supplier, there's a huge swing in the variability of the orders. We also observed the development of spurious correlations among orders. Something that we don't believe has been observed before in the literature. We observed a persistence of variation. After a one time change in the supply network, the system continues to ring for about a year of simulated time.

Finally the thing that I'm going to talk to you about today is variation generation. This supply network can generate internal oscillations under stable boundary conditions. It's interesting to note that the top three of these are linear phenomena. The model was operating within a linear domain. The oscillatory phenomenon, though, is nonlinear.

Here's how we set up the system with stable boundary conditions. The customer on the top has constant demand. There's not even any stochastic noise in it. Constant level of demand at the top. The supplier at the bottom has infinite capacity, and always provides what is being requested by its immediate customer. But the intermediate sites see oscillating inventory activity. The configuration of the system is that we're doing everything in batch sizes of one, so there's no threshold nonlinearity associated with batching. However, sites two and three have a capacity limitation. They can produce a hundred units per time period. That represents a threshold nonlinearity in the face of demand, if demand goes higher than one hundred. We set consumer demand at varying levels to see what would happen.

Here's what happens. Up until time twenty here, we had supply and demand balanced at a hundred units per time period. At time twenty, the consumers' demand went to a hundred and ten. And what happened was the system entered a sawtooth regime of oscillation. What you're looking at in the red and blue lines are the inventory levels at site two, and site three. The sawtooth climbs rapidly, and then drops off gradually. When we increase the demand at the top level consumer to a hundred and fifty, now we're in a situation, after we get out of the original sawtooth, with a much finer frequency sawtooth. You'll notice that a shift takes place right about here. Before this point, the oscillators are out of phase with each other, but here sites two and three lock into phase with each other.

M: You said variation correlation. What's that again?

Mr. Parunak: The top level, the demand, is IID, Independent Identically Distributed Galcian Variates. At the low level supplier, you see correlations in the time series. In other words, they're not there in the consumers' demand, but they are there in the real world.

Now what's really getting interesting is when I jack my demand up to two-twenty over a hundred, I have a biperiodic oscillation. There is a lower frequency sawtooth on which is modulated higher frequency spikes. Some of you would say this is unrealistic. Nobody would ever impose more demand on their supplier than the supplier's capacity. In that case you haven't lived in manufacturing. Customers often do this to lock in suppliers and make sure that the supplier has capacity, and is expecting business from them in a pinch. So they'll intentionally over-represent their demand; and we actually have concrete data from the automotive industry to back up that behavior. It's very realistic behavior.

Well, what's the mechanism in this? What happens is if you have a safety stock level, and your demand coming in is greater than what you're producing each time period, you have to steal a little bit from your safety stock each time period to make up the defect. When the safety stock gets to zero you can no longer fulfill an order. What you have to do is backlog the order that you received at that time period, and stockpile the production that came in that time period. Those backlogged orders can delay the impact of changed orders. There's a memory effect. When demand drops below capacity, if you've got backlogged orders, the oscillator continues to function, but it happens in the reverse direction. The sawteeth change their slope, their orientation. That's the simple agent-based model.

We then said this is interesting. Why don't we try to model this using a system dynamics approach? Here's the differential equations that underlie the system. There are only four time- varying variables in the system. There are the two finished goods levels, and the whip levels that feed into them. In the model, material comes into whip, it's aged there for a period of time to model production time, and then it's gated through site capacity into finished goods. You're probably more used to seeing system dynamics models in the form of a diagram like this. That's a Vensim model that we ran the system on.

There's an even simpler model that you can use. It turns out that the fundamental dynamics that we're looking at here don't depend upon a two-stage production. You can see them in a single stage production. So you can represent the whole thing with two difference equations. One says ship at time T if the inventory at the beginning of that time period, plus what I produce during that time period, is greater than or equal to the order size. Otherwise, don't ship anything. My inventory next time is equal to my inventory at time T, plus what I produce, minus the shipment. You could put that in a two column spreadsheet, and graph the thing; and it's a very simple equation-based model of the behavior that we see.

Here is the agents in Swarm. There's the double sawtooth that I showed you. There's what we got out of a Vensim model.

You can see it's qualitatively the same behavior. The first observation is who needs agents, I can do it in Vensim? Of course you could take it to the next stage and say who needs system dynamics? I can do it in Excel. Well, obviously, both of those conclusions are inappropriate. There are many things that you can do in a stock and flow model, such as a system dynamics model, that would be very awkward to do in Excel. It just happened that in this case the fundamental dynamics could be represented in a couple of interlocked difference equations which we could capture that environment. It's also true that there are a number of phenomena that we saw appear that we could not duplicate in a system dynamics model; largely having to do with the replication of the PPIC algorithms that were involved, which we simply could not find a way to represent.

If anybody here knows of a realistic MRP or PPIC algorithm in a system dynamics formalism in captivity anywhere, I would love to know about it. Some of you, I know, are specialists in this field. Here's an example where things aren't quite so clean. Here's a moderate overload, a hundred and fifty over a hundred. Here's the Swarm model. You will remember that we were out of phase, then we locked into phase. There's none of that complexity in the system dynamics model. The system dynamics model just immediately jumps into a steady oscillatory state, and sits there. You don't see the transition effects that were captured down here.

Now we went one step further. As we studied the outputs from these models, we observed certain irregularities in them. Being scientifically minded, we wanted to understand what was going on. You'll notice here we've got sort of two levels of decreasing things. There's a local maximum here, and then it starts to go down again. There seem to be certain limits on this. We began to ask how do these regularities emerge? What lies behind them? We were able to prove a number of theorems about this oscillator.

The assumptions here are that the demand and the production have no common factors, so we divide out all common factors to get the variables in terms which the theorems are expressed. We measure inventory in those same units. So we have to divide out a factor of fifty that's common to both supply and demand. We also divided out of inventory. And we define a supplementary variable, which is the minimum of production, and demand minus production. Here are the theorems that we got.

One is we can show that this region, inventory between zero and demand, is an attractor. If you start the system outside of that, it will be attracted into the region, and once the inventory is in that region, it stays there. That's a theorem, that's provable. Second, we can prove that the system is periodic in D, which is the demand level. That number of time periods brings you back to the inventory you had before. And it's a deterministic system, so once you hit that state, you're there. Third, we can prove that over the course of that period, the inventory will assume every value in that range. We can prove that local minima and maxima are adjacent, that is one time step apart. The system goes from a local maximum to a local minimum, or from a local minimum to a local maximum; and the patterns are predictable, and they depend on that H variable.

We can prove that there are H local maxima in one period of the system. We can prove that the local maxima have two different lengths. They are not of the same length. If you have two different local maxima, the periodicity (the subperiods) of those are of different lengths, and we can prove that the less common type of extremum, are always separated by instances of the more common type. Those are all theorems that come out of an analytical theory of the inventory oscillator, which is available in our report.

Now, one way to summarize all those things, which my colleague Bob Savit came up with, is that those statements are equivalent to observing that the dynamics of this oscillator live on a torus. Here's a simple little plot of one of these oscillators. We observed that if you take the slope of this line, and continued it up, and then picked it up at the bottom, and brought it up to here again, you'd run into this point. In other words, if you took this, and wrapped it around from there, and down there, it would continue. You could effectively join the top and bottom of this diagram together, and form a cylinder. Then, because it's periodic, because we're coming back to our starting point at the end, if I wrap the ends of this diagram together like this, and connect them, I have a torus. The dynamics live on a torus. What does that mean? There are two implications. One is that if we had an irrational D over P, we'd have a quasiperiodic structure, which would be very complicated-looking. However, because the dynamics is restricted to the surface of a two-torus, it cannot become chaotic in the limit of continual flows (unless you introduce other dynamical effects). Now this system can become chaotic, but within the scope that I have described to you, our analytical model allows us to reach the conclusion that under these particular conditions a system cannot become periodic. I could never have showed you that from the agent-based model. I could never have showed you that from the system dynamics model. To get that kind of a result, I have to be able to abstract it to the point of this kind of a proof.

Lessons that we derive here. When your only tool is a hammer, every problem looks like a nail. We want to learn, as system modelers, when we should use other tools as well. Screwdrivers, wrenches, saws. Let's build a simple taxonomy of models. We started with an agent-based model in which agents were related together by behaviors. Then I showed you two instances. A system dynamics model, and an Excel spreadsheet model in which observables were related together by equations. I think it's important to emphasize here that there's a very fundamental shift that's taken place in modeling between those two categories of models. In one case, the fundamental entities of the models are the individuals in the problem domain itself. In the other case, they are abstract quantities which we as observers associate with those individuals.

I'm going to argue that there is a logical priority of an agent- based or an individual-based model over an equation-based model. One deals more closely with the reality of the domain itself. The other requires a step of abstraction to say I'm interested in these observable quantities, and in the relationships among the quantities, rather than in these entities, and the behaviors of those entities as they impinge on one another.

There's a third step that we then took. We went to our analytical model in which we could prove theorems, and now we have even higher level abstractions that are related to one another by various logical principles, such as modus ponens. It's pretty clear that we have taken yet another step of abstraction as we have moved in the dynamics of our model the manipulation of the model from an approach of simulation, to an approach of analysis and proof. These two categories of models are used by running them. We turn them on, and let them go, and see what they do over time. This category of model down here is manipulated in a very different way, according to a set of logical inference rules.

The strength of agent-based modeling over equation-based modeling is that the model structure, being divided up according to entities in the world, as opposed to observables, is often advantageous in a commercial environment. It permits the protection of proprietary interests, because the units of my model are the firms that I'm dealing with here, as opposed to variables that may span firms. It's a better foundation for implementing distributed control.

We found a tremendous set of advantages in system representation such as ease of construction and being able to represent, in a straightforward fashion, the decision algorithms that were being used.

There's often an issue of verisimilitude. It's very tempting, in an equation-based model, to use certain lumped parameters or averages over individuals. When you're modeling individuals as separate agents, it's quite straightforward to associate with each agent its own local variables. And so there's less of a temptation to average things out. The deviation of individual agents from a system-wide average can often be the small difference in initial conditions that leads to divergence in behavior down the road. That's why people are finding, in fields as diverse as insect ecology and traffic models, that individual-based or agent-based models are often superior to equation-based models; because of that avoidance of lumped parameters.

Weaknesses. This is the least explicit type of model, in terms of system level analysis. What you see is what you get. There's no explicit setting forth of the relationships among the things that we can actually observe, and it's the most demanding computationally. It seems to be the category of model best suited, in our experience, for exploration and discovery. Here's a domain, let's build a computer model, then we'll play with it, and see what we see. But the level of understanding will be substantially enhanced if I could start to write down some equations about this. If I can write down a set of equations, I'm making system level dependencies explicit. And I've got a model now that's more computationally efficient. I can run it on a slower computer. I can run it on my Palm Pilot if I want to. There are weaknesses. The three weaknesses correspond to the three strengths of the agent-based model; and I'll refer you to our paper for that. But I think equation- based models are powerful, in connection with an agent-based model, to test hypotheses about integrated formulations.

Rob has done some very important work on the idea of docking models. The fundamental idea is model something in several different ways, and then compare the results of those models to each other. You will be tremendously instructed by the differences that you see, and what's required to get those models into line with each other. Some of you may be aware of Wilson's work in animal ecology, docking an equation-based model with an individual-based model. It is a very, very instructive exercise. It also helps us to formulate insights for communication.

I can now say here's the relationship among these variables. I wrote down an equation, and it captures it. In the analytical model, the strength is that this provides the greatest depth of understanding. Now I can say with some confidence that under appropriate conditions this system will not go chaotic, for example. I can formulate testable predictions using this. However, it is rare that I can in fact produce an analytical model for a complex system. When I can do it, it's only going to be in a very limited subdomain, as in the case here. In general, these are not computational. I can't give it to a computer to run. I have to sit down, and think real, real hard about it, and wrestle with it to tease out what the fundamental principles are, and how they're related. Going back and forth to the data from the simulations. This is tremendously valuable for packaging specific results for further consumption understanding, when you can do it. But it's rare that you can do it.

I've outlined to you the basic problem of supply networks, I've exhibited to you three different models that we constructed of one behavior of several that we observed, that is the inventory oscillator, and suggested to you that we don't just have a hammer. We're very fortunate in having an increasing armamentarium of modeling tools at our disposal. Agent-based models, in my mind, hold pride of place. That's where we start, because it's the closest to the domain, the most straightforward. Equation-based approaches help us to refine what we observe in the agent-based model, help us to begin to capture it in some kind of a closed form that we can communicate. If we are fortunate enough to be able to formulate an analytical model of some subdomain, then we have a very powerful tool in our hands for making testable predictions, and communicating our results more broadly.

Mr. Fulkerson: I would like to invite Win Farrell to join us. He makes his living with agent-based models, and it would be interesting to have another component, which would be the commercial view of things.

Mr. Win Farrell: I'm Win Farrell. I'm with PricewaterhouseCoopers. We do make agent-based models for commercial purposes. I've recently written a book called How Hits Happen. We do a fair amount of work in forecasting. I'm interested in some of the points that Van made about trust with the forecast. But anyway, the ideas of forecasting for very volatile industries such as CDs, motion pictures, toys, perfumes, and the like. I just want to make a couple of comments about what we do with agent-based systems that I don't know if I heard exactly today. I just wanted to make the point when we do build these commercial based systems, we started off a few years ago building what people would call toy-based systems. I don't know if that's a derogatory comment or not, but I think it's an important kind of role of these systems, from the point of view of instruction.

What we're doing now is we're trying to do things to help, in terms of decision support for real people that have real problems. With the forecasts, I'm interested in this trust factor, because it's always fun to wonder if people do trust you in that sense. We look at vast quantities of information - the data warehouse terabyte size of information that feeds our models. We work on agent populations that are in the hundreds of thousands to millions. We make, if you would, predictions or forecasts that are typically a week or so in duration. No longer than that.

We look at a vast number of adjacent probable conditions. We call it visiting the future. And we visualize that in sometimes metaphors, and visual metaphors that go all the way to the photorealistic.

That's what we do in order to commercialize this, what we consider to be a very important set of technologies and constructs. That's one approach to make this stuff commercial.

One question that I have, can we expect that we could find some of these scaling laws that can give us some guideposts as to whether or not we can look at agent models which have scores of agents, versus hundreds of thousands or millions? I know it's problem domain-specific. How deeply do we have to go into these agencies, in terms of their individual cognitive skills, and their rule bases?

Mr. Axtell: That's not an easy question, but I can take a stab at it, Win. I actually have a grant to study a problem that's very similar, which is what is the tradeoff between agent complexity, versus number of agents in a computational system? For example, if you have X number of clock cycles to execute, is it better to have a small number of complex agents, or a lot of very simple agents? As you said, it's going to be domain- specific. What I find to be a useful way to motivate this is to look at the data, and to see what the target is you're trying to hit. If the target you're trying to hit is going to be a distribution function at the macro level, then maybe you actually need to have a lot of agents to get robust statistics. If the target you're trying to hit is just some stylized behavior about how individuals act, then maybe a small number of agents are fine. You may need to have some more complicated kind of strategic behaving agents. I think it's always domain- specific. My personal sense is that having a large number of simpler agents is an easier problem than a relatively small number of complicated agents. As you know so very well, one of the powerful features about this modeling approach, that perhaps has not been touched on today in any detail, is that you can, in a very small amount of code, control a lot of execution. In essence the agents are just replicas of one another. They're heterogeneous, but they're going to be replicas to some extent. There's a funny sense in which a few thousand lines of code can control a few million bytes of data, or maybe hundreds of millions of bytes. There's a sense in which those simple agents, in vast numbers, are useful. Particularly in domains when you don't know how the agents are behaving, a large number of agents can cancel each other out, and you can get robust statistics, even though you're not sure how the real microworld looks.

M: You guys want to talk about scaling at all?

Mr. Parunak: I think it's important to recognize that the behavioral space of these agents is a very complicated one. And traditional methods of varying parameters to explore that space can be very inefficient. I think methodologically it's extremely important to look at using genetic mechanisms. My favorite approach involves fairly simple agents that are parameterized, and then taking the parameter sets for the agents as the chromosomes in some genetic approach, and using that to explore the space; because it can make a much more efficient search of what turns out not only to be a very large space, but also a space whose topology is extremely complex.

Mr. Deshmukh: As far as the scaling goes, I think the performance that you're trying to measure really plays a big part in how well a society will scale out. One of the things that we have been looking at is the type of the performance measure being monitored, in the sense to see if the performance is additive in the participants. If the formalism of the final performance is an additive, or an aggregation mechanism, then your predictions are easier to make than if it is some kind of a multiplicative or a nonlinear function. It really depends on the performance that you're trying to monitor.

Mr. Farrell: Just another point I wanted to make - the level of intelligence of a particular agent as sort of an adjunct to this, in terms of numbers of agents you can AKA stuff into a computer. You can just get a bigger computer, I suppose. But the issue does come that what we're facing now - those of you who know my work know that we're working on extremely primitive forms of learning systems. They're basically taken from some of the work that Bernardo Huberman has done. But essentially, one of the questions I have is that how can we generate agents that switch modes of learning and invoke either a higher resolution aspect of learning and information gathering, or a lower resolution, just based upon the context. I don't know if you guys have thought about this, but we're struggling with this now. Because we have for the duration of the simulation, agents that have a single form of learning. We try to build the simulation into that mode of learning, and then modify that single mode of learning. Now we're saying well, that doesn't do it for all the cases. We'd like to switch from one to another to another, depending upon the context. Have you guys faced this situation?

Mr. Deshmukh: The stuff that we have looked at, in terms of learning, has been with the use of analysis packages in a design environment. Let me give an example. We've been looking at design of aerospace components using agents. The question is when you are designing, say, a nozzle for a big aircraft, there are different analysis packages that you can use to do flow dynamics analysis. So you could do 1D, 2D, 3D. Which analysis do you want to use at what point? That's not exactly learning, but it relates to how at what stage in your design process you're in dictates the use of a particular mechanism. You may not know what the solution is, but if you know what domain you are searching, then that dictates how refined your search, or your learning process has to be. And that's the approach we have been using. It's not perfect, but it works.

Mr. Parunak: I'd like to make three points with respect to learning, particularly in agents that are intended for industrial deployment, as yours and mine are; and to a lesser extent Abhi's. The first is that while there are some very powerful learning mechanisms out there in the AI world, if I can't make fairly clear to a user what's going on inside an agent, that agent will never be trusted in industry. I have a very strong bias in favor of simple agents, and that includes learning mechanisms, as well as everything else. So I'm very much in favor of things, like reinforcement learning, that are quite straightforward. Now, but that doesn't mean you can't do complicated things.

The lesson of agents in societies is that complex system level behaviors emerge from very simple individual level behaviors. My second point is that this is true of learning as it is of everything else. There's system-level learning that takes place. The construction of a pheromone path in an ant colony is a memory that belongs to the entire colony, and can be accessed by the individuals in the colony by fairly simple mechanisms. The second point is when you're thinking about techniques, think about ways that the society can learn. Think about ways to use the environment to hold the memory, as opposed to relying just on what goes on inside a single agent's head.

The third point, stimulated by some remarks that Eric Bonabeau made a few weeks ago at the University of Michigan when he was discussing some of the ant models that you talked about. He came into this area from telecommunications, and some ant-like algorithms are now being used extensively in telecommunications for routing applications, and distributed networks. They have found that to make the systems work right, they have to have the pheromones evaporate more rapidly than they do in nature. Even in nature, they evaporate; but Bonabeau claims that functionally that's not very important. The observation here is that almost as important as learning in an agent world is forgetting. If we're dealing with a changing environment, the stuff I learned yesterday needs to go away. When you think of learning, think simple, so people can understand it. Think system. Think how can I learn as a system? And think symmetry, that is, don't forget the forgetting side of learning. That's not much studied in AI, but I'm persuaded that agents that can forget appropriately are going to have similar performance boosts to those that can learn appropriately.

Mr. Axtell: I have one follow-up point on that. There are very powerful theorems, as some of you may know, in game theory. For any game which does not have cyclic strategies, you can get social learning to happen as long as agents make mistakes. Making mistakes is crucial to kicking the system out of some kind of minimum local suboptimal position. What happens is you can learn any Nash equilibrium as long as agents make mistakes. You can learn at the social level. With no agent trying to compute it, it can be learned.

M: It's really funny, because it was yesterday I was having lunch with Tom Bach at Harvard. He's one of the really nice graduate students there, and he went to the computational school at Santa Fe. We were talking about lock in of mistakes, and how are we going to handle that because we're not really building strict game theoretic constructs, we're building these simulations that traverse this infinite multidimensional space. But there is this sort of error lock-in that occurs, and gets built into a system. And then as we face, because we have emotionally charged agents, this sort of fear of embarrassment, we put a lot of anthropomorphic terms on them. It ends up we only do it because it's useful for us, in terms of a shorthand of explaining what we see in the simulations. But it certainly seems to be important.

M: What exists today to make modeling agents easy? Are there some standard packages that you like, or would suggest that people pick from as they leave here today? And are those standard packages that you might recommend capable of helping you identify some of these pitfalls?

Mr. Farrell: I just want to make a point. I notice already in the back of the hall systems that are relatively straightforward to produce. If there's any sort of advertisement for how easy it is to build stuff in Starlogo, let this be it: as we were discussing dendrites, Yuri put together a simulation. I think it took about fifteen minutes to put together the thirty or so lines of code that looked at dendrites. Is that correct? About that time. So in one case it's simple. I will tell you from my experience that some of these systems are relatively simple to set up, but they're devilishly hard to calibrate.

Mr. Axtell: I would agree with that, and I would say there are two approaches for handling that problem. One is when the system you're trying to model is actually up and running, instead of feeling as if you're done, you should feel like you're about ten percent done, because you have to then do all the analysis of the thing. And oftentimes, as Van mentioned, he thought he was going to add a lot more features to his simulation, and never got around to it. Usually it's all- consuming to just analyze what you have running. But I think there is a sense in which - and I want to come out on the positive side here - there's a sense in which these systems are very easy to build. Particularly when the agents are relatively simple, and the population of agents is a mere replica of relatively simple individuals. In that sense, for example, when it takes a team at Microsoft a year to build Microsoft Word, and it's five megabytes of code, and it runs at about five megabytes on your computer, there's a strong correlation between, in conventional software, the amount of code you write, and the amount of space it takes up. In this world, that need not be the case. And for example, we have a model of exchange where agents just engage in very simple trade behavior. It's under a thousand lines of code, but it fills up a gigabyte on a workstation just by replicating, by instantiating a hundred million agents. The hard part is the analysis.

Mr. Pernak: I'd like to put my comment in context. I was speaking of the simplicity of modeling the particular domain that we were working with in agents, as opposed to modeling it with differential equations. And the difficulty of representing a PPSE algorithm in differential equations. Clearly, if I were trying to model the propagation of radio waves, I would be much better advised to use differential equations than to use agents to model that. It's a domain-specific claim, first of all. The ease here has to do with modeling systems that are naturally distributed, consist of lots of different pieces following various decision algorithms. I find it easier to cast that in rules or computational expressions than in differential equation formalism.

In terms of packages, I'm a great fan of the Swarm environment. I am a tremendous admirer of Starlogo. For reasons of interfacing with industrial people, I have been exiled to spend my life in the Wintel land, and so we don't yet have Starlogo available. But I'd like to share an anecdote with you around Swarm. And I don't know how many here know that Deer and Corporation, under Bill's leadership, has made significant investment in the Swarm package. So there's really a vested interest on Deer's part in seeing it advance. It's now available for Windows '95, as well as Windows NT. And, of course, for UNIX systems, as it has been for years. We have a contract with two companies. A major semi-conductor firm, and a major software engineering firm, to participate with them in building an agent-based scheduling system for the semi-conductor world. As part of working out the algorithms, one of the tasks for the team was to produce a simulation of the fab, so that we could try out different agent-based scheduling techniques in the fab. We said we ought to do this in Swarm, and the software firm said we're not so sure. We think this sounds like some academic toy. We're not sure it's a good thing to play with. Why don't we do it in Java, which we know real well? We're Java experts. So they turned six Java programmers loose on building a model of the fab. We had one programmer who, at the beginning of the exercise, had never seen Swarm before in his life. And our model of the fab was up and running before theirs was. Finally, they said you take it, and run with it now. We're not going to spend any more time on this. That says a lot about the pre-packaged facilities in something like Starlogo, or like Swarm where you've got support already for the notion of agency. The kinds of encapsulation that are necessary. You already have scheduling mechanisms to invoke the different agents. And different scheduling mechanisms, so that you can guard against the artifacts that may arise from always following one particular polling policy. There is a whole passel of random number generators. Again, one can avoid artifactual results that might stem from idiosyncratic selection of one particular random number generator approach. There's a whole box full of widgets to give you graphical displays of the results. If you want to form the kind of plots that I had up on the screen, you drop it in place, and it's all there. It's a wonderful tool kit for constructing these things. So if you're in a domain that is discrete, and has discrete-based behaviors, you want to use agents rather than equations. I think that's fairly straightforward. And you're well-advised, if it's a very simple problem, you're playing with it, and you're fortunate enough to live in MacLand, Starlogo is a very easy entry point. If you're willing to study a little bit more for a more industrial strength tool, I think Swarm is well along the way to proving itself to be just such an industrial strength platform.

Mr. Deshmukh: I see this as two problems. One is the modeling and simulation of agent-based systems. The second problem is implementing agent-based systems. Now these are not exactly the same. The needs for modeling are different than the needs for actually implementing them so that they are used in a real world situation. For the modeling issue, there are packages that are available. Swarm, Starlogo, and others that are available that give you a good handle on developing systems, and also make it easy to implement the models. As far as implementing these systems in the real world, I think we are still away from an implementation that truly interfaces with any system. There are systems that are used that can interface with a certain set of systems that are available - and when I say systems I'm talking about computer systems. But there aren't really a generic set of agent-based systems that are available that can interface with almost any software that is available out there. That's an issue. I want to close my comments with a question to the panel. You can technically model radio waves as agents. The question is, are we going to use these systems for kinds of domains where the explicit knowledge about the domain is limited as an exploratory tool, and then graduate to an area where we have a science-based knowledge - by science-based I mean analytical knowledge - and then use those as we get more information? Or is there a fundamental difference in systems that can never be analytically modeled, and that's where these agent-based systems are going to be used? Is this a stepping stone, or is it a truly different modeling methodology?

Mr. Parunak: There's some formal evidence that agent-based systems escaped the Turing bounds on computational complexity. I haven't examined those theorems in detail. But to the extent that those claims are true, agents can compute things that are fundamentally non-computable by other mechanisms. However, that's not very useful in a practical sense. I think from a practical perspective, we start with the real world, and we build models of the real world, using whatever formalism is closest at hand. In many interesting domains that's going to be agents. Whether we're fortunate enough to be able to get to an analytical model in our lifetime is another matter. I'm not making a claim ontologically about whether the methodologies are equivalent. I'm making an existential claim that I think there will always be some things that all we know how to do is model with agents; and there will probably be other things that yield to formalism representations.

Mr. Axtell: My perspective is that when it comes to modeling any system involving humans, it seems like agents are the way to go. That's not a theorem, it's more of a qualitative feeling about it - a gut reaction. There's a little paper by Feigenbaum about chaos wherein she talks about using a computer to solve a very nonlinear field equation. Iterating a bunch of local densities forward in time perpetually she calls a foolish double-regress. The only important information that's in the equation, and that you're numerically integrating is the scaling information. My sense is that when it comes to the agent model, the way that social science has progressed to date, which is idealize the human behavior in some axiomatic way, and then formalize that, and idealize it further through assumptions about aggregation. Then write down the equations for it, and solve the equations. Then make some inference about the real world. It's kind of a foolish double-regress again. Because we can just start out with the agent representation, heterogeneous agents in software. We can spin that thing forward in time, and you get an answer.

Mr. Farrell: A perspective I would have is that when we put our sort of agent-corrective lens on a standard numerical process, we were able to increase the accuracy well beyond the standard numerical approaches. This stuff works.

M: What is the state of the art in terms of actually involving humans in the simulation itself?

Mr. Axtell: My understanding is there's not been very much use of having, for example, real human agents involved as playing against artificial agents. A much more common thing we have going on right now, particularly in economics, is a big wave of so-called experimental economics, where you put human subjects into environments, and you actually give them real payoffs for behaving in various ways. Those experiments show that people fail to have strictly convex preferences. They do not maximize expected utility. They have all these deviations. In fact, they actually care about reciprocity, and they actually are willing to engage in punishment of others, say in public goods games, and not just maximize their own utility. To build those subrational - maybe even non-neoclassical - behaviors into agents is kind of the current state of the art. There has been very little work though on actually putting these real subjects in the multi-agent world. At least very little I'm familiar with.

M: Let's thank Abhi, Van, Rob, and Win for their preparation and their presentations today.

Complexity and Management

Tom Petzinger
Wall Street Journal
William McKelvey
University of California, Los Angeles
Peter Senge
MIT Sloan School of Management and Society for Organizational Learning
Laurence Prusak
IBM
Henry Mintzberg
McGill University
Ron Schultz
Senn-Delaney Leadership Consulting Group

M: It's my great pleasure to welcome someone who most of us read, if you're in the business community, every Friday without fail, looking to see what pearl of wisdom we would have neglected if we had not read him. Welcome Tom Petzinger.

Mr. Tom Petzinger: Thank you. Did you read this morning's column?

M: Yes.

Mr. Petzinger: Okay. Then you've passed the test. Those of you who read comic strips growing up are familiar with the Chester Gould strip Dick Tracy. Perhaps his most famous strip appeared in 1945 after all these dastardly characters that he had introduced into American folklore were top of mind among American comic strip viewers. In this particular strip, the climax in the final frame is all of these terrible bad guys, a half a dozen of them, lined up, ready to pounce on Dick Tracy all at once. These characters had only ever appeared one at a time. But they were all united now suddenly in this one frame, and Chester Gould draws Dick Tracy with a panicky look on his face. And the bubble says "Hey, what's the big idea?" Well, I'm here to say to Yaneer, what's the big idea? Here I am like Dick Tracy with a half a dozen of the leading figures in their field - they're not bad guys, but they are towering figures - all brought together in the same frame for what may actually be a somewhat historic, and certainly very interesting gathering.

We have with us Bill McKelvey. Bill is the Director on Strategy and Organization Science at the Anderson School at UCLA. I don't think anyone in the last three years has applied evolutionary dynamics to firms more deeply, and more rigorously than Bill McKelvey. In fact, he may be the only true complexity theorist, per se, on our group.

We have Peter Senge. Peter built the Center for Organization Learning at MIT into what is undoubtedly one of the most influential and admired groups of its kind anywhere on the planet. He is the author of the book, The Fifth Discipline.We have Larry Prusak. Larry is Managing Principal in the IBM consulting group. He's right here in the Boston area, and he is the editor of a very influential book, known by the title Knowledge in Organizations. Let the record show that as editor he included his own essay last; which is probably a first.

We have Henry Mintzberg, who again, though not explicitly rooted in the complexity tradition, probably appears in more bibliographies and citations at this conference than just about anybody else. Henry is a Management Professor in North America and Europe, and the author of many seminal books, the most recent and perhaps important of which is The Rise and Fall of Strategic Planning. The book of books on strategic planning.

My colleague, Ron Schultz, is here. Ron is a writer, a speaker, a broadcast producer. He is the Director of Publishing, and a whole number of other writing-type activities at Sendelaney Leadership Consulting in Santa Fe; and is the co-author, with Howard Sherman, of a book called Open Boundaries.Now in brief, let me suggest that history is very humbling when ideas that we take as new have preceded us under different names, and under earlier intellectual eras. But history is also inspiring - it might be better to say history is comforting when we find that the ideas we thought were new were actually timeless. That they're actually verities that again and again reappear, reasserting their authenticity. I'm going to submit that complexity is one such meme.

Anyone recognize this woman? That's Mary Parker Follett. Mary Parker Follet was a management theorist back when there was no such thing. This is pre-cyberneticists, pre-Norbert Weiner, pre- general systems theory. "No one can understand the labor movement, the farmer movement, or international situations unless he is watching the internal stimuli, and the responses to the environment." This writing is circa 1920. She called this circular response: "We cannot watch the strikers, and then the mill owners. Trade unionism today is not a response to capitalism, it is a response to the relation between itself and capitalism. What about authority? Where does it emanate from? It is not something from the top. It comes from the intermingling of all. Of my work fitting into yours, and yours into mine, and from that intermingling of forces a power is created which will control these forces. Authority is a self- generating process."

The Austrians were not complexifiers, per se. But they used the very language that we now take as recent. The emphasis in this quote is in the original, the emergence of new patterns as a result of this increase in the number of elements means that this larger structure as a whole will possess certain general abstract features which recur independently of the particular values of the individual data.

Anyone recognize this fellow? This is Abraham Maslow, the great humanist psychologist best known for postulating the hierarchy of human motivations. Less well known is that Maslow also spent a sabbatical in an industrial plant in Southern California. He spent the summer walking around with a tape recorder in his hand, and having his thoughts transcribed. To my knowledge the only time he was ever in a business or a firm-type setting. And he walked out with these thoughts on such subjects as holistic business. "A business in which everything is related to everything else. Not like a chain of links of causes and effects, but rather a spiderweb, or geodesic dome, in which every part is related to every other part." These writings, incidentally, are circa 1962. As for creativity, he noted that it is correlated with the ability to withstand lack of structure, lack of predictability of control, the tolerance for ambiguity, for planlessness. Concepts that are redolent of many of the presentations throughout this week.

The cyberneticists brought us a little bit closer to where we are today. We have been learning to see social processes as the links tying open systems into large and interconnected networks of systems, such that the output from one becomes inputs to others.

A few years pass, and we have the likes of Mintzberg theorizing about the firm itself with the notion of emergence beginning to creep into the concept in a big way. And although not explicity rooted in this tradition, before long Mintzberg is actually coining complexity-like terminology, such as emergent strategy. Senge in 1990 urges us to look to nature's templates for the patterns that control events in organizations. He popularizes systems thinking, which he defines in many ways, but the most concise of which is seeing patterns where others see only events - even postulating the use or the invention of microworlds.

This brings us to a time where we begin encountering the explicit use of complexity in management. The folks in Santa Fe develop a business network. Some of the consulting practices spring up. We have theorists like our own Bill McKelvey, Jeffrey Goldstein, and Stuart Kauffman. We have consultants - Ernst and Young, McKinsey, Casa, the Santa Fe Center for Emerging Strategies. We have the toolmakers - Win Farrell, Bill Fulkerson of Deer. We have authors - many of whom are present with us today. We also have organizations that have erupted. NECSI, the complexity in management mail list, and of course the Santa Fe Institute.

I'm going to conclude by posing a problem. The problem being that in the course of naming something, or in explicitly linking so much deep thinking from across sciences, or many fields, or many intellectual traditions. In the course of naming it, you run the danger of oversimplifying it. In oversimplifying it, you draw in the corrupters, the charlatans, people seeking pat solutions. This forces us to use the F word. Is this a management fad? Is it possible that the management folks will wreck this concept, even for the scientists perhaps? To contemplate this, I invite you to consider this quote from the business world. "In the world we enter, chaos is order. Evolution is revolution. Adaptation is survival. ...[O]rder, truth - perhaps, even beauty - will emerge from the constructive chaos of disruptive change." How many buzzwords per sentence in that one?

(general laughter)

Mr. Petzinger: We run the risk of wisdom becoming confused with word salad. I don't mean to pick on anyone in particular, but furry animals know how much snow to expect in the coming winter, and dress accordingly. They communicate superbly. Nothing we have created in any human organization comes close. Certainly not in my family where we were told to put on our coats every morning before going out.

Here's, from one authority, words of warning, which I take just a little bit personally. "As soon as one management fad disappears, another is waiting in the wings to replace it. What? You didn't catch on to quality circles? That's okay. The big guy read an article about chaos science in the Wall Street Journal, and wants to implement it throughout our North American operations right away."

My wife, whom I describe as a complexity widow, was reading Elle, or Vogue, or some fashion magazine, and encountered a double page ad for a new scent from Donna Karan. Beautiful photograph, obviously, and on the facing page we get the name of the scent. [Chaos.]

Now, before going to the first of our speakers, I'm going to boldly assert that we are in very little danger of creating a management fad for these reasons. Complexity describes the world as it is. It is not an idealization tool. It emphasizes removing behaviors over adding behaviors. It's not a program. It's not something you can start on Monday morning. It's behavior you have to stop doing on Friday. It rewards our humility over our conceits. It defies methodology in packaging. It has to be customized. It is the situation that complexity addresses. It is science, rather than a fad. With that, I'd like to introduce the first of our scientists, Bill McKelvey.

Mr. Bill McKelvey: When I was at MIT, I was a student of Warren Bennis, one of the leading gurus on leadership. In thirty years this is the first time I've had a paper that had the L word in it. This is a new venture for me.

Yaneer set me up very nicely in his opening remarks. He said, if a single component controls a collective behavior (not the individual behaviors of all the components of a system) then the collective behavior cannot be more complex than the individual behavior; i.e., there is no emergent complexity. If we don't have CEOs and leaders in the firm which cannot create complexity, there's no reason for any of you to exist.

This is largely a conference of complexity and computational models. What I'm interested in is how do we deal with complexity in firms. In thinking about that, I started from my embeddedness in Kaufman's NK model. I was thinking gee, how do I talk about a firm, and think about a firm, and worry about the complexity of the social network in a firm getting to the point where we have complexity catastrophe? Then I thought oh my God, what if the firm doesn't even have a social network, and it doesn't have any executives that could create complexity in the first place - which is what Yaneer's little comment is pointing to.

Then I found out I was going to be on a panel with gurus; Henry Mintzberg, and Mike Hammer, and Peter Senge. I was in Harvard Coop a couple weeks ago, and I found this book. It's called Rethinking the Future. And it's got gurus in it. I mean, it's got their pictures. I thought well, I'm not a guru, what am I going to do on a panel like this? I'm just an academic. I'm from Los Angeles. I was thinking of Mark Twain's, Connecticut Yankee in King Arthur's Court. What you've got here is a Tinseltown academic in GuruLand. With that, I will keep going.

The guru view of leadership. I'm going to start right at the top beyond the efficient frontier; which is basically Michael Porter's thing these days. If you read the guru books - rapid change, hyper-competition, nonlinear dynamics, chaos, human social capital, learning infrastructure, network dynamics - these are all themes that appear in the guru talks in this book. What does this mean? Let's go to Porter.

In his 1996 Harvard Business School Review paper, Porter tells us that he's joined the resource-based view of strategy, and he recognizes that the things that strategists worked on for the first twenty years or so are more or less irrelevant, if you have a firm that's looking for economic rents. Where are we going in this? If you look at the early days of consulting firms and strategy, the problem was how do we get firms up to the efficiency curve? If you're the head of General Motors, you still haven't achieved that yet. But here's the struggle. How do we get GM up to the efficiency curve? Then the second thing happened is we began to worry about what position should you be on the efficiency curve in order to get rents? Should you be on the low cost end, or the high quality end? I could spend an hour just talking about this, and how the Japanese do it. But really the action is now in this third thing - new novelty planes.

The idea is how do you get the product out there first? You have to stay ahead of this constant evolutionary process as your competitors move toward the efficiency curve. Once you're on the curve, rents disappear, and you get normal profits. The problem we have now is how do we keep firms moving fast enough in a rapidly changing world that they can try to get rents? This is really what the strategic question is in this day.

Strategy people are mostly economists, although that's changing. Economists worry about the production function. So you've got some owner up there. He's the guy with all the money. The economist's theory of the firm is that you have a lot of capital, and you have labor. As Porter points out in his paper, what people have been doing is we go to Taiwan, we build a plant, and we hire a bunch of labor at twenty cents an hour, and we have an advantage. He points out that that lasts about a year. Your competitor can do exactly the same thing. If he's smart he'll go to China, and eventually they'll go to Bangladesh, and they can get labor at seven cents an hour. That doesn't get you very far, and it doesn't deal with what the current problem is in strategy, which is how do you get rents?

Now what you see is that we have an additional piece to the production function. I'm calling it O right now. There actually is a paper called "Organizational Capital". It takes a lot of dead economists to get a new thing onto this equation. You all know economics moves one funeral at a time. We're looking at a lot of dead economists here. What's happened is this. I put O down here just so I wouldn't ruin their equation by putting in non-independent variables.

Gary Becker introduced human capital a long time ago. That's what the H is. So for economists we now have capital, labor, and human capital. Then Ron Bird, who was a professor at Chicago, introduced S, which is social capital. That gets us into networks, and that gets us into complexity. Now you see how this is working. I just grouped all these things on the bottom under O. The D really goes back to Chandler, Williamson, and all those org theorists that talked about the structure of an organization.

It's going to take a gymnasium full of dead economists before we get the D and the S into that production function. I just want you to know that we're talking about human and social capital, and then what happens after we get that going.

Back to the metaphors. My mentor was Bennis. His talk in this guru book talks about visionary, heroic leadership. That's where it stands in the leadership literature. He talks about this as herding cats.

Bennis's view of how you get organizations to work in firms is essentially irrelevant to the modern age. There's no capital, there's no labor, there's no human capital, and there's no social capital in that statement. It all depends on having this visionary leader. As Yaneer says, if you have a visionary leader that can't get leadership down into the lower parts, you don't have any complexity. We're dead if we follow Bennis. But the gurus are getting there.

Peter Senge sets a very high standard for knowledge. He says there's information and knowledge, and knowledge is "how to". There is no knowledge in this whole book by that definition. Zip. I'm not even sure there's any knowledge in my talk. There's no knowledge in this conference, by that definition, unless you're a modeler, and you're worrying about how to do another model. But if you're worried about dealing with executives, there's no knowledge anywhere in this room. "How to" is how do we get executives in firms to deal with these kinds of issues that the gurus, and everybody else is worried about?

What I'm here to talk about is distributed intelligence. There actually is a body of literature on this - the brain. We have a lot of brain science going on, and we know that brains work with neural networks. Neurons are basically on/off switches, and as one brain writer says, intelligence is the network. It's not in the nodes, it's the network.

One of the things you can think about are what I call the thin and thick models. Normally in decision science we have thick models. We have a really complicated algorithm or a model of how a particular executive or purchasing agent makes a decision. All of that is in the node. We all use AI models. In these kinds of models, all the action is in the network, and the nodes are treated as real simple.

In fact, in the management world, we actually have some combination of both. When you have people in a firm, it's hard to imagine that they act all day long like simple on/off switches, but we also have to worry about the network. We now have a large body of literature on social network theory. Ron Bird is a network sociologist. He says that for strategy, all the intelligence in a firm is in the network, it's not in the nodes. It's not in the human capital or the people. Think about that. There's no network in this room, so by that definition there's no intelligence or knowledge in this room.

Here we are - a lot of the gurus. Bennis, and most of the leadership people, talk about visionary, heroic leadership. They're going to guide the organizational ship toward the good future. Yaneer has said, and I agree, that if you'd have that view of it, you'd have no complexity in the firm, so then all of you people, and all of us, are out of business.

What I think we ought to talk about, if we're in the complexity business, is do we have human capital in firms? How do we get it? Do we have social capital in firms? Do we have intelligence distributed on some combination between these two things? That's the next thing.

The disconnect is most of the work on leadership, and all of this talk in the guru books. The information gets us to the point where we have firms dealing in a rapidly changing environment, we have hyper-competition. The only response we have is that the leaders should be visionary. How do you have a leader direct an emergent system, if we're in fact going to talk about emergence in organizations? If the leader really leads, according to the current conception, he or she shuts down emergent structure.

How do we do leadership in a way that fosters emergent structure in a firm without the leader somehow creating a bunch of passive followers following some vision. Further, in a rapidly changing world, what chance is there that the leader has the right vision at the time, or at the right level of technology, and so forth? Very little chance.

That is when I turn to complexity theory for an idea. Critical complexity. There's kind of a progression here. Stochastic microstates, subcritical complexity, Newtonian simple rules, deterministic chaos, simple rules again. What we're really focusing on is this zone of complexity in between chaos, and sort of low energy conduction or something, if we're dealing with fluid dynamics, or something like that.

What I'm interested in here is critical value dynamics. The storm cell is a good example, where if you hit below the first critical value you get conduction from one molecule to another, and not much happens. If you go above the second critical value, you get chaotic behavior, oscillating. We have tornadoes, and things, and then they disappear, and they come and go. In between, however, you have bulk air currents. And so you get more efficient, and more rapid response to energy differentials.

Now think of the organizational context. We have a firm facing an energy differential. It's out of date, it's obsolete, it's not keeping up with the rapidly changing world. It's under a lot of adaptive tension. It's like some kind of generalized energy differential. The question is, how do you get the firm to move in between these two critical values?

You'll see it with a lot of M and A activity. We buy a small firm in New England. We get rid of management, we change the culture, we change the accounting system, we change the IS systems. Right away we create a lot of chaos. If we don't do all of that, if we just passively buy the firm, and hope for the best, this little firm just stays the way it is, and not much of any good happens. It's another example of how do you move a firm into the critical value zone where you begin to get emergent structure that's significant, that's in the bulk category, that actually develops ideas to move the firm in a more adaptive position more rapidly.

I started this whole line of reasoning with Kauffman's work, and the idea of complexity catastrophe. When you stop to think about human and social capital in firms, there are a lot of what Kauffman refers to as "epistatic links". If I have a node that has some adaptive improvement that gets attached to it, what is the yield that we get out of that? Usually in an interdependency network, if I begin having to deal with all sorts of other people, they say you can't do that, we don't have enough money for that, it's not really the right time, you don't have the right people. The idea gets shut down, and the yield value isn't very good.

Labor actually has brains - it's not all just muscle. If you're going to imagine an executive that's trying to deal with a set of scientists, or other intelligent people, how do you get that executive coached into the frame of mind where he or she can create enough adaptive tension that energizes emergent structure, creating the focus, and the vision, and all these other top down leadership attributes, without creating the dictatorial perception? How do we get an executive to generate the right kind of adaptive tension? The books don't tell us.

It's not by running out in front and saying come on. How do we manage the critical values? How do you know when your system is above the first critical value? How do you know when it shoots over the second critical value? What are the diagnostics? We don't have some digit in the second decimal place that flips, and you can tell. It's more complicated than that. If you think the visionary leadership is essentially a point attractor. What we're trying to do is to get managers to set up strange attractors, so that you get relevant behavior, without somehow identifying a point in advance you want the system to generate where the thing should go. Clearly there's a lot of stress. I tell them my doctoral students: I try to run you at the edge of dysfunctional stress. You don't want them to be dysfunctional, but I don't want them to just sit there, and think they've got forever to get through the program. You've got to keep pushing them along.

Managing the agency problem. The owner is going to say, we're going to pay these people to do whatever they want to do? How do we know it's for the good of the organization? How do we know it's for the good of the shareholders or the owners? You have an agency problem here.

Finally, I have a friend, Ron Sanchez, who writes a lot of papers on modular design, which is very popular in a lot of firms now. It fits the complexity notion very well. Modular design is essentially the opposite of letting engineers design an efficient total system where all the interdependencies are worked out. With modular design you simply identify models, define the interfaces, and the modules are essentially autonomous. Each module can change, and move in its direction. Sony's Walkman is a classic example of modular design. It allows them to move much faster in a competitive world than the standard decision scientists' optimization designs.

The modular design really fits with the notion of emergent structure in human and social capital. That's all I'm going to do right now.

M: I have the unfortunate role of living in Big Company

Land, and have to convince executives - this is the way I earn my keep. Invest in social networks, invest in human capital, be serious about these ideas, because they seem to be truer than some of the older models.

This is a world of heuristics and persuasion. These are the skills you need to do this. You need heuristics - little things you invent that seem to work - and powers of rhetoric and persuasion; which have great force in large companies. Deirdre McCloskey recently estimated that persuasion accounts for about twenty-eight percent of the GNP in the United States. This is in the AAR, this is serious stuff, and there's a lot of truth to that. Many of us in this room are living through persuasion. With the correct use of rhetoric, persuasion, and heuristics, you can make some progress. How do you get firms to seriously invest in, and take action, vis a vis nonlinear dynamics? I'd like to elaborate on little on that later. Thanks.

M: I have a question for you, Bill. You talked about leadership in two respects, and you said that there's no kind of "How To" knowledge in the room. But in one respect you did talk about leadership with regard to How To. You talked about leadership mostly with regard to the hype, and the overdone, and the sort of great leader intervening, and all that. All of which I agreed with. But then you talked about leadership in a second respect, with regard to How To. Namely how you lead your own doctoral students. And you talked about how you push them along. That sounds inconsistent with everything else you said about leadership. I'm curious how you reconcile that.

Mr. McKelvey: Well, it's a good question. I have a doctoral student right now - she's a theoretical economist. I didn't plan it that way. I have another one who has gone off into computational models, and neural networks, and network sociology. And I have a third one who's main skill is econometrics. She's quite good at it. There is no pattern among these; although all are studying human and social capital, as it turns out, and that's an emergent thing. They all started this two or three years ago, and we only discovered these terms a little while ago. I'm not giving them much intellectual direction, but I am setting the bar. I don't tell them too much about what they have to do to get over it. I just keep talking about well, if you do that, you might not get hired. And if you do this, you might not get through the program. And if you do this, you wouldn't be very competitive. I'm just keeping the pots boiling. They're smart enough that they figure it out.

M: But take that back to the other leaders, and they're doing the same thing. They're managing the process, rather than the content. They're doing what you do.

Mr. McKelvey: They may be, but I don't see that in the literature. Maybe that's what you do when you talk to them in the firm, and that's the proprietary side of it. Which it could very well be. But I don't see any cues in this book, or some of the others. There's a chapter in the Handbook of Organizational Studies that reviews current leadership, or the history of leadership, and the 1980's was largely the Bennis, heroic leadership stuff. His comment is that in the 1990's you begin to see more attention on dispersed leadership. A lot of what he talked about was teams. Well, who's forming the teams? The top management is forming the teams. These teams are not left to an emergent structure. Or if they are, I don't see rules, or procedures, or processes that leaderships, or CEOs are being given that tell them how to foster bulk, emergent structure in firms. Now maybe you do that in your consulting.

M: No, no, no. I'm not talking about myself. I can't stand that leadership literature you're talking about for the same reasons you can't. All the hype and everything. We're just obsessed. Leadership has become an absolute cult in American society. Not around the world, certainly not in Japan, but in America leadership is an absolute cult. It's almost perverse. I've just done an article called "Managing Quietly", which is very consistent with the things you're saying. It seems to me that a dimension of all that crap literature is about inspiration. That's essentially what you're talking about doing with your own doctoral students. Your role is to inspire, and show them the constraints, and keep pushing them, and so on. In a sense, there is a reconciliation.

Mr. McKelvey: I don't have any problem with that.

M: Let me take the talking stick from you Bill, and ask Peter Senge to speak. My grievance with the Bennis book is the extended discussion of how one particular motion picture director finally got Barbra Streisand to cry. Page after page on that very topic.

Mr. Peter Senge: I'm not exactly sure where to start. I'm inclined to try to make a comment that connects to one of the themes that Bill talked about. Since we already touched on the problems with heroic leadership, let's come back to the whole business of knowledge for just a minute. As I'm sure a lot of you know, knowledge management would qualify as a fad today. It is remarkable to me how much time, and money, and all sorts of resources we can invest in something that people can't even define. There is a very strong tendency for us to define knowledge as information, or the other way around, for us to regard information as if it's knowledge. I don't know what the world's greatest definition of knowledge is. I can only tell you what's been helpful for us in our work.

I've always found a useful definition of knowledge to be, "the capacity for effective action". Learning in that sense is about the enhancement of capacity for effective action. While that's very simple, and undoubtedly has some limitations, it really gets you out of the confusion that knowledge is somehow information, but a little bit more so. Which is what I find again and again in the world of business today. Everybody is saying well, yes, knowledge isn't quite information, it's like really important information, or like really big information, or like really... - there's something different about it. I think any of those definitions ultimately end up not adding much value. It is very challenging, and I appreciated you raising the bar higher. I hadn't thought of it that way, but in practical terms, it is very challenging if you think of knowledge as the capacity for effective action. All the things we might be interested in, whatever common denominators, and common territory there are amongst us clearly has to do with highly complex, interdependent realities, and how people build knowledge. That is, not get ideas, not just even build better theory - although I think theory is quite essential - but actually develop enhanced capacities for effective action.

Quite a few of you probably heard John Sterman's presentation. John and I have a very similar background at MIT. We both had the same mentor - a man named Jay Forrester - who instilled in us a certain perspective for appreciating nonlinear dynamical feedback systems. It is the engineer's view. It has strengths and limitations.

The thing above all that I value from that particular background is, when all is said and done, Jay was a kind of penultimate engineer. He believed deeply in tools that help people do something. While he is a significant intellectual figure, what he cared most about was the consequences of his work. He was a builder. He was an engineer who thought that ultimately you had to produce something, or you hadn't really been an engineer. That spirit has pervaded the work I've been involved with for a long time in the Center for Organization Learning.

This group, a consortium of corporations, has been reorganized over the last few years. We felt it ultimately had outgrown becoming organized as part of MIT. It's now an independent nonprofit, membership-owned and organized, and called the Society for Organization Learning. MIT is a member of SOL, as are Ford, AT&T, Shell Oil, and some other Fortune 100 companies. This is not because we thought they were the most important, it was just a coincidence that those were the organizations we were working with. More than that, it was strategic in the sense that we felt if you could produce some significant learning, or generation of knowledge in those kind of enterprises, then that was quite a challenge. The problem with looking at all the startup organizations, and all the organizations that look so new, and different, and distinctive is that unfortunately very often they will become, if they're successful, the same stodgy, bureaucratic, centrally controlled, power-oriented, politically determined organizations that are the type I mentioned. We figured we might as well start where it's hardest. That's why we've always concentrated on those organizations. Although today, SOL is now starting to grow as a global network, and I suspect a lot of the new SOLs that are organizing will organize around small organizations, and nonprofit organizations, and public/private organizations, and hopefully educational organizations.

About ten years ago, when we were trying to explain at an academic conference what we did, George Roth and I came up with this picture, which points to the notion that there are two types of complexity that we're wrestling with in a complex human or social system.

Most of what's talked about in a conference like this is along that vertical axis - dynamic complexity. By dynamic complexity I mean things like emergence. I mean situations where cause and effect are not close in time and space. In the behavior of a complex, non-linear system, the areas of most significance are very often very distant from the symptoms of problems. Managers in organizations think if we're losing market share we ought to crank up a marketing intervention. Or if product development is somehow not up to snuff, we ought to re-organize the product development organization. Very often these are symptoms of the interactions of the enterprise as a whole. The greatest leverage might lie in very distant parts of the organization.

That's always been a cornerstone idea in system dynamics. Quite a few of us became aware of certain deep limitations in the field of system dynamics. As somebody once said in one of our meetings about fifteen years ago, "The only problem with these human systems is they're full of human beings." Other than that, we could capture them in our models. I have been modeling for a long time. The discipline of system dynamics is a discipline of simulating nonlinear feedback systems as continuous systems. It's a very particular perspective. The problem was that there was no foundations in the development of our students. There was no real serious attention in practice to the fact that you weren't trying to build a model, ultimately, about the system. You were trying to help a group of human beings be more effective in accomplishing something they really cared about. And then the model, if it was the world's greatest model, was only great insofar as it enabled the human beings to do that. Now I should extend that statement by one more step and lay the foundation that could help other human beings also be more effective. I do think theory matters.

The theory that I became most interested in was a theory constructed by the practitioners. To talk about theory in the world of business is usually a way to have yourself shown to the door, but not necessarily. I think it's more because we're not very skillful at it than because it doesn't make sense. I think the best practitioners are theorists. They are thinking about what my assumptions are behind these actions. Why does one strategy make more sense than another, in terms of the world view that it is based upon? And in so doing, if they're really good, they are quite reflective about their assumptions about that world view. They know they're just constructing a view upon which their decisions are based, and ultimately the continual inquiry into that view is quite critical to their work. Deming used to say: "no theory, no learning". I think he's right. However, theory doesn't produce learning, because theory can very easily then just become more information. It was these kind of questions that led me to be interested in the subject of learning, and it took many years. Obviously, we're all interested in learning. We all enjoy it when we're learning. When we're in a learning mode, life is a little more exciting. It started to become evident that these were puzzles that had to do with the fundamental nature of learning processes, individually and collectively. These are processes whereby human beings become both more capable of effective action, and more appreciative of the world in which they are acting. They are processes that integrate thinking and acting. The essence to me of what all learning is about is the effective integration of thinking and acting.

In a human world, that means you have to deal with human complexity. You have to deal with the fact that people are different. They see the world differently. They construct different realities. They have very different assumptions. Oftentimes they have different values. This dimension, of course, has traditionally been an area that people have focused on a lot in organization behavior, in team building, and in negotiation theory.

In a sense, we've been trying to pose the question, how do you make progress in this domain? There are a lot of tools for generating insight about complex systems which don't necessarily enable the human actors in those systems to become more effective. There are a lot of tools and methods for helping people listen to one another, appreciate one another, really slow down enough so that we can start to see that the person who appears crazy is maybe not crazy. If I could hear the process they have gone through to come to their view, I might learn something about the reality that I'm facing. And so in a funny, simple way, everything we have been doing has been trying to integrate tools, and theories from these two different dimensions.

When all is said and done I have no interest in anything that doesn't allow human beings to be more effective in their action. The work we've been doing should be judged by how effective it has been, not just how intellectually intriguing it has been. In organizing this consortium many years ago, that was the aim. Could we generate more momentum by creating a critical mass of people in different organizations?

Most of what we've done for the last ten years is get involved in long term projects. Most of the projects have been at the level of what I would call a team. I don't agree with the notion that CEOs create teams. That's what's happened a lot, but teams is the way work has always been done. My definition of a team is a group of people who need one another to get something done. Teams have always been there, and they have always been embedded in social networks. The real question is can we see them better, and can we nurture and support their effectiveness and their evolution better? Most of the work has been at the level of teams. Product development team, sales team, manufacturing units. In the last three or four years, increasingly at the level of larger organization.

I'll give you an example from a team setting. I brought along an example of some of the research documentation that gets created in this process. This is called a learning history.This is one from a product development team, in an automobile company. It's a real company; take my word for it. A team was developing a new model year car. The project, or their program, as it's called within the company, was a five to six year program with about a billion and a half dollar budget. It's a big program. It involved over a thousand full-time equivalent engineers at its peak. These teams tend to ramp up over several years, and then tend to ramp down before a launch.

I'll read you an exchange. It's the best way I know to give you a feeling for what actually goes on when you're trying to build knowledge around these kind of issues. This is a staff meeting, about two years into the project, in which a group of engineering team leaders are gathered with one of the senior managers in the team, who they call a business planning manager. He has a lot of influence in the launch process, but he's part of the team the whole time. In any case, some of the engineers are complaining that they have a lot of meetings. One of them says, we're coming to meetings all the time. My biggest pet peeve is that we are wasting our time in sometimes four and five change control meetings a week. This is not unique at Epsilon (Epsilon is the code word for the team). This has been going on for years at Auto Co. (the code name for the company). "Look", he finally said, "you're making our lives miserable. You're making our jobs difficult because you're trying to control us. I can't get anything proved around here without coming to you, and getting permission from you to get it approved. Why do we need a system that's so cumbersome?" By the way, the data for this are tape recordings of meetings, and tape recordings of interviews with people after meetings. Some of it's real time, and some of it's retrospective, but these are quotes from the person who said that. Now this is a quote from the launch manager, who I call the business planning manager in an interview afterwards. "Lo and behold, I found myself saying 'Because I don't trust you.' "

Here's another interview with an internal consultant who was sitting in on the same meeting reflecting when the launch manager said that - and he actually shouted it. "There was an uncomfortable silence in the room. What went through our minds was that we always suspected he didn't trust us, now he's telling us as much. Then the launch manager proceeded to say, 'And let me tell you why I don't trust you. If I did nothing to pressure you to meet your deadlines, you wouldn't meet your deadlines.' " I offer that in the spirit of giving you a walk into the middle of something like this.

What proceeds then on the next two or three pages, is a variety of both quotes from the meeting, and reflections of people after the meeting of what happened. The bottom line is that nobody was surprised. They were surprised by the fact that he said it. "We all knew he didn't trust us.", as the one gentleman comments. Well, after this, a remarkable thing happened. Because he had said that, and because they all then responded that they all knew he didn't trust them, and they wondered why all along, in a paradoxical way a certain quality of trust started to develop.

Some of you may know the work of Chris Argyris. Chris has had a big influence on me, and a lot of us along this dimension. One of the things that Chris has talked about for years is what he calls the left-hand column. One of the tools and methods we use in exercise in these projects is a left-hand column. A left- hand column is you get people who are having a difficult issue in a meeting to write down a little bit of what's being said. I said, she said, I said, she said. And they write that in the right-hand column. In the left-hand column of the same piece of paper, each person writes everything they're thinking but won't say. When a team is far enough along, you sit down, and together talk about your left-hand column cases. This group had actually done a fair amount of that, so it started to happen over time that spontaneously stuff would start to come out from their left-hand columns.

One thing that's interesting about this to me is that it's always dangerous when you pick a little tiny piece of an example. You get a thread, but you don't get the weave. There is a weave here. These people are spending a lot of time talking about why they don't get parts done on time. Why they don't reach a launch date, or a prototype date on time. Talking a lot, and conceptualizing a lot the kind of interdependent world they live in.

In another very heated meeting, the head of the program and a finance manager of a program had a shouting match. It came out that the head of the program said the finance manager only cares about cost, and the finance manager said the head of the program only cares about engineering goals. "You have no discipline to meet costs." It was the finance manager who helped them see where the key leverage was in this particular system, as they had diagrammed it.

Here is an interesting example of human complexity, and dynamic complexity meeting. The openness, candor, and trust they have started to develop, combined with their ability to try to understand the system, started to help them deal with what Ian Mitchoff termed "wicked problems", which are problems characterized by high human complexity, and high technical, or dynamic, or conceptual complexity.

M: This audience is familiar with the concept that in a complex, diverse, highly interlinked system, a small set of simple rules produces massively rich possibilities for behavior, and you set a standard for action in theory. I wouldn't expect you to enumerate a half a dozen rules or design principles for all organizations, but are there design principles for the design principles? We agree that the leader has to bring something to the organization, if not charisma, or vision, or any of these other 80's style characteristics. He brings rules, or at least the context in which rules are developed. What are the design principles for design principles?

Mr. Senge: Could I be a little bit rude?

M: Sure.

Mr. Senge: You're illustrating something that I was hearing in several levels before. Two things are really problematic for me in that question. The use of the word "the leader", and then of course the use of "he" referring to the leader. But here's the real point: there are no executives in this team. Now there are bosses, there is a hierarchy. But I know how the bosses on this team got involved, and it didn't come because of their enlightenment. It came because some internal consultants saw some possibility for something being done very differently.

One of the things that's a huge limitation in our thinking about, and taking more effective action in this whole area is this notion that leader is boss, and boss is leader. It's everywhere in our language. We use leader as a synonym for boss. I would suggest to you if we do that, we are actually saying two things, which we don't intend, probably; but we're saying them very clearly. The first is that other people aren't leaders; and the second is we have no definition of leadership, right? We don't need two words to describe the same thing.

M: There's someone that precedes others in the organization.

Mr. Senge: Yes. There is leadership. It's a word we should suspend for ten or twenty years. It may be so screwed up in its connotations that you can't use it in a productive fashion. I'm very open to that idea. But a rose by any other name... There are people who stick their necks out. There are people who step forward. There are people who initiate, and then there are people who work to sustain significant forces for change; all of which is what I would consider the kind of core work of leadership. It's not just initiating, because when you initiate, you bring a lot of difficult stuff out of the closet. Then a lot of times people who initiate aren't very effective at dealing with that difficult stuff. The main point is that these people are all over the place. We found at least three different types of people playing critical leadership roles.

Now what are their rules of thumb? I don't know. It's the kind of question that I have to tell you I'm not too used to thinking about. All the things that came to mind are timeless verities. Commitment to the truth. That doesn't mean truth with a capital "T", but as I was trying to illustrate, a commitment to talk about what is so for us, even though it might put ourselves at risk. An appreciation of our interdependence. That may be a bad phrase to use, but I mean as Ben Franklin said, "We must all hang together, or most assuredly we shall all hang separately." That's the kind of timeless verity. People as individuals do not create anything. Creation, or bringing something new into being is always a product of human communities. The closest word that comes to mind on this is love. The real appreciation of the other. The appreciation of the quality of our relationship. As far as I'm concerned, the quality of thinking in organizations is very, very strongly influenced by the quality of relationship. These are not very new or novel ideas; and I'm sure it's much too simple, but those are things that came to mind.

M: As a reaction to what you said, do leaders necessarily initiate, or do they recognize the initiatives of others?

M: Either way they're articulating.

M: Okay. But that's very different.

Mr. Senge: It is very different. That is where we have to suspend the word "leader". There are folks initiating. We can find them, we can see them. Then there are people doing lots of very, very important moves as a result of the initiating. Now I would prefer myself to call all of that part of the phenomenon of leadership, and define leadership as the way human communities shape their future. I don't think there's a snowball's chance in hell of redefining leadership in this day and age.

We found some of the most effective local line leaders, who I regard as critical, who will stick their neck out, and try something that is completely counterproductive in the larger organizational context. They shoot themselves in the foot time and time again, because they don't understand the political dynamics of a larger corporation. Consequently, no larger learning occurs. They're vital as initiators, and they're very counterproductive as diffusers. They become zealots, they become true believers.

This is a public story, so I can tell you this one. Taurus, the most successful car development team probably in the last twenty years at Ford, every single key person on Team Taurus had left Ford within a year after the Taurus was launched. There is no evidence whatsoever it had any impact on how Ford develops cars.

M: Whose responsibility would it have been, if anyone's, to create the culture or conditions in which those people might have stayed?

Mr. Senge: You need to be careful when you use language like that. "Culture" I would be real careful in using, but let's pick "conditions", because it's a little more neutral term. Clearly you would like to have executives who are always wrestling with those questions. You need line managers really initiating and trying new ideas, because that's where value is being generated, and that's where you've got to improve the value-creating process. You need folks who are concerned about how we operate and how we learn from ourselves.

Let's pose the counterpoint. How would Ford have not lost that knowledge base? (I would argue there was a knowledge base in Taurus.) How would they have not lost it? One way might be if somebody felt personal accountability, in the sense that part of their job was to study the innovators. If there was some research ethic built in. There are very few examples of that in corporations today. There are one or two, but very, very few. I don't think you can expect the line people, who are focused on their business accountability, and their customers, and so on, to be stepping back, and saying, how can we conceptualize what Taurus did that nobody else has done before? They broke a lot of rules. How could we sustain an inquiry into whether or not some of those rules shouldn't be changed? That's tough stuff. That's politically sensitive stuff. The people who led Team Taurus were hated. They were incredibly controversial. You need someone who can rise above that controversy, and again sustain that kind of inquiry.

The Marine Corps came up with a notion of this that they called the creative contributor. That within the unit they identified who this creative contributor. They identified that they were killed very easily, and that they had to be -

Mr. Senge: I think you mean metaphorically killed.

M: In this case both ways, but to a certain degree metaphorically. They could be burned out very quickly, and they had to protect them. It was proved in Vietnam that it was the creative contributor who was getting people home alive.

Mr. Senge: I know. The US Army is member of SOL. And I have to say, in some ways they are more sophisticated than any corporation around some aspects of learning. They'll all trace it back to Vietnam, and the incredible breakdowns that occurred, and the need to fundamentally rethink their values and their processes. I've found that people in the military are very sophisticated in their thinking, and their practice.

M: The Marine Corps have openly embraced complexity thinking, and have brought SFI gurus into Quantico.

Mr. Larry Prusak: I come as an emissary, as I said before, from Big Company Land. My own background, quite different than my colleagues, is that I was an academic in an earlier life, studying the history of ideas. Eventually I got bored with academia. I wanted to be in the world, and so I went into the world as a management consultant.

I specialize in knowledge. I think I invented the term "knowledge management". Mea culpa. We did this because we're paid to invent terms. There is a real subject, in economics and sociology, of what people know, and what organizations know, and how it's manifested. This is a real thing, it has a lineage in epistemology and other subjects too. I tried to study a great deal of that, and then did some writing on knowledge management. You can't manage knowledge, because you can't see it. This is a given. The world adopted that word. I think Mike Hammer would say this about reengineering if he were here. That the world adopted the word.

The increasing unremitting vulgarization and commercialization are terms that Tom Pettinger referred to. DeToucville noticed this in 1830. P.T. Barnum should have statues put up to him in every town square. This is what American commercial society does. Nothing wrong with it. It gives us freedom, and a lot of wealth. It also tends to take things, and turn them into something I can sell to you on a disk. All the consulting firms I know sell something called "packaged enabled business Transformation". Something that makes no sense whatsoever if you look at the words, but they're making billions of dollars on this. This is absolutely true.

I work for IBM, a very big firm. They have big consulting and services types of practices. They talk a lot about knowledge. They work with information, which is sort of a message that's bounded by repositories. Information and messages, me to you. Guernica is a message. Schubert's music is a message. Emily Dickinson's poems are messages, and so are memos. Knowledge, as I think our two speakers said, are what people know, and what groups of people know. It's the space between them, very often, rather than the people themselves - the social capital and human networks role.

IBM and all the firms want to sell you something about knowledge. They go to these large firms, and say we want to do something around knowledge in order for you to buy services, software, and hardware. I get called in to say, let's do something about knowledge. It's an interesting job. You end up at large firms, talking to executives. People who have the budgets and the power to do something. They say, what are we going to do about knowledge?

I'll tell you how I try to answer that, within the realm of maintaining some integrity, and still being able to pay my mortgage. This is based on real stuff. First thing I try to do is find out where the knowledge is in the organization. The best book I know about the confluence of knowledge and economics is Nelson and Winters book, An Evolutionary Theory of Economic Change, which is a very serious book about how knowledge manifests in routines, in social networks, and in other places. Where is the knowledge in an organization? Is it in people? Is it in processes? Is it in routines? Is it in documents and databases? Is it in the social networks that our two speakers were talking about?

Once we begin to talk about that, with whoever wants to listen, we begin to ask what is a unit of analysis? When you talk about knowledge in an organization, what is it? What are you going to analyze? An event? A group? An individual? The entire enterprise? Its past history? A GM is interested in analyzing decisions. The decision is the unit of analysis.

We heard something about the Ford Motor Company. Not long ago they did a little experiment in this area, and found they had five hundred and sixty some odd people across the firm who were passionate about the technology of brakes. This might not turn you or me on, but if you're Ford, it's big time stuff. What do you do when you learn something? You have five hundred people across the world, across functions, across boundaries, across rank. What do you do with that? Do you give them money? Do you give them space and time? Do you acknowledge that they're valuable to the firm? They talk to each other, they take that extra step, they read something, and send it to you, when it's not in their job description. That's a real issue. Is there an intervention you can make in an emerging group, a community. People who are not organized by the hierarchy, but have self- organized themselves around a passion. What do you do with those people? It's a legitimate question. You want to help them. You want to make an improvement. How do you improve their behavior? Maybe you should just say, live and be well, as my grandmother would say. If we can identify networks, and communities, what can you do about them?

The other thing I want to say is that people ask all the time, what's the return on knowledge? If we make this investment, what's the return? This is so insipid, but it happens to everyone I know who tries to do things in firms. Unless you speak with a position of immense cognitive authority. Peter Senge might actually have this. He's really very well-regarded. Peter Drucker can do this, or people like that.

There's probably only one or two answers for this. One certainly I did with my first consulting job. I came to this when I was in my mid-thirties. One of the biggest food companies in America, which exists in the wilds of Delaware and Pennsylvania, wanted to build a central library so that they could keep stuff, and keep documents, and find out where they are (this was radical stuff). They said, we can't get a project approved unless you can prove to us the internal rate of return. How much will we get internally for the dollars we spend on this library? And it has to be over thirty percent. In my ignorance towards what I actually thought consultants did, I tried to work out a methodology that could do that. This, of course, is just absurd. It was silly. But I spent time doing this. One night, in a fit of integrity, I woke up at two a.m. and said, this is bullshit. I came in the next day, and told this company, which is a very large firm, thirty-six point eight percent. They wrote it down, brought it to their management, got the project approved; and I had made that number up. Totally made it up. It didn't matter. They needed the library. I did what I was paid for. You couldn't disprove it would be thirty-eight percent. It's a Karl Popper Memorial - it couldn't be disproved, therefore it was proved.

But what is the return? This gets to very big issues. I think Peter Senge was talking about this. Comfort with ambiguity: a sense of understanding that's beyond raw, reductive linear models. This is the hardest act I have to play. As near as I can tell, about seventy-four percent of senior executives have engineering or accounting degrees. Now what does this mean? It makes my life very difficult, because they think there's one right answer, and it's a linear approach, and you can get there. You do A, B follows. I build a library, we get this internal return. We help networks across Ford, give them a little money, a little time, and we'll get better brakes. Maybe yes, maybe no.

What I used to do is carry around with me a little picture of Martin Luther. I'm not a Lutheran, to put it mildly. Luther talked about justification by faith, and justification by good works. I say, read Martin Luther, he talked about this. They built the whole Reformation on these processes. It doesn't always work, but it's something. This false quantification, this false science, this obsession with numerology really is killing a lot of firms. This has been talked about for many decades previous to me. Eventually you rub up against it - this sense that everything can be measured. What's remarkable to me is people don't live their lives that way. They live their company life that way, but what's the return on having children? I defy someone to tell me this...

M: Thirty-six point eight percent.

Mr. Prusak: I can't disprove it. You can't prove it. It's a dead heat. What's the return on marriage? What's the return on piety? Stuff we all live with. Patriotism, piety, community sense. No one can measure that. We all live with it. When you go into firms, you've got to measure it. It's a remarkable change.

I've seen about a hundred and fifty knowledge projects now over time. In a very raw sense, what we're talking about is that knowledge, what people and groups of people know, is local, contextual, and sticky. There's tons of books about it. I just red a book by a political scientist at Yale, James Scott, called Seeing Like a State. I just finished it on my way up here. Scott wrote about why large scale experiments like Brasilia fail. It's a fascinating book on the difference between local knowledge - what people know who have to do things - and planning knowledge.

It's very difficult to transfer knowledge in organizations because the knowledge is embedded in social capital, and social networks. Knowledge is not Cartesian. It's not out there. It's not an individual. It's embedded, embodied in groups of people who share common goals, common ideas, and common emotions. That is a very elusive subject. Ron Bert did fine research on this. I'm writing a book on this myself. What do you do when knowledge is embedded in social capital? It's the space between people. What can you do? What's an interventionist stance towards that? One thing Professor Nonoka wrote about, and something I've absolutely grown to agree with, is the need for space.

A lot of firms are run with mechanical models. The models they have in mind are of a lean machine. This is ridiculous, but it's still very current - it's the way people think. Machines don't need knowledge, they just need energy and direction. If you think of an organization as a machine, you're not going to make much progress here. We fight against this.

Nothing happens without reflection, and there's no reflection without space. I'll give you an example. I did some work for the Monsanto company. Very fine company. Smart, good firm. They had a big library in their R and D center. It was a room not much smaller than this filled with chem abstracts, and physical abstracts, and books about science. The scientists would go there, and use the library. Some administrator in Finance had this brilliant idea: since almost everything in that library can be found on a CD-ROM which can be put on a server, and we can bring that the desktops of the scientists, we can eliminate the space for the library. This is technically true, there was no argument. He knew he was right. I was in the room, and he said let's just go ahead with it, and he was going to write off that library. I said wait, wait, wait. Ask the scientists, ask them. He said no, no, no, we don't have to. It's a no-brainer; a phrase you hear all the time. People who say that usually have no brains, I'd add. Even though I'm a consultant I said I will interview and I will spend an extra day here no charge. I will not bill my time. But let me please ask a few of these scientists what they think of this brilliant idea. Of course the reaction was overwhelming. Kill that library, and I'm leaving this firm. You kill that library, and I will transfer out of Monsanto. I have a resume, I have an application. These were the people who make that firm go. These are the scientists. Talk about creating knowledge. There is no Monsanto without them, believe me. The passion, and the power of their reaction really surprised me. You can deliver most of the stuff to a desktop. You can buy all that stuff. They said we know that. Do you think we're stupid? We understand that. We need a place to go to that's unstructured, where no one finds us, where there are no telephones, and where we can talk and think. They had no such place but the library. Nothing happens without reflection, and nothing happens without space - cognitive space.

I have a friend who works for Arthur D. Little. He's in his early seventies, and one time he found two hundred annual reports of American firms over a couple of years which mention in some way, in some form, that they believe in learning. God love them. They believe in learning in some way. He called people in some of these firms up (he knew a lot of people) and said, did you not believe in it before? What makes you a learning firm? Is this just something you're saying? He'd ask them one question: if you sat all day at your desk, and you had an open office, or left the door open, and read a book all day, what would be the social reaction? What would be the reaction if all day you didn't answer the phone, you didn't boot up the computer; you read a book? Mike Porter brought a new book out on strategy, and you read it all day; and you're in corporate strategy, so it's relevant. Although other things are equally, if not more. No one said they'd be comfortable doing that. No one. We're a learning organization.

Space, cognitive space. If you don't have space to learn, what are we talking about? Physical space, cognitive space. You can get all the technology you want. You can get money for technology. It pays off my debts, it will pay off yours. Sell them all - you can't get space, and you can't get attention. These two things are the most valuable for doing knowledge stuff, or learning stuff in, and you can't get them. They're becoming a scarcer and scarcer commodity.

Another key thing, along with space, is trust. Ken Arrow wrote a book twenty or thirty years ago, Limits Of Organization, where he talked about trust being the most efficient economic tool since you don't need to negotiate when you have trust. You don't have to barter or bargain, if you trust someone, you trust them. Trust is anticipated reciprocity. It's an options model. I will help you, with the understanding, tacit or covert, that you will help me in time, and that you have something that I would want to have help with. It's a long way of saying it makes a tremendous difference.

There are things that can be done to enhance trust in organizations. Charles Sable at MIT wrote a very interesting paper on how, if nations could re-create their myths, you'd have less wars. Companies can sometimes change stories. That's why I like learning history. Sometimes if you change stories about the past, it will help engender trust.

I'm at IBM. We own Lotus. We didn't merge with Lotus, we bought Lotus, and we're eating it.

M: Is that off the record too?

Mr. Prusak: You can put that on the record. The Lotus Eaters. One of the things we're doing is changing the stories they tell about IBM, which of course are not dissimilar to stories I heard about Germans in the 1940s from my own parents. We're changing the stories, and it changes the legends, the myths, the way people understand; and it will help build trust. Sable was right. There's a lot of very interesting literature emerging, in economics and political science, about the value of trust. Again, this is not mechanical, this is why some of the models don't take that into account - the human aspect of trust. It's a big, big deal. The third very important element, which again is not science so much as heuristics, is perceived equity, which is a form of trust. If you sense in an organization that you're not getting your due, that there's wealth being created not being accruing to you. Not only financial wealth, but social wealth, intellectual wealth. You're going to under-optimize, you're going to under-perform, and it kills trust. Perceived equity is a key ingredient in social capital. Networks will occur, but within these social networks that we're talking about within the growth of social capital, it creates a tremendous constraint if you have a sense that someone is going to misuse what you give. They'll be thrown out of the network. Not formally, but slowly their calls don't get answered. It's a slow process. If the whole firm has a sense of this, it's deadly.

There are firms that have low levels of social capital, very low levels, but make so much money it swamps it. It doesn't matter. Some of the Wall Street firms have no social capital, but enough wealth that people say okay, I'll take it. It doesn't matter. I'm a star, I'll work by myself, or find a few friends. Most large firms are more process, and less star-driven.

The last thing I'd say about social capital and knowledge that's embedded is an investment model. One needs to get across in these organizations a sense that there are interventions that can be made, investments which lower transaction costs, and enable cooperation so knowledge can be moved and can be generated. Someone wrote a book in systems theory, Colin Eden, I think, called Messing About With Problems. I'd like to write Messing About With Knowledge. There are people in firms who know how to do things, and what they know how to do can sometimes be made more valuable if more people knew who knew had a little time, a little space, a little money, a little technology, a little pat on the back. Large firms in America are still monstrously hierarchical, and monstrously stressful. Don't let anyone kid you. Everyone salutes, and drinks Maalox. But make no mistake about it, there's changes occurring. With the diffusion of technology, you get a sense that knowledge gets more visible in organizations. As it becomes more visible, because of intranets and web sites, people get a sense of who knows what. There's a real disparity in organizations between who knows, and who has the money, and who has the power. It's beginning to look a little like France in 1780, where some of the bourgeoisie creating wealth said, now tell me again, what do those monks and aristocrats actually do? It's a very interesting sense of social stress. Knowledge isn't free, it isn't cheap, it's expensive, and it takes interventions.

The last point I wanted to make was the role of symbols and signals. The people who control the budgets, the people who allocate the resources - what they do is watched by everyone else. The way they symbolize their activity is very important.

Before I joined IBM, I worked for another great firm, Ernst and Young. I had a friend there, Tom Davenport, who wrote a book on process engineering. This book made Ernst and Young, hundreds of millions of dollars. You don't need to know the causality. Take my word for it, it established them as knowing something. Although Tom knew it, they said he works for us, so we know it, and they made a ton of money reengineering firms. Tom never heard one sentence from the people who ran Ernst and Young. Not a sentence. He got paid very well, this is interesting, but he never heard a word, and it hurt him, it bothered him. How come I never hear from the top?

The firm shifted management. The top moved on, another group came. Then Tom gave a small interview to some magazine - a pretty small level magazine.

The new managing partner read that interview, and wrote him a handwritten note. He took out a pen, and said, hey I liked reading this. Next time you're in New York, let's talk about this. I was on the plane with Tom when he read that note. He got his mail, he was reading it on the plane. It really moved him, it really mattered to him. It helped him stay in the firm. It created an impression on him, as a unity. Not in his brain, not in his pocketbook; him. That's really important stuff. It's like your Taurus story. It's not wildly different from that.

We need, again, a big element of social capital. Children need it, adults need it. You need to be recognized. You need a sense of identity. There's a real thing called identity. Identity in this country, for better or worse, is often wrapped up in where you work, and what you do. Whether it's right or wrong, it is. You can't divorce a company from the environment it lives in. We live in a society. I've never heard it said better than C.B. Macpherson - possessive individualism. You're not different when you enter a firm door. It's possessive, and people want to possess something. Their sense of identity is strongly tied to what they do, and how they're rewarded for it. It's a big part of the social capital issue. It's not enough to know things. It's enough to know, and be recognized for what you know, and somewhat rewarded.

It's a very odd group today. I've never spoken to a group like this, and it's sort of interesting for me to hear all this, and I hope this helped you somewhat. Thank you.

M: I'll share one tiny anecdote. We've been talking a lot about large organizations, but now about a very small organization, a fifty million dollar a year organization. A fast food franchising operation. European hearth style bread that's called Great Harvest Baking.

M: I love that. Oh yes.

M: A great place.

M: You say it's a great place.

M: Well, at least their bread is good.

M: It is. The bread's wonderful. They're continually changing recipes. They're continually innovating marketing strategies, pricing strategies, dealing with local officials, employment, recruiting, training. It's a knowledge organization. One of the rules this organization has is that any franchisee anywhere in the country who wants to go to the expense of traveling to see any other franchisee, or to invite one to his or her place, has a fifty percent reduction on the cost on the payment back to the parent company, on the royalties.

Mr. Prusak: That's a great idea - that's exactly what I'm talking about.

M: The people that run the parent company - I say run, but I should say the people who actually own it, because the franchising company is owned by a family - promulgate almost no rules. This is a franchise operation. The cover page of the franchise agreement says anything not expressly prohibited by the terms of this agreement is allowed. They consider themselves innovators. They swap this knowledge around, and they're way ahead of all competitors in that space.

Mr. Henry Mintzberg: I'm going to sit up here. If I stood, you wouldn't notice the difference, because I'm not that tall. I have the same feelings about being in front of a group that I don't know much about. I wasn't even sure A) about complexity theory; or B) about whether I fitted this or not when I came down here. I put together some notes of some things I might talk about, but I really wanted to listen, and react to what's going on. Let me do a bit of that. The first word I learned, or probably the word I learned most emphatically yesterday, was clockwork. I got a sense that clockwork means the same thing as machine organizations, or bureaucracy, or structure, or what have you. The second thing I learned is that this conference is running like clockwork. I mean that in both senses of the word. We talk about something running like clockwork meaning it's running very orderly; but that of course is the same meaning as the word clockwork has been used, so far as I gather, in complexity theory. It kind of bothered me quite a bit.

I attended two sessions yesterday, one of which was on health care, and it ran more or less like clockwork, except there was time for questions. Then I attended a session in the evening on organizational issues, and that ran like clockwork, but there was almost no time for discussion. I thought well, doesn't there have to be a consistency between the content of the presentations, and the style or approach of the presentations? I'm being quite serious. Linear is kind of bandying around here as a kind of dirty word. What we've all been doing is linear. I am giving you a talk that's linear. It started with my first word, it will end with my last word, and every word follows in sequence, and you're not going to interrupt me; although I wish you did. I do wish you did. But it's linear. In fact the conference is very much like clockwork. A lot of it runs linear. Now we swarm - another word I learned (I mean, I actually knew what clockwork was before I came, and I know what a swarm is because I'm interested in bees, but leaving that aside). We swarm when we go out for coffee, but we don't swarm within this room. And Bill sort of alluded to something similar. I'm going to try and keep my comments brief, so that we can do a little swarming, because I think that's the purpose of why we're here, and not clockwork. In other words, I really believe the things you've been saying, even if many of you don't.

Some random thoughts. I also heard the word "shotgun". Somebody talked yesterday about shooting a bird with a rifle, or shooting a bird with a shotgun. The point made was if you shoot the bird with the shotgun, the pellets that miss are not wasted, because you weren't sure exactly where the bird was going to be, so you had to spray the pellets in a certain area; and the fact that some miss is fine, as long as some hit. I am going to shotgun my presentation. This is not linear and clearly conceived, it's just a bunch of random thoughts; half of which, maybe most of which, will be irrelevant or silly, or what have you. But if one of them is like the butterfly in Rio, then maybe it will create some kind of turbulence in the room; which is after all not so far away as Rio is from wherever it is; New York, or somewhere. So let me just say a few things.

First of all, one of the things I discovered here is that I've been a complexity theorist for a long time. Probably before anybody even used the word "complexity theory". The first thing I did was study managers. After being imbibed with the old view of managers as planning, organizing, coordinating, and controlling. Now going back to Fayole.

Think of what the world would be if we had listened to Follett, and not to Fayole. Nobody heard anything much from Follett until decades after she died. Whereas we were enamored, one way or another, with Fayole, either directly or through Urwick and Gulick. Imagine if we had had a world of cooperation, instead of a world of control, as our model of management. As you just finished saying, we may believe we're into the age of Follet finally, with cooperation, instead of the age of control, but it's a crock. The world out there is still clockwork, just like the world in here is still clockwork.

I went out to observe managers, and I discovered that all kinds of funny things were going on in the office, and they were being interrupted all the time. They were doing short things. A journalist described managerial work as calculated chaos; which I kind of like. I think it is a kind of calculated chaos.

Managers have to act. That's what management is about. When Bill deals with his doctoral students, in fact he acts. When he writes management theory maybe he doesn't act, but when he deals with his management students he raises the bar. Bill is a perfectly action-oriented manager when it comes to managing the things he has to manage for results; which is to get his doctoral students out the door. By the same token, Peter writes things that are theoretical, and conceptual too, and not necessarily actions. There's not so much of a difference; but there was a difference in the way they talked about things. It seems to me that because managers have to act, or intervene, or change things, they have to simplify. No matter how complicated the world is, managers have to act. The worst thing a manager can possibly do is not act. Doing nothing is far worse for a manager than doing something, no matter what it is. The best thing is to do the right thing, the worst thing is to do nothing, and the second best thing is to do something, even if it's the wrong thing. At least you can correct it, as long as it's not one of those big things we were talking about before. A little thing you can get started, and corrected, and so on. If you've read Karl White, you know what the conceptual underpinning of this idea is.

That leads to something I called emergent strategy. I've been using that word at least since the early eighties, or late seventies. The notion of emergence is something I'm sympathetic to. An emergent strategy is based on the fact that you don't make a strategy as some kind of Moses-like process where you walk down from the mountain, and present the tablets, and everybody else runs around implementing. Instead you make a strategy by trying something, and testing, and changing. Other people take the initiatives. There's a three-part process for what management does. There are initiatives that people take. Those can happen anywhere in an organization; and often down on the ground, where people know what's going on. There's a kind of championing or promoting role, which is somebody who recognizes the initiative for more than it looks to the person who started it, and promulgates it, to some extent. Then there's an acceptance role. Part of the promoting role is to seek some kind of convergence in the initiatives. Strategy becomes simply the consistency or pattern that develops in behaviors. That is what management is really about.

That leads me to the conclusion. I just did an article called "Managing Quietly". What managing quietly is about is the fact that the hype gets in the way, and the big initiatives get in the way. These big turnarounds where people come in, and do dramatic things to organizations, and drive everybody nuts, get in the way because they don't know what's going on. Managing quietly basically means building up the system quietly, slowly, and low-key, by which you encourage people to take initiatives.

Decision making. I'm doing a paper called "Decision Making: Enough Thinking?" In it I contrast a thinking first model of decision making, which is the standard kind (define the problem, diagnose the situation, generate solutions, etc) from a see first view of decision making, which is a bit the Gestalt notion of preparation, incubation, illumination, verification. You don't understand a problem until you see it. Like the chimpanzee in the cage who wanted to get the banana, and then suddenly realized that the box in the corner could be put under the banana. He climbs onto it, and he gets it. When he saw the problem, he solved the problem. The third approach is the act first. You act in order to think. It seems to me that the more complex something becomes, the more you have to act first, and the less thinking is actually useful; at least initially, as a starting point. It's a bit like coming here. I had to act and see before I thought, because I always misjudge audiences that I don't know. It's no use trying to write speeches or talks before I come to an audience I don't know well, because I always get it wrong. Sometimes you don't have time to adjust, and sometimes the audience is less tolerant than you are.

One other comment about decision making concerns the view of decision making as a chaotic process. Organized anarchy. The thing that gets raised in my mind about describing decision making as kind of a garbage can, anarchistic process is: is this the complexity of the process we're talking about, or is this the inability of the observer to understand what's going on? Anything we don't understand is, by definition in our minds, chaotic.

For example, a lot of the talk about turbulence. I hate that word "turbulence". I think it applies well to hurricanes, literally, but I don't think it applies well to anything in management. I've read things from people like Kahn who wrote in the seventies of the view that we live in a turbulent world. They said, "Nothing mankind has experienced is quite like the turbulence we're experiencing now." True, there was a Depression, and a couple of wars in this century, but "nothing is like the turbulence we're experiencing now." A lot of that, claims about turbulence, is just an effort to pat ourselves on the back, and say we're really important because big things are happening now. I don't see any turbulence in this room. I don't see any turbulence outside this room on the highway. I don't see any turbulence in this hotel. I see very little turbulence anywhere. If you want to take turbulence to be something like the siege of Leningrad, if you lived through the siege of Leningrad, I grant you that was turbulent. I don't mean to be callous, but the use of the word is absolute and utter nonsense most of the time. We pat ourselves on the back.

I accept the notion of complexity, and complex systems. But how much of the complexity we supposedly see around is real complexity, and how much is simply confusion because we don't have the right theories or approaches to understand things; which may in fact be a lot simpler than we think? How much of the complexity is the unexplained variance, and how much of the complexity is real, true complexity?

M: In that case, I'm going to ask Ron Schultz, in his role as my co-facilitator, to share a few thoughts, and to do some summarizing before we start swarming.

Mr. Ron Schultz: I have an unenviable task here of trying to summarize what has gone before here. I will begin by quoting Thomas Pynchon, "Do not underestimate the shallowness of my understanding," in attempting to deal with this. Henry brought up something, and this will make that statement evident. I need some help with something here, and I wonder if you could all do this. If you could hold out your arms, and just kind of do this for me please? Could all of you? Great. I just want to really screw up that butterfly in Rio. It's making me pretty damn sick.

What I want to talk to you about today is in terms of summarizing some of these ideas. Peter began with this idea that knowledge is the capacity for effective action. What that is really saying to me, on one level, is that our ideas are shaping our behaviors, so that what we think shapes what we do. When we get stuck in our old ways of thinking, we limit the possibilities available to us. We see this kind of lock even in our language. Bill was talking about it, a couple of other people were talking about it. We've been talking all day about the firm, which is a real mechanical statement to describe this organization. And we seem, even in our language, to get locked into this place.

Henry was talking about our mechanical perspective as well, and how we do get locked in. What happens when we get locked in is we limit our possibilities. When we limit our possibilities, we limit our ability to respond to unexpected occurrences. And when we limit our responses to strategies, we become vulnerable to catastrophic events. And when we are vulnerable to catastrophes, an unexpected occurrence can wipe out the system. Which in my estimation is why it's important for us to understand how ideas shape behaviors.

Now all of you may think oh, this is just wonderful. This is all well and good. But the question that you really have burning in your minds is can this guy do balloon animals? This is all a way of summarizing what has been going on here today. And -

M: You bring balloons along just in case?

Mr. Schultz: Just in case, you never know. The answer is yes, he can do balloon animals. Now the question that I have is, can these guys do balloon animals?

M: No.

M: You've had practice.

Mr. Schultz: Yes I know. It's cheating. Now I'll put these guys on the spot. Let's see if they can do this, and if they will. The point of this is that what they are showing to you is that they are not getting stuck in their old ways of thinking. At least they are attempting to try and do this. Oftentimes, if you were to give a new task to somebody in management, they would not be willing to attempt this, right? Don't hyperventilate.

Mr. Schultz: I cheated. I'll tell you now. But let me give you a little story about this whole notion of getting stuck, and what happens when we do get stuck. There's a little place here in New Hampshire called Lake Sunapee. It's a little north of here. It's a beautiful little lake. They were getting a new minister in the town, and it made these three guys rather happy, because they always took the new minister out on the lake to go fishing. So they arrive at the lake, and the minister arrives; and to their surprise, the minister is a woman. And they thought well, that's okay. That's fine. We can take her out fishing, she'll enjoy it. They get in the boat, they go out to the middle of the lake, and they're fishing, and they're talking, they're having a wonderful time; the discussion is great. And the sun starts to go down, and she starts to get a little cold. So she says I'll be right back. So she throws her feet over the side of the boat, walks across the water back to her car, gets her jacket out, puts it on, closes the car, walks back across the water, gets back in the boat. Nobody says anything, but the next day the village is abuzz. Did you hear? We have a new minister, and it's a woman, and she doesn't know how to swim! What happens here is when we get locked in to our old ways of thinking, we miss the inconceivable.

What I want to provide for you here is what I call a sustainable model for inconceivable development. Which I think is a little bit of the How To, but then again, it might just be more information for you. The process is very simple. There are only three steps in this whole process. It begins by understanding the system.

As soon as we have this new understanding, as with complexity, it requires us to do something. It requires us to adjust our relationships according to that new understanding. We go in, and we adjust all our relationships with people that we work with. And what happens? Something emerges out of that interaction of our understanding and that adjustment of relationship.

Now we have a new understanding because of what's emerged. We have to go back in, and create a new understanding of how the system works. Well, that means that we have to go back, and we have to adjust our relationships according to that new understanding. And we do that and something new emerges. Which means we then have to go back in and readjust our understanding of the system, and the system continues on down.

What happens in an organization when we do this is that by the time you get three levels down in this process, what emerges was absolutely inconceivable. What usually happens within our organizations is after the first emergence, we stop. We don't go back in, and readjust our understanding, and continue the process on down; and the system dies. We see this a lot in organizations, in terms of continuing the process.

Before I tell this last story, I want to make a couple statements here. Whether it's true or not I'll be interested to hear. On one level, I was thinking that complexity is really an outcome. It arises out of the very active organizing within an ever-changing environment. In business it comes out of our interactions with each other, and that environment; and the interpretations of those interactions are all based on one common experience. That one common experience is our thinking. When we don't get stuck in our outmoded ways of thinking, we can greatly affect the way in which we operate. If our thinking is hectic, and overly busy, the way in which we operate is hectic, and overly busy. If our thinking is quiet and calm, so too are our interactions.

Now I would like to provide one more story about this whole process of what happens when we get stuck in our old ways of thinking, and why this kind of thinking is so important. And I apologize to anyone here of Irish distraction.

Paddy is taking his donkey to market, and he comes to this low overhanging bridge. The donkey's ears are too long to fit under the bridge, so Paddy gets this idea, and he takes out his chisel, and he starts chiseling two grooves down the underside of the bridge. When his friend Erin comes by and says "Paddy, what are you doing?" He says "Well, the ears of my donkey are too long, so I'm putting these grooves in." He says "Paddy, Paddy. Why don't you take your shovel, and dig a trench?" And he said "Erin, you're not listening to me, are you? I didn't say his legs were too long, I said his ears were too long!"

Enough thinking, and enough talking. I hope that gives you a summary of the afternoon. Thank you.

Mr. Bar-Yam: I think the importance of putting the human being back in the organization, also in the context of thinking about organizations in the context of complex systems, is a very important one. I wanted to mention something in this regard at a higher level of abstraction which at least suggests that complex systems thinking should not be understood to neglect it entirely. That is in contrast to the usual discussion of chaos, and the concept of edge of chaos, and so on, where people are being pushed to the limit, in order to be able to exist in a complex environment. The discussion that is relevant here was a discussion that Herb Simon gave at the banquet talk last year. For those of you who were there, I will just remind you.

He spoke about the concept of homeostasis, which was very important in the previous generation of complex systems thinking. It is missing in much of the current thinking about complex systems. The idea is that there is a complexity that the organism exists in, in the context of the environment; but the parts of the organism are not in the complexity of the environment outside, because the organism is shielding, and protecting its own components. Realizing that a complex system is protecting and shielding its components from the complexity that they otherwise would encounter, I think is an important part of understanding how complex systems thinking should be applied in the context of human organizations.

Mr. Mintzberg: I want to make one comment, which is while I agree with the comment, I hate the term "social capital" more than I hate the term "human resource".

M: How about in relation to turbulence?

Mr. Mintzberg: About the same. I'm not social capital. I'm not a human resource. All those things belittle human beings as individuals. If we're going to take your comment seriously, Yaneer, we should stop using those labels and start talking about real live human beings.

Mr. Bar-Yam: Let me answer that comment. One of the things that we have to recognize is that it is important both to have abstract, larger scale models, as well as models that allow us to deal with the emotional, or individual human content. Recognizing that there are all those different levels of treating a system is important. While one should not be out of balance with the other, it is not unreasonable to have the more abstract models as well.

Mr. Mintzberg: The trouble is that companies grab onto these things. Where it used to be called "personnel", which was a perfectly innocuous label for dealing with human beings, for a decade or more it's now been called "human resources", and I just think it's a horrible, horrible expression, that's all. It translates into practice, and it remains at that level of abstraction, and dehumanizes. The best managers I know don't like human resources, but they like people.

M: We're talking about languaging here, which is the importance of the issue of knowledge management. Your mea culpa is well taken. It's a very descriptive turn of the phrase. But it may tend to tell managers, the packagers, the charlatans, that knowledge is indeed something that can be managed, and quantified. We all know that it's like Schrodinger's cat. When we do it, we affect it, and we hurt it, and we may deaden it, which your examples very nicely brought to life.

F: My comment is directed to Bill. I have some reservations about the edge of chaos model, in terms of its use as the next single solution to our business problems. The way I see the edge of chaos model is that we're trying to achieve an opening of the degrees of freedom of a system to allow for creativity. But once a manager has created those conditions, a selection process must occur; therefore, the degrees of freedom must be reduced. In any learning environment, the degrees of freedom have to be reduced even further to get that behavior integrated into the organism as a whole.

In effect, the manager has to know not only that they need to open the degrees of freedom, but when to open the degrees of freedom, and when to pull the system back more toward an equilibrium place. I don't think any system can continuously operate at the edge of chaos. It's just too much stress. I don't think anyone can even manage or maintain those levels of creativity.

Mr. McKelvey: The French have a term, "dragueur". Which is usually applied to men who try to pick up young women. We are all being here intellectual draguers, in the sense that some of my colleagues are busy trying to drag managers into a world of more complexity than what they're comfortable with. Some of the rest of us are trying to drag academic text in the same direction. What Tom started with in his opening remarks was the realization that we have had chaos, and complexity, and nonlinear events in organizations for many decades, and that we're only now beginning to develop a language that gets us out from under linear thinking and hierarchical order. While I agree with you that nobody's going to want a firm to live on the edge of chaos; on the other hand, I think we just struggle with an enduring, persistent climate, which is constantly dragging firms toward hierarchy, and stability, and an inability to respond to the kind of product development challenges they have.

F: I think Peter Senge would agree that the real challenge is to teach managers that they must do both, the appropriateness of the timing, and the skills involved in going from one state to another.

Mr. McKelvey: I have a paper that got published last year called "Quasi-natural Organizations". And I'm not the only one. I was reading a thing in Gleick's book on chaos. This is an old idea that again we struggle between the naturally emergent phenomena in firms that are outside the control of managers, and at the same time it's clear that managers have tremendous responsibilities and abilities to make changes instantly. Jack Welsh can sell half of GE and buy another half overnight.

And yet what's interesting, and what we haven't really gotten to, is the intersection between intentional behavior by whomever - top management, or some jerk at the bottom line who decides to put a screwdriver in the machine - and the naturally emergent, idiosyncratic, stochastic microevents that go on all the time that provide the basis of emergent complexity in firms. Until we recognize and develop a language that allows us to talk pretty cleverly about emergent phenomena in conjunction with intentional behavior, I don't think we're going to be much use.

Mr. Prusak: I would completely agree with especially the latter part of what Bill said. Language, and the fact that if you don't have words for things, things don't happen. Edward Sapir wrote about this in the turn of the century, and it's been proven again and again - if there's no vocabulary, and narrative around the vocabulary, things don't happen. They happen naturally, as you're saying, but they don't happen intentionally. There is no intentionality. One of the reasons that I use words like social capital is that if we don't insert words into organizations, i.e., the learning organization. We can fool around with the use of fads in language, but if there aren't words, these organizations default. They default to what they did ten years ago, twenty years ago, what they were taught in graduate school. John Maynard Keynes has a wonderful sentence, he said that the most hard-shelled, pragmatic American executive is often the slave of the ideas of dead economists. This is exactly right. People say I don't care about theory, I don't care about ideas. I'm into action. These people are the slaves of Samuel's Third Edition. They're the slaves of the textbooks that we fool around with.

Unless you put in new words as strange attractors, as focal points, as something that creates new energy, it won't happen.

Mr. McKelvey: I did have a comment on this point with Henry. One of the reasons we use social capital is that we're trying to talk to economists. People do not get into that production function. The theory of the firm does not include people. The only way you can get an economist to respond to the idea that people are not all uniform, or that they count for something, or that brains count, or that networks count is to call it something capital, because that's what's in their equation. If you don't do that, it doesn't exist. Whether you like social capital or not, if you want to sell to economists, you're stuck with their language. It is a language game.

M: Life's too short to sell anything to economists.

M: They're all dying anyway.

M: Not enough of them have died yet.

M: I've never wasted a day of my life trying to convince economists of anything - it's not worth it. On the edge of chaos, it's a lot of nonsense. Most of what we want from organizations is stability. The reason we're in this hotel is because it's predictable. If it wasn't predictable, we'd go to another hotel. If we go out for lunch at 12:30, and it's not ready until 1:00, we don't come back. If we go up to our room, and we lift up our pillow, and a Jack-in-the-box jumps up and says surprise we say I'm not amused, but my wakeup call came three minutes late. We want our airlines to leave on time, and arrive on time, and provide service precisely the way we want it, with minor variations. If the chocolate chips in the cookies are a little bigger, we rave about that airline. Most of the services we want from organizations are stability and predictability.

The most change-resistant organization, in a certain respect, is the university, and academia, and academic conferences, and academic behavior. We can talk about the edge of chaos all we like, but we don't experience that at all. The reason most of us, certainly speaking for myself, are academics is because it's a nice, quiet, predictable life. The edge of chaos is nice if you're developing software, or you're in one of these highly dynamic industries. But there aren't that many of them, when you get right down to it.

Mr. LeBaron: I have a comment and a question. At least it's my observation, over an extraordinarily small sample size, that the limitation in the description of having a very nice emergent system is the individual. The individual does not want to live at the edge of chaos, or near to it. When you introduce a system which periodically or continuously is emergent, and adaptive, nonforecastable by a linear or hierarchical sense, it appears to be out of control. The individual will do whatever it can to freeze it up. Now periodically you can get some pulses through. You can send through some shocks. That's what a merger or an acquisition is - a shock. Or you can change people around, shuffle them around. You can't do that very often, but they sort of restart the clock. But you can't keep it up continuously.

I'm concerned with the implementation side on the individual. How do you have the individual who's comfortable, want to be comfortable in the emergent systems which you will build?

Mr. McKelvey: What is your name?

F: Lynn.

Mr. McKelvey: Lynn? I'd like to start with you in saying that you went from stability to the edge of chaos. When you did that, you left out what is the subject of this conference, which is complexity, which is the zone between the two critical values. It's important to recognize it. In statistical mechanics

Coppen talks about it as the melting zone between frozen agents. Our problem is to try to make this zone as large as possible. If it's minuscule, then you're right. We go from bureaucracy, hierarchy, homeostasis - all these dampening things - to chaos. Nobody wants that. Our problem is to make this liquid zone as large as it can be, because that's where you get the entrepreneurship, and the innovation. With luck, we'll get more than just one entrepreneur here or here, but emergent networks that create the bulk movement. I don't think individuals have much trouble living in the region of what we really want to think about as complexity. They have trouble living in a chaotic world. And a lot of people get upset by having to live in a hierarchically driven world that's run by managers who are thinking in a mechanistic way.

The emergent complexity area that this conference is about is where we'd like firms be, where they need to be, and where most people would like to be. I live in a university. Henry is absolutely right - it's a disaster. It's full of people who are dying to behave in an emergent, networky kind of way, and we're treated like passive agents.

M: There's been a number of aspects of chaos and complexity discussed throughout this week. It's very interesting, and people have to adjust their perspectives. One of the issues that is the grounding of the edge of chaos stuff is the differential between rigidity of systems - that carry on in a very predictable way, but cannot adjust to changes in their environment - and a system which progresses as normal, but is still adaptable, such that when there are perturbations, it can move and change. This is the crux of the edge of chaos, as opposed to deep in chaos, where it's essentially unpredictable, or deep in linear systems, where it works perfectly, and will carry on its little clockwork motion, even if the world changes. I want my airplane to leave at a predictable hour. But if I want to change my flight, I want the airline to be a little adaptable with my ticket reservations.

There's a breadth of concepts here. To say people don't want to live on the edge of chaos is almost to say they want to live in a very rigid, and nonadaptable environment. Certainly they want predictability, but they want, and need adaptability.

M: The original notion of the edge of chaos is where a self- organization occurs. As Chris Langton says, it's where information gets its foot in the door of the physical world. It's not where things break down, it's actually where they come together.

M: I'd like to go back to the discussion on language and complexity. And Henry, I admire you for the passion with which you express your shock and distaste for language which is very popular in organizations. We use "human resources", we use "social capital", and we don't even understand what impact it has on human behavior. I believe it has an impact, but nobody really cares. Nobody is prepared to do anything about it. My own feeling is that we cannot take the concepts of complexity to our front line people, or to the managers unless we have a language which goes with that. And Ron, I was reading your book yesterday, and I found a chapter which is on language and metaphors. I would like to know from you and Peter, you have written about language that goes with system thinking. You can't have just system thinking planted, there's a language you need to develop.

My question to the panel is, could you please give us some hints on how we go about creating that language, and making that language acceptable in our organizations?

Mr. Schultz: One of the things that I touched on earlier too is this notion that we first have to recognize how locked in to the old language we are, and how we get stuck in the mechanical terminology to describe the world as it is. Just in the notion of the name of this organization: the New England Complex Systems Institute. The word "institute" comes from the root "sta", that has the notion of stasis, of static, of stable. In the very description of our complex system, we have stability built in. That's part of what we need to do, in terms of languaging. Also, the notion that we were hearing earlier about providing the space for that to develop is important, in terms of allowing that language to develop as well. My sense of what she was saying about the edge of chaos was that there was a certain freneticism to that, and that it caused companies to really charge ahead. What was missing from that definition for me was this notion of fallowness - of being able to have the open space so that we can reflect on these ideas, and also reflect on what we're saying so that we're not stuck in our old ways of thinking.

Mr. Senge: When I was listening to the question I was thinking a bit more along the lines of Ron's comment, in that I guess I'm a little less interested in creating new language than I am in helping all of us to continually pause and consider the language we're using. I worry a lot about reengineering our language. If we only got the terms right, everything would go fine. What is to me often much more generative is when you just stop and go, now what did we just say? I've been needling people for years about "human resources", but to the point of just saying, have you thought about what the word "resource" means, and the whole world that you bring forth when you use the word "resource"?

I'm quite confident that people will come up with language that will support them. I'm much more concerned about the thoughtlessness and the lack of awareness with which we use language. If I'm going to advocate for something, it will be very much for the space, and maybe the nudge that we all need to give each other - which we do give each other all the time, but we often miss it. When someone says something we don't understand, we miss it. That's the opportunity right there. Now would you please slow down, and say that again for me, because there's something you're saying that makes no sense to me. Which is to say there's somewhere in your world that I am not, but I might be. And of course that's what conversation is.

Mr. Prusak: There's a wonderful example of what you just said. Have any of you read Diane Vaughn's book on the Challenger disaster? Diane Vaughan's a sociologist at BC. It's called The Challenger Disaster. At the last minute, they have to make this decision to send the rocket up, and it's distributed. The people that have to make that decision are in three different places, and they're talking via speaker phone. The hierarchy of NASA is as strong as Stalin's Russia, so they couldn't really give pause. They couldn't look at each other, and give signals that I don't really want to send it up, but I can't say no. No one intervened when people said oh, that's okay, the O ring. There wasn't intervention. There was no pause. There was a momentum sense, and they couldn't get the body language, because they were separated. It's a wonderful three or four page description - and I see this all the time in firms - of where no one gives each other space. Well, what do you mean by that?

I was at a very large financial institution in New York City. One of the big places down there told me they had a nine hundred and forty million dollar, four year project underway to increase the productivity of financial planners who work there. When I asked, how would you define "productivity"? No one ever thought what that word meant. They never stopped to think. I am not making this up. You could buy that much technology, and that much consulting, but no one said what do you mean by "productivity"? Output per hour? Value of assets? There's a million ways this could be done. No one ever thought that.

Mr. Prusak: Capital implies an investment model. I find that dialogue with economists is often valuable. Capital implies you invest for a return. That's an expected rate of return. You put in effort, or money, and get an expected rate of return. When I talk about social capital I mean things you do, that cost time or money, that would enhance the social value of a firm. Lower transaction costs so people can behave towards each other without a big effort, and enable cooperation. Somewthing like a room where you can go and talk.

Let me give you a clear example - Daichi Pharmaceuticals, the Merck of Japan. I did a tour there one time, and the tour guide, a Japanese woman who was one of my students a long time ago, took me into a room about this size, and there were futons, and green tea, and people sitting around chatting. She said, this is our talk room. T-A-L-K. I'm from New England, and you have to spell that word out because it could be pronounced many ways. The talk room. I said Yoko, what do you mean? It's not a conference room. She goes no, no, it's a talk room - I said what the hell is a talk room? It's not a lounge. There weren't couches. And she said every day it is expected - social norms - it is expected that people doing research would go into this room, have some tea, and talk to whomever they meet about what they did that day. They'd say oh, I just read a new article on a certain molecule, or I'm working on this project. It's sort of a Brownian motion theory of knowledge generation. The random discussions would pay off. It's an investment. It's a big room. Tokyo real estate is very expensive. It's an investment where the return would be people would get to know each other. You'd talk to different people. There would be some knowledge created.

M: A number of people who have been here this week have said that they only realized they were doing complexity theory in retrospect. It was things they had always been doing that somehow now came under this umbrella. Tom mentioned that these ideas are not necessarily new. It would be interesting to hear if there are particular ideas or concepts, or models, or techniques that come from this portfolio of ideas we're putting under complexity theory that you've found particularly inspirational, or useful in your work, or you think will be useful going forward? Where do you see the most value coming from, and what areas would you like to see in complexity theory develope more, and move forward, and be incorporated into the organizations that you work with? What are the top ten, top five, top two complexity ideas that you see being valuable?

M: My number one is in terms of relationships, and looking at the interactions, and interpretations out of which we form our relationships.

Mr. McKelvey: I don't think there's any question that my inspiration has been Stu Kauffman. The reason is that I've always thought of myself as an evolutionist. I've spent a lot of time reading in biology, and I think a lot of economists and a lot of my colleagues basically see the organizational world as firms co-evolving toward a more fit state in a changing world. What was intriguing to me was the discovery of Kauffman's work where he said that there's a limit. There's places where order is not the result of evolutionary processes, but it's really the result of complexity, and so the natural selection process has been thwarted. In the world that I live in, what could be more fundamental than a book and an idea that says natural selection comes to a stop? The question is, when does that happen in organizations? If there's no evolution in organizations to begin with because you've got a dead firm run by a huge hierarchy, then there's no reason to worry about what Kauffman is worried about - what Campbell calls blind variation, selection, and retention. If that fundamental process never gets going very adequately in the first place, we don't have to worry about Kauffman's concerns. How do you get variation and selection going in a firm in ways that help it to respond to a changing world? And of course having done that, then I got into statistical physics.

The second thing then is the melting region. It's not the edge of chaos, it's a region at the edge of chaos. When you look at statistical physics, the models Weinberger, the model flips from stability to chaos almost instantly. The region is less than one decimal place. Imagine being in a firm where you either are running like a dinosaur, or over the edge into chaos. Nobody can live with that. Our problem is how do you make this melting zone large enough so that you can have emergent structure and all these good things that we want?

Mr. Prusak: I had dinner with Murray Gelmon once, and read two books, that's quite literally the extent of my knowledge. However, I was very interested too in the evolutionary theory of the firm, and evolutionary economics. I would share that interest, and realize that that is somewhat self-limiting. I would also put to you a human example of something that I lived through, and I think your theories would help to explain. This is the failure of information to change organizations, to change a system. I worked as a consultant to IBM for many years before I joined them. I was there in the early nineties. IBM had bought, had owned, had in systems every scintilla of codified information on this planet on the nature of IT markets. There was nothing you could produce that was for sale that they didn't buy or have in systems that were accessible; and yet they almost folded. With all that information available to the people who ran the firm, and lower down, they came very close to closing up shop, and why? What happened? It began to strike me so vividly of the failure of information acquisition, of bringing in signals, the failure to change. What was this failure? Some of course Chris Argyris and others have talked about for years. But the cognitive constraints, the social constraints, the political constraints, the dominant logic - all these various models - begin to come across in some of the more semi-popular things that Gulick and others write about. Complexity and chaos theory. That's my answer.

Mr. Mayer-Kress This is a follow-up question to what was asked before concerning the integration of chaos theory and your specific domain. There are some specific results in one of Stu Kauffman's books wherein he is talking about patches as some kind of a guidance for how big you want to have a team so that they're most effective within a company. I would like to get your feedback on this. The second aspect that is very universal in complex systems is that you have feedback. That was also indirectly mentioned here. That you need some kind of a feedback - some reward, some punishment system - within the company so that you can see that those in a team who contribute to the solution are reinforced. That in the next team that they are participating that their contribution is more predominant. That's something that you see very typically in co-evolving systems, and artificial intelligence systems. I would be interested to see if there is any kind of evidence of applications of these two examples in firms.

Mr. Senge: Can you say a bit more about what patches are?

Mr. Mayer-Kress: Stu Kauffman has done a number of simulations trying to figure out if it's better that you have large teams in a big organization with a little communication between the teams, or if its better to have a number of teams where you have optimized within the team, and then communicate, and then do the optimization for the organization based on the results of the large team. Essentially, how much do you want to subdivide tasks in order to solve a big problem?

Mr. Prusak: I spent a little time with Stu Kauffman. He was working at Ernst and Young in some commercial endeavors when I was there, and we had talks about this. Some of this sounds like operations research, which has been going on since Hector was a pup. What I see in organizations is large scale virtual teaming failing. We can talk about systems, but if you live in a system, it's different then when you have a system described to you. This is something I'm doing research on. I spend my time concerned with "virtual teams" of fifty to one hundred people. These virtual communities, virtual teams, make some sense in saving money on real estate. They give people a laptop, and say, go forth and do something valuable. This stuff stinks, to use very unscientific language. The people hate it, it doesn't work, and nothing good comes from it. Nothing innovative, productive, or socially valuable. And we're getting a ton of this. I'd be interested in how much time you have to write into a budget or meetings where you have to sniff each other out like animals. Face to face, get all the subtext like animals.

Mr. Senge: Like animals? Just like animals.

Mr. Prusak: Yes, right. Isn't it?

Mr. Senge: That's what we are.

Mr. Prusak: We are animals. You're right. Thank you. We forget. No, it's true. You're very much correct. I work in an organization that pushes virtuality, because they sell technology to so-called enable this. Yet I constantly get called down to the headquarters of IBM to come to a meeting and discuss things. And because people use the hierarchy, they can do it. They can get me to come down. I said, well, why can't I give you an article? Why can't we do it over the phone? Why can't we try some of these other methods? No, no. We want you to be here. Why is that? I kept pushing. Why? Why do I have to get on a plane, spend six hundred dollars, go to Westchester, and do all that? Because we want your presence here. You can't push them much further. They can't conceptualize further. It's the same reason dogs run in packs, I would say. That's a fairly unscientific answer to patches. But you see this constantly in firms and organizations. The idea of the pushback from people who need other people. I sound like Barbra Streisand. We are social - you can have an over-socialized or an under-socialized view of life, as Durkheim talked about. We are social enough. It's very hard to produce stuff in vacuums. All of you come here. You could read our books, we could all do this by email. No one does that. The planes are full. Anyone who flies a lot, like I do, knows that the planes are full. It's not all sales.

M: I'm going to close the session by returning to Marshall's question, which I'm reminded to do by Peter's insight that we are animals. Marshall said that the aspect of complexity that is most in form of what I do is the realization that as valuable as living systems may be as metaphors for organizations, it's not nearly as valuable as the knowledge that complexity teaches us, which is that human organizations are biological systems. We are biology. The organizations we create, just like the organizations that exist in the natural world. We aren't the natural world, but nevertheless, we operate according to - if not identical, then very similar - dynamics that scale from species to species and from organism to organization. There's a lot of science, including quantitative science, in the life sciences and the physical sciences. There's less in our organizations, but we can learn a lot from them.

With that, I would like to thank our panelists. It was a pleasure being with you all the last few days.

Mr. Bar-Yam: I am going to close the conference at this point, and thank you all for joining us, and participating in this activity. As I close, I wanted to make a couple of comments. One is that there was a lot of discussion in this session about what is good, and what is bad. Is it chaos? Is it structure? Is it order? What is good? It is essential to remember that the system is only created in response to the environmental demands that are placed upon it. The nature of what is good in the system is only in response to the nature of what is happening around it.

This is very relevant, not just to this session, but to this conference, and more generally to the understanding of why complex systems has become relevant to us. Something has happened. There is a reason why the science of complex systems is arising in these times. That has to do with the same reason that it is found to be so relevant in thinking about how human organizations are functioning, and that it is of such great interest, both in popular and scientific communities. That is because the environment in which we are operating, the human environment, is in some ways becoming complex. We have to deal with it on an individual level, as well as on an organization level, and we are now trying also to understand it from a scientific level.

Thank you.

Part II

Papers

Evolutionary Algorithms on a Self-Organized Dynamic Lattice

Paul Halpern
Department of Math, Physics and Computer Science
University of the Sciences in Philadelphia
600 S. 43rd St., Philadelphia, Pa. 19104
p.halper@usip.edu

We investigate a new optimization model, one which places an Evolutionary Algorithm with some, but not all, of the features of the Evolution Strategy approach of Rechenberg and Schwefel on a changing lattice of site connections – namely a structurally dynamic cellular automaton. After seeding 1000-site lattices with real values, one per site, we observe these site values' behavior over time, subject to evolutionary survival of the fittest. In addition to the standard Darwinian optimization processes of selection, breeding, mutation and replacement, we add the feature of a self-selective neighborhood structure. Each generation, the link structure of the lattice is organized by the sites themselves, based upon fitness criteria. In our model, we have found, for various parameters, an overall tendency to greater average fitness. We have discovered, as well, for particular mutation rates, evidence of phase transitions in lattice link structure.

1 Introduction

In recent decades a number of optimization techniques have been developed that exploit Darwinian notions of natural selection in an effort to arrive most efficiently at the fittest solution to a problem. Falling under the general heading of Evolutionary Algorithms, these include the Genetic Algorithm approach [6] and Evolution Strategies schemes [1, 9, 10]. These methods have achieved considerable success in providing speedy pathways toward approaching optimal or near-optimal solutions. Optimization occurs in each of these models through specific processes of selection, recombination, mutation and replacement (though in some models not all of these steps are undertaken).

Mühlenbein has pointed out, in the case of Genetic Algorithms, the dangers of premature convergence to sub-optimal solutions [8]. This difficulty arises in situations in which a population evolves toward a smaller peak in the fitness landscape, becomes more and more uniform and consequently fails to notice the largest peak. To avoid this situation, he suggests the introduction of a spatial population structure. Spatially isolated populations, he feels, would have the opportunity of developing toward different peaks. This would greatly improve the chances that a segment of the population would sample, and eventually hone in on, the fittest solution to the problem (the highest peak of the landscape). Mühlenbein's Parallel Genetic Algorithm model [8] provides a means of allowing for such spatial isolation.

We take a somewhat different approach to the issue of the spatial organization of the population. We place our Evolutionary Algorithm within the structure of a flexible lattice of self-selective neighborhood connections, namely a Topological Cellular Automaton[3, 7, 5]. A Topological Cellular Automaton is a variation of the standard Cellular Automaton model [11] in which link connections between lattice sites are added or removed each generation based upon the values of neighboring sites. The advantage we hope to gain is to allow each organism the choice of communities at any given time step. The choice of neighborhoods in our model is therefore dynamic, rather than fixed, determined exclusively by the participants.

One of the motivations for our inclusion of dynamically changing neighborhoods is to model the tendency for animals to choose mates from neighboring groups, but not from their own immediate families. A well-studied example of this is the propensity for song-birds to select partners with similar, but not identical songs [2]. Following this strategy, they hope to avoid mating with family members, while increasing their odds of finding fit neighbors with whom to procreate. As this principle works in biological systems to increase species fitness, we hope that it will similarly serve well for computational models.

2 Lattice-based Evolutionary Algorithms

We now present the dynamics of our model for an Evolutionary Algorithm (with some, but not all, of the features of an Evolution Strategy) on a dynamic lattice. Our algorithm can be described as follows:

STEP 0: A representation of the problem as a vector space and an associated fitness function mapping each vector onto a number are each defined.

STEP 1: An initial lattice configuration is created by randomly connecting a set of N sites with links, where N is an even integer. Isolated nodes are prevented by guaranteeing that each site has at least one link. Thus, disconnected graphs are avoided. The connectivity between sites is encoded in the connection matrix C_{ij}, where i and j each range from 1 to N. Each matrix element is equal to 1 if and only if the two sites i and j are directly connected; otherwise it is equal to 0. In this manner, all neighborhoods are symmetrically established.

STEP 2: Each site i is seeded with a single vector of length L, encoded by x_{im}, where m ranges from 1 to L.

STEP 3: Based on the fitness function, the fitness of each site is evaluated. (Although strictly speaking we have determined the fitness of the vector located on a given site, for simplicity, we refer to this value as the "site's fitness".)

STEP 4: In this step, based upon fitness criterea, old connections are either maintained or discontinued, and new connections are established.

Consider a given site i. Site i's nearest neighbors (link distance 1) are ranked according to fitness. Based upon the decoupling parameter set initially, the site disconnects with the least fit fraction of its nearest neighbors. The value of the decoupling parameter determines how large this fraction is.

Simultaneously, site i's next-nearest neighbors (link distance 2) are ranked according to fitness. Based upon the coupling parameter set initially, the site forms new direct connections with the fittest fraction of its next-nearest neighbors. (That is, these especially fit next-nearest neighbors are singled out to become nearest neighbors.) The value of the coupling parameter determines how large this fraction is.

In other words, each site forms a new neighborhood based upon fitness criteria. To avoid ambiguity, the updates are not performed until the entire lattice has been evaluated.

STEP 5: Each site randomly mates with one, and only one, of its nearest neighbors (as determined by the updated connection scheme). This mating takes place according to the procedure of intermediate recombination [**?**]; namely, each offspring consists of the vector average of its two parents. Therefore, the number of offspring is $N/2$.

STEP 6: With small probability (chosen as one of the initial conditions) each offspring has the possibility of becoming mutated. This mutation consists of a small random increment added to or subtracted from the vector. (Note, in contrast, that for an Evolution Strategy, all offspring are mutated, with their mutations following a normal distribution.) Periodic boundary conditions are applied here. If the mutation is such that it would take the vector out of its designated range, then a suitable correction is either added or subtracted until the vector lies once again within the appropriate span of values.

STEP 7: The set of $N/2$ offspring replaces the $N/2$ least fit members of the general population. These offspring vectors are randomly placed at the sites of the least fit members they have replaced.

STEP 8: Repeat steps 3 through 7, for a fixed number of time steps, or until a suitable convergence criterion is met.

The following parameters are set initially before each run begins:

N: Size of lattice
L: Length of each vector
M: Initial number of links per site
C: Connection fraction – Fraction of fittest next-nearest neighbors with which each site must connect each time step.
D: Disconnect fraction – Fraction of least fit neighbors with which each site must disconnect each time step.
P: Probability of mutation – Chance that a given offspring is subject to mutation
T: Total number of time steps

For our trial runs on DEC workstations we used lattices of N= 1000 sites each. For the purpose of simplicity, we chose vectors of dimension $n = 1$, with values ranging from 0 to 1.

To facilitate understanding of the model, we utilized a simple fitness function with a single peak:

$$f(x) = 1 - x^2 \tag{1.1}$$

We set the amount of mutation to range from -.05 to +.05, for any given time step. Periodic boundary conditions were established such that if a mutation threatened to bring a vector's value outside the range of 0 to 1 then a unit value was either added or subtracted as appropriate.

The mutation rate, fixed for an entire run, was generally set to be P= .001. The reason the mutation rate was chosen to be so low (one mutation per step, on average) was to understand better the effects of the connectivity dynamics. As much as possible, we wanted to separately examine the distinct consequences of changing neighborhood connections. The mutation rate was varied in some runs, however, for the sake of comparison.

The decoupling and coupling criteria were generally set at D = C = 0.1. (10% of the least fit nearest neighbors were disconnected, and 10% of the fittest next-nearest neighbors were disconnected per site at any given time step.) Once again, though, these rates were altered for particular runs for the sake of comparison.

We ran this algorithm for T= 1000 time steps per run, with 10 different randomly chosen initial configurations.

3 Site Value Behavior and Connectivity

As in the case of (non-lattice based) Evolution Strategies and Parallel Genetic Algorithms, we found that our model served to optimize the chosen fitness function. This is not surprising, however, since our fitness function is fairly simple.

For our default initial parameters, listed above, we found that our algorithm brought the average fitness to its maximum value within 200 time steps. Varying the parameters from their initial values seemed in most cases to have a deleterious effect. For example, as the disconnect fraction D was increased, the time until the model reached its fitness peak grew longer (see fig. 1).

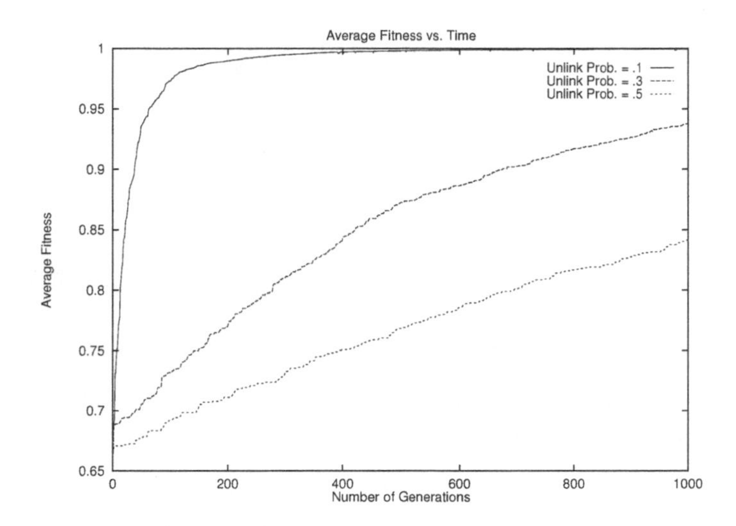

Figure 1: Here we set the link probability equal to .1 for a 1000 site lattice, and vary the unlink probability between .1 and .5. Note that lower probabilities of decoupling seem to result in a faster progression toward greater fitness. The mutation probability is set to be .01

We speculate that the reason for this change in performance stems from the fact that, for higher values of D, neighborhood connections are broken down too quickly. For optimum performance, one needs to build up links first before breaking them down.

When D=C, since each site starts off with many more next-nearest neighbors (link distance 2) than it has nearest neighbors (link distance 1) it initially creates more connections per time step than it severs. Eventually, however, the connectivity of the lattice saturates at its maximum value, and number (not the rate) of disconnections catches up to the number of new connections per time step.

Of more interest than the fitness growth itself is the connection between model performance and lattice connectivity. We found that the fastest hillclimbing took place when lattice connectivity increased, then decreased, and finally levelled off. The first stage, of increasing connections, allows for the establishment of the fittest possible communities. But then, once these fit neighborhoods are established, connections can be severed with the poorer performers.

We found two different sorts of approaches toward high connectivity. The first sort, for low mutation probabilities, is monotonic growth over time (see fig.

2).

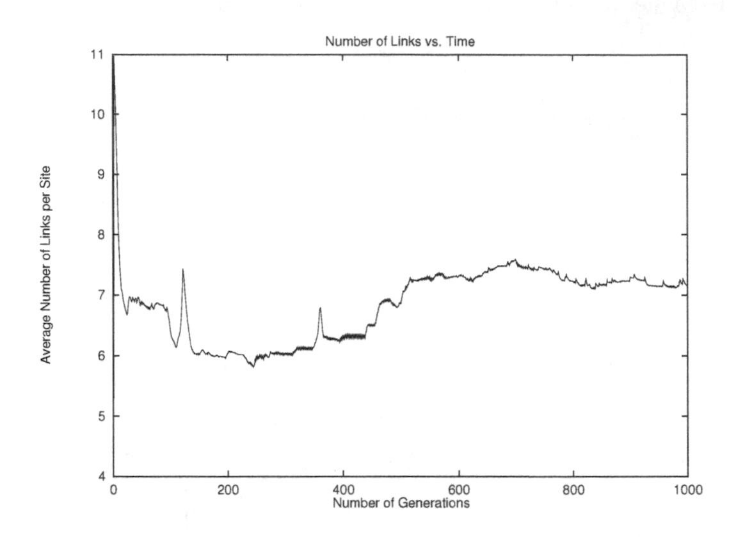

Figure 2: Here we plot the structural evolution of a lattice, typified by the value of the average number of links per site over time. If one sets the mutation probability to .01, the link coupling rate to .1 and the decoupling rate to .1, one observes a reasonably smooth approach to a static link structure.

The other type of behavior, of more interest to those studying complex systems, is that of a catastrophic "phase transition" between lower and higher degrees of connectivity. We have found this sort of behavior to occur when P is set to be approximately 0.05 (and within a small window of this value). There is a sharp increase in number of links per site between generations 350 and 450. (See fig. 3).

Examining the overall link distribution over time, we obtain a sense of why this "phase transition" occurs. During the early part of the model's evolution, most sites have few links. Therefore, there is still considerable room for growth to occur.

As lattice growth continues to take place, each site has a greater selection of potential partners. Thus, whenever an especially fit site arises, many other sites notice it as a next-nearest neighbor and want to be directly connected with it. Over time, connections to this especially fit site grow dramatically.

Why would a site suddenly become especially fit? This occurs because of mutation. Occasionally, the periodic boundary conditions make it possible for mutations to take place that radically alter the fitness of particular offspring. These offspring suddenly jump in fitness. Because of this increase, suddenly many sites want to be connected to these fitter vectors. The level of connectivity dramatically increases, as many sites assume new connections. In short, a

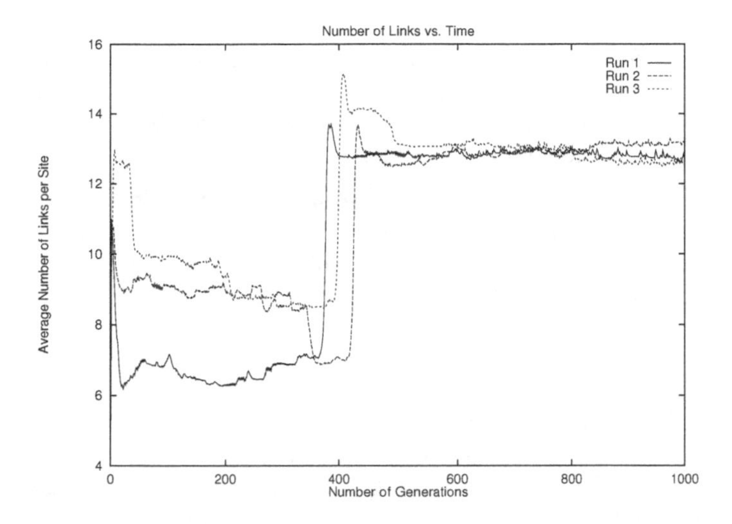

Figure 3: In contrast to Figure 2, we now increase the mutation probability to 0.05. Curiously, we observe a qualitatively different behavior in the link structure dynamics. We notice a sharp increase in connectivity as the model evolves from generations 350 to 450. This is characterized by what appears to be a 'phase transition' in the average number of links per site. Fixing these parameters, we detect this behavior in every run. Three sample runs are shown here.

bifurcation occurs in which some of the sites maintain their original level of connectivity, while others increase theirs drastically. Thus, the periodic boundary conditions seem to lead to a "phase transition" in the average connectivity.

4 Conclusion

In an earlier version of this model [4], a lattice-based variation of Genetic Algorithms, we noted that lattice connectivity can lead to an improved rate of optimization. Here we note that placing a special sort of Evolutionary Algorithm (with some, but not all, features in common to an Evolution Strategy) on a lattice can serve to organize a physical space in addition to a vector space, potentially leading to an overall increase in fitness. Furthermore, it can result in an interesting dynamics, including the possibility of phase transitions.

Possible applications of this model include cases in which flexible connectivity would benefit optimization. One natural arena in which such a procedure might be found useful is in the domain of parallel computing. Several units, working on a particular problem, might use fitness criteria to determine whether or not it would be best to make or sever a connection. Then optimization might proceed through the best networking of these units.

Because this prelimary study was designed mainly to examine the basic features of the model, the optimization problem chosen was very straightforward, with a simple fitness function. In future studies of this model, we plan to look at a more complex range of fitness functions. We will continue to examine how placing an evolutionary algorithm on a lattice affects the overall performance of the algorithm and the connectivity of the lattice.

Acknowledgments: We wish to thank Werner Ebeling, Helge Rosé and Torsten Asselmeyer for their help, suggestions, and for many useful discussions.

This project was supported by the German Fulbright Commission and Institute for Theoretical Physics, Humboldt University, Berlin.

Bibliography

[1] BÄCK, T., F. HOFFMEISTER, and H. -P. SCHWEFEL, "A survey of evolution strategies." In R. K. Belew and L. B. Booker, editors, *Proceedings of the Fourth International Conference on Genetic Algorithms and their Applications*, Morgan Kaufmann (1991), 2–9.

[2] BAKER, M. and M. CUNNINGHAM, "The Biology of Bird-Song Dialects," *Behavioral and Brain Sciences 8* (1985), 85–133.

[3] HALPERN, P., "Sticks and stones: A guide to structurally dynamic cellular automata," *American Journal of Physics 57* (1989), 405–408.

[4] HALPERN, P., "Community-Based Genetic Algorithms," *DYNAMIK - EVOLUTION - STRUKTUREN*, Verlag Dr. Koester (1996),11–18.

[5] HALPERN, P. and G. CALTAGIRONE, "Behavior of Topological Cellular Automata," *Complex Systems 4* (1990), 623–651.

[6] HOLLAND, J. H., *Adaptation in Natural and Artificial Systems*, University of Michigan Press (1975).

[7] ILACHINSKI, A. and P. HALPERN, "Structurally Dynamic Cellular Automata," *Complex Systems 1* (1987), 503–528.

[8] MÜHLENBEIN, H., "Evolution in time and space – the parallel genetic algorithm." In G. Rawlins, editor, *Foundations of Genetic Algorithms*, Morgan Kaufmann (1991), 316–337.

[9] RECHENBERG, I., *Evolutionsstrategie: Optimierung technischer Systeme nach Prinzipien der biologischen Evolution*, Frommann-Holzboog (1973).

[10] SCHWEFEL, H. -P., *Evolutionstrategie und numerische Optimierung*, Dissertation, Technical University of Berlin (1975).

[11] WOLFRAM, S., *Theory and Practice of Cellular Automata*, World Scientific (1986).

The Heart as a Complex Adaptive System

J. Yasha Kresh, Igor Izrailtyan and Andrew S. Wechsler
Depts. of Cardiothoracic Surgery and Medicine
MCP-Hahnemann University School of Medicine
Philadelphia, PA

There are many systems in nature in which a large assembly of autonomous parts (agents) interacting locally, in the absence of a high-level global controller, can give rise to highly coordinated and optimized behavior. The complex adaptive behavior of global-level structures that emerges is a consequence of nonlinear spatio-temporal interactions of local-level processes or subsystems. This form of nested co-optation (across levels of organization) constitutes isolated cells, organisms, societies and ecologies. Systems of this type are governed by universal principles of adaptation and self-organization, in which control and order is emergent rather than predetermined and have come to be known as complex adaptive systems (CAS). Consistent with the basic rules governing CAS, the integrative action of living organisms can not be derived simply from their concatenated fractions but evolves "relationally", i.e., it emanates from emergent internal requirements of the constitutive "parts within parts".

There is mounting evidence that the heart is a system onto itself and that it is intimately intertwined with the nervous and endocrine system residing within its borders. How these regulatory subsystems communicate and how uncoupling of the hierarchical organization results in loss of adaptive "fitness" remains a challenge in human biology. The traditionally and long held view has been that functionally the heart is a muscular organ pump responsive and adaptive to short term physiologic stimuli (physical demand, emotional/psychological disturbances). It is also known, that the heart is involved in adaptive accommodation to altered environmental conditions, such as prolonged exposure to high altitude and weightlessness (microgravity) in outer space. This adaptive propensity is made evident in the ability of the heart to increase the delivery of oxygenated blood by augmenting the beating frequency (heart rate-HR) and velocity of muscle fiber shortening. Most change in complex systems is emergent

In the past, the study of cardiac function (mechanical, electrical and neuro-hormonal) has been disjointed with minimal overlap and unification. It is becoming increasingly apparent that the heart is not merely a muscular pump, the function of which is regulated by external stimuli. The heart is a complex organ-system with a built-in capacity for self-regulation and adaptation. In particular, regulation of contractile proteins in response to mechanical overload, alteration in autocrine / paracrine regulation in metabolically stressed hearts, "learned" tolerance (preconditioning) to deprivation of muscle blood perfusion are examples of extent of the cardiac "plasticity".

1. Cardiac Autoregulation

The concept of "self-regulation" is based on the axiom that the heart is a regulatory system, integrating many internal and external functional units, including endothelium-mediated control and afferent/efferent neural mechanisms and thereby generating feedback of its beat-to-beat performance as a muscular pump (Fig. 1). Signaling involved in cellular response to stress contribute to cardiac adaptation at the organ level. Dynamic "crosstalk" between the components of the intrinsic cardiac neuroendocrine system underlies the adaptive organ-response to the imposed the environmental stress. Complex network of signaling pathways enables cells to respond to specific patterns of extracellular perturbations. In principle, the highly interconnected network of intracellular signaling reactions can be "trained" to recognize certain input patterns of stimuli and to execute a specific output pattern of responses.

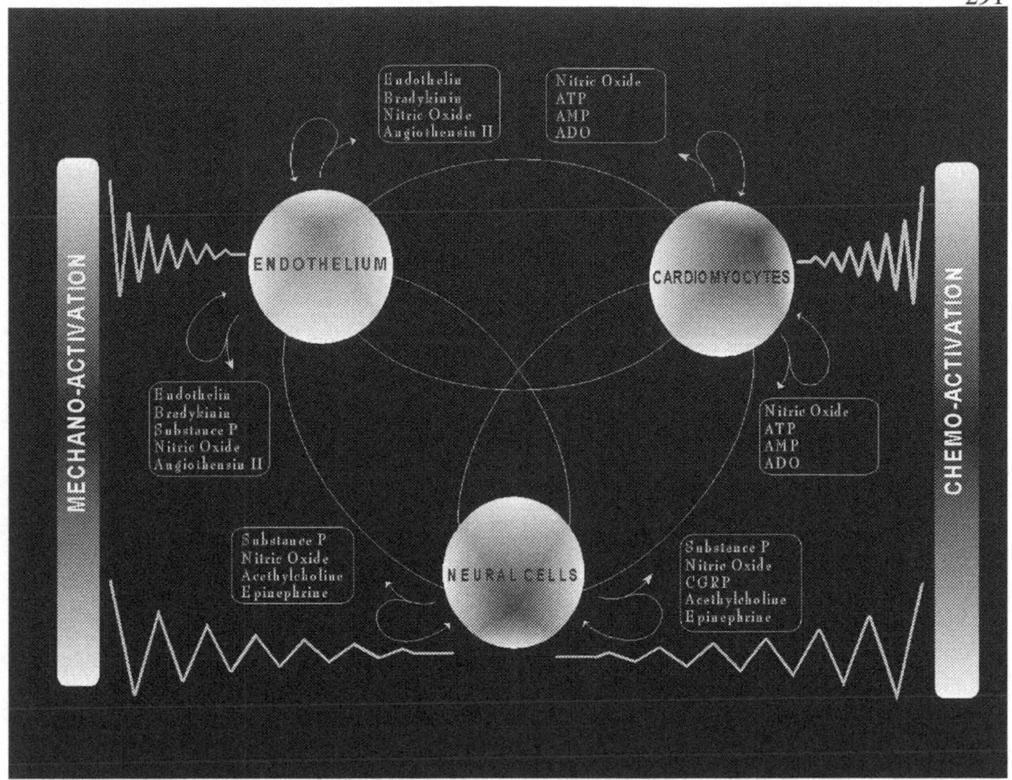

Figure 1. The heart as a complex adaptive organ-system: homeodynamic regulatory interactions.

Cardiac self-regulation is vital for coping with myocardial ischemia, a dynamic process associated with both destructive and protective cellular response mechanism [1]. Myocardial ischemia and reperfusion is commonly encountered in cardiovascular medicine and surgery. Acute ischemia results in a spectrum of deleterious effects, ranging from transient depression of contractile function (myocardial "stunning" and "hibernation") to severe abnormalities, such as reperfusion-induced lethal arrhythmias and accelerated myocardial cell death (both necrosis and apoptosis). The mechanisms involved in myocardial dysfunction during ischemia and subsequent reperfusion are complex and depend on a number of cellular signal transduction pathways [1]. The capacity of the heart to adapt to ischemic stress is manifested by a phenomenon known as "ischemic preconditioning" [1, 2]. During a preconditioning mode, a brief ischemic stress rapidly induces tolerance to a later prolonged ischemic challenge.

Attenuation of free radical production, activation of adenosine receptors and potassium (

K_{ATP}) channels, and induction of heat shock proteins are known responses to stimuli of preconditioning [2]. Extracellular stimuli elicit signal transduction pathways (Fig. 2) utilizing a number of integrating intracellular enzymes (e.g., phosphatases and kinases). The pattern of the resulting signaling determines whether myocardial injury or adaptation to ischemia is the eventual outcome.

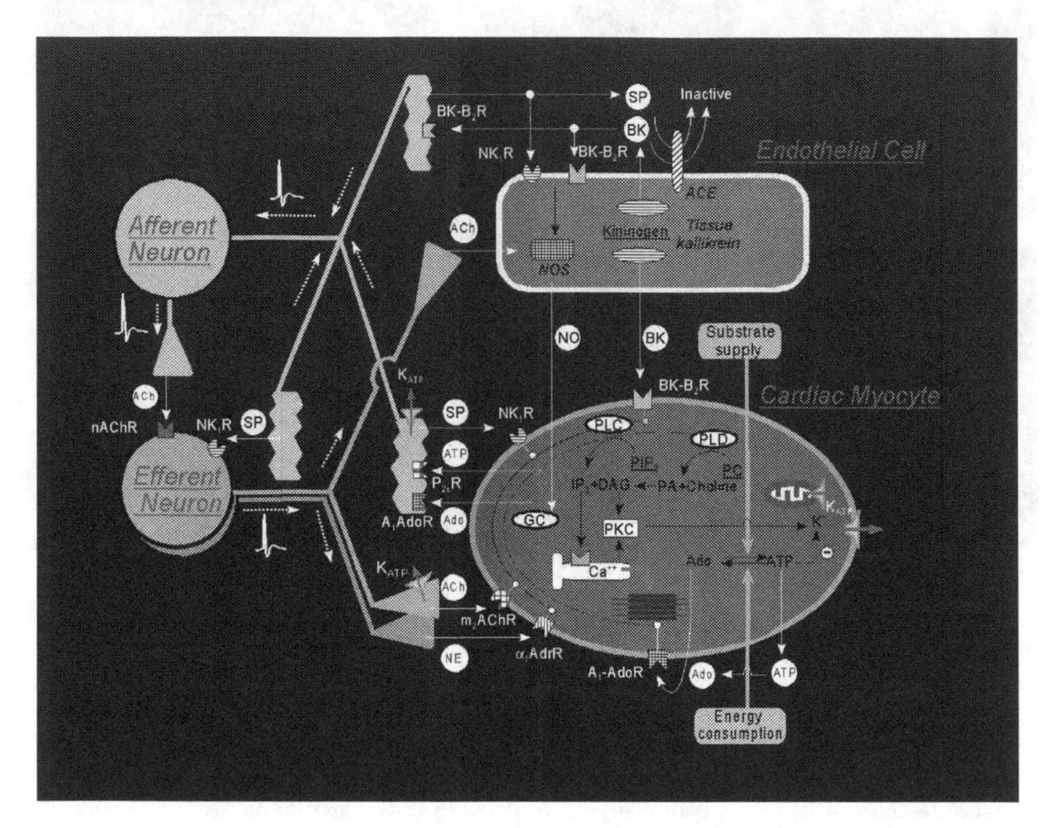

Figure 2. Neuro-endothelio-cardiocyte regulatory unit signaling cardiac adaptation to ischemic stress. Many extracellular mediators of ischemic preconditioning, including adenosine, bradykinin, nitric oxide, endothelin, norepinephrine, cytokines and several other growth factors and hormones have been identified [3] as components of signaling pathways underlying the adaptive mechanism induced by preconditioning. The role of intracellular protein kinase C (PKC) dependent cascades in cellular signaling has been implicated [4]. PKC is activated via stimulation of phospholipase C (PLC), which subsequently degrades membrane phospholipids to diacylglycerol (DAG), an important PKC cofactor. Certain receptor agonists (e.g., adenosine) may also be linked to

PKC signaling via phospholipase D (PLD) mediated production of DAG. In addition to PKC, the importance of concomitant action of other intracellular enzymes, e.g., ecto-5'-nucleotidase [5], tyrosine kinase [6], stress activated protein kinase, e.g., p38/SAPK [7], or extracellular regulators, e.g., cellular stretch [8], can not be ignored. The phosphorylation of heat shock protein 27 (Hsp27), downstream to p38/SAPK, appears to enhance the polymerization of actin and is thought to help repair the actin microfilament network which becomes disrupted during cellular stress and thereby may aid in cell survival. Activation of heat shock protein production may be involved in mediating a "second window" of myocardial protection following (24 hrs) a brief period of ischemic preconditioning. While these signal transduction pathways may function synergistically with respect to PKC, their action may not be simply linear or additive.

It is important to recognize that cellular adaptation cannot be viewed simply in terms of biochemical events and/or processes. In particular, cellular architecture, i.e., cytoskeletal structure and mechanics, self-organize to stabilize the functional integrity of the myocardial tissue. Importantly, many of the stress-activated enzymes and other substances that control protein synthesis (polymerization) are co-localized with the cytoskeletal structures. Consequently, the interaction between the biochemical reactions and cytoskeletal geometry/mechanics can effectively dictate the adaptive strategy of the cell, including the activation of genes that direct the synthesis of the structural and contractile proteins. Thus, the loss of cellular "tensional integrity"[9] may trigger a cascade of mechanical restructuring, responsible in part for activating a cell death program, i.e., apoptosis, and ultimately dictating the fate of the heart as a complex organ-system.

The assumption that the intrinsic cardiac nervous system has the capacity for independent regulatory action, intimately involved in the adaptive processes of the heart was tested in isolated in-vitro perfused hearts. It was demonstrated that a brief cessation of myocardial perfusion (preconditioning, i.e., "learning" event) prevented as much as 50% of cardiac systolic function deterioration in subsequent exposure to a prolonged ischemia/reperfusion insult [10]. In addition, a nearly 10-fold preservation in heart chamber filling properties was observed in the ischemically "adapted" hearts. Moreover, hearts that exhibited a significant deterioration in contractile function had a more pronounced derangement in the electrical activity and more frequent incidence of premature beats and tachyarrhythmias. This type of susceptibility to dysrhythmias was more often seen in hearts pretreated with neuro-transmission blockers (e.g., hexamethonium-HEX). Pharmacologic probes were used to "dissect" specific signaling circuits of the intrinsic neural cholinergic pathways, interacting with cardiac tissue kallikrein-kinin system, constitute a self-regulatory mechanism, involved in mediating the ischemic myocardial adaptive process.

The extensive network by which intrinsic cardiac neurons are "functionally wired" to all cardiac tissue elements (e.g., myocytes, endothelial cells) enables a selective and directed modulation of the constituent parts, giving rise to a set number of regulatory states. The heart is goal-seeking and purposeful organ that can adapt/select a course of action out of many possible strategies so as to optimize its functional integrity in response to imposed "environmental" stresses. With respect to cardiac function this implies homeodynamic selection process. Indeed, there may be a parallel between the brain acting as a self-organizing system and the intrinsic cardiac nervous system of the heart [11, 12].

To reveal the functional correlates of intrinsic neural regulatory activity and the state of "ischemic adaptation", produced by a brief cessation in coronary flow, we analyzed the dynamic behavior of sinus node automaticity (RR interval time-series). Illustrated are data (Fig. 3) from two studies, using an isolated perfused heart preparation, in which myocardial perfusion was stopped for a 5-min period in presence (left panel) or absence of HEX (right panel). The onsets of ischemia (intervals A-B; spanning ~ 3-min) and the reperfusion (intervals C-D; spanning ~ 5-min) are identified accordingly. In the upper panels, the RR interval (heart rate period) time-series are plotted. The lower panels depict the corresponding changes in computed point correlation dimension (PD2 [13]) for the respective time-series. The ischemia/reperfusion stimulus caused multi-modal changes in the dynamics of heart rhythm generating system. In the experiment without HEX, relatively small but discernable changes in heart rate variability (HRV) were observed at the early stages of ischemia (interval A). The late phase of ischemia (interval B) is characterized by an abrupt increase in the "dimensional state" which was sustained until the perfusion was reestablished (interval C). The reperfusion phase (interval D) was characterized by a diminished dynamics of the system. However, this newly established dynamics of HRV did not return to the preischemic state. HEX antagonized both the early ischemic changes in HRV (A) and the sustained increase in heart-rate dynamics during the reperfusion phase (D). The fact that the dimensional changes preceded significant alterations in HR (interval A) suggests that the heart rhythm generating system may have "reorganized" to a new operating mode (i.e., phase transition). The sustained higher level of HRV (interval D) implies that the heart can be "adaptively learned" such that it retains a "memory" of the preceding ischemic event. The observed phase transition and apparent "dynamic preconditioning" presumably is a result of the intrinsic cardiac neuroendocrine function. At this time it is unknown what specific mechanism(s) is/are responsible for this dynamic adaptive changes. The underlying cellular plasticity (e.g., prolonged potentiation of synaptic junction activity, second messenger neuromodulation, etc.) could be involved in this phenomenon.

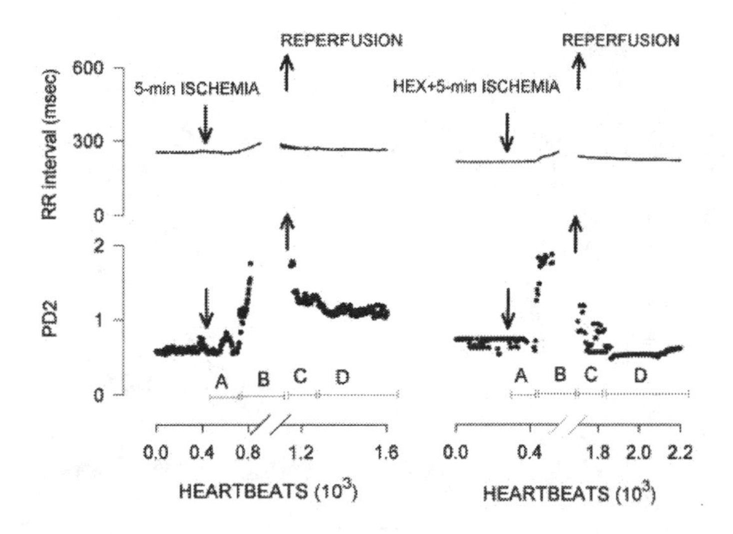

Figure 3. Heart rate variability (R-R) and dimensional analysis (PD2) in isolated rabbit heart. Shown are the effects of ischemia (A-C intervals) and reperfusion (D-interval).

The experimental evidence strongly suggest that the intrinsic cardiac nervous system is not merely a passive assembly of "relay stations", involved in the central command and control, but is robust in its ability to express independent regulatory action in response to imposed changes in cardiac state. Bradykinin (BK), a known mediator of ischemic tissue response was used to probe the dynamic attributes of intrinsic cardiac regulation. As seen in Figure 4, the metronome-like D2~1) of the isolated heart exhibited a significant phase shift (PD2 ~ 3) in response to a BK-stimulus. This effect of BK on heart rate variability was markedly attenuated by HEX (PD2 ~ 1.5) and abolished completely by tetrodotoxin, an inhibitor of neuronal electrical activity.

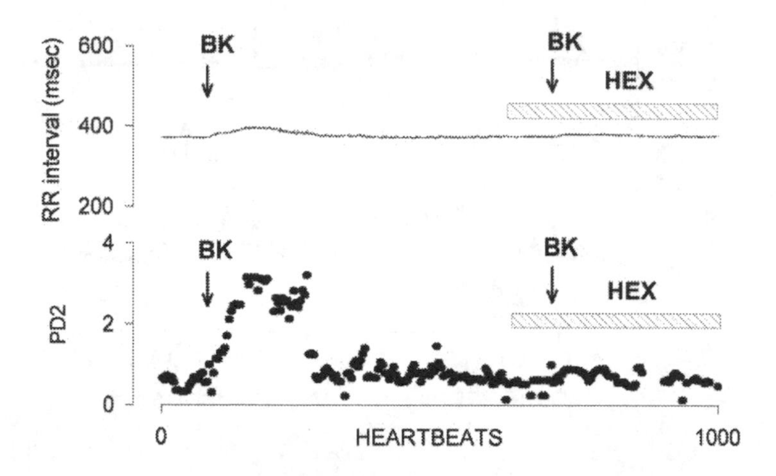

Figure 4. Dynamic changes of R-R intervals following bradykinin (BK)-injections: before and after hexamethonium (HEX). Note that the BK-mediated dimensional changes (PD2- point correlation dimension) are markedly attenuated by HEX.

2. Emergent Properties of the Heart

The capacity of self-organized systems to adapt is embodied in the functional organization of intrinsic control mechanisms. Evolution in heart rate variability dynamics was used as measure of the capacity of the transplanted human heart to express newly emergent regulatory order. In a cross-sectional study consisting of 100-patients post (0-10 yrs) heart transplantation (HTX), heart dynamics was assessed using pointwise correlation dimension (PD2) analysis [14].

In general, the number of variables required to characterize the time-dependent events of a dynamic system determines its dimensionality. PD2 of 4-5 in normal subjects is expected to reflect the minimum number of independent control variables, i.e., degrees of freedom that define the dynamics of the cardiac pacemaker. How precisely the interaction of the components comprising the autonomic nervous system translate into a system dimension that is greater than the mere number of known pacemaker control loops (attributable to the parasympathetic and sympathetic system) remains to be elucidated. Nevertheless, the number of control variables involved in the "regulation" of the HR-dynamics is considerable less than the number of neuromodulators known to alter sinus node automaticity. This is consistent with the view that a dissipative system having a large number of state variables do not transverse the whole of its state-space but operates within a bounded subset, defined by only few variables.

Commencing with the acute event of transplantation, the dynamics of cardiac rhythm formation exhibited a
number of phase transitions. Shortly after implantation, the donor heart manifested a metronome-like behavior (PD2~1.0). The dimensional trajectory of HRV reached a peak value of PD2~2.0 at 11 to 100 days post-HTX. The subsequent dimensional collapse to PD2~1.0 at 20-30 month post-HTX was followed by a progressive near linear gain in the functional order of the rhythm generating system, reaching PD2~3.0 at 7-10 years post-HTX.

These time dependent transitions are illustrated in Figure 5, where HRV-dynamics are confined within a two-dimensional phase-portrait. The observed return maps are assumed to be the projections of a multi-dimensional phase-space trajectory of an "attractor", underlying the HR generator dynamics. Figure 5 depicts the dynamic patterns of HRV for a normal volunteer and three patients studied at different intervals post-HTX. Clearly, the shapes the shapes of the Poincare' plots are distinctly different. The attractor in normally innervated heart displays a wandering pattern. 1-week post-HTX, the trajectory of the attractor is grossly simplified, taking-on an appearance of convergence to a point in space. It would appear that the dynamics-pattern of HRV at 1-month post-HTX attains a geometrically ordered temporal "cone-like" structure, signifying an emergence of a new dynamic state of the heart rhythm generator. At 7.5 yrs post-HTX, the trajectory in the system dimension nears the dynamics of the normal innervated heart. Clearly,"rebuilding" biological hierarchical organization takes time.

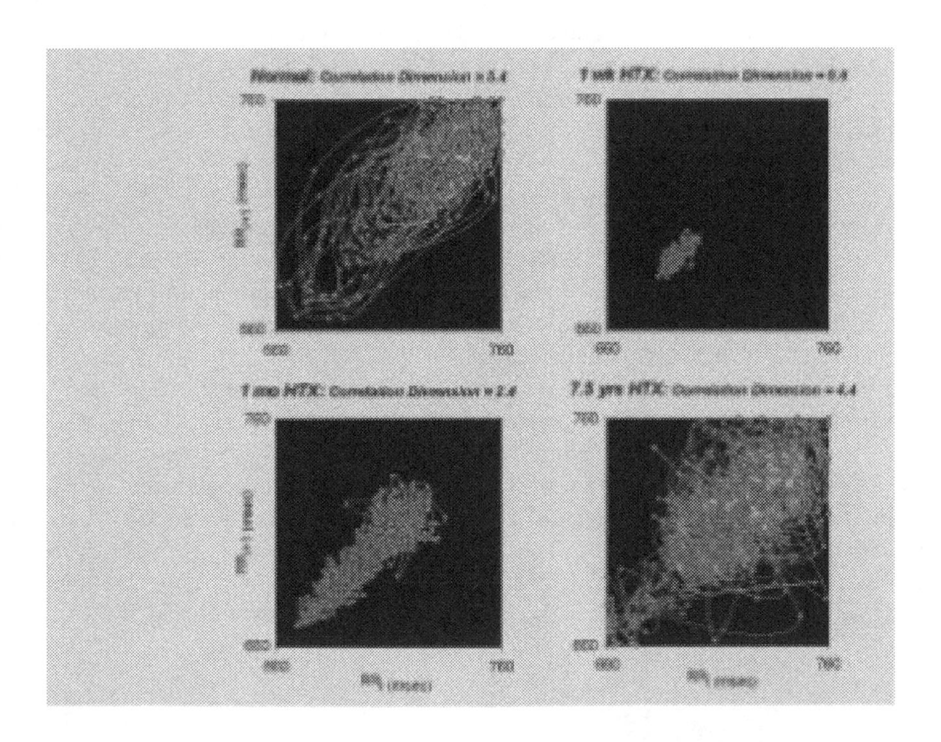

Figure 5. Phase-plane portraits (Poincare' plots) of cardiac allograft RR-interval time-series, in route to/from complex dynamics with its respective attractor dimension (point correlation dimension-PD2).

The "dynamic reorganization" of the rhythm generating system of the transplanted heart, seen in the first 100 days can be attributed to adaptive capacity of intrinsic control mechanisms carried together with the donor organ. It is important to emphasize that the HRV dynamics in early stage of post-HTX (within first 100 days) are in fact generated by a system devoid of central autonomic control, thus implicating a causative role to the intrinsic cardiac regulation. The fact that a finite dimensionality associated with the heart rhythm generator was observed implies that the allograft is not merely a passive participant in the assimilation process within the host "landscape". Importantly, the decentralized heart is capable of expressing new patterns of self-regulation. The fact that the heart can restitute the multi-dimensional dynamics, relatively independent of external command and control, is an impetus for reassessing the prevailing paradigm of cardiac regulation and adaptation.

The phase transition in the evolution of heart rate dynamics seen at 2 years post-HTX is an indication of a withdrawal of a dominant intrinsic control. This time frame is coincident with early signs of partial sympathetic reinnervation [15], thus giving rise to a dominant "sympathetic" mode (attractor). The newly derived evidence based upon the power spectral analysis supports these observations [14]. In particular, the period 20-30 months post-HTX is characterized by loss of spectral reserve or reduced complexity of the HRV-dynamics, associated with the observed rise in low to high frequency ratio, which is often used as an index of sympatho-vagal balance. This provides additional evidence in support of the hypothesis that a dominant sympathetic control resurfaces during this period. In non-transplant patients, predominance of sympathetic (e.g., post-infarct patients) tone is accompanied by a loss of HR generator dimensionality [16]. While the specific mechanism for these phenomena is not identified, a reasoned speculation can be offered. Sympathetic overdrive is a prominent component of a generalized stress response. To achieve this "unison" of action and body response would require each of the participating elements to be "phased" or "locked", to perform a limited and directed task. This coordinated effort would necessitate a collapse in the overall degrees of freedom. A remaining plausible explanation is related to the process by which the reestablishments of extracardiac sympathetic control downregulates the preexisting intrinsic regulatory activity. The
interactions between the intrinsic and extrinsic autonomic regulatory loops have been identified as functional parts of the intrinsic cardiac nervous system [11, 12].

The subsequent long-term progressive rise in dimensional complex of HRV (>2 yr post-HTX) can be attributed to the restoration of a functional order patterning, patterning parasympathetic control. Interestingly, mutual interaction of several feedback systems with delays can produce and show multistability (periodic, quasiperiodic and chaotic dynamics). Predominance of a sympathetic feedback control is usually associated with one regulatory mode (low dimension) [16], while the parasympathetic accentuation is shown to be poly-modal, characterized by rise in HRV dimension [17]. One plausible mechanistic explanation of how parasympathetic drive contributes to apparently greater HRV-dimensionality can be ascribed to the inherent delay in the feedback loop associated with the dispersion/integration of the signal within the intrinsic neurons of the heart, a property of cholinergic nervous system.

The biological implications associated with observed HRV-changes post-HTX go beyond the simple characterization of sinus node pacemaker function. In a centrally denervated heart, the modulatory response of cardiac rhythm dynamics is indicative of regulatory function of intrinsic cardiac neuroendocrine mechanisms [18, 19]. The highly distributed nature of the intrinsic cardiac nervous system enables it to integrate changes in the chemical milieu [10, 13, 18] and mechanical state [11] of the myocardium. The organization of complex intrinsic feedback systems [11, 12, 18] constitutes a functional "heart-brain" that may participate actively in the cardiac response to environmental changes. The dynamic interactions among functional components intrinsic to the heart, e.g., neurons and endothelium, give rise to a complex self-regulating organ-system. The transplanted heart, in its encounter with the host, manifests many of the attributes characterizing complex adaptive systems: network of interacting components, organizational order, multi-functionality, and fluctuation in system parameters. An early surge in dimensional order (PD2) of heart-rate generator proved to be a characteristic feature of donor heart-recipient dynamic interaction. Structural and functional uncoupling resulting from an imposed allostatic load, e.g., organ rejection [20], can trigger a loss/decline in network (hypercycles) of regulatory organization, manifested as a reduction in the dimension of HRV state-space attractor.

3. Summary

The denervated heart offers a unique experimental model to study the reassembly of the mechanisms responsible for a complex HR dynamics. The transplanted heart cannot benefit from fixed structural arrangement of feedback mechanisms, i.e., homeostatic goal directed behavior. Viewing the heart as an open system may prove to be useful in the understanding of organizational forces involved in restituting a system that exhibits a higher order of behavior. The progressive gain in HRV-dimension indicates some degree of restitution of "functional capacity" to cope with perturbations as they arise. Otherwise, the system would get "stuck in a rut" once it settles in stable environment, where the dynamics of operating point that is not restricted in time to a fixed state space attractor. The observed gain in dimensional complexity may be a product of newly evolved interactions and/or reinforcement of existing local control mechanisms that may facilitate new modes of regulatory function, i.e., emergent order. Specifically, upregulation of existing (dormant, less active) modes of sinus node modulation, such as mechanical stretch or activation of intrinsic neuroendocrine system are possible candidates involved in the emergence of new regulatory patterns.

The transition in the heart-rate generator "dimension" (gain/loss) may help in quantifying at least in some reproducible way the functional reserves and adaptive capacity of the intrinsic control mechanisms. This conceptual framework goes beyond the view that fluctuations in HRV are simply random events. Adaptation of the heart to its host is an impetus for greater level of functional integrity, i.e., higher system dimension, emergence of pattern and order.

i. From a system theory standpoint, a system that is endowed with greater number of degrees of freedom is
more robust and has a greater ability to accommodate imposed disturbances. In general, biological systems, independent of hierarchical organization (molecular to multi-cellular), normally operate such that a finite number of regulatory modes can be invoked. Chaotic systems are very susceptible to changes in initial conditions, i.e., small changes in a parameter of chaotic system can produce very large changes in the output, i.e., poised at the "edge of chaos" [21, 22]. This allows the system to switch from one state to another rather quickly. It may be that chaotic regime enables the heart to exert its function such that regulatory changes can be achieved with minimal external input, reminiscent of "self-organized criticality" seen in other physical phenomena. From a standpoint of economy of performance (energy use, responsiveness) there must be some upper limit set on the number of active degrees of freedom (control variables) that can or need be summoned. It seems that most of physiological time-series data are restricted in dimensional complexity to 3-6 degrees of freedom.
ii. One must appreciate the "whole-part" dualistic nature of the heart, comprised of multiple nested loops of nonlinear interacting regulators (homeostats) making it especially suitable/prone to chaotic behavior and thus amenable to finer/rapid control/adaptation. Sustained periodicities (predictability) are often associated with "unhealthy" event (epileptic seizures, tachycardia / bradyarrhythmia). Moreover, healthy young person shows greater complexity, (i.e., higher dimension) in HRV than older person. Greater loss of complexity, i.e., reduction to almost constant rhythm may be premonitory to fatal event [23]. Diminution in dimensionality of HR-dynamics is often associated with changes in the adaptive plasticity of the heart to cope with adverse cardiac events:

- The loss of total spectral power as well as reduction in correlation dimension has been shown to be sensitive early markers of risk to sudden death [16, 24].
- As much as 80% of sudden cardiac death in cardiac transplant patients as well as increase incidence of arrhythmia is confined to the period starting at the second year post-HTX [25], an interval which was shown in our paper to be associated with the loss of HRV. In patients exhibiting acute rejection, the dimensional complexity of HRV was suppressed [20].
- The noted progressive long-term increase in the degree of HRV in HTX patients is consistent with other studies that documented partial reestablishment of autonomic regulation [15, 26]. Importantly, in those patients in whom the reappearance of neurologic markers was documented there was a consistent exaggerating of HRV

features [27]. Increasing "fitness" of the heart within the host seems to be associated with the complexification of the regulatory order. This provides supporting evidence for advancing the proposition that adaptive reinnervation process that evolves over time (rebuilding hierarchical organization) is associated with enhancement in dimensionality of the system, i.e., "adaptation builds complexity."

The explanatory model of the newly emergent functional order attributable to graft-host interaction may benefit by evoking organizing principles of "co-evolution". The heart is an organ endowed with an adaptive plasticity (genotypic/phenotypic memory) and capacity to assimilate ("fitness capacity") within the host and in the process modify the environment determining the fate of the body system as whole. The principles by which "emergent properties" and functional order of self-organizing system, such as the heart, achieve (homeo) dynamic stability provide a non-reductionist framework for understanding how biological system adapt to imposed internal and external stresses, i.e., ischemia, organ/tissue replacement. The newly emergent dynamics of cardiac rhythm arising after heart transplantation may represent a more stable, versatile and adaptive (as per "law of requisite variety") bipartite whole. The integrative action of the living organism cannot be gotten from their concatenated fractions [28, 29] but is evolved "relationally", i.e., emanate from emergent internal requirements of the constitutive parts.

References

1. Jennings RB, Reimer KA. The cell biology of acute myocardial ischemia. Annu Rev Med42:225-246, 1991
2. Abd-Elfattah AS, Wechsler AS. Myocardial preconditioning: a model or a phenomenon? J Card Surg. 10(4 Suppl):381-8, 1995
3. Schaper W. Ischemic preconditioning, remembrances of things past and future. Basic Res Cardiol. 91(1):8-11, 1996
4. Ytrehus K, Liu Y, Downey JM. Preconditioning protects ischemic rabbit heart by protein kinase C activation. Am J Physiol 266:H1145-H1152, 1994
5. Kitakaze M, Minamino T, Node K, Komamura K, Hori M. Activation of ecto-5'-nucleotidase and cardioprotection ischemic preconditioning. Basic Res Cardiol. 91(1):23-6, 1996
6. Maulik N, Watanabe M, Zu YL, Huang CK, Cordis GA, Schley JA, Das DK. Ischemic preconditioning triggers the activation of MAP kinases and MAPKAP kinase 2 in rat hearts. FEBS Lett 396(2-3):233-7, 1996
7. Mocanu MM, Baxter GF, Yue Y, Critz SD, Yellon DM. The p38 MAPK inhibitor, SB203580, abrogates ischaemic preconditioning in rat heart but timing of administration is critical. Basic Res Cardiol. 95(6):472-8, 2000
8. Whittaker P. An alternative perspective on ischemic preconditioning derived from mathematical modeling. Basic Res Cardiol. 91(1):47-9, 1996
9. Ingber DE. The Architecture of Life. Sci American 278:48-57, 1998

10. Izrailtyan I, Narula J, Schwartz AB, Kresh JY. Intrinsic Cardiac Neuroendocrine System Mediates Ischemic Myocardial Adaptation [Abstract] JACC 31(2)Suppl:134A-135A, 1998
11. Kositskii GI. The Afferent Systems of the Heart Moscow: Meditsina; 1975, 208 p.
12. Kresh JY, Armour JA. The heart as a self-regulating system: integration of homeodynamic mechanisms. Technology Health Care 5(1):159-169, 1997
13. Skinner JE, Wolf SG, Kresh JY, Izrailtyan I, Armour JA, Huang MH. Application of chaos theory to a model biological system: evidence of self-organization in the intrinsic cardiac nervous system. Integr Physiol Behav Science 31(2):122-146, 1996
14. Kresh JY, Izrailtyan I. Evolution in Functional Complexity of Heart Rate Dynamics: A Measure of Allograft Adaptability. Am J Physiol 275(3):R720-R727, 1998
15. Wilson, R.F., Christensen, B.V., Olivari, M.T., Simon, A., C.W. White, and D.D. Laxson. Evidence for structural sympathetic reinnervation after orthotopic cardiac transplantation in humans. Circulation 83(4):1210-1220, 1991
16. Elbert T, Ray WJ, Kowalik ZJ, Skinner JE, Graf KE, Birbaumer N. Chaos and physiology: deterministic chaos in excitable cell assemblies. Physiol Rev 74:1-47, 1994
17. Vaughn BV, Quint SR, Messenheimer JA, Robertson KR. Heart period variability in sleep. Electroencephalogr Clin Neurophysiol 94(3):155-62, 1995
18. Izrailtyan I, Kresh JY. Bradykinin modulation of isolated rabbit heart function is mediated by intrinsic cardiac neurons. Cardiovasc Res 33:641-649, 1997
19. Izrailtyan I, Kresh JY, Williams J, Morris RJ, Brockman SK. Heart transplantation induced modulation of cardiac rhythm dynamics: the role of intrinsic neuroendocrine control [Abstract]. Circulation 96(8)Suppl:I-69, 1997
20. Izrailtyan I, Kresh JY, Brozena SC, Morris RJ, Kutalek SP, Wechsler AS. Early detection of acute allograft rejection by linear and nonlinear analysis of heart rate variability. J Thorac Cardiovasc Surg. 120(4):737-45, 2000
21. Huberman, B.A. The adaptation of complex systems. In: Theoretical Biology, eds. B.Goodwin and P.Saunders. Baltimore & London: The John Hopkins University Press, 1992, p.124-133
22. Kauffman SA. The origin of order. Self-organization and selection in Evolution. New York - Oxford: Oxford University Press, 1993, 709 p.
23. Lipsitz LA, Mietus J, Moody GB, Goldberger AL. Spectral characteristics of heart rate variability before and during postural tilt. Relations to aging and risk of syncope. Circulation 81:1803-10, 1990
24. Malik M, Camm J. Electrophysiology, pacing, and arrhythmia. Clin Cardiol 13:570-576, 1990
25. Patel VS, Lim M, Massin EK, Jonsyn GP, Ates P, Abou-Awdi NL, McAllister HA, Radovancevic B, Frazier OH. Sudden cardiac death in cardiac transplant

recipients. Circulation 94(9 Suppl):II273-7, 1996

26. Ramaekers D, Ector H, Vanhaecke J, van Cleemput J, van de Werf F. Heart rate variability after cardiac transplantation in humans. Pacing Clin Electrophysiol 19(12 Pt.1):2112-9, 1996

27. Ziegler SI, Frey AW, Uberfuhr P, Dambacher M, Watzlowik P, Nekolla S, Wieland DM, Reichart B, Schwaiger M. Assessment of myocardial reinnervation in cardiac transplants by positron emission tomography: functional significance tested by heart rate variability. Clin Sci 91 Suppl:126-8, 1996

28. Weinberg GM. An introduction to general systems thinking. New York: John Wiley & Sons, 1975,279 p.

29. The relevance of general systems theory. Ed. Laszlo E., New York: George Braziller, 1972

The DASCh Experience: How to Model a Supply Chain

H. Van Dyke Parunak
Center for Electronic Commerce, ERIM
3600 Green Court, Suite 550
Ann Arbor, MI 48105
vparunak@erim.org

1. Introduction

Nonlinear dynamical systems have been a fertile field for the application of simulation techniques. Since the 1960's, System Dynamics has studied such problems by integrating systems of ordinary differential equations (ODE's) over time. More recently, increases in computer power have permitted the broad application of agent-based (or individual-based) modeling. In our work on supply chain modeling, we have found agent-based modeling to be more flexible than ODE models for basic exploration. One phenomenon we discovered, the inventory oscillator, can also be modeled in ODE's, an approach that permits more rapid manipulation in a spreadsheet environment. Further study permits derivation of a closed-form analytical model as well, which makes explicit a number of interesting structural features of the oscillator.

This paper does not pretend to enrich the repertoire of nontrivial behaviors known to complexity researchers. Mathematically, the behavior we observe is not particularly sophisticated: the inventory oscillator turns out to be computing a modulus function. Its intended contribution is twofold. First, and primarily, we seek to highlight the differences among agent-based, equation-based, and analytical system modeling, in terms of when they can be applied and the results one can expect to derive. The comparative simplicity of our system is what makes the analytical treatment possible at all. Second, manufacturing engineers find the potential for inventory fluctuation under stable boundary conditions counterintuitive and of great practical import. Its reducibility to the modulus function, far from making the results trivial, suggests that similar threshold nonlinearities may be responsible for other unexpected time-varying manufacturing measurements, and thus points the way to stabilize these important commercial systems.

Section 2 of this paper describes the supply chain problem. Section 3 reports the three models that we constructed. Section 4 reviews the roles of each model and recommendations for their deployment. Section 5 summarizes our conclusions.

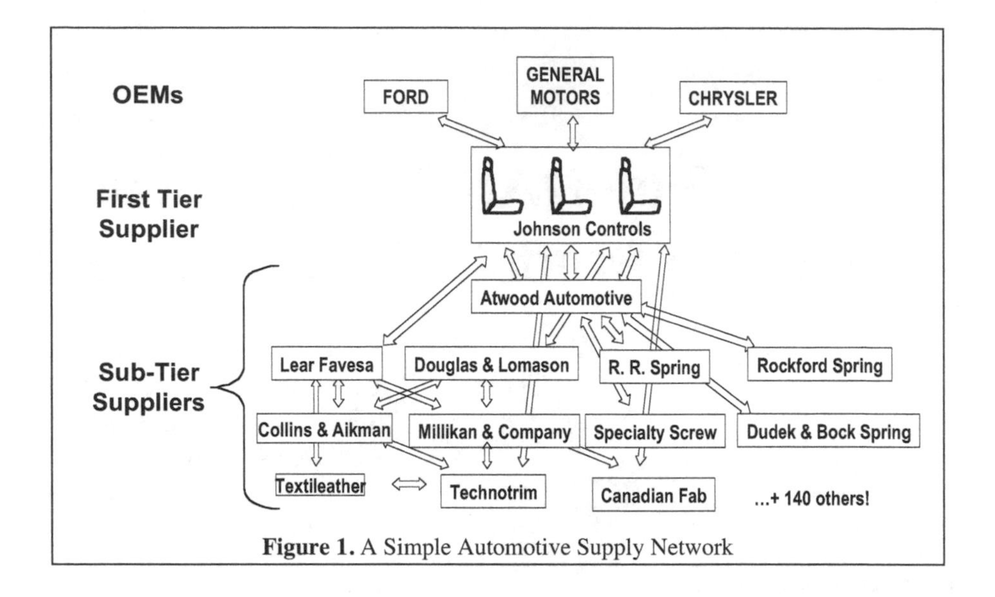

Figure 1. A Simple Automotive Supply Network

2. The Supply Chain Challenge

Modern industrial strategists are developing the vision of the "virtual enterprise," formed for a particular market opportunity from independent firms with well-defined core competencies [4]. The manufacturer of a complex product (the original equipment manufacturer, or "OEM") may purchase half or even more of the content in the product from other firms. For example, an automotive manufacturer might buy seats from one company, brake systems from another, air conditioning from a third, and electrical systems from a fourth, and manufacture only the chassis, body, and powertrain in its own facilities. The suppliers of major subsystems (such as seats) in turn purchase much of their content from still other companies. As a result, the "production line" that turns raw materials into a vehicle is a "supply network" (more commonly though less precisely called a "supply chain") of many different firms.

Figure 1 illustrates part of a simple supply network [1, 3]. Johnson Controls supplies seating systems to Ford, General Motors, and Chrysler, and purchases components and subassemblies either directly or indirectly from over 150 other companies, some of which also supply one another. Product design and production schedule must be managed across all these firms to produce quality vehicles on time and at reasonable cost. Historically, this vision has been frustrated by unexpected behavior of the supply network, such as large swings in orders and inventories and unreliable information. Our research explores these problems from a dynamical systems perspective.

3. Three Models

We have modeled one aspect of supply chain behavior using three different approaches. Our initial agent-based model exhibited internal inventory oscillations under stable conditions at the chain's boundaries. We replicated much of this behavior in an equation-based model using ODE's. Then we developed an analytical model in which we could prove certain empirically observed characteristics of the oscillator.

3.1. Agent-Based Model

The DASCh project (Dynamical Analysis of Supply Chains) [5, 6] includes three species of agents. *Company agents* represent the different firms that trade with one another in a supply network. They consume inputs from their suppliers and transform them into outputs that they send to their customers. *PPIC agents* model the Production Planning and Inventory Control algorithms used by company agents to determine what inputs to order from their suppliers, based on the orders they have received from their customers. These PPIC agents currently support a simple material requirements planning (MRP) model.[1] *Shipping agents* model the delay and uncertainty involved in the movement of both material and information between trading partners.

The initial DASCh experiments involve a supply chain with four company agents (Figure 2: a boundary supplier, a boundary consumer, and two intermediate firms producing a product with neither assembly nor disassembly). Each intermediate company agent has a PPIC agent. Shipping agents move both material and information among company agents.

We expected this simple structure to exhibit relatively uninteresting behavior, on which the impact of successive modifications could be studied. In fact, it shows a range of interesting behaviors in terms of the variability in orders and inventories of the various company agents.

We found four different behaviors in the model: amplification of variance in the order stream as one moves away from the customer, induction of spurious correlations in the order stream, persistence of disturbances long after a single change in orders has been made, and generation of variation in inventory levels in the system when the boundary conditions are held constant. Details of these behaviors are discussed in [6].

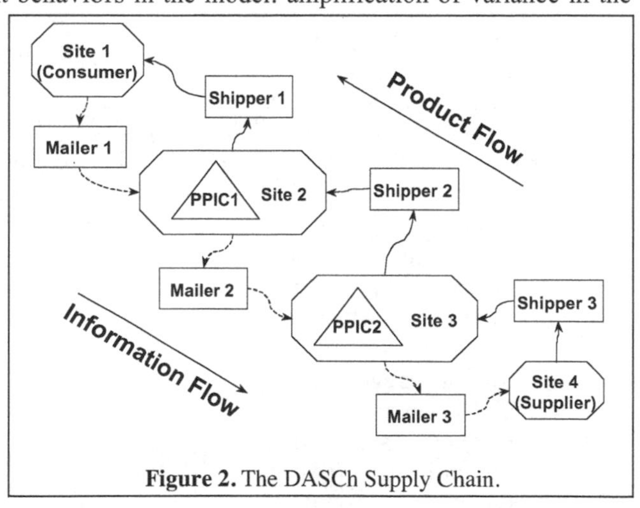

Figure 2. The DASCh Supply Chain.

---

[1] The basic MRP algorithm includes developing a forecast of future demand based either on past demand or on customer forecast (depending on location in the hourglass), estimating inventory changes through time due to processing, deliveries, and shipments, determining when inventory is in danger of falling below specified levels, and placing orders to replenish inventory early enough to allow for estimated delivery times of suppliers.

This report focuses on the last effect, the generation of inventory variation. Even when top-level demand is constant and bottom-level supply is completely reliable, intermediate sites can generate complex oscillations in inventory levels, including phase locking and multiperiodicity, as a result of capacity limitations.

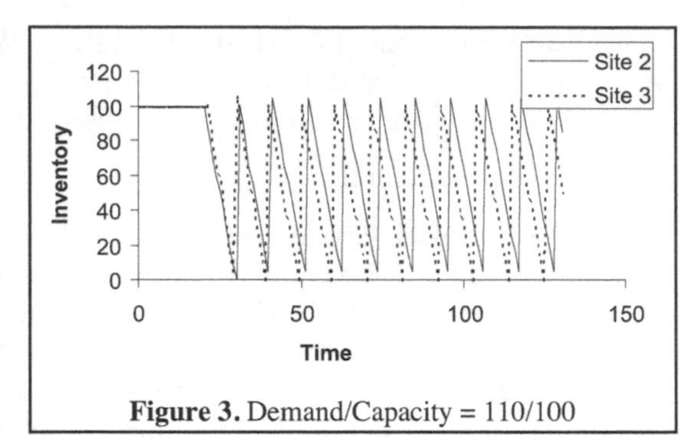

Figure 3. Demand/Capacity = 110/100

The consumer has a steady demand with no superimposed noise. The bottom-level supplier makes every shipment exactly when promised, exactly in the amount promised. Batch sizes are 1, but we impose a capacity limit on sites 2 and 3: at each time step they can process only 100 parts, a threshhold nonlinearity. As long as the consumer's demand is below the capacity of the producers, the system quickly stabilizes to constant ordering levels and inventory throughout the chain. When the consumer demand exceeds the capacity of the producers, inventory levels in those sites begin to oscillate. The basic dynamic is that filling orders draws down inventory to make up a shortfall in production. When inventory falls too low, the current order is backlogged and the current production run provides a new inventory.

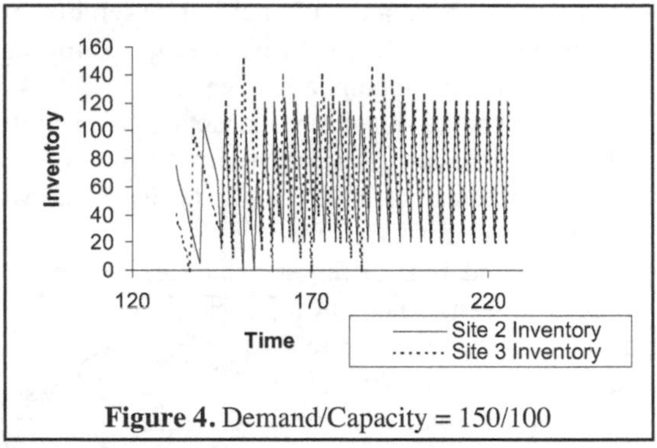

Figure 4. Demand/Capacity = 150/100

Figure 3 shows the behavior when demand exceeds capacity by 10%. Site inventories oscillate out of phase with one another, in a sawtooth that rises rapidly and then drops off gradually. The inventory variation ranges from near-zero to the level of demand, much greater than the excess of demand over capacity

Figure 4 shows the dynamics after increasing consumer demand to 150. The inventories follow a sawtooth of shorter period. Now one cycle's production of 100 can support only two orders, leading to a period-three oscillation. The inventories of sites 2 and 3, out of synch when Demand/Capacity = 110/100, become synchronized and in phase after a transition period.

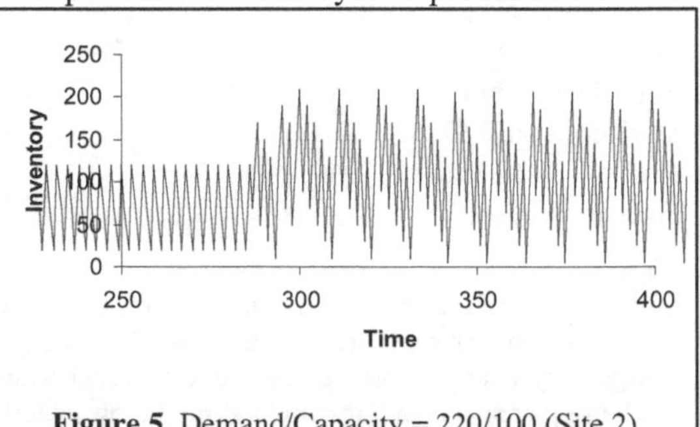

Figure 5. Demand/Capacity = 220/100 (Site 2)

The transition period is actually longer than appears from Figure 4. The increase from 110 to 150 takes place at time 133, but the first evidence of it in site 2's dynamics appears at time 145. The delay is due to the backlog of over-capacity orders at the 110 level, which must be cleared before the new larger orders can be processed.

Figure 5 shows the result of increasing the overload even further. (Because of the increased detail in the dynamics, we show only the inventory level for site 2.) Now the consumer is ordering 220 units per time period. Again, backlogged orders at the previous level delay the appearance of the new dynamics: demand changes at time 228, but appears in the dynamics first at time 288, and the dynamics finally stabilize at time 300.

This degree of overload generates qualitatively new dynamical behavior. Instead of a single sawtooth, the inventories at sites 2 and 3 exhibit biperiodic oscillation, a broad sawtooth with a period of eleven, modulated with a period-two oscillation. This behavior is phenomenologically similar to bifurcations observed in nonlinear systems such as the logistic map, but does not lead to chaos in our model with the parameter settings used here. The occurrence of multiple frequencies is stimulated not by the absolute difference of demand over capacity, but by their incommensurability.

3.2. Equation-Based Model

Following the pioneering work of Jay Forrester and the System Dynamics movement [2], virtually all simulation work to date on supply chains integrates a set of ordinary differential equations (ODE's) over time. It is customary in this community to represent these models graphically, using a notation that suggests a series of tanks connected by pipes with valves. The dynamics of our simple model can be represented by the following set of ODE's:

$$d(WIP3)/dt = orderRate - min(capacity, WIP3/productionTime)$$

$$d(Finished3)/dt = min(capacity, WIP3/productionTime) - \alpha$$

$$d(WIP2)/dt = \alpha - min(capacity, WIP2/productionTime)$$

$$d(Finished2)/dt = min(capacity, WIP2/productionTime) - \beta$$

where

$\alpha = orderRate$ if $Finished3/orderPeriod + capacity > orderRate$, otherwise 0;

$\beta = orderRate$ if $Finished2/orderPeriod + capacity > orderRate$, otherwise 0

WIP{2,3} is work in process inventory at site 2 or 3, respectively;

Finished{2,3} is finished goods inventory at site 2 or 3, respectively;

orderRate is the rate of consumer orders to the chain;

productionTime is the time needed at site 2 or site 3 to turn WIP to finished goods;

capacity is the amount of WIP that site 2 or site 3 can

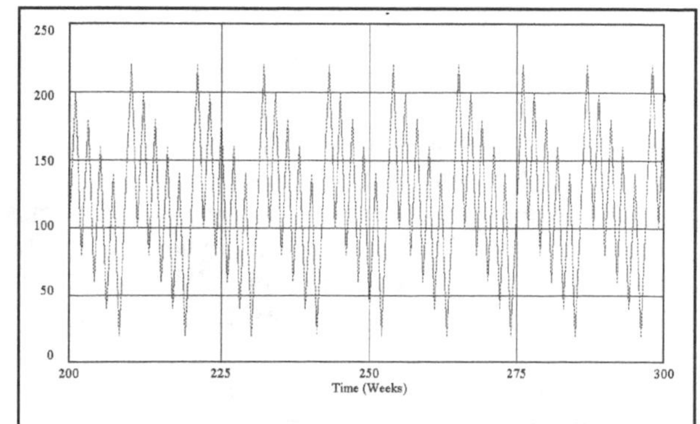

Figure 6. Inventory Oscillation in an ODE Model

turn into finished goods each time step.

This model does not support many of the behaviors in the agent-based model. In particular, amplification, correlation, and persistence of variation depend on the PPIC (Production Planning and Inventory Control) algorithm in DASCh, which is extremely difficult to capture in an ODE formalism [8]. However, the ODE model does demonstrate oscillations comparable to those in the DASCh model. For example, Figure 6 shows the biperiodic oscillations for Demand/Capacity = 220/100, generated by the VenSim® simulation environment. The system dynamics model shows the same periodicities as the agent-based model, though it does not show the transitional dynamics or phase locking behavior seen in Figure 4, because it has abstracted away the PPIC algorithm.

3.3. Analytical Model

If we further abstract away the dynamical behavior of production and shipping that generates the observed behavior, an even a simpler model is available. Since each time step generates new inventory of *capacity* and outstanding orders ship everything in excess of *order*, the inventory at the *n*th time step is just *mod((n-1)*capacity, order)*, where *mod()* is the modulo function, the essence of a threshold nonlinearity. This level of abstraction permits us to prove a number of interesting relations among the *Inventory(t)* at a site, the constant *Demand* (order rate) from its customer, and its constant *Production* (capacity level). Critical derived quantities include *D* and *P* (the smallest integers such that *D/P = Demand/Production*), *I(t)* (*Inventory(t)* in the same units as *D* and *P*), *H* (the minimum of *P* and *D – P*), *Period* (the minimum *n* such that *I(t) = I(t+n)*, and *Sequence* (the shortest sequence of steps-to-next-local-maximum over the course of a single period). Many of these results are well-known characteristics of the modulo function. Proofs are available in [6]. For example:

Attractor.–If the system is initiated with *Inventory* ≥ *Demand*, it will enter the region *0 ≤ Inventory < Demand*, and then remain there.

Scaling.–If we multiply *Demand* and *Production* by the same integer factor, or if we divide out common integer factors, the *Inventory(t)* is multiplied or divided by the same integer factor, but *Sequence* and *Period* are unaffected. This result motivates the use of *D* and *P*, from which all common factors have been removed, as a unique representation of a given ratio *Demand/Production*.

Period.–For any *I(t)* in the region *0 ≤ I < D,* the system will return to the same inventory level at time *t+D,* so that *Period* = *D*. By the previous result, *Period* = *D* not only for systems in the *(D,P,I)* units, but for arbitrarily scaled *(Demand, Production, Inventory)* units.

Coverage.–Between *t* and *t + Period, I* assumes every value in the attracting range. This result holds only for the reduced units *(D, P, I)*, since it concerns units of parts produced. For systems in which *Demand* and *Production* have a common factor *k*, there will be bands of inventory values of width *k* that the system will never visit once it is in the attracting region.

Length.–The number of items in a *Sequence*, corresponding to the number of intermediate maxima between maxima of the same size (counting one of the ends), is *H*.

In addition, the pattern by which $I(t)$ moves between local minima and local maxima in the attracting region, the proportion of long and short subperiods, and the number of monotonic subsequences in the overall *Sequence* depend on H in ways defined more precisely in [6].

These results are consistent with a concise geometrical model of the dynamics, familiar to those acquainted with the behavior of the modulo function. The complete dynamics can be represented in a square of D units on a side. The left edge of the square corresponds to time t, the right edge to time $t+D$, the bottom to inventory 0, and the top to inventory D. The system trajectories behave as though this square were formed into a two-torus. In our manufacturing domain, D and P are integer parameters, so D/P is rational by construction. However, the torus model supports irrational D/P as well. In this case, we would have quasiperiodicity, and the orbit on the torus would never retrace itself. Since the surface of a 2-torus is two-dimensional, this interpretation shows that the dynamics of the oscillator can be embedded in two dimensions. Thus in the limit of continuous time, and under the rules we explored, the oscillator can never go chaotic.

4. The Right Tool for the Job

Each of the three modeling approaches offers distinctive contributions to our understanding of the dynamics of the inventory oscillator.

Each agent in the agent-based model maps directly to an entity in the problem domain. It is straightforward to represent the PPIC algorithm in such a model, so we did, and were able to discover a much wider range of interesting behaviors than in the ODE model, which lacks such an algorithm. Even for the oscillator, it supports some behaviors (transition effects and phase locking) that simpler models do not show. Elsewhere [7] we discuss in depth the advantages of agent-based models over the equation-based models of system dynamics. However, the agent-based model offers no *a priori* characterization of the relationships among the model observables.

The equation-based model makes these relationships explicit. However, its construction requires deciding in advance what observables to study, and demands that the relations among them be expressed in closed functional forms. The inventory oscillator lends itself to such expression. Other important features of the supply network (such as interacting PPIC algorithms) do not.

The analytical model offers a detailed characterization of the oscillator that is not available to either of the other approaches. It shows clearly why the oscillator cannot enter the formally chaotic regime without introducing some other complication. However, it is the most limited of the models. It depends on the reducibility of the dynamics to a simple function, it applies only to the oscillator, and then only to an abstraction in which common factors are removed from the values for demand and production.

5. Conclusion

The three modeling methods explored in this paper can be compared in several ways. The agent model offers the most natural representation and greatest breadth of potential behavior, followed first by the equation-based model and then by the analytical model. However, the explicitness of the relationships among system observables is greatest in the analytical model, followed by the equation-based model and then by the agent-based model. It is unlikely that we could have developed the analytical model without first of all discovering the oscillatory behavior in one of the other two models, and the ease of manipulation of the equation-based model in a spreadsheet form was a great help in testing hypotheses that led to the formulation of the theorems in the analytical model. The equation-based model, in turn, is only possible because these particular system observables and behaviors lend themselves to representation in closed functional forms. Other behaviors observed in the agent-based model could not be duplicated in the equation-based model, and would not have been discovered if we had begun with that form of model.

Thus our experience recommends that system modeling begin with a formalism as close as possible to the entities in the problem domain (that is, with an agent-based model). In some cases, experience with this model may permit the construction of a second, equation-based model that may be useful in generating large numbers of test cases quickly. Inspection of such results may (in simple cases) suggest analytical formalisms for specific behaviors.

Acknowledgments

DASCh was funded by DARPA under contract F33615-96-C-5511, and administered through the AF ManTech program at Wright Laboratories under the direction of James Poindexter. The DASCh team includes Steve Clark and Van Parunak of ERIM's Center for Electronic Commerce, and Robert Savit and Rick Riolo of the University of Michigan's Program for the Study of Complex Systems.

References

[1] AIAG, "Manufacturing Assembly Pilot (MAP) Project Final Report", *M-4*, Automotive Industry Action Group (1997).

[2] Forrester, J. W. *Industrial Dynamics*, Cambridge, MA, MIT Press (1961).

[3] Hoy, T. "The Manufacturing Assembly Pilot (MAP): A Breakthrough in Information System Design",*EDI Forum*, 10(1996), 26-28.

[4] Nagel, R. N. and R. Dove. *21st Century Manufacturing Enterprise Strategy*, Bethlehem, PA, Agility Forum (1991).

[5] Parunak, H. V. D., "DASCh: Dynamic Analysis of Supply Chains", http://www.erim.org/~van/dasch (1997).

[6] Parunak, H. V. D., R. Savit, R. Riolo, and S. Clark, "Dynamical Analysis of Supply Chains", http://www.erim.org/cec/projects/dasch.htm, ERIM (1998). Available at http://www.erim.org/cec/projects/dasch.htm.

[7] Parunak, H. V. D., R. Savit, and R. L. Riolo, "Agent-Based Modeling vs. Equation-Based Modeling: A Case Study and Users' Guide", *Proceedings of Workshop on Multi-agent systems and Agent-based Simulation (MABS'98)*, Springer (1998), Available at http://www.erim.org/~van/mabs98.pdf.

[8] Warkentin, M. E., "MRP and JIT: Teaching the Dynamics of Information Flows and Material Flows with System Dynamics Modeling", *Proceedings of The 1985 International Conference of the Systems Dynamics Society*, International System Dynamics Society (1985), 1017-1028.

Some Properties of Multicellular Chemical Systems Response to the environment and emergence

Jerzy Maselko Michael Anderson
Department of Chemistry, University of Alaska,
Anchorage AK 99508

1. Summary

The interactions of multicellular chemical structures with the environment are studied numerically. These chemical structures develop from a single perturbation in a 2-D array of coupled, by diffusion, chemical cells forming very complex, highly organized and beautiful patterns. During interactions with the environment, these structures exhibit a self-repairing property, learning capacity and "chemotaxis". It is a new approach that will allow us to study very high dimensional systems.

2. Response to the environment

Chemical reactions occurring in a chemical cell, in the case of two reacting chemical species, can be described by the following general equation [1].

$$dx/dt = k_0 +- k_1x + k_2y +- k_3xy - k_4x^2 + k_5y^2$$
$$dy/dt = b_0 +- b_1y + b_2x +- b_3xy - b_4y^2 + b_5x^2$$

Where the reactions

1) Outside → x
2) A → x

are described by the parameter k_0, whereas k_1 describes the following reactions:

3) A + x → 2x
4) x → outside

Reactions 2 and 4 represent the interaction of a chemical cell with the environment. Parameters b_0 and b_1 play the same role for the chemical component y. The other parameters k and b are the rate constants for reactions occurring inside the chemical cell. The surroundings or the environment may influence the chemical cells in two ways.

1. The state of the chemical system may change in the response to the changes in the environment represented by parameter k_0 and b_0 . This response may be studied by a bifurcation diagram where parameter space is divided into areas where attractors exist and areas where they do not. By changing system parameters, the chemical system may undergo a bifurcation, attractor collapses, and system moves from one attractor to another.
2. The chemical system may respond to perturbations in the values of concentrations of chemical species. This type of response is best studied by a phase portrait. The phase portrait represents all attractors and their boundaries.

For the multicellular chemical system the approach discussed above becomes futile. The number of attractors may reach thousands, even millions as will the number of variables. Therefore, a new approach is necessary.

In this paper we will explore the response of multicellular chemical systems to perturbations and the interaction of these systems with the environment.

3. Numerical calculations:

The Gray-Scott model was used for numerical calculations [2,3]. This system of chemical reactions is a simplified version of a mechanism in the iodate-sulfide-hexacyanoferrate(II) reaction [4,5]. It suggests that the generated patterns and properties could be realized experimentally. The patterns formed in multicellular chemical systems have been studied, until now, in few cell arrangements [6,7,8]. The Gray-Scott model for the one-dimensional multicellular chemical system is described by the following simple set of ordinary differential equations. Following Gray and Scott we are using (a) and (b) as the chemical species.

$$da_i/dt = (a_0-a_I)/tr - a_i b_i^2 + d_a(a_{I-1} + a_{I+1} - 2a_1)$$
$$db_i/dt = (b_0-b_I)/tr + a_i b_i^2 - kb_i + d_b(b_{I-1} + b_{I+1} - 2b_I)$$

In the case of two dimensions, there are four neighbors, and in the case of three dimensions, six neighbors are included. The coupling strength between connecting cells for species (a) and species (b), respectively, are represented by d_a and d_b

4. Numerical results

For the parameter range studied in this paper, the Gray-Scott model has a unique stable steady state at a=1 and b=0 which is excitable. Fig.1 represents the phase portrait for the system.

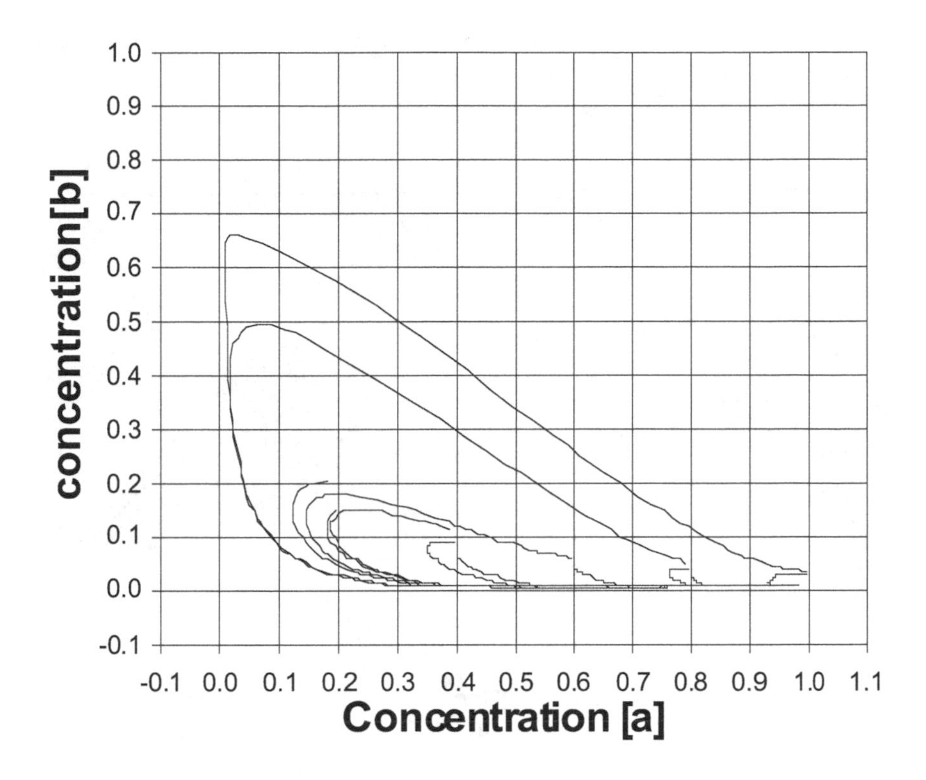

Figure 1. The phase portrait for the Gray-Scott model of chemical reaction.
$k=0.034$, $tr=313$, $b_0 = 0.066$, $a_0 = 0$

Above and very close to the stationary steady state there is a saddle point. By perturbing the cell with component (b), the chemical system will move further from steady state, increasing the value of (b) substantially, through an autocatalytic process, before returning to the excitable steady state. If the cells are connected by diffusion, then after the perturbation with chemical species (b) a chemical wave will be generated in the following way. The component (b) diffuses from the excited cell to the surrounding excitable cells, increasing the (b) concentration above the excitability threshold and these cells become excited. If the coupling of (a) is greater than that of (b) then a new phenomena emerges. The concentration of the (a) component in the excited cell is smaller than in surrounding cells and component (a) diffuses inward to the excited cell, decreasing the concentration in the surroundings. Because of the properties of a phase portrait as is illustrated in Fig.1, the decrease in the value of (a) causes an increase in the threshold value, which makes excitation impossible. The wave can not propagate and a stationary pattern develops.

The examples of multicellular chemical structures that develop from single perturbations are represented on Fig.2 (below)

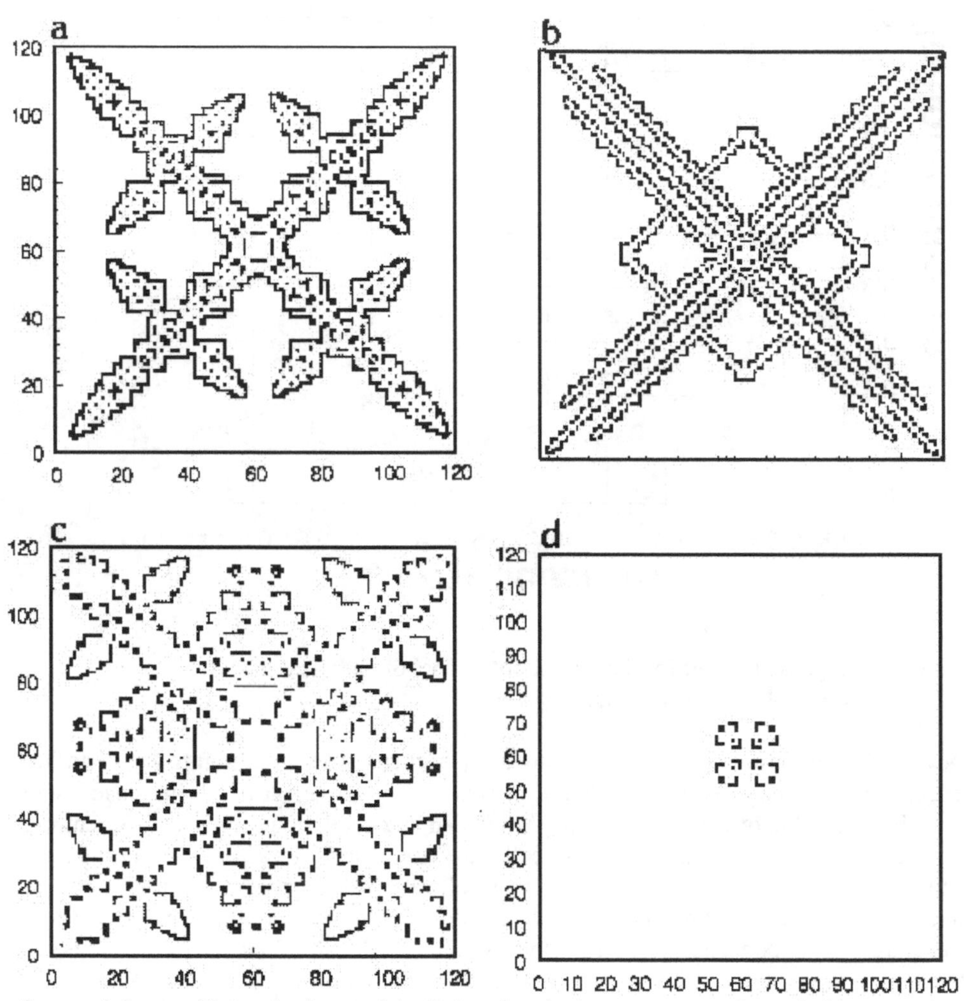

Some of these self-developing multicellular chemical structures are bounded in space whereas another grows to the system boundary. The characteristic properties are the growth of branches, sub-branches, and formation of closed compartment. Under these system parameters, the developed structures preserve symmetry of cellular arrangement.

5. Response to the environment in multicellular systems.

The response of multicellular chemical systems to the environment may be studied similarly to the single cell system. However, due to the enormous number of attractors and possible perturbations, the new approach is necessary. In this paper we concentrated on the new emergent phenomena that can not be seen in singe cell chemical systems.
Fig. 3 represents the response of the self-developing multicellular chemical system to the cutting of a growing branch. Three cases have been found.

a) The cut branch will regenerate with the same shape.
b) The branch will regenerate with a different shape (fig. 3a).
c) No regeneration will be observed (fig, 3b).

In case of b, a special perturbation may cause the branch to regenerate, as is represented in Fig.3 c.

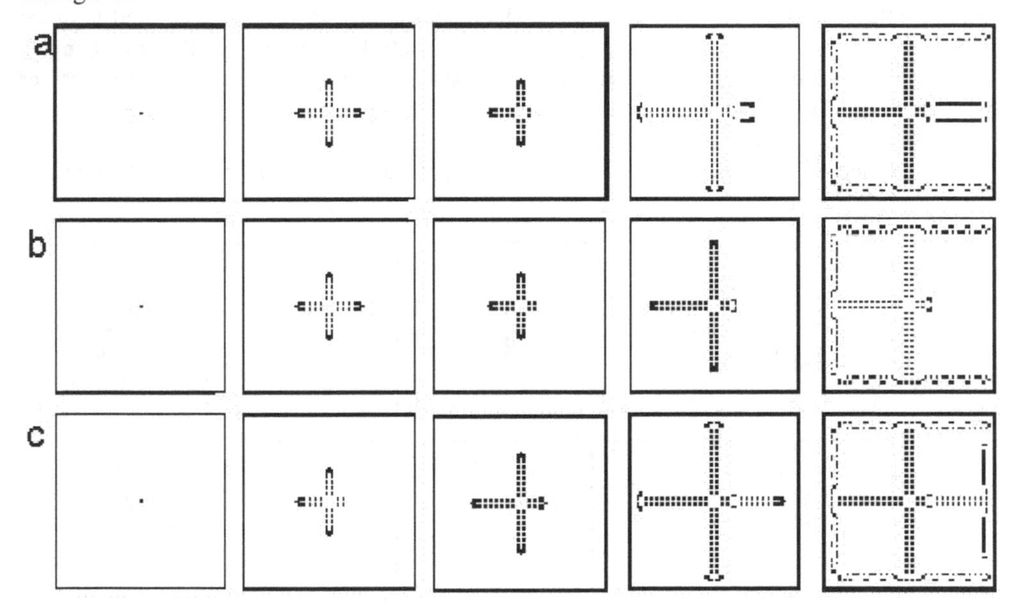

Figure 3. Response of multicellular chemical systems to branch removal.

The response of the structure to the perturbation of a single cell is presented in Fig.4. After the structure develops (Fig.4b), the single central cell is perturbed with the component (b) (see Fig. 4.c).

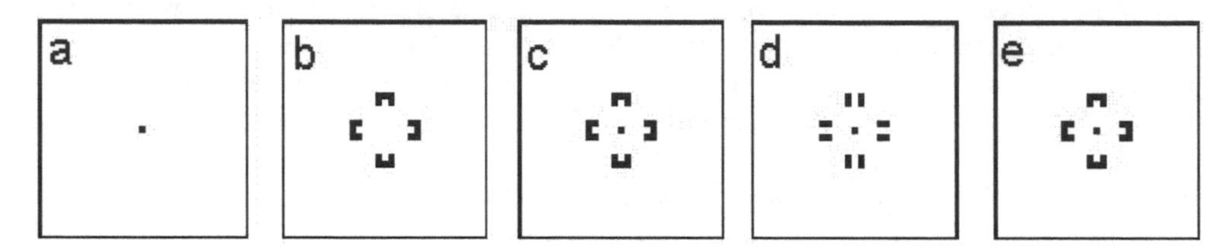

Figure 4. Development of " immunity". d_a =0.01835, d_b = 0.004

The structure changes its shape and the concentrations in the central perturbed cell are different (Fig.4e). Following perturbations to the central cell do not cause any changes. After the first perturbation, the system adjusts its structure and becomes immune to further perturbations. This behavior may be considered as a learning.

The influence of the multicellular chemical system on its surroundings is presented in Fig.5.

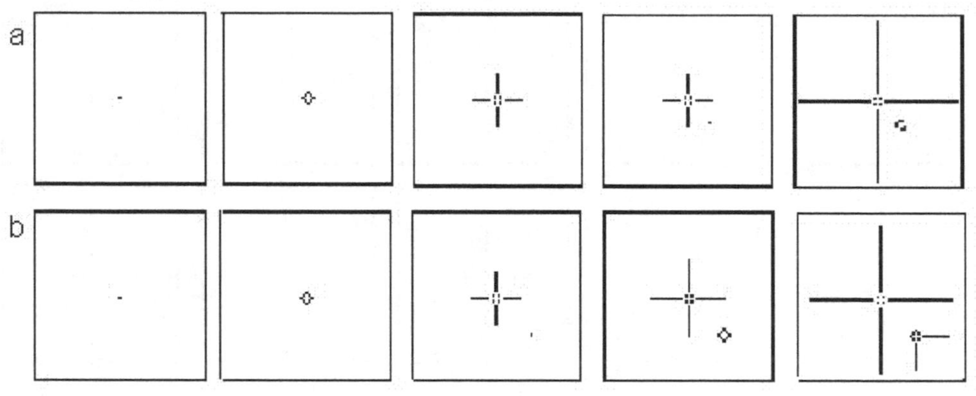

Figure 5. Influence on environment. k=0.03, tr=330, d_a =0.01835, d_b = 0.0007

After the multicellular chemical structure develops, the surroundings are perturbed with the same perturbation that causes the original structure to grow. If the perturbation is close to the structure, the new structure will develop but without the branches. For more distant perturbations, branches will grow only in directions away from the primary structure. This "behavior" of the multicellular chemical structure is similar to that of the chemotaxis observed in biological systems.

The interaction of the growing branch with the impermeable boundary has been discussed in paper [9].

Depending on system parameters, the growing branch may:

1. Be stopped by the boundary
2. Split close to the boundary and develop along the boundary, and stop development in the corner.
3. After meeting in the corner, branches may change their direction of growth by 45° and move to the middle of the pattern.
4. After meeting in the corner, the branches may stop growing and develop growing sub-branches.

From thousands of possible perturbations of chemical systems, we presented a few when the emergence of a new phenomena, not observed in single cell systems, are obviously seen. All of these phenomena are very similar to those observed until now only in biological systems. Usually the language of mathematics and chemistry is used to explain the biological phenomena. Here we are using the language of biology to describe phenomena observed in chemical systems.

The response of multicellular chemical structures to the environment may fulfill some function. The applications in materials science and particularly in synthesis of highly structured and smart materials have been discussed in [10]. One of the important properties of smart materials is the self-repairing property that is rather common in multicellular chemical structures. Other structures may be applied to the formation of reinforced materials. For example, the structures presented in Fig.2 may form reinforcing skeletons. Three-dimensional structures may form neural networks, complicated structure of tubes or 3-D computer chips.

6. Toward the experimental realization

The most important project is obviously the experimental realization of self-developing patterns in multicellular chemical systems. The one possible idea, already explored, involves the use of ion exchange beads that may support chemical reactions as a chemical cell [11,12]. Other directions already under study involve the construction of inverse micelles with chemical reactions that may support formation of Turing patterns. The most difficult and challenging project will be to construct chemical systems with a growing numbers of cells.

Literature

1. Tyson J., Light J., 1973 J. Chem Phys. 59, 4164
2. Gray P., Scott S., K., 1984 Chem. Eng. Sci. 39, 1087
3. Gray P., Scott S., K., 1985 J. Chem. Phys. 89, 22

4. Edblom E., C., Orban M., Epstein I., R., 1986 J. Am. Chem. Soc. 108, 2826
5. Gaspar V., Showalter K., 1987 J. Am. Chem. Soc. 109, 4869
6. Crowley M., Epstein I., 1989 J. Phys. Chem. 93, 2497
7. Fei Z., Kelley R.G., Hudson J.L., 1996 J. Chem. Phys. 100, 18987
8. Wang Y., Hudson J.L. 1992, J. Phys. Chem. 96, 8667
9. Maselko J., 1996 J.Chem.Soc., Faraday Trans. 92, 2881
10. Maselko J., 1996, Materials Science & Engineering C4, 199
11. Maselko J., Showalter K., 1991 Physica D 49, 21
12. Maselko J., Showalter K., 1989 Nature 339, 6226

Hierarchical Genetic Programming using Local Modules

Wolfgang Banzhaf
Dept. of Computer Science, University of Dortmund, Germany
Dirk Banscherus
Quantum GmbH, Dortmund, Germany
and Peter Dittrich
Dept. of Computer Science, University of Dortmund, Germany

This paper presents a new modular approach to Genetic Programming, hierarchical GP (hGP) based on the introduction of local modules. A module in a hGP program is context-dependent and should not be expected to improve all programs of a population but rather a very specific subset providing the same context. This new modular approach allows for a natural recursiveness in that local modules themselves may define local sub-modules.

1 Introduction

Genetic Programming is the development of computer programs by evolutionary means [9, 5]. A population of randomly generated programs is subjected to mechanisms of variation and selection in order to arrive at behavior specified by an explicit or implicit fitness function. Over the course of the development, programs are generated that more and more approach the desired behavior.

The mechanisms used to vary and select computer programs are similar to those in other areas of evolutionary computation [6], and employ stochastic events as the main driving force for innovation. Mutation and crossover are operators used for variation, proportional or tournament selection are frequently used as selection operators to direct the search process.

As in other fields of evolutionary computation, the representation of the problem is an important aspect of its solution. Genetic Programming originally started with the tree representation of computer programs. Program trees are easy to manipulate by mutation and crossover, and until today they are the most frequently used representation in GP.

Genetic Programming is able to solve an impressive variety of problems from different problem domains [5]. However, it is well known that there are performance problems with Genetic Programming when tasks grow complex. In such a case, human programmers would rely on a modularization technique allowing them to decompose the task into sub-tasks which are subsequently solved independently, to arrive at a solution by recomposing the solutions of sub-tasks. Some modularization techniques have been proposed for Genetic Programming. Koza has suggested automatically defined functions [4], recently augmented by architecture altering operations [11]. Angeline and Pollack suggest libraries of functions [2], Rosca and Ballard adaptive representations [17]. It seems, however, that the real break-through for modular Genetic Programming is not yet made.

This paper presents a new modular approach to Genetic Programming (hGP - standing for hierarchical GP) which is based on the introduction of local modules. In contrast to other approaches, our notion of a module in a program is that the context of the module in the calling program is of great importance. A module should not be expected to improve all programs of a population but rather a very specific subset providing the same context. At the same time, our modular approach allows for a natural hierarchy in that local modules themselves may define local sub-modules.

Modules are allowed to evolve at a much slower rate than programs reflecting the need of programs to rely on their modules for improving their function. We discuss this principle which seems to be at work in other natural and artificial modular systems.

Results are presented on a set of discrete and continuous problems, including comparison with regular Genetic Programming. More details can be found in [4].

2 Modular Concepts in Genetic Programming

2.1 The Problem

One of the important issues in Genetic Programming is whether GP is able to scale up. Although there are a number of interesting applications of GP already (see [5], chapter 12), real world applications suffer from a complexity threshold. It seems that programs of small size may be readily evolvable, but as soon as one gets into hundreds or even thousands of nodes[1], GP becomes less and less effective as a means to generate the targeted function.

[1]three nodes in a tree usually correspond to one line of code

A natural method to improve GP performance is therefore the introduction of sub-programs. Partitioning of a problem into sub-problems which can be solved independently is one of the most powerful and general approaches to problem solving that we have developed [1]. In Computer Science in particular, where problems of large complexity are solved daily, modularization is a key enabling technology for progress. Many of the biggest steps in software and hardware development over the last decades may be traced back to the introduction of modularization / hierarchization techniques.

Thus, one of the big challenges for genetic programming may be formulated as this: Is it possible for a Genetic Programming system to evolve modular solutions to problems *automatically*? Note the emphasis on "automatically". It is clear that a manual specification of sub-problems will work, provided the sub-problem complexity is sufficiently small to be treated by regular GP. However, will it be possible to delegate the structuring of the problem to an automatic process like GP?

2.2 Existing Approaches

Mainly three approaches have been proposed in the course of the last decade to solve the problem of modularization by Genetic Programming: "automatically defined functions" (ADFs) [10], evolutionary module acquisition [2] and adaptive representation [17]. A more detailed discussion can be found in [5, 4]

2.3 Local, context-sensitive modules: hGP

We introduce another general method for specifying modules. The idea of local and context-sensitive modules is motivated by the success of Gruau's work on cellular encoding [8]. At the surface, cellular encoding is about making graphs available for use with genetic programming. Gruau develops neural networks, other researchers develop other graph-like applications, e.g. electric circuits [12].

The aspect interesting here, however, is that of hierarchical evolution. We use a number of hierarchical levels of evolution, with a population on each of them. On the highest level, individuals of the population evolve their functionality. On the lower levels, modules of level 1 ... n evolve through the same mechanisms of variation and selection. Figure 1 depicts the situation. Modules on higher levels (including the individuals on the highest level) are able to call modules of the next lower level as subprograms. Each level has its own set of terminals and functions.

As in ADFs, the newly defined modules are local to an individual. They are not available to the population as a whole but only to the one individual which has called them. Thus, an individual has to evolve a good choice of modules completely for itself, only taking help through crossover of material from other individuals having defined modules at the same level. Much as an entire GP system has global convergence to a solution, so do the local modules have a tendency to converge, even without being able to be accessed by all individuals.

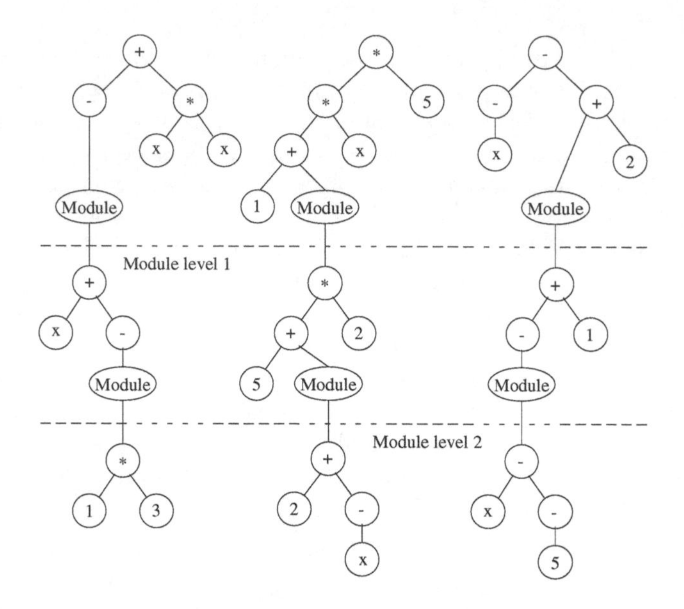

Figure 1: Example of three hierarchical levels of evolution in hGP. Modules on each level evolve in their own level and are called from the next higher level.

Arbitrary crossover of material is forbidden in this method. Rather, modules at the same level of description are able to exchange material. Koza has called this method structure-preserving crossover [9]. ADFs make use of this method, too, since the two types of branches in an ADF are only allowed to be crossed over with their kin.

Apart from the potential difference in terminal and function sets, what is the difference, then, between modules on different levels? An important difference lies in the fact that modules on different levels evolve with different speed. Similar to the compress operation in module acquisition, which explicitly forbids further evolution of material that has been compressed, speed of evolution is the key difference. The radical step of freezing the compressed material completely is substituted, however, by a less radical, but more general step: to decrease the speed of evolution. The lower in the hierarchy a module is located, the slower it is allowed to evolve. Although this is somewhat counter-intuitive at first glance, it is indeed the method which Nature used when evolving modules. The more fundamental the modules are the less evolution Nature allows at that level. The appearance of the genetic code is a typical example of this phenomenon [14], the development of repair mechanisms in the replication of genetic material is another [7].

Thus, our method to evolve modules at different levels will be to adjust the speed of evolution. The lower in the hierarchy, the less crossover and mutation events will hit them. In a nutshell, higher level modules can be discerned from lower level modules by their larger speed of evolution. Interestingly if we turn this argument around, another observation in Nature seems to fit in very well with this picture: Higher level modules, i.e. modules commanding higher complexity must be faster in evolution if there is no way to reduce evolution speed in lower levels, i.e. to stabilize developments there.

The generation of a lower level module in hGP is done during the evolution of the higher level individual: After crossover, modules are identified in the best individuals of a population only. Modules are formed by search for valuable sub-trees in these individuals. The general method for finding valuable subtrees is to compute the differential fitness [16, 18] with and without the subtree under discussion. Ranking selection is then applied to identify the best subtrees and to generate a module for the next lower level. Various parameters determine this procedure, like e.g. maximum number of modules per individual, maximal depth for computation of differential fitness, etc.

Since on the lower level evolution should progress, too, a fitness must be assigned to each of the newly created modules. In hGP, the fitness of a module is exactly the same as the fitness of the individual which is calling it in the next higher level. Thus, a good program will automatically transfer its high fitness to the module used by it.

Crossover and mutation on lower module levels work similar as on higher levels. hGP also allows different variants, e.g. based on homology and quality of subtrees. hGP was implemented as an extension of gpc++0.4.

In pseudo code the algorithm executed for each generation in hGP reads:

```
FOR level := 0 TO maxLevel DO
  DO popSize(pop[level])*evolutionSpeed[level] TIMES
    (mum, dad, child) := Selection(pop[level])
    Crossover(mum, dad, child)
    Mutate(child, mutationStrength[level])
    IF level < maxLevel - 1
      ModuleList := searchModules(child)
      AddModules(pop[level], ModuleList)
    FI
  OD
OD
```

3 First Results with Hierarchical Genetic Programming (hGP)

For the following experiments hGP has been substantially restricted. Two variants referred to as *hGPminor* and *hGP* are tested with the following restrictions:

- Number of modular levels: Only level 1 modules allowed
- Number of modules per calling individual: Only 1 module allowed
- Mutation only on highest level allowed

In *hGPminor* evolution on module level is not allowed. In this case the generation of modules works mainly as a protection of valuable code agains mutation and crossover. In *hGP* evolution on the module level is allowed. The settings for the evolution on the module level are:

- Crossover variant: replace a bad subtree by a randomly selected subtree
- Crossover probability on module level: 33% (i.e., evolution speed on module level is 1/3).

Surprisingly, the crossover variant "replace a bad subtree by a good subtree" has led to significantly worse results. Experiments have also confirmed that hGP is robust concerning the setting of the crossover probability on module level (up to 50 %).

We studied six test problems that have been used to compare the performance of hGP with standard GP: 4 continuous problems from function regression and two instances of the discrete even-N-parity problem [9] with $N = 5$ and $N = 7$. The time measurement performance was based on the number of node evaluations, not on the number of individuals or generations evaluated (see [4]).

Problem	Type	Symbol	Regression function
1	continuous	f_1	randomly selected y-values
2	continuous	f_2	steps
3	continuous	f_3	$x^6 - 4x^5 - 3x^4 + 4x^3 - 2x^2 - x + 4$
4	continuous	f_4	$\frac{x^3-x^2-x+3}{x+\frac{5}{9}}$
5	discrete		even-5-parity
6	discrete		even-7-parity

Even-5-parity, even-7-parity and regression on f_4 have been used during the development process of hGP and extensive experiments have been carried out based on these problems [3]. Regression problems on f_1, f_2, and f_3 are used after development of hGP for validation. Table 1 gives the run parameters. Due to a lack of space we report here only on the even-7-parity problem. The reader may compare [4] for the performance on other problems.

We compared standard GP, i.e. Genetic Programming without modules, and hGP, without (hGPminor) and with evolution on the module level. In preliminary experiments a reduction of evolution speed to about 1/3 that at the level of individuals turned out to be efficient, although different applications shall require different module evolution speed. In another application, we were successful with a speed of 1/10 of that of the higher level programs [13]

4 Discussion and Conclusion

Figures 2-5 show the performance of standard GP, hGPminor and hGP for the test problems 6 (even-7 parity). In addition to best, average, and worst fitness

Parameter	Setting
population size	3 000
selection	(10,1)-tournament
generation equivalents	100
crossover-frequency on top level	100 %
crossover-frequency on module level	0 % (hGPminor) 33 % (hGP)
mutation-frequency on top level maximum tree depth	2 % 17
maximum initial tree depth	6
initialization	ramping half and half
maximum number of modules per individual	1
problem	even-N-parity, N = 7
raw fitness	$\phi = 100*$ (number of mismatches)
parsimony term	$100 \cdot (1 - \frac{10}{10+\kappa(a_i)}$
terminal set	$T = \{D_0, D_1, ..., D_N\}$
function set	$F = \{AND, OR, NAND, NOR\}$
termination-criterion	exceeding the maximum number of generations

Table 1: The Koza tableau of parameter settings for the even-N-parity problem in hGP. Comparison with standard GP containing no modules. $\kappa(a_i)$ is the expanded structural complexity of the individual a_i [15].

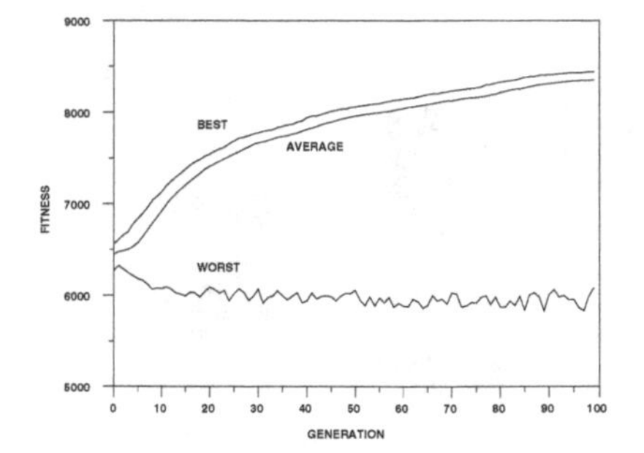

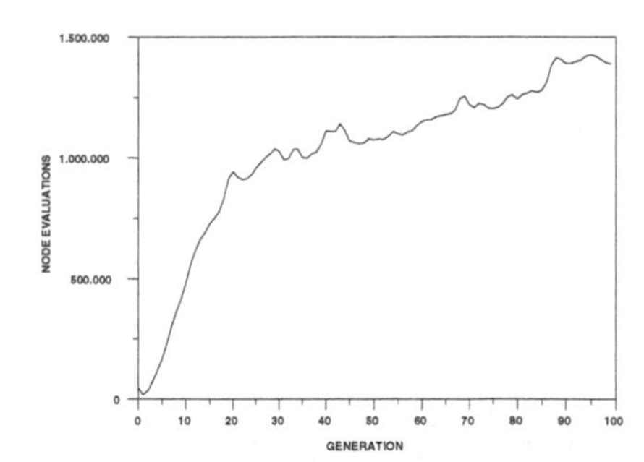

Figure 2: Even-7-parity Standard GP. Left: Best, average and worst fitness over time. Right: Number of node evaluations accumulated over time. Average of 50 runs.

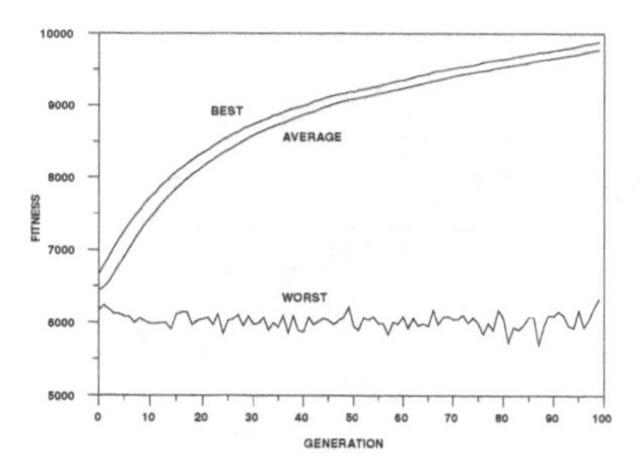

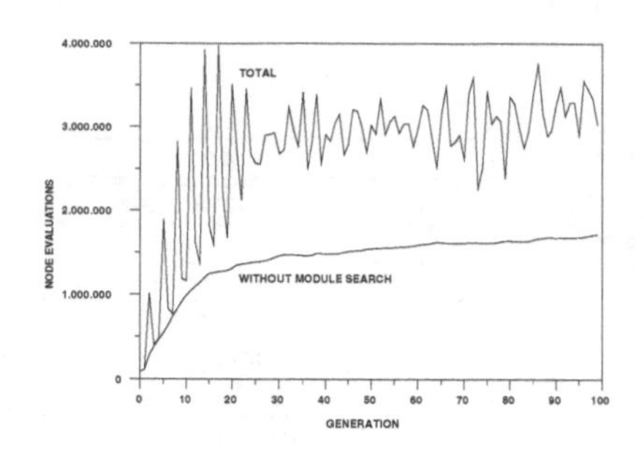

Figure 3: Even-7-parity hGP minor. 50 runs. Left: best, average, and worst fitness over time (measured in generation). Right: node evaluations per generation. Lower curve shows the node evaluation needed only for fitness evaluation. The upper curve shows the node evaluation needed for fitness evaluation and module search. The area between the upper and the lower curve represents the additional effort which is spent for searching good modules.

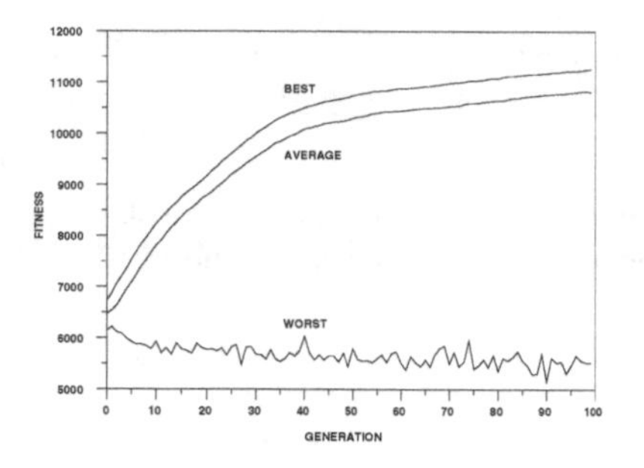

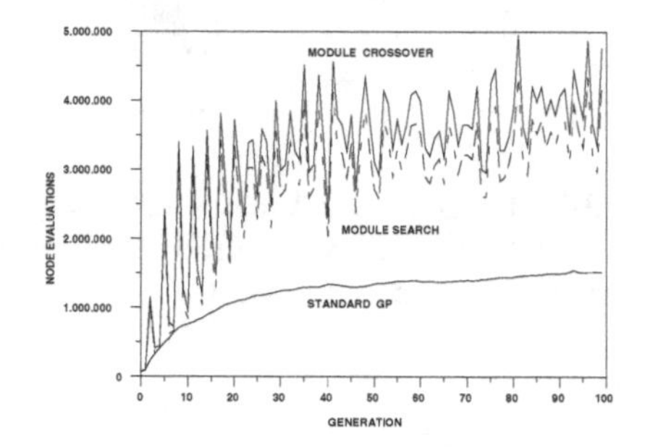

Figure 4: Even-7-parity hGP. 50 runs. Same as above.

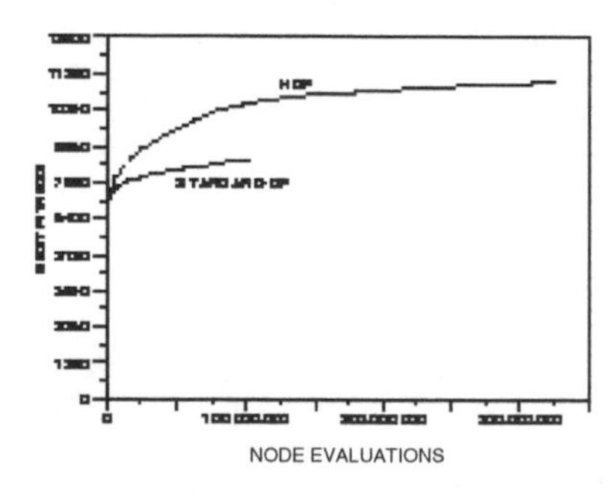

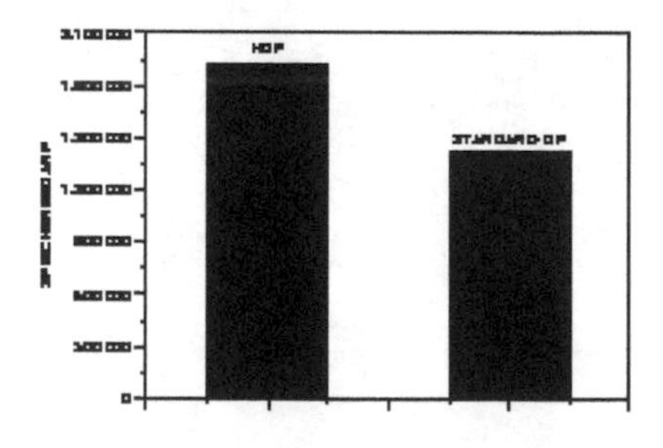

Figure 5: Even-7-Parity, GP vs hGP. 30 runs each. Left: Fitness progress over number of node evaluations. Right: Memory consumption compared between standard GP and hGP.

the nodes evaluated per generation is depicted in the right figures. It can be seen that hGPminor outperforms standard GP and hGP outperforms hGPminor. The number of nodes evaluated per generation increases which reflects the fact that average individual length is growing. The growing process is bounded because a parsimony pressure is activated. Note, that the pasimony pressure is very weak. For the parity problem it has only an effect if two individuals represent exactly the same function.

For a fair comparison of convergence speed in Fig. 5 time is now measured in node evaluations. For all 6 test problems (5 not shown here) hGP outperforms standard GP [4]. The performance gain depends on the problem. Figure 5 compares also the memory consumption of standard GP vs. hGP. In general, hGP does not consume significantly more memory than standard GP. In many cases its memory consumption is even considerably smaller. Even-7 parity is an exception from this rule.

hGP shows good performance even when a more detailed time model – number of node evaluation – is applied. The performance gain is based on efficient module search techniques which are based on the differential fitness calculated by replacing the designated module by a neutral structure. Whether a larger number of levels increases the performance of hGP is still an open question and should be a subjects for future investigations.

ACKNOWLEDGMENT

Support has been provided by the Deutsche Forschungsgemeinschaft within the Sonderforschungsbereich 531, project B2.

Bibliography

[1] ABELSON, A., and G. SUSSMANN, *Structure and Interpretation of Computer Programs*, MIT Press Cambridge, MA (1985).

[2] ANGELINE, P. J., and J. B. POLLACK, "The evolutionary induction of subroutines", *Proceedings of the Fourteenth Annual Conference of the Cognitive Science Society*, Lawrence Erlbaum (1992).

[3] BANSCHERUS, D., "Hierarchische Genetische Programmierung mit lokalen Modulen" (1998), Diploma Thesis.

[4] BANZHAF, W., D. BANSCHERUS, and P. DITTRICH, "Hierarchical genetic programming using local modules", Technical Report *50/98*, University of Dortmund, Dortmund, Germany (1998).

[5] BANZHAF, W., P. NORDIN, R KELLER, and F. FRANCONE, *Genetic Programming — An Introduction*, dpunkt/Morgan Kaufmann Heidelberg/San Francisco (1998).

[6] FOGEL, D., *Evolutionary Computation*, IEEE Press Piscataway, NY (1995).

[7] FRIEDBERG, E., G. WALKER, and W. SIEDE, *DNA Repair and Mutagenesis*, ASM Press New York (1995).

[8] GRUAU, F., "Genetic synthesis of modular neural networks", *Proceedings of the 5th International Conference on Genetic Algorithms, ICGA-93* (University of Illinois at Urbana-Champaign,) (S. FORREST ed.), Morgan Kaufmann (17-21 July 1993), 318–325.

[9] KOZA, J., *Genetic Programming – On the Programming of Computers by Means of Natural Selection*, MIT Press Cambridge, MA (1992).

[10] KOZA, J., *Genetic Programming II*, MIT Press Cambridge, MA (1994).

[11] KOZA, J., D. ANDRE, F. BENNETT, and A. KEANE, *Genetic Programming III*, Morgan Kaufmann San Francisco, CA (1999).

[12] KOZA, J., F. BENNETT III, D. ANDRE, M. KEANE, and F. DUNLAP, "Automated synthesis of analog electrical circuits by means of genetic programming", *IEEE Transactions on Evolutionary Computation* **1**, 2 (July 1997), 109–128.

[13] OLMER, M., W. BANZHAF, and P. NORDIN, "Evolving real-time behavior modules for a real robot with genetic programming", *Proceedings of the international symposium on robotics and manufacturing* (Montpellier, France,), (May 1996).

[14] OSAWA, S., *Evolution of the Genetic Code*, Oxford University Press Oxford (1995).

[15] ROSCA, J., "An analysis of hierarchical genetic programming", Technical Report *566*, University of Rochester, Rochester, NY, USA (1995).

[16] ROSCA, J., and D. H. BALLARD, "Evolution-based discovery of hierarchical behaviors", *Proceedings of the Thirteenth National Conference on Artificial Intelligence (AAAI-96)*, AAAI / The MIT Press (1996).

[17] ROSCA, J. P., and D. H. BALLARD, "Learning by adapting representations in genetic programming", *Proceedings of the 1994 IEEE World Congress on Computational Intelligence, Orlando, Florida, USA* (Orlando, Florida, USA,), IEEE Press (27-29 June 1994).

[18] ROSEN, R., *Life Itself*, Columbia University Press New York (1995).

Dealing With Complex Human Systems From An Interpretive Systemic Perspective[1]

Hernán López Garay and Joaquín Contreras
Departamento de Sistemología Interpretativa
Escuela de Sistemas,
Universidad de Los Andes, Mérida, Venezuela
hlopezg@ula.ve

1. TWO NOTIONS OF COMPLEXITY

The investigation of the nature of complexity and how to deal with it has been a major focus of research in the systems community. Joining this stream of research, we would like to introduce in this paper a new way of looking at the issue of complexity and illustrate briefly with a case study how to approach it. We will start by stating what we consider is the predominant notion of complexity and the constellation of concepts associated with it (concepts such as: system, emergence, environment) and then proceed to do something similar with the second notion. The contrast between the two will help us to understand better each conception of complexity.

1.1. Organismic Complexity

Complexity is usually defined in terms of the number of system's components and their interrelationships. However, when systems are seen from an organismic perspective (i.e., systems are like organisms in a given environment) is the complexity of system's behavior in a given environment, what is important to model. Accordingly, considerations such as whether the system's components are dynamic or static, goal-seeking or adaptive, and whether the system is in a turbulent or calm environment, are vital to understand system's behavior. Then, our first notion of

[1] We would like to acknowledge the financial support given to the first author by the New England Complex Systems Institute (NECSI) and the Consejo de Desarrollo Científico Humanistico y Tecnológico (CDCHT), to attend the International Conference on Complex Systems organized by NECSI. This conference provided a rich environment for the present paper.

complexity has to do with the *amount of information necessary to describe the behavior of the system* (i.e., number of internal states, inputs, outputs, the environment's states, etc.). Hence, complex systems are not necessarily systems that have a large number of components and interrelationships but those which require more information to characterize their behavior.

1.2 Interpretive Complexity: A Conceptual Framework[2]

1.2.1 A Phenomenological Conception of Systems

A second notion of complexity arises from a change in the way we conceptualize systems. Suppose that we stop thinking that: A. Systems are units independent from the observer and composed by a set of interrelated parts from which the system's properties and behavior *emerge*. B. Systems are hierarchically organized; C. Systems are adaptive wholes immersed in environments with which they trade resources and information. Suppose, further, we admit just the opposite: Systems are neither purposeful nor purposeless. Systems are acts of *distinction*. The act of distinction is the act of drawing the boundary of the system. In such an act, an "outside" and an "inside" are created. Equivalently, a figure and its background are created. Drawing the boundary constitutes a *system-background unity*. Moreover, the boundary is a boundary for someone. This means that drawing the system's boundary brings forth *both, the system and the observer of the system*. Thus, a *system-observer* unit is created. It can be shown [***SP***, pp.449-490; López-Garay 1986] that the observer is the ground for the system and vice versa. According to these assumptions, systems are more like a *figure-background* unity than like sets of interrelated parts, hierarchically organized with communication and control processes and immersed in an environment. Note this is not simply a change of labels. In other words, we are not merely replacing the notion of, say, system's environment with that of *background*. "Environment" is something independent from the system and is composed of other systems and entities. On the contrary, the "background" of a figure is the undifferentiated *ground* at the *back* of the figure which is both constitutive of the system and constituted by it. In fact, if we try to examine and articulate the background, we immediately discover all we can say about it is that it is the ground of this figure. Nevertheless, if we insist, we find that we cannot point to anything concrete without turning the background into a new distinction or figure, which again has a background, and so on, and so forth. This characteristic of the background we call it its "unindicability".

Now, how would one approach systems or phenomena thus conceived? Certainly, the basic idea is not to decompose it into a figure and a background and then proceed to analyze each of them. This would destroy the essence of the phenomenon we want to study. On the other hand we have to deal with the background in some way, in order to understand the mutual constitution figure-background. In [***SP***, 473-490], and [López-Garay 1986] have been shown that it is

---

[2] The following sections of this paper are based, mainly, on notions introduced and developed in ***Systems Practice***, Volume4, Number 5, October 1991, and ***Systemic Practice and Action Research***, Volume 12, Number 1, February, 1999. We will refer to these particular journals and issues as ***SP*** and ***SPAR*** respectively To avoid repetition, the main authors to be cited from these journals are from ***SP*** (Ramsés Fuenmayor) and from ***SPAR*** (Ramsés and Akbar Fuenmayor, Hernán López-Garay and Tomasz Suárez).

possible to increase one's awareness of some relevant regions of the background (i.e., without turning it into a new figure) by means of an interpretive method in which designed *contexts of meaning* (or interpretive contexts) are contrasted with each other[3]. The idea here is that a figure-ground relationship is similar to a *word-con-text* relationship. To understand the meaning of a word we have to remember that a word becomes a word (i.e., means something) only within a text (e.g., "hero" in the Iliad means something different than hero in a Batman's comic strip) or a context (e.g., "kill him" in a football game does not mean the same than in a war). Equally, a better comprehension of the word is obtained if we unfold its use in different contexts (i.e., if we begin to see the unity underlying the different word's meanings).

We shall call *systemic understanding* the outcome of building these contexts of meaning and *interpreting* in each of them the phenomenon under study. By *interpreting* we mean the investigation of what sense the phenomenon makes when "placed" in each interpretive context (recall our metaphor word-con-text). The purpose of these theoretical tools and the unfolding of different interpretations is not simply to explore different regions of the background but also to understand how they form together with the system a unity. The variety of interpretations unfolded, tell us something about the figure-background unity. However, the study does not finish here. Recall that every system study is seeking to understand the holistic or unitary (emergent?) character of the phenomenon under study. Therefore, the next step is to *search for this unity in the diversity of boundary drawings unfolded.* This is done by orchestrating a debate between these various interpretations. The outcome of this debate is a clearer sense (not necessarily a tangible product) of a possible answer to the question: What is the unity underlying the different interpretations? Equivalently, it is an answer to the question of emergence: what emerges is precisely a better comprehension of the system's unity. The answer to this question is never a tangible product because, as it can be shown, the background is inexhaustible [***SP***, pp.473-490; López & Suárez in ***SPAR***]. We shall call this second intangible outcome *systemic comprehension.*

In sum, the second notion of system described above is strongly related to a figure-background idea of *wholeness* or *emergence*. Following, we will expand a little further these concepts and show the notion of complexity that arises from them[4].

1.2.2. The *Historical* Nature of the Background and its Relation to Individuals and Society

So far, our description of a new way of understanding systems has made little reference to how the notion of *system-background* is related to the notion of *system-observer*. In other words, we have not dealt with the concrete realities of the situation which involves the act of drawing a system's boundary. In particular, we have said nothing about the human individual and his/her *involvement* with circumstances. Let us explain. We have said that every distinction brings forth not only that which is to

---

[3] Formally, we call *context of meaning* or *interpretive context (see **SP**,* pp.485-487*)* a coherent theoretical structure centered around a fundamental idea (*generative concept, seminal idea* or *geno-idea*) from which the rest of the structure is derived in logical terms (which is the reason why we also call them *logic-based* contexts of meaning. More about it later).

[4] Similar ideas are beginning to take root in some streams of the systems movement. See for example [Senge, 1998] and his concept of "wolism". Also see [Flood, 1998, pp.2-3].

be "observed" but also the "observer". But who or what is the observer? Certainly, it is not a disengaged being watching reality from some privileged, neutral stand. The observer is born with every distinction and hence is engaged from the very first second of its existence to what is distinguished. Moreover, the observer's engagement transcends the particular physical situation within which it emerges. For instance, as [Schrag 1969] has pointed out, the events in the life of a person are always positioned against a background of remembered happenings (p.30). This means that the subject is constituted historically in at least two senses. One sense is the personal events, what one might call its personal history. The other sense is related to cultural events. Because of this, **"...the experiencer is already lodged in history and in some manner and degree constituted by it." (ibid., p.207). What is distinguished occurs within a context in which some sense of the *historical* is already in force (p.31, italics added).** Notice, however, that Schrag's conception of history and past is not that of a collection of fixed and structured events (i.e., the background of a distinction is not a set of recollections waiting to be retrieved from memory). The notion of past ventilated here is akin to that of a sediment of lived experiences which is structured in every present, according to (among other things) the cultural and historical conditions. In sum, there is no distinction ---no drawing of a system's boundary--- without bringing forth the "ground" of a personal and cultural context. It is in these contexts which the distinction is placed. In this sense we can say that what is distinguished *is* an *interpretation* i.e., it is not something that stands by itself (as words do not stand by themselves but depend on the contexts in which they appear). Its meaning changes with the context.

1.3. *Interpretive* Complexity

What notion of complexity can we derive from this *interpretive* conception of systems? Let us recall the roots of the word complexity. It has to do with that which resists being embraced or encircled. In fact, complexity comes from the Latin *complexus* which means encircling, embracing. Its common use, however, has qualified this meaning as synonymous with what is difficult to understand (stand-under!). As we saw above, the background is elusive, it can never be made explicit, i.e., we can never embrace or encircle it completely. Yet it is there, it lurks in the back-ground of whatever is the case, not as something fixed but as something whose character is that it is always undetermined. As mentioned in previous sections, in an interpretive systemic study one has to disclose the background, through the fabrication of different interpretive contexts. This research activity generates a number of interpretive contexts which unfold the object of study from different perspectives. However, in so doing, we generate a new problem, namely, the problem of how to gain some sense of unity from such diversity of views unfolded. "Embracing" this interpretive variety is a sort of "complexity" different from that of understanding the behavior of the system in terms of its states, inputs, outputs, etc. We may call it *interpretive* variety.

1.4. Two Levels of Complexity

Notice we have here a different type of complexity. *Organismic* complexity has to do with the organismic variety of system's components and their interrelationships, both among themselves and with the environment. "Embracing" this variety of components and interrelationships means to be able to find their underlying unity or *function*. In the case of *interpretive variety* the idea is to find the unity underlying the *variety of interpretations*. Notice that the two types of complexity are related thus. Every distinction (i.e., every interpretation) gives rise to a system's boundary and its environment. Further distinctions give rise to the system's components and their interconnections. In this sense, organismic complexity is first level complexity whereas interpretive complexity is second level.

Before going on and illustrate with a case study how to deal with organismic and interpretive complexity in human systems, let us present one more concept we need in the conceptual framework which supports our notions of systems and complexity. It is the concept of *social reality*. We will use the metaphor of an iceberg to explain this notion. The metaphor will serve also to illustrate the organismic conception of social reality. The contrast between these two conceptions of reality will help to clarify further our interpretive conception of the world and how to conduct research on complex human systems.

2. THE ORGANISMIC AND THE INTERPRETIVE CONCEPTIONS OF SOCIAL REALITY: THE *ICEBERG*

Let us imagine that social reality can be represented by means of an iceberg. At the tip of the iceberg we have the concrete *manifestations* of human action. The second layer, that of patterns, rules and social practices in general is not visible, although it is very close to the surface. The third layer, is interpreted differently by organismic and interpretive approaches. To the former, this layer is the *structural* layer; i.e., that layer, which exists apart from the researcher, and that generates systems behavior [Jackson, 1992]. On the contrary, for the latter, this is the *constitutive-meaning* layer, i.e., the fundamental "ground" which breathes meaning into social practices and human action. In either case, this layer is the deepest most hidden part of the iceberg.

Accordingly, an organismic systems study requires to start at the tip of the iceberg, where the *phenotypical* manifestations of the phenomenon appear, and then proceed to discover and formalize its *geno-structure*, i.e., that deep, fundamental structure which generates the phenomenon's visible manifestations. When we imagine how one could apply this approach to the study of the prison system in Venezuela, then we see that we could start describing the phenotypical manifestations (e.g., riots, violation of human rights, high rates of recidivism, etc.). Next, we could continue our investigation unfolding the cultural patterns and social norms which make possible such manifestations. Finally, we would go deeper, looking for the underlying, permanent structures (and their mathematical formalization) which explain high recidivism, violent behavior, etc. But the interpretive systemic approach differs from this interpretation of the iceberg, particularly in relation to the third layer. One difference lies in that we are not after any permanent, fundamental structure (there is none!), nor are we looking for some laws which explain the transformation from one structure to another. In effect, as we disclose the background and interpretive variety emerges, we find that there is no system's structure waiting to be discovered by the researcher. This does not mean that an interpretive systemic researcher does not make use of formal interpretive contexts (i.e., theoretical constructions) which could be seen as "tools" to *freeze*, temporarily, the background in some form of "structure" (i.e., meaning structure). *Nevertheless, their real use is to help us become aware of precisely the impossibility of freezing the background in one single structure. By articulating the background in several contexts of meaning, we realize how limited are these contexts and how enigmatic is the background and the system's unity*. This interpretation is then closer to what [Piaget, 1970, section 21] calls *structuralism without structures*. But there is also another reason why to attach this label to this approach. The latter is beginning to unravel what it means to say that social phenomena are historically constituted. [Fuenmayor, A & Fuenmayor, R., in ***SPAR***]'s notion of *history* clarifies this matter. *History* is the narration of **"...the series of cultural contexts that have led us to experience reality....in the way we do at present." (ibid., section 3.2)**. In other words, *history* is the narration of how we have arrived at our current major cultural context of meaning, the *Weltanschauung* (in Dilthey's sense) which orders and gives meaning to our past, present and future. It could be seen as some sort of *historical a priori*, inasmuch as these basic contexts of meaning are like conditions of possibility of the constitutive meanings of a culture. Accordingly, they form and shape social action and are embodied in it (i.e., they do not exist independently from social practices. They manifest themselves only in concrete specific situations). However, unlike the Kantian *a priori* of experience, they apply only for a limited period, hence they are neither a-historical nor universal. Moreover, these historical a priori are not affiliated with another in causal or "genetical" terms. Hence we cannot talk of a historical law that can predict their changes. Nor can we talk about some transcendental onto-epistemological structure underlying them. Therefore, what is most basic is not a fixed structure but a fluid narrative, the "history" of a continuos becoming which orders and gives sense to what we are and what we do. In sum, we humans are constituted by some form of *historical a priori*. Therefore, and contrary to classical scientific research, our inquiries are not guided by the principle of "objectivity", if one understands by it, the effort to get rid in our investigations from our "subjectivity" (precisely our and the object's historical constitution). What we are after is to bring forth such a "subjectivity" and show that it is precisely "it" that makes possible any knowledge or experience in a particular situation.

3. DEALING WITH SECOND-LEVEL COMPLEXITY: AN INTERPRETIVE SYSTEMIC STUDY OF PRISON SYSTEMS IN LATIN AMERICA[5]

3.1. The Venezuelan Prison: A "narrow" opening of the "Background"(Logic-based contexts)

Now, let us see how the above conception of social reality is investigated in our study of the prison. We also start at the tip of the iceberg, with the phenomenal manifestations of the prison. We notice, first, that the complexity of the prison system in Venezuela arises not only because the prison is seen as belonging to a

---

[5] Although we will deal mainly with prisons in Venezuela, there is plenty of empirical evidence that prisons in Latin America are very much similar institutions.

complex set of institutions dealing with justice, but also because in the Venezuelan cultural context the prison is opened to a variety of interpretations, some of which display their paradoxical character. For instance, formally speaking, according to the Venezuelan constitution, prisons are in the "business" of rehabilitation. This means that prisons should play an educational role in society, making sure prisoners are transformed in useful members of society. But this is not happening. Prisons are ridden with frequent brutal and bloody riots, continuous violation of the prisoners' human rights, widespread corruption of the judicial system, and other ailments.

What is the "background" on which such *distinctions* about prisons are made in the Venezuelan society? Our first attempt to disclose the background of prisons demanded questions such as: are prisons centers of punishment or rehabilitation? If both, how does punishment and rehabilitation mix together? Why is there such an abysmal distance between the official discourse on prisons and actual fact? To pursue these questions, we built what we call *logic-based* interpretive contexts. These contexts have the form of a "theory", with a central concept or, as mentioned before, a *geno-idea* (e.g. "punishment") developed in a rational conceptual system bonded together exclusively by logical relationships. In this connection we design modern and post-modern contexts [López-Garay; Suárez, in ***SPAR***]. Our aim was to see how our current social order (as it is currently perceived) "fits" in these contexts. The outcome was appalling since we seem to view prisons as centers to inflict pain and continuously violate human rights. Such a behavior is in open contradiction with basic principles of modernity and also postmodernity. However, could it be that this behavior corresponds to some sort of special social punishment proper of these contexts? But then, how have we come to see and administer social punishment in this way? What about the Western cultures, which are the creators of modernity, why their conception of social punishment has not become like the Venezuelan? The latter questions are leading the way to a "historical" opening of the background.

3.2. A "wide" opening of the "Background": Towards a *History* of Punishment in Venezuela

Following we summarize the first steps we gave to answer these questions taking into account the historical nature of the background (see [Contreras 1998] for a full account). Considering that Venezuela has transplanted the prison institution from Western cultures, our research task was aimed at building a *historical* account of how prisons came to be seen and experienced as the dominant mode of punishment in Western cultures, and then examine our historical records to see if our institutions could be interpreted in a similar fashion. The differences between the Western's and our institutions will indicate where and to which extent we need to reform the western historical context in order to make sense of our own historical situation. The work that guided our research was Michele [Foucault's 1977] classical study of the rise of the prison (as the dominant mode of punishment) in France.

A main outcome of our research was that while prisons in France may be seen as fulfilling the role of masks in a disciplinary society[6], the Venezuelan prisons,

---

[6] Foucault shows that the continuos failure of prisons in France (e.g., high percentages of recidivism) makes sense in a disciplinary society. In effect, prisons are the only institutions that have as an explicit purpose to "discipline" people. Its

on the contrary, are rather like *mirrors which reflect the contradictions of modernity.* In effect, prisons in Venezuela are *two-faced.* One face is modern and is manifested not only in the formal statements and plans of the state, where it is stated that their mission is to rehabilitate the prisoners, but also in the country's constitution. The other face is continuously subverting this formal mission. How is this possible? As Foucault, we wanted to find out what is the "logic" of Venezuelan prisons, i.e., what makes them possible the way they are. A new light was brought up to this problem when we realized that such "strange" behavior was happening not only with prisons but with a whole set of Western institutions transplanted to Venezuela since the fifteen century onwards. The fact is that this whole project of transplanting modernity was from the very beginning contradictory. If as [Fuenmayor, A & Fuenmayor, R., in ***SPAR***] have argued, the essence of modernity was not only the liberation from tradition but the *will to invent* a new social order, then it goes against its own essence the attempts to transplant (i.e., make a copy of) the products of this process (i.e., its institutions and social order in general). Hence, we need to build a context of meaning that can help us understand how have we come to appreciate the world as "moderns" of a particular kind, namely, as people tutored by the real modern ones (i.e., the Western cultures)! It is in the light of this historical-ontological context that we have to re-open the background and start a new research cycle asking not only for the sense of our prisons in such a context but also for what could be done.

4. SUMMARY: ON DEALING WITH SECOND LEVEL COMPLEXITY

A new concept of systems complexity has been presented and its application briefly illustrated. Dealing with interpretive complexity requires a spiral-like process aimed at opening the system's background. Such a process is dialectic inasmuch as every opening brings forth interpretive variety which, consequently, forces us to "synthesize" it by searching its underlying unity. Therefore, the question may arise: Where does it stop this process? Furthermore, what is gained by this attempt to "complexify" the study of complex human systems? A simple answer is this: we gain a better understanding of the enigmatic nature of complexity itself! To say it otherwise: **"What is being gained, ultimately, is a two-sided experience: one side is composed of the *enigmatic variety* and *limitations* that the different ways of articulating the sense of a phenomenon have, confronted with, on the other side, the *enigmatic unity* and *limitlessness* of the *unarticulated* presence of that same sense."** [López-Garay & Suárez in ***SPAR***]. Inasmuch as the systems movement in general, and the study of complex human systems in particular have been focused mainly on what in this paper is called organismic complexity, and taking into account the relation of this notion to interpretive complexity, then we think our work can help not only to widen the notion of complexity but also to open new ways of dealing with it, particularly in the context of complex human systems.

REFERENCES

---

failure brings the attention of the public to this fact, thus keeping their eye away from the on going process of disciplinarization in society as a whole.

Contreras, J. J., 1998, El Sentido Histórico de la Prisión Rehabilitadora en Venezuela. M.Sc thesis, Graduate program in Interpretive Systemology, Universidad de Los Andes, Venezuela.

Flood, R. L., 1998, Editorial. *Systemic Practice and Action Research*, 11(1): 1-8.

Flood, R., Ulrich, W., 1990, Testament to conversations on critical systems thinking between two systems practitioners, *Systems Practice*, 3: 1.

Foucault, M., 1977. *Discipline and Punish. The Birth of the Prison*. Alen Lane, London.

López Garay, H., 1986, *A Holistic Interpretive Approach of Systems Design*, Ph.D. thesis, The Wharton School, University of Pennsylvania, Philadelphia.

Piaget, J., 1970, *Structuralism*. Harper, New York.

SP: *Ssystems Practice*, V.4, N.5, October, 1991

SPAR: *Systemic Practice and Action research*, V.12., N.1, February, 1999.

Senge, P., 1990, *The Fifth Discipline*, Doubleday, New York.

Senge, P., 1998, Some Thoughts at the Boundaries of Classical System Dynamics: Structuration and "Wholism". *Proceedings* of the Sixteenth International Conference of the System Dynamics Society, Quebec, Canada.

Schrag, C., 1969, *Experience and Being*, Northwestern University Press, Evanston.

Connected Science: Learning Biology through Constructing and Testing Computational Theories

Uri Wilensky
Center for Connected Learning and Computer-Based Modeling
Northwestern University
uriw@media.mit.edu

Kenneth Reisman
Department of Philosophy
Stanford University
kreisman@csli.stanford.edu

1. Introduction

There is a sharp contrast between the picture of the field of biology as studied in school settings and the picture that emerges from the practice of current biology research. While the two pictures are linked by similar content and the objects of study are recognizably the same, the *processes* involved in the two activities are quite different.

In school settings, typical instruction emphasizes the memorization of classification schemas and established theories. In middle school, classification may take the form of learning the names of the bones of the body, the names and shapes of different plant leaves or the phyla in the animal kingdom. In high school and early undergraduate studies, the content broadens to include unseen phenomena such as parts of the cell or types of protozoa, but the processes of memorizing classifications remains essentially the same. Even in cases where the theories are not yet established, such as the extinction of the dinosaurs, the alternative theories are presented as competing stories to be memorized. And even when students are exposed to research techniques in laboratory work, the emphasis is on following a prescribed procedure rather than reasoning from the evidence gathered in the procedure.

This picture contrasts sharply with the picture that emerges from the recent biology research literature. In this picture, the participants are active theorizers. They devise new evidence gathering methods to test their theories. Instead of accepting classifications as given, they are seen as provisional theories that are constantly reassessed and reconstructed in light of the dialogue between theory and evidence. They reason both forwards, by constructing theories that are consistent with the known evidence and backwards by deducing consequences of theories and searching for confirming/disconfirming evidence. In constructing or assessing an account of a biological phenomenon, they focus on the plausibility of the *mechanism* proposed—can it achieve the task assigned it in a biologically feasible manner? This assessment of the mechanism often involves reasoning across a range of levels—they ask: is the mechanism constrained by the structure at the molecular, the cellular, the organismic and/or the ecological level?

The contrast between the processes in which these two communities are engaged leads biology students to form a misleading picture of the biological research enterprise. Students form beliefs that biology is a discipline in which observation and classification dominate and reasoning about theories is rare. Furthermore, they believe that learning biology consists of absorbing the theories of experts and that constructing and testing their own theories is out of reach[1]. In this paper, we present an approach that attempts to narrow the gap between school biology and research biology. The approach centers on the use of innovative computer modeling tools that enable students to learn biology through processes of constructing and testing theories.

In recent years, several educational research projects (Jackson et al, 1996; Ogborn, 1999; Roberts et al, 1983, 1988) have employed computer modeling tools in science instruction. The approach taken herein differs from these approaches in its use of object-based modeling languages that enable students to model biological elements at the level of the individual (e.g., individual wolf/sheep) as opposed to aggregate (differential-equation based) modeling languages that model at the level of the population (wolf/sheep populations). This technical advance in modeling languages enables students to employ their knowledge of the behavior of individual organisms (or molecules, cells, genes...) in the construction of theories about the behavior of populations of organisms. They embody their theories of individual behavior in a computational agent. This ability to model individual behavior enables students to employ their personal experience with sensing and locomoting in the world as initial elements in their models of other organisms.

In previous work, the authors and other object-based modeling projects (Repenning, 1994; Resnick, 1994; Smith et al, 1994; Wilensky, 1995; 1999; Wilensky & Resnick, 1999) have described the "embodied modeling" approach in a

[1] For an extended account of the gap between school and research in the domain of physics, see (Hammer, 1994).

broad inter-disciplinary context. In this paper, we explore the use of this approach, specifically, in biology instruction. We begin, in the following section, by describing our embodied approach to biological modeling and the object-based parallel modeling language, StarLogoT[2] (Wilensky, 1997a), in which the models are constructed. In section three, we illustrate this approach by developing embodied models of predator-prey population fluctuations. In section four, we develop a computational model of synchronously flashing fireflies to frame a discussion of the student modeling process and the relationship of this process to modeling within science. Finally, in our concluding remarks we summarize the major points of the paper.

2. The StarLogoT Modeling Language

StarLogoT is a general-purpose (domain independent) modeling language that facilitates the modeling of complex systems. It works by providing the modeler with a framework to represent or "embody" the basic elements—the smallest constituents—of a system, and then provides a way to simulate the interactions between these elements. With StarLogoT, students write rules for hundreds or thousands of these basic elements specifying how they should behave and interact with one another. These individual elements are referred to as "turtles". (StarLogoT owes the turtle object to the Logo computer language). Turtles are situated on a two dimensional grid on which they can move around. Each cell on the grid is called a "patch", and patches may also execute instructions and interact with turtles and other patches. Some typical commands for a turtle are, to move in a given direction, to change color, to set a variable according to some value, to "hatch" new turtles, or to look at the properties (variables) of other turtles. Turtles can also generate random values, so that they can, for example, execute a sequence of commands with a fixed probability. Patches can execute similar commands, though they cannot change location. The wide range of commands executable by turtles and patches makes it possible to use them to represent many different systems. For example, turtles can be made to represent molecules, cells, or individual organisms, while patches represent the medium (whatever it may be) in which they interact. Time in StarLogoT is represented as a discrete sequence of "clock-ticks". At each clock-tick, each turtle and patch is called upon to execute the rules that have been written for it.

The modeling approach we describe—instantiating the individual elements of a system and simulating their interactions—is not unique to StarLogoT. Such models have been used across a wide variety of domains and have been referred to by several different labels, including: object-based parallel models (Wilensky, 1995; 1997b), agent-based models (Beer, 1990; Maes, 1990; Repenning, 1993; Epstein & Axtell, 1996), and individual-based models (Huston, 1988; Judson, 1994). These "new wave" modeling approaches have transformed biology research practice enabling researchers to model increasingly complex multi-leveled biological systems (Forrest, 1989; Langton, 1993; Keen & Spain, 1990; Taylor et al, 1989). For the remainder of this

[2] The StarLogoT modeling language and associated materials can be downloaded from: www.ccl.sesp.northwestern.edu/cm/ .

paper, we will employ the term "embodied modeling" to refer to this general approach. While the term, "object-based parallel modeling", which we have used in the past is, perhaps, a more accurate description of the technical workings of StarLogoT, the "embodied modeling" label more closely matches the experience of a biology modeler who is actively engaged in understanding and embodying the behavior of individual biological elements.

In the following two sections of the paper, we will illustrate the embodied modeling approach in biology with two examples of modeling biological phenomena. We intend these examples to illustrate both how such an approach can 1) facilitate the creation of predictive multi-level models in biology and 2) enable biology students to create more powerful explanations of and deepen their understanding of biological phenomena.

3. Modeling Predator-Prey Population Dynamics[3]

The dynamics of interacting populations of predators and their prey have long been a topic of interest in population biology. Comparisons of a number of case studies have revealed similar dynamics between such populations, regardless of the specific species under study and the details of their interactions (Elton, 1966). Notably, when the sizes of the predator and prey populations are compared over many generations, we tend to find regular oscillations in these sizes which are out of phase; where one increases, the other tends to decline, and vice-versa (figure 1). In this section, we will present embodied models, developed by students, that reproduce these dynamics.

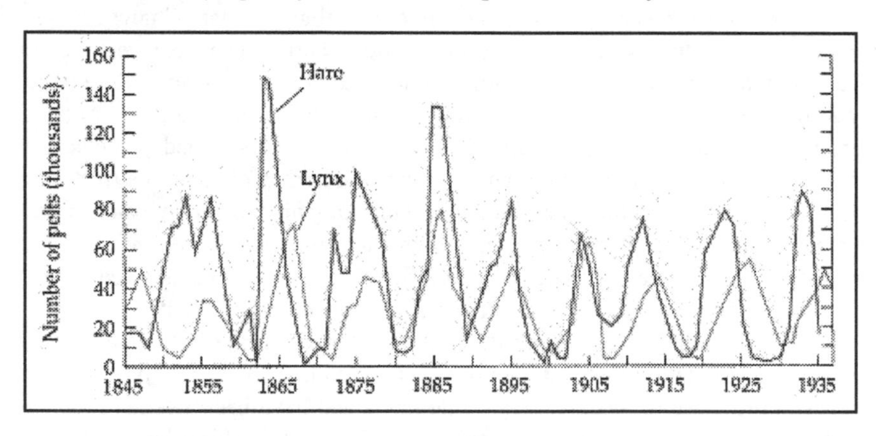

Figure 1. Fluctuations of the sizes of predatory lynx and prey hare populations in Northern Canada from 1845-1935 (Purves et al, 1992).

3.1. The embodied approach

Using embodied tools, such as StarLogoT, students, typically, first approach the modeling problem by asking what kinds of rules an individual predator or individual

[3] The predator-prey model and other models (collectively known as "Connected Models" (Wilensky, 1998)) can be downloaded from http://ccl.sesp.northwestern.edu/cm/models

prey must follow so that, when allowed to interact in large numbers, populations of such individuals will exhibit population oscillations. It may seem to readers that one would need to be highly familiar with the phenomenon being modeled and with current theories in order to develop a working set of rules, but our experience indicates otherwise. In the Making Sense of Complex Phenomena project, we have found that students are often able to develop solid explanatory models of various phenomena, with only a small amount of background knowledge. We generally encourage modelers to try and make sense of a problem on their own before seeking external resources, and often they are quite surprised at how far they are able to get. To help convey a sense of this process, we will describe the development of a StarLogoT predation model from the standpoint of a student, Talia. While Talia's case has its own unique characteristics, the topic of predator/prey interactions is popular with high school students and, thus, we have observed many "average" students going through a modeling process quite similar to Talia's.

3.2. Finding rules for wolves—an initial model

Talia's task was to formulate a plausible set of rules for a typical predator and a typical prey. Recall that the characteristic properties of predator-prey interactions are invariant across many species and many different conditions. Rather then be specific, then, these rules needed to point to general behaviors that all such species can be said to perform in one way or another. In her first attempt, she described a predator (say, a wolf) as moving about in the StarLogoT world and looking for prey (say, sheep). As the wolf needs energy to live (which it must obtain by eating sheep), she decided that each step in the world will cost the wolf energy. Running out of energy will cause the wolf to die, and the only way to gain energy is by eating sheep. Here is a simple rule-set for a wolf based on the Talia's description (stated in summary form here):

Rule-set W1: wolf
at each clock-tick:
1. move randomly to an adjacent patch and decrease energy by E_1
2. if on the same patch as one or more sheep, then eat a sheep and increase energy by E_2
3. if energy ≤ 0 then die
4. with probability R_1 reproduce

Talia gave the sheep a simpler rule-set. Sheep only move about and reproduce, though they may still be eaten by the wolves:

Rule-set S1: sheep
at each clock-tick:
1. move randomly to an adjacent patch
2. with a probability of R_2, reproduce

3.3. Running the model

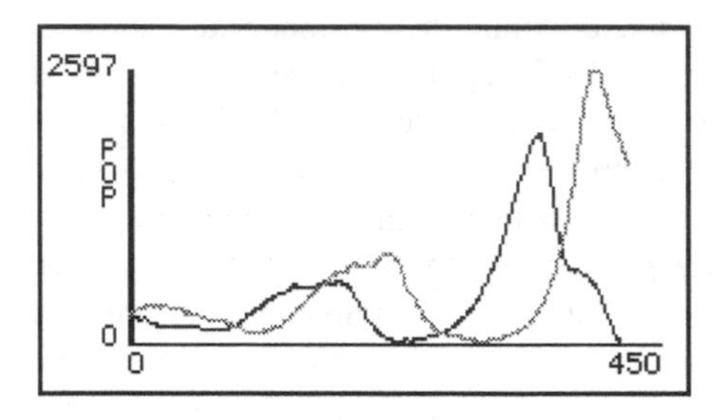

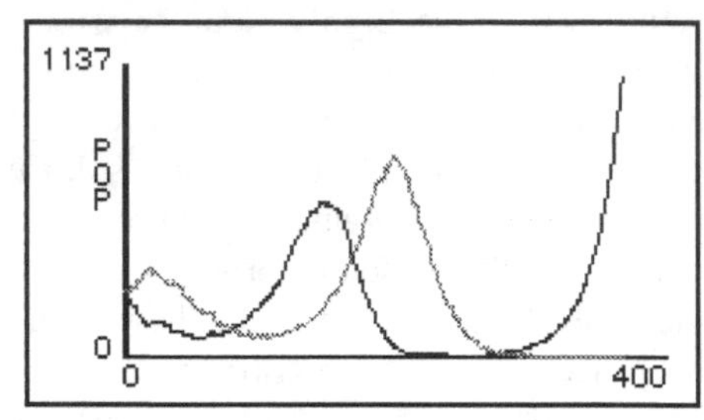

Figures 2a (left) **and 2b** (right). Two different outcomes from rule-sets W1 + S1. Gray represents wolf population size and black represents sheep population size.

After Talia ran 1her model several times, she noticed that one of two general outcomes would always occur. Either there were oscillations until all the sheep were eaten, whereupon the wolves died from starvation (figure 2a). Or, there were oscillations until the number of sheep dipped too low and the wolves all died off, at which point the sheep reproduced exponentially (figure 2b). Her simple rule-set thus succeeded in producing oscillations, but the pattern was unstable. The next logical step in the modeling process was for her to determine the cause of this instability and correct her model. In order to do this, Talia engaged in a process of successive revision—she would repeatedly devise some variation of her rule-set, and then program it and observe its effects.

3.4. Researching the relevant biological literature

Research into scientific literature is often a part of the model development process. This can help amend any errors in a student's knowledge of the phenomenon or reveal any important facts that the student might be overlooking. After experiencing difficulty devising a rule-set that would lead to stable oscillations, Talia did some research to determine the source of the problem. Notably, she read that when such systems were first created in the laboratory the findings were very similar to her StarLogoT model: either the predators ate all the prey and then starved, or the predators first died, and then the prey multiplied to the carrying capacity of the environment (Gause, 1934). Several differences between the natural and experimental/model settings could account for this discrepancy. The most significant such factor was the lack of constraints on the growth of the prey population in the experimental settings. In nature, the size and rate of growth of the prey population are constrained by several factors, including limits on the food resources available to prey and limits on their maximum density (Luckinbill, 1973). The laboratory experiments and Talia's model, however, included abundant food for the prey, and no other adversities in the system but the possibility of predation.

3.5. Revising the model

The major disparity between the experimental setting and the natural case study was the lack of constraints on the growth of the prey population. Talia addressed this by including a third species within her model, grass, which the sheep must feed on, and which is available in limited supply. There were then two ways in which the prey could die: either by being eaten or by starving. This resulted in an updated rule-set for sheep and a new rule-set for grass:

Rule-set S2: sheep
at each clock-tick:
1. move randomly to an adjacent patch and decrease energy by E_3
2. if on grassy patch, then eat "grass" and increase energy by E_4
3. if energy < 0 then die
4. with probability R_1 reproduce

Rule-set P1: patches
at each clock-tick:
1. If green, then do nothing
2. If brown, then wait X_1 clock-ticks and turn green

Shown in figure 3 is a typical outcome from rule-sets W1 + S2 + P1. This plot shows the population levels of the predators, the prey, and the grass.

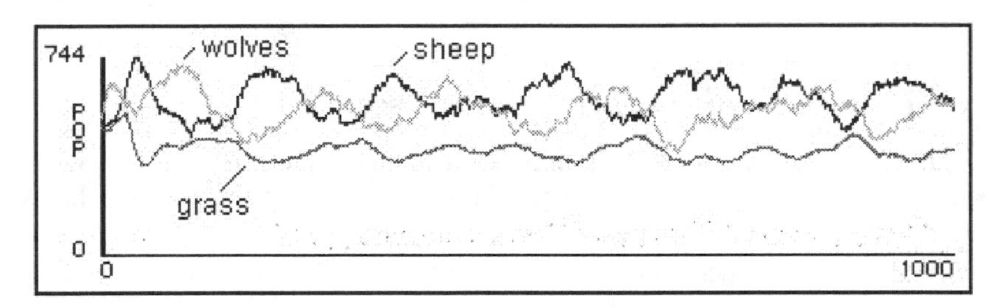

Figure 3. A typical outcome from rule-sets W1+ S2 + P1. Light gray represents wolf population size, black represents sheep population size, and dark gray represents relative amount of grass.

Talia was surprised to find this model much more stable. Oddly enough, she saw that limiting the resources of the prey population actually *increased* their chances of survival. Conversely, by allowing the prey population unlimited food, they were actually more likely to die off! This surprising result, which holds in the natural world just as it does in the StarLogoT world, is known in the scientific literature as the "paradox of enrichment" (Rosenzweig, 1971). Talia was pleased to find that her revised model had correctly predicted this result.

4. Modeling Synchronized Fireflies

In this section, we present a second example of how students can use StarLogoT as a laboratory for exploring biological mechanisms. Our example follows the inquiry of an undergraduate student, Paul, a humanities major whose formal biology instruction consisted solely of high school biology courses. Paul had heard of the phenomena of synchronously flashing fireflies and was intrigued. The following paragraph will provide some background.

For centuries, travelers along Thailand's Chao Phraya River have returned with stories of the astonishing flashing fireflies that inhabit the mangrove trees along the banks. Come nightfall, the fireflies in a given tree will pulse their lights, on and off, in near perfect synchrony with one another. There are several species of firefly that are known to do this, such as the Southeast Asian *Pteroptyx Malacae* and *Pteroptyx Cribellata*. When one such firefly is isolated, it will typically emit flashes at regular intervals. When two or more such fireflies are placed together they *entrain* to each other—that is, they gradually converge upon the same rhythm, until the group is flashing in synchrony (Buck, 1988).

How do the fireflies achieve this coordinated behavior? When we think about how behavior is coordinated in our daily lives, we tend to think of schedules and elaborate plans. Paul was perplexed at how creatures that seem to have little capacity for intelligent planning are nonetheless capable of such coordination. It was Paul's suspicion that there must be a simple mechanism behind the feat of the synchronizing fireflies. His goal was to try to understand this mechanism by building a model of it in StarLogoT.

4.1. Approaching the problem—initial assumptions

To begin, Paul made several working assumptions about these fireflies (he was prepared to revise them later if necessary). First, he decided that the mechanism of coordination was almost certainly a distributed mechanism. That is, the fireflies were not all looking to a leader firefly (or some other stimulus) for "flashing orders", but rather were achieving their coordination through passing and/or receiving messages from other fireflies. From his previous experience with StarLogoT, he had learned that not all coordinated group behavior requires a leader to direct the group (see Resnick, 1996; Wilensky & Resnick, 1999)[4]. A second assumption, following the first, was that the system could be modeled with only one set of firefly rules; that is, with every firefly in the system following the same set of rules. Although he recognized that this assumption might have been too strong, just as ant and bee populations do divide roles among their groups, he decided to first try out the simpler hypothesis of undifferentiated fireflies. Yet a third assumption Paul made concerned the movement of the fireflies: that it was not necessary to model this movement as coordinated or as governed by deterministic rules, but rather that it could be modeled as random flights and turns. From experience with other StarLogoT models, he had come to appreciate

[4] Note that some of the first theories proposed to explain synchronously flashing fireflies were, in fact, "leader" theories (Morse, 1916; Hudson, 1918).

the role of randomness in enabling coordination (Wilensky, 1997b; 1999). In a wide variety of domains, ranging from the movements of particles in a gas, to the schooling of fish and the growth of plant roots, Paul had seen how stable organization could emerge from non-deterministic underlying rules. A final assumption was that the behavior of the fireflies could be modeled in two dimensions[5].

These assumptions transformed Paul's task into one of finding a plausible set of rules for a typical firefly. Since Paul knew that a firefly left to its own would simply continue to flash at a regular pace, he reasoned that there were now two principle questions that he had to answer. The first was to determine what would trigger a firefly to change the timing of its flash, and the second was to determine what the response of the firefly would be to this trigger stimulus.

4.2. Thinking like a firefly

Often when building a model, students find it helpful to identify with the individuals within the model and to view phenomenon from their perspective. In order to attack the first of his two questions, Paul began to "think like a firefly". He reasoned along the following lines: If I were a firefly in that situation, what information would *I* have to go on? It would be dark, and I probably wouldn't be able to see the other fireflies. I probably wouldn't have much capacity for hearing or sensing the other fireflies either. I would, however, be able to see their flashes. Perhaps, then, I could look to see who else is flashing and then use this information to adjust my own flashing pattern. From this, Paul provisionally concluded that the trigger stimulus was probably the flashing of other fireflies.

Next, he considered what the response of a firefly would be to this stimulus. Paul could think of several possibilities: a firefly could flash in response, a firefly could increase or decrease the delay until the next flash, or perhaps the firefly could modulate this delay depending on the intensity of the stimulus. He found it was difficult to intuit the effect of each possibility—none of them result in synchronization in any obvious fashion. The next step was to test each mechanism by coding it up as StarLogoT model and observing the effects.

4.3 Putting it together—an initial model

Paul consolidated his decisions and assembled an initial model to test out various mechanisms. To regulate rhythmic flashing behavior, the model firefly incorporates a timer that continually counts down from a parameter R to zero and then resets to R and begins the cycle again. Each time the timer reaches zero, it causes the firefly to flash. That is, the firefly changes its color to yellow for one clock-tick (rules F1.1-F1.3). The firefly also "flies" around by moving in a random direction at each clock-

[5] While the first three assumptions were derived, to a great extent, from Paul's understanding of plausible biological mechanisms, this last assumption was primarily driven by the limitations of computer displays and of the StarLogoT language itself. Since building a three-dimensional model is more difficult in StarLogoT, Paul was, essentially, hoping that the three dimensionality of the firefly world was the not the key factor in enabling their coordination.

tick (rule F1.4). Finally, rule F1.5 is the flash reset rule. Paul began with a flash reset rule would, in response to any flash seen at an adjacent patch, cause the firefly to flash immediately and reset its timer to R.

Rule-set F1: firefly
to initialize:
0. set timer with random value between 0 and R
at each clock-tick:
1. if color is yellow (flash is on), then change color to black (flash is off)
2. if timer is zero, then change color to yellow and reset timer to R
3. decrement countdown timer by one
4. move randomly to an adjacent patch
5. if there is a yellow firefly within one patch, then change color to yellow (flash is on) and reset the timer to R

When Paul coded up this rule-set in the StarLogoT language and executed it, he found that it didn't quite work. In fact, it did cause flashing behavior to propagate through the population, but the flashes wouldn't terminate. That is, the fireflies perpetually caused each other to flash until the result was one persistent flash. In response, Paul experimented with several other flash reset mechanisms and variations to his rules. Though he observed the emergence of several interesting patterns, he was unable to find a plausible rule-set to give him the precise behavior he was interested in.

4.4. Researching the relevant biological literature

Notice how far Paul was able to get without reference to detailed research on real fireflies. From his initial goal to model "whatever" was going on, he was able to reason to the point where he was seeking a very particular sort of mechanism. At this point, Paul did some research into the scientific literature in order to obtain information about real fireflies (Buck, 1988; Buck, 1976; Buck, 1968). From this research, Paul discovered that many of his design decision were, in fact, biologically plausible. This includes his focus on a distributed synchronization mechanism, his assumption of undifferentiated fireflies, and his conclusion that flashing serves as the means of communication for synchronization. He also learned that the mechanisms of synchronization he had tested out were not far from the actual mechanisms in fireflies. Though different methods of synchronization are seen across species, one strategy that is similar to Paul's strategy is called "phase-delay synchronization". In this strategy, when a firefly perceives a flash it delays its next flash so that it will occur one flashing period after the perceived flash. This is the strategy known to be employed by the Southeast Asian fireflies of the *Pteroptyx* genus. Paul's next step was to alter his existing model in order to determine whether this strategy would actually work.

4.5. Revising the model

Paul's rule set needed only slight modification in order to test the phase-delay synchronization strategy. He altered it as follows:

> Rule-set F2: firefly
> 0-4. Identical to rule-set F1
> 5. if there is a yellow firefly within one patch, then reset the timer to R

Paul executed rule-set F2 using 1000 fireflies, and was amazed to see the model fireflies converge upon a single rhythm before his eyes (figure 4). Paul was encouraged by the initial results of his investigation, and was left with even more questions to consider and to research. For example, he was intrigued by the ability of some fireflies to adapt not only the timing of their flash, but also the duration between flashes. The papers he had looked at gave no complete theory of how this could be done.

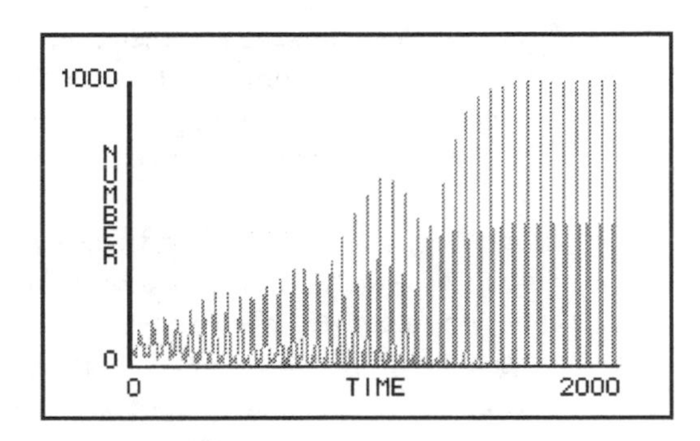

Figure 4. Typical plot of the number of flashes in a population at a given time under rule-set F2.

5. Concluding Remarks

The embodied modeling approach we have presented and illustrated herein can be the basis of a modeling-centered biology curriculum in secondary and post-secondary contexts. By removing the barriers of formal mathematical requirements, it enables students to meaningfully engage the dynamics of complex biological systems. They are able to construct models of such systems, reason about the mechanisms that underlie them and predict their future behavior. Because they are able to use their knowledge of the individual elements in the system to construct their model, they are provided with an incremental path to constructing robust models. When their knowledge of the individual biological elements is combined with their knowledge of their own embodiment, their own point of view, they are enabled to think like a wolf, a sheep or a firefly.

The above examples have, we hope, demonstrated the power of the embodied modeling approach to enable students to construct robust models and engage in exciting scientific inquiry. The modeling-based classroom is dramatically different from most venues of classroom practice. Rather than passively receiving an authority's explanation of science and mathematics concepts, students seek out and consider these concepts on their own. Rather than carry out the directions for predetermined lab studies, students engage in new investigations. What underlies this approach is our deep conviction of the value of *reasoning about scientific order*. In both the predation and firefly examples presented above, students were encouraged to reason through a problem, creating and testing their own theories and hypotheses, before reaching for the established literature.

There are, however, significant practical obstacles to employing the embodied modeling approach on a large scale at this point in time. Foremost amongst these is the unfamiliarity, to most teachers, of the modes of operation of the modeling-based classroom as described above. Overcoming this obstacle will necessitate significant teacher education efforts on the changed role of the teacher in the classroom. Another significant obstacle is the relative sparseness of personal computers in classrooms. StarLogoT requires the computational power of typical current personal computers. To take full advantage of the approach, a class needs to have at least one of these per three students. Many classrooms do not meet this requirement. Finally, the modeling approach requires teachers and students to be comfortable with algorithmic thinking and debugging, topics that are optimally learned through longitudinal developmental strands. As of now, there is not significant commitment to these topics and strands in K-16 education.

Despite these significant obstacles, there are good reasons to believe that the approach we describe will come to be widely incorporated into the science and social science curriculum. Most of the obstacles described above are short-term obstacles. The eventual proliferation of computers in classrooms combined with educational studies describing the effectiveness of the modeling-based classroom and the overwhelming adoption of modeling tools by practicing scientists will put pressure on teacher education programs to adopt these new methods and tools. Indeed, it is already happening in school districts and schools of education that are located near high technology centers.

Our approach promotes several processes of reasoning that are central to science: developing original hypotheses, formalizing ideas, researching existing solutions, and critical analysis of results. We believe that experience with these processes will be of significant advantage to *all* students as they seek to understand science and, more generally, the world around them. Few students will go on to become scientists. To the ones that don't, we owe them more than an introductory glimpse of current theories—we owe them the tools with which to appreciate scientific evidence and to reason in a thorough, scientific manner on their own. For the ones that do, we owe them a framework within which they will be better prepared to absorb and appreciate the myriad facts they will encounter for years to come. Thus, it is our hope that the approach we are developing will serve as a framework for all students to be able to work and think like scientists.

The preparation of this paper was supported by the National Science Foundation (Grants RED-9552950, REC-9632612). The ideas expressed here do not necessarily reflect the positions of the supporting agency.

References

Beer, R.D., 1990, *Intelligence As Adaptive Behavior: An Experiment in Computational Neuroethology*, Academic Press (Cambridge).

Buck, J., 1988, Synchronous Rhythmic Flashing of Fireflies II, *Quarterly Review of Biology*, **63**, 265-287.

Buck, J. & Buck, E., 1976, Synchronous Fireflies, *Scientific American*, **234**, 74-85.

Buck, J. & Buck, E., 1968, Mechanism of Rhythmic Synchronous Flashing of Fireflies, *Science*, **159**, 1319-1327.

Elton, C., 1966, *Animal Ecology*, October House (New York).

Epstein, J. & Axtell, R., 1996, *Growing Artificial Societies: Social Science from the Bottom Up*, Brookings Institution Press (Washington).

Forrest, S., 1989, *Emergent Computation: Self-Organizing, Collective, and Cooperative Phenomena in Natural and Competing Networks*, North Holland Pub. Co. (Amsterdam).

Gause, G.F., 1934, *The Struggle for Existence*, Williams and Wilkins (Baltimore).

Hammer, D., 1994, Epistemological Beliefs in Introductory Physics, *Cognition and Instruction,* **12** (2), 151-183.

Hudson, G., 1918, Concerted Flashing of Fireflies, *Science*, **48**, 573-575.

Huffaker, C., 1958, Experimental Studies on Predation: Dispersion Factors and Predator-Prey Oscillations, *Hilgardia,* **27**, 343-383.

Huston, M., DeAngelis, D., Post, W., 1988, New Computer Models Unify Ecological Theory, *BioScience*, **38** (10), 682-691.

Jackson, S., Stratford, S., Krajcik, J., & Soloway, E., 1996, A Learner-Centered Tool for Students Building Models, *Communications of the ACM*, **39** (4), 48-49.

Judson, O., 1994, The Rise of the Individual-based Model in Ecology, *TREE*, **9** (1), 9-14.

Keen, R. & Spain, J., 1992, *Computer Simulation in Biology*, Wiley-Liss (New York).

Langton, C., 1993, *Artificial Life III*, Addison Wesley (Reading).

Langton, C. & Burkhardt, G., 1997, *Swarm,* Santa Fe Institute (Santa Fe).

Luckinbill, L., 1973, Coexistence in Laboratory Populations of Paramecium Aurelia and its Predator Didnium Nasutum, *Ecology*, **54** (6), 1320-1327.

Maes, P., 1990, *Designing Autonomous Agents: Theory and Practice from Biology to Engineering and Back*, MIT Press (Cambridge).

Morse, E., 1916, Fireflies Flashing in Unison, *Science*, **43**, 169-170.

Ogborn, J., 1999, Modeling Clay for Thinking and Learning, in *Computer Modeling and Simulation in Science Education*, edited by Roberts, N., Feurzeig, W. & Hunter, B., Springer-Verlag (Berlin).

Papert, S., 1980, *Mindstorms: Children, Computers, and Powerful Ideas*, Basic Books (New York).

Purves, W.K., Orians, G.H. & Heller, H.C., 1992, *Life: The Science of Biology*, W.H. Freeman (Salt Lake City).

Repenning, A., 1993, AgentSheets: A Tool for Building Domain-oriented Dynamic, Visual Environments, Unpublished doctoral dissertation, Dept. of Computer Science, University of Colorado, Boulder.

Repenning, A., 1994, Programming Substrates to Create Interactive Learning Environments, *Interactive Learning Environments*, **4** (1), 45-74.

Resnick, M., 1994, *Turtles, Termites and Traffic Jams: Explorations in Massively Parallel Microworlds*, MIT Press (Cambridge).

Resnick, M., 1996, Beyond the Centralized Mindset, *Journal of the Learning Sciences*, **5** (1), 1-22.

Resnick, M., & Wilensky, U., 1998, Diving into Complexity: Developing Probabilistic Decentralized Thinking Through Role-Playing Activities, *Journal of the Learning Sciences*, **7** (2), 153-171.

Roberts, N., Anderson, D., Deal, R., Garet, M., & Shaffer, W., 1983, *Introduction to Computer Simulations: A Systems Dynamics Modeling Approach,* Addison Wesley (Reading).

Roberts, N. & Barclay, T., 1988, Teaching Model Building to High School Students: Theory and Reality, *Journal of Computers in Mathematics and Science Teaching*, **7** (Fall), 13 - 24.

Rosenzweig, M., 1971, Paradox of Enrichment: Destabilization of Exploitation Ecosystems in Ecological Time, *Science*, **171**, 385-387.

Smith, D. C., Cypher, A., & Spohrer, J., 1994, Kidsim: Programming Agents Without a Programming Language, *Communications of the ACM*, **37** (7), 55-67.

Smith, J.P., diSessa, A.A., & Roschelle, J., 1994, Reconceiving Misconceptions: A Constructivist Analysis of Knowledge in Transition, *Journal of the Learning Sciences*, **3**, 115-163.

Taylor, C, Jefferson, D, Turner, S & Goldman, S., 1989, RAM: Artifical Life for the Exploration of Complex Biological Systems, in *Artificial Life*, edited by C. Langton, Addison Wesley (Reading).

Wilensky, U. & Resnick, M., 1999, Thinking in Levels: A Dynamic Systems Perspective to Making Sense of the World, *Journal of Science Education and Technology*, **8** (1).

Wilensky, U., 1999, GasLab—an Extensible Modeling Toolkit for Exploring Micro- and Macro- Views of Gases, in *Computer Modeling and Simulation in Science Education*, edited by Roberts, N., Feurzeig, W. & Hunter, B., Springer-Verlag (Berlin).

Wilensky, U., 1997a, *StarLogoT*. Center for Connected Learning and Computer-based Modeling. Northwestern University. www.ccl.sesp.northwestern.edu/cm/ .

Wilensky, U., 1997b, What is Normal Anyway? Therapy for Epistemological Anxiety, *Educational Studies in Mathematics*, **33** (2), 171-202.

Wilensky, U., 1996, Modeling Rugby: Kick First, Generalize Later?, *International Journal of Computers for Mathematical Learning*, **1** (1), 125-131.

Wilensky, U., 1995, Learning Probability through Building Computational Models, Proceedings of the Nineteenth International Conference on the Psychology of Mathematics Education, Recife, Brazil.

Wilensky, U., 1993, Connected Mathematics: Building Concrete Relationships with Mathematical Knowledge, Doctoral dissertation, Media Laboratory, MIT (Cambridge).

Diffuse Rationality in Complex Systems

Ron Cottam, Willy Ranson and Roger Vounckx
The Evolutionary Processing Group (EVOL)
Laboratory for Opto and Micro Electronics
VUB-IMEC Electronics Division (ETRO)
The University of Brussels (VUB)
http://etro.vub.ac.be/evol
evol@etro.vub.ac.be

1. Introduction

We formulate a new overall framework of rationality capable of supporting not only the reductionist scientific description of nature, but also the implications of an evolutionary development of structure. The classical clearly observable difference between animate and inanimate entities emerges naturally from this framework, which has the form of a generalized precursor to quantum mechanics and within which approximative Newtonian metastates develop through a primitive evolution.

The described computational model is consistent with the presupposition that survival by the maintenance of organizational integrity is the main individual aim of living entities. The principal survival criterion can be described by the relationship between the time-scale of an internal reactive process and that of the external reaction-demanding environment, and evolutionary success then becomes primarily a question of ensuring phenomenological computability. Survival requires the prioritized identification and objectivization of dimensions within which an entity is most fragile or most at risk [Cottam et al 1998a]; the evolutionary co-development of a selectively dimensioned environment [Szamosi 1986] brings with it enhanced computability and consequently competitive advantage.

Timely computation of an entity's reactions to unpredicted environments necessitates the availability of simplified representations of the interaction between an entity and its surroundings in their combined multi-dimensional phase space. Metastates characterizing these objectivizations correspond to regions of the phase space where its contextual description can be reduced reasonably accurately to a small number of parameters or degrees of freedom, and they are equivalent to the visualizable limited-parametric provisional stages of an evolutionary computation.

Starting from this description of adaptive real-time environmental-reaction survival computation in a natural temporally demanding medium [Langloh et al 1993], we present a more general model of the evolutionary context itself as a computational machine [Cottam et al 1998a]. The effect of this development is to replace deterministic logic by a modified form which exhibits a continuous range of dimensional fractal diffuseness between the isolation of perfectly ordered localization and the extended communication associated with nonlocality as represented by pure causal chaos.

2. An Adaptable Real Time Processor

We can obtain processor response satisfying the double requirement of both fast and accurate environmental responses by using a query-reflection structure [Langloh et al 1993].

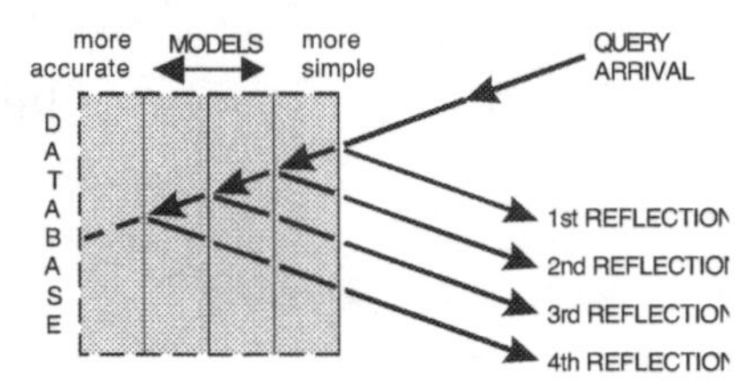

Figure. 1. Multiple reflections in a query-reflection structure, showing the availability of both fast and accurate responses.

We arrange a suitable hierarchy of models organized into a series of reflecting planes, as shown in Figure 1 with the simplest at the right and the most complicated one at the left. When we fire a query at the structure the immediate reflection is from the simplest model, so the response to the query can be very rapid. The second reflection is from a model which is more complicated, so its processing demands more time. Progressively the successively returned reflections take longer and longer to appear, but they are more and more accurate with elapsed time.

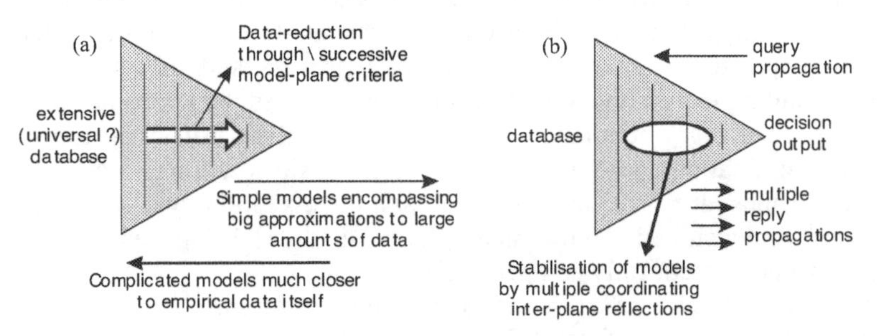

Figure. 2. (a) Decision-making by data-reduction through sequential criteria, and (b) combining query-reflection and pyramidal decision-making.

Decision-making usually proceeds by the reduction of an available data set through progressive applications of sets of rules to a final binary conclusion. In its most simple form this has the shape of a pyramid (Figure 2a). Combination of this query-reflection idea with pyramidal decision-making gives us the basic form (Figure 2b) of an adaptable real-time processor ["AQuARIUM": Langloh et al 1993]. Data-like queries propagate from right to left, provoking data-like multiple reflections from the sequential model-planes in the body of the structure. Recursive inter-model-plane reflections stabilize the relationships between consecutive model-planes, resulting in a coherent progressive change between the simplest and the most complicated data representations. The data is not now confined to the originally described database at the left-hand side, but is implicitly distributed throughout the model-plane assembly in the associated rule-sets.

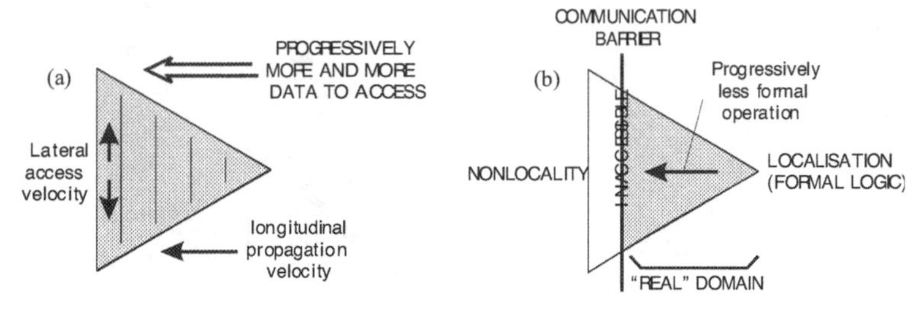

Figure. 3. (a) Query velocity limitation resulting from increasing lateral data access, and (b) the general symbolic representation of evolution through computation.

By waiting as long as possible we can obtain the best response which is feasible within the time available; but now there is a problem. It is the external stimulus *itself* which defines the permitted response time scale, and as the "lateral" query velocity (Figure 3a) must increase as the size of the accessed data-set increases we finally run out of "longitudinal" query (group) velocity, and model-planes at the extreme left hand side become completely inaccessible.

3. A Transferable Computational Model

We can extend this albeit crude computational system into not only a useful model for information processing, but also finally into a generalized description of the evolutionary appearance of localized entities in a universal context, as in Figure 3b. The imposition of a perceptional system on information processing implies the reduction of possibly unstructured information to a structured perceivable form related to symbolic representation. The model-planes which appear in the computational description as stabilized partial computational results are functionally equivalent to the metastates which are found in the evolution of complexity. We include explicitly the necessary condition of a "real" communication limit for the existence of causality in any environment [Prigogine & Stengers 1984].

In the resulting symbolic representation of a generalized environment shown in Figure 3b, an equivalence is made in the longitudinal direction between {pure communication, pure nonlocality, complete disorder and pure "causal" chaos} at the left hand side, and between {complete lack of communication, complete localization, complete order and degenerate formal logic} on the right.

The "real" domain now spreads from the imposed communication barrier to the pyramid apex on the right (Figure 3b), and movement from the apex in the direction of the left hand side is equivalent to a progressive change away from formal logic into a more partially rational state. The similarity between this symbolic portrayal and the simple chaotic logistic plot shown in Figure 4 is not accidental: both are describing similar relationships between unstructured information and its structured representation. The white vertical lines which appear in the chaotic zone of the logistic plot are regions within which low-dimensional representations are locally valid. These correspond to the metastatic model-planes which appear in our survival-computational description.

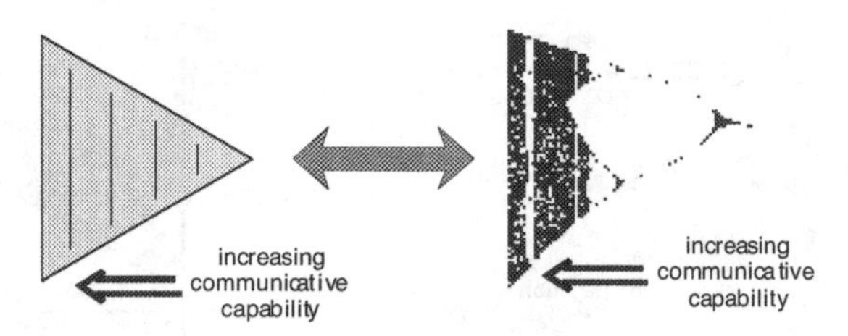

Figure. 4. A comparison between the forms of query-reflection and the logistic plot.

Nicolis has pointed out in 1993 that chaos enables a system to explore its phase space. This corresponds to the necessary in-plane data access required for the model-assembly stabilization described above. "Deterministic" chaos seems to be nature's best attempt at penetrating an imposed communication barrier, and represents a meta-model of the underlying "causal" chaos which is equivalent to communication itself.

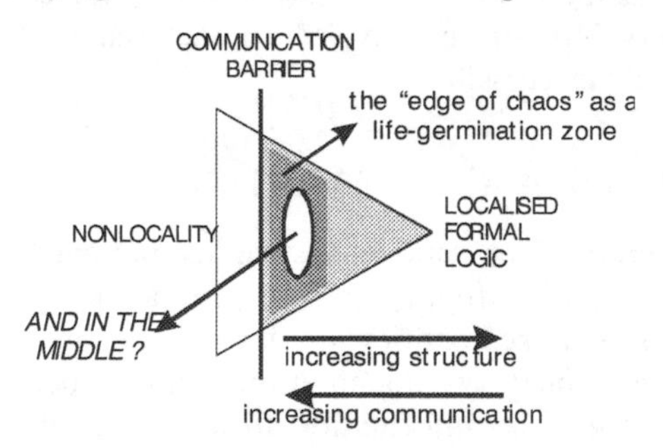

Figure. 5. Location of the edge of chaos region in the symbolic representation.

If we now add chaos itself into the symbolic description of Figure 3b, we can identify the location of the "edge-of-chaos" described in 1990 by Langton, some way to the right of the causally-imposed communication barrier (Figure 5). This region exhibits the balance between communication (towards the left in our representation) and structure (towards the right) necessary for the evolutionary germination of complexity [Langton 1990] and in all probability "life" itself.

4. A "Psychological" Potential Well

Above a certain level of perception of complexity we no longer have the formal mental tools available to approach solution of a problem, and we abandon the attempt or try non-formal methods. Below some arbitrary level of perceived complexity we make blind use of pre-existing rule structures, and it is only when our perception of the complexity of a situation passes this threshold that we begin to "think" about how to deal with the situation. If, when we superimpose these phenomena, there is a gap between the two, then this implies that there is a region of perceived complexity within which there is the capability to react in a consequent manner to the context in which we find ourselves: this is a necessary condition of, and *also* the result of, consciousness itself, and we can locate this "consciousness-capability" gap in the general symbolic model as an emergent metastate [Cottam et al 1998b].

Evolutionary growth of this "consciousness-capability" gap appears to be equivalent to the development of "conscious life" as we conventionally describe it; Deacon (1998) has described "Our self-experience of intentions and 'will'" as "what we should expect an evolutionlike process to feel like!". Evolutionary development of this kind would be the case not only for "living entities", but for ***all*** perceptional localizations, consistently with, for example, David Bohm's assertion [Weber 1987] of the (presumably) low-level "consciousness" of electrons.

The development of localizations of this kind demands evolution along the lines proposed in 1986 by Szamosi as a modification of Darwin's proposals, in which a dimensional domain evolves into reality *coincidentally* with the dimensionally embedded structure itself.

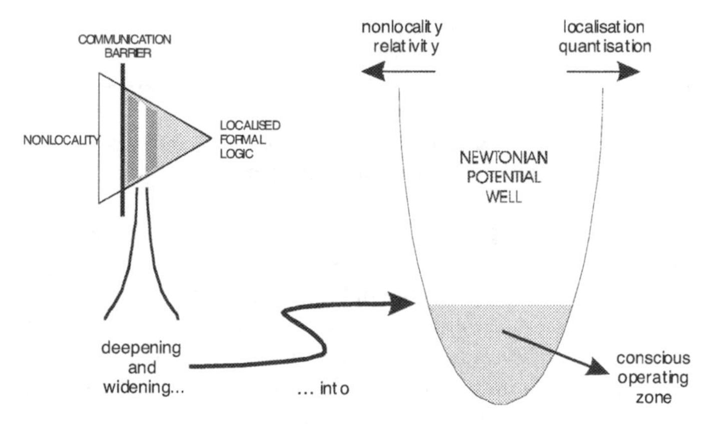

Figure. 6. Widening of the consciousness-gap into a Newtonian potential well.

It is reasonable to suggest that the dimensions which would become the most perceptionally organized (or objectivized) would be those within which the developing structure is the most fragile, viz. apparently in space and time. We could expect the "gap" to develop in both width and depth (Figure 6) to constitute a *containing* " potential well" for consciousness, centered around the "easiest" or "most-computable" representational form of an approximate Newtonian objectivization. We should explicitly point out that metastates are not just computational properties of some system, they are potentially alive *themselves*: a metastate *is* the entity it represents [Cottam et al 1998b].

5. Nonlocality and Quantized Perception

Quasi-particle emergence and stabilization in this scheme consistently depend on "local" criteria, as does electron-positron emergence in "free space", for example. The modified evolutionary description fits in well with super-string models [Green et al 1987], where there is also the requirement for co-generation of strings and their background space-time. Nonlocality appears in this context as a description of perfect communication, and as such it also appears in the real zone in a partial manner (Figure 7a), rather than remaining an esoteric phenomenon uniquely associated with quantum physics.

The most general view of the relationship we describe between the "global" and the "local" is shown schematically in Figure 7b as a simultaneous symmetrical reductionism between nonlocality and localization.

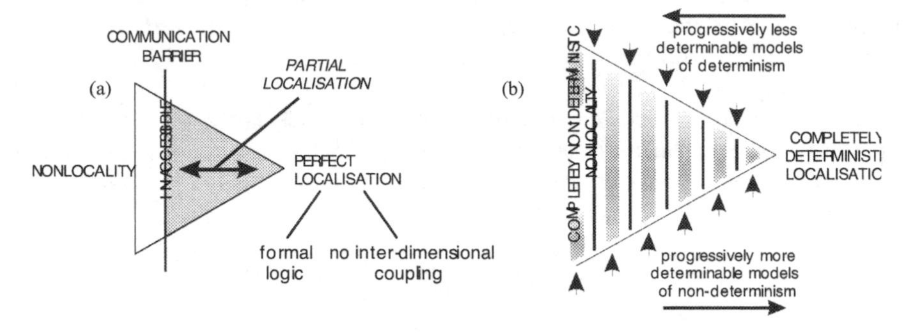

Figure. 7. (a) The appearance of partial nonlocality in the "real" zone, and (b) the general form of a symmetrical reductionism.

Any parameterized description of universal phenomena will then exhibit a similar form: "deterministic" chaos is a low-dimensional representation of high-dimensional causal chaos; quantum mechanics is the deterministic representation of a higher-level more general process description; physics is the ground state of an extended quantum-consciousness model of the universe; energy is an inanimate description of consciousness; the parameters of a low-dimensional model are simplified representations of metastatic agents. Perceptional access to any system-descriptive level relies on relativistic process-dependent effects and the degree to which the local representational diffuseness is *sufficiently computable*.

It is vitally necessary to apply a "correct" mathematical approach to describing inter-dimensional processes. We are most concerned with modeling relationships and not simply functions, and here classical probability suffers from a lot of difficulties, particularly if the relations are not simply one-to-one. A better approach is to use upper and lower bounded probabilities as described by Dempster in 1967 and Schafer in 1976, where there is no imposed "deterministic" central value to the range of probabilities describing an event. When this probabilistic conversion is carried out in a recursive manner, it is capable of generating the diffuseness [Cottam et al 1998a] associated with a progressive change from localization towards nonlocality in a multi-dimensional system.

Coupling between different localized dimensions in this scheme always takes place more or less through the "chaotic" database (Figure 8); there is no coupling *at all* directly between completely localized dimensions, and consequently dimensional interactions are always to some extent irreversible.

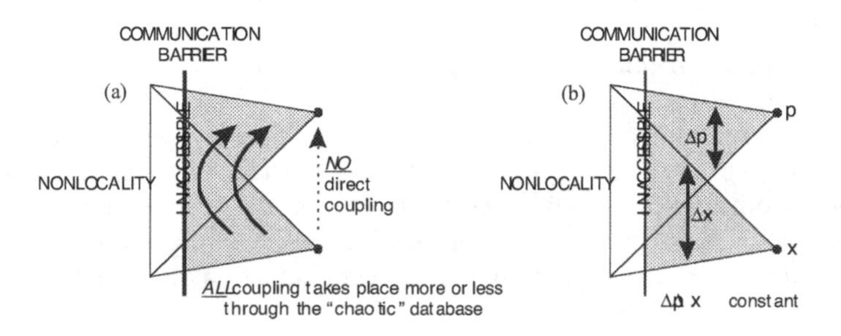

Figure. 8. Inter-dimensional coupling is
(a) never direct, but only through the "chaotic" database, and
(b) consistent with the limitation indicated by Heisenberg's uncertainty.

Heisenberg's uncertainty principle appears as a special case of coupling between two dimensions which are linked by a constant coupling factor: if one dimension is completely localized, the other is completely delocalized and therefore unknown.

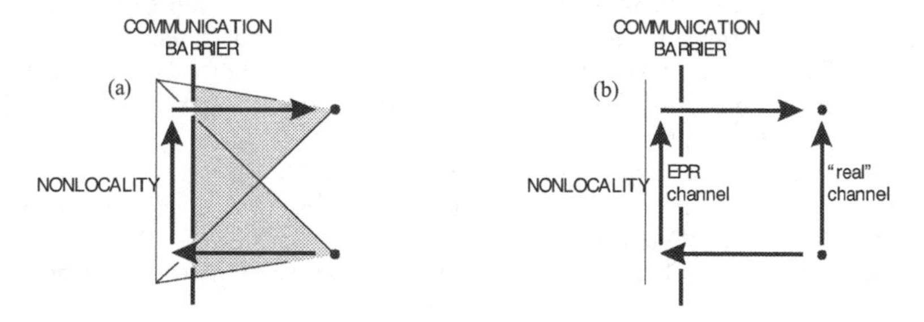

Figure. 9. The computational paths implied by
(a) nonlocal communication without using a "real" channel, and
(b) Bennett et al's "teleportation" through combined EPR and "real" channels.

Finally, what about the apparent possibility of coupling two dimensions directly through the nonlocal communicative background, as some sort of telepathy? Bennett et al presented in 1993 a scheme for the uncertainty-free transport ("teleportation") of a quantum state between different locations through the use of an Einstein-Podolsky-Rosen nonlocal channel (Figure 9). Closer examination reveals this to be a single-quantum version of the more general propositions of this paper, and it is clear that it is fundamentally necessary that there is not only communication via an EPR channel, but by a "real-world" channel as well: for a "real" communicative result we still need "real" communication, although the combination of the two channels may only be "partially real".

6. Conclusions

We conclude that this generalized quasi-model framework, which is presented here only in its most primitive form, is capable of describing *any* localized entities, including perceptions themselves. The formulation is entirely consistent with David Bohm's description in 1980 of nature in terms of implicate and explicate order. The proposition of the appearance of partial or diffuse rationality with movement away from perfect localization is confirmed by Ioannes Antoniou's work in 1996 on the extension of quantum mechanics to large systems, where he finds that logical completeness breaks down.

We believe that the quantum tradition of this century can be complemented by treating consciousness itself as a system of interacting quasi-particles as agents, and similarly for all other (real) localized entities. The prefix "quasi-" then accrues a quantitative sense in terms of the degree of localization involved, and all structure is seen to be part of a universally correlated hierarchical interaction of metastatic agents.

References

Antoniou, I., 1996, Extension of the Conventional Quantum Theory and Logic for Large Systems, in *Einstein Meets Magritte*, VUB Press (Brussels).

Bennett, C.H., Brassard, G., Crépeau, C., Jozsa, R., Peres, A. & Wootters, W.K., 1993, Teleporting an Unknown Quantum State via Dual Classical and Einstein-Podolsky-Rosen Channels, *Physical Review Letters*, **70**, 1895-1899.

Bohm, D., 1980, *Wholeness and the Implicate Order*, Routledge & Kegan Paul (London).

Cottam, R., Langloh, N., Ranson, W., & Vounckx R., 1998a, *Partial Comprehension in a Quasi-Particulate Universe : A Framework for Evolution*, in preparation.

Cottam, R., Ranson, W., & Vounckx, R., 1998b, Emergence: Half a Quantum Jump? *Acta Polytechnica Scandinavica: Emergence, Complexity, Hierarchy, Order*, Finnish Academy of Technology (Espoo), 12-19.

Deacon, T., 1998, *The Symbolic Species*, Penguin (London), 458.

Dempster, A.P., 1967, Upper and Lower Probabilities Induced by a Multivalued Mapping, *Annals of Mathematical Statistics*, **38**, 325-339.

Green, M.B., Schwarz, J.H., & Witten, E., 1987, *Superstring Theory*. Cambridge U. P. (Cambridge).

Langloh, N., Cottam, R., Vounckx, R., & Cornelis, J., 1993, Towards Distributed Statistical Processing - "AQuARIUM", *in ESPRIT Basic Research Series, Optical Information Technology*, edited by S.D. Smith & R.F. Neale, Springer-Verlag (Berlin), 303-319.

Langton, C.G., 1990, Computation at the Edge of Chaos: Phase Transitions and Emergent Computation, *Physica D (Special Issue on Emergent Computation)*, **42**, 12-37.

Nicolis, G., 1993, in *ECAL 93, Self-Organization & Life: from Simple Rules to Global Complexity*, Centre for Non-Linear Phenomena and Complex Systems, ULB (Brussels).

Prigogine, I., & Stengers, I., 1984, *Order out of Chaos: Man's New Dialog with Nature*, Flamingo-Harper Collins (London).

Schafer, G.A., 1976, *Mathematical Theory of Evidence*, Princeton U. P. (Princeton).

Szamosi, G., 1986, *The Twin Dimensions: Inventing Time and Space*, McGraw-Hill (New York).

Weber, W., 1987, Meaning as Being in the Implicate Order Philosophy of David Bohm: a Conversation, in *Quantum Implications: Essays in Honour of David Bohm*, edited by B.J. Hiley and F.D. Peat, Routledge & Kegan Paul (London), 440.

Spontaneous evolution of self-reproducing loops on cellular automata

Hiroki Sayama
New England Complex Systems Institute
sayama@necsi.org

A simple evolutionary system "*evoloop*" implemented on a deterministic nine-state five-neighbor cellular automata (CA) space is introduced. This model was realized by improving the structurally dissolvable self-reproducing loop I had previously contrived after Langton's self-reproducing loop. The principal role of this improvement is to enhance the adaptability (a degree of the variety of situations in which structures in the CA space can operate regularly) of the self-reproductive mechanism of loops. The experiment with *evoloop* met with the intriguing result that the loops spontaneously varied through direct interaction of their phenotypes, smaller individuals were naturally selected, and the whole population gradually evolved toward the smallest ones. This result shows that it is possible to construct evolutionary systems on such a simple mathematical medium as a CA space by introducing the mortality of individuals, their interaction, and their robustness to variations into the model.

1 Introduction

This article gives an affirmative answer to the question whether it is possible to construct an *evolutionary process*—here I view this phrase as a process in which self-replicators vary and fitter individuals are naturally selected to proliferate in the colony—by utilizing and tuning up a simple deterministic cellular automata (CA) space.

In this article, a simple evolutionary system "*evoloop*" implemented on a deterministic nine-state five-neighbor cellular automata (CA) space is introduced. This model was realized by improving the structurally dissolvable self-reproducing (SDSR) loop I had previously contrived[12] after Langton's self-reproducing (SR) loop[6]. The principal role of this improvement is to enhance the adaptability (a degree of the variety of situations in which structures in the CA space can operate regularly) of the self-reproductive mechanism of loops, besides a slight modification of initial structure of the loop.

The experiment with the *evoloop* met with the intriguing result that, though no mechanism was explicitly provided to promote evolution, some evolutionary process emerged in the CA space, where loops varied by direct interaction of their phenotypes, smaller individuals were naturally selected thanks to their quicker self-reproductive ability, and the whole population gradually evolved toward the smallest ones. It is characteristic that in this result genotypical variation was caused by precedent phenotypical variation, which is quite different from the idea of mutation usually considered. This result shows that it is possible to construct evolutionary systems on such a simple mathematical medium as a CA space by introducing the mortality of individuals, their interaction, and their robustness to variations into the model. This implies that, in the future, we will be able to create extraordinary large-scale evolutionary systems in a fine-grained superparallel machine environment by using a very simple algorithm with neither explicit management of living individuals nor generation of random numbers for stochastic mutation of genotype.

2 Former works

Langton's SR loop[6] is one of the most famous models of self-reproduction on CA. It was implemented on a simple eight-state, five-neighbor CA space. Fig. 1 shows the manner of self-reproduction of the SR loop. This loop contains several signal states '4' and '7' in its Q-shaped tube enclosed by sheath states '2'. Each signal travels along the tube counterclockwise and splits into two identical signals at the T-junction of the tube. One of them circulates into the loop again and the other goes down toward the tip of a construction arm that is thrust outward from the loop. When a signal reaches the tip of the arm, translation from genotype to phenotype will occur, such as straight growth or left turning of the arm. When the tip of the arm reaches its own root after it has turned left three times, the tip and the root bond together to form a new offspring loop, and then the connection between parent and offspring—which Langton called the "umbilical cord"—disappears. The SR loop reproduces itself in such a way in just 151 updates and will try to do the same again in the same way but rotated by 90 degrees counterclockwise, until its self-reproductive ability halts because of a shortage of space. Langton's SR loop was such a useful model in studying self-reproduction on CA that various modifications were invented after it[2, 3, 4, 8, 10, 14, 15].

After this SR loop, I previously contrived the SDSR loop capable of structural dissolution (a form of death) as well as self-reproduction, where a new *dissolving state* '8' was introduced into the set of states of the CA while exactly preserving

Figure 1: Self-reproduction of Langton's SR loop.

states '0'–'7' and all state-transition rules relevant to them[12]. The dissolving state was granted with an ability to travel along a tube and dissolve neighboring structures so that once a site takes on the dissolving state, a continuous structure that includes that site will be extinguished quickly. The SDSR loop shows several characteristic behaviors that were never seen in the SR loop world, such as continuous self-reproduction in finite space, production of many 'merged' loops through collision of two or more normal loops, competitive exclusion between loops of different sizes living in the same finite space, and so on. However, the SDSR loop could not actually evolve, which is the very problem resolved in the following sections.

3 Evoloop: an evolving SDSR loop

3.1 Reconstructing the state-transition rules

The reason the SDSR loop did not show any apparent evolvability is that its state-transition rules, which designated all mechanisms necessary for self-reproduction, were specialized only for a set of particular situations that appeared in an ordinary self-reproductive process of the original SR loop. Thus even a slight fluctuation such as a one-site discrepancy in propagation of signals could easily ruin the self-reproductive process of the loop. For example, when the form of the arm of the SDSR loop is altered by force during its self-reproductive process, it cannot reproduce any self-reproductive offspring; it either generates a dissolving state (Fig. 2, left) or falls into a sterile structure (Fig. 2, right). In such cases, neither connection of the tip of the arm and its root nor dissolution of the umbilical cord between parent and offspring occurs correctly, because in these cases the location of genes near a bonding T-junction is different from the situation expected by Langton's state-transition rules. Such rigidness of rules seems to have prohibited evolution of the SDSR loop.

Figure 2: What happens if the form of the arm of the SDSR loop is altered by force during self-reproduction.

To resolve this problem, it was necessary to make the self-reproductive mechanism described by the state-transition rules more "adaptable." The word "adaptability" used here means a degree of the variety of situations in which structures in the CA space can retain their regular operations. To enhance the adaptability of the state-transition rules, I reconstructed mechanisms of its self-reproduction carefully, while keeping fundamental behaviors of signals as is. I first defined general rules concerned with sustenance of sheath structures and propagation of signals. Next, to clarify what behaviors must be realized in the state-transition rules for self-reproduction, I divided a self-reproductive process of the loop into the following six phases: (1) Lengthen the construction arm, (2) turn the tip of the arm left, (3) bond the tip and the root of the arm together, (4) dissolve the umbilical cord between parent and offspring, (5) germinate a new sprout of the arm, and (6) lengthen the new sprout of the arm. Then, I manually refined each part of the state-transition rules relevant to each of the six phases to make it adaptable to a greater variety of situations than before.

On granting adaptability to the self-reproductive mechanism of the SDSR loop, some inadvertent complication of the old state-transition rules became a nuisance. Specifically, in the CA of the SR/SDSR loops, rules concerned with bonding of the tip and the root of the arm and germination of a new sprout of the arm in the parent loop were constructed in such a heuristic way completely dependent on some specific situations that they defied any modification. In addition, old rules had some redundancy in that the location of a new sprout of the parent's arm was pointed by a messenger '5' traveling on the sheath while that of the offspring's was pointed by a different messenger '6' traveling in the tube.

I conducted a thorough revision of the state-transition rules to fix these problems. For example, the mechanism for germination and growth of a new sprout was made to be identical in both parent and offspring. To equalize the length of the parent's sprout with that of the offspring, I let the sprout be explicitly stimulated to grow by all of signal '7's contained in the loop in any case. As a result, the length of the umbilical cord became longer than that in the SR/SDSR loops. For dissolution of such a lengthened umbilical cord, signal '6' was reassigned to be a special umbilical cord dissolver much more powerful than that in the old rules. Functions formerly possessed by signal '6' were reassigned to '3', '4', and '5' in new rules. These reassignments made it much easier to refine the state-transition rules to make them keep their regular operations in a greater variety of situations. After these above-mentioned operations, a dissolving state '8' was introduced into the set of states of the CA in the same way as in the SDSR loop.

I eventually obtained a new loop that was extremely resistant to fluctuation of environmental conditions with neither increase in number of both states and neighborhood sites of the CA nor alteration of the basic structure of the loop. I named this "*evoloop*." Fig. 3 depicts general behaviors of refined phases of the self-reproductive process of the *evoloop*. Mechanisms concerned with phase 3, 4, 5 and 6 are reconstructed this time from scratch, while those concerned with 1 and 2 are exactly the same as in the SR/SDSR loops. The complete state-transition rule set of the *evoloop* is presented in the other literature[11, 13].

Self-reproduction of the *evoloop* is shown in Fig. 4. Since the sprout of the

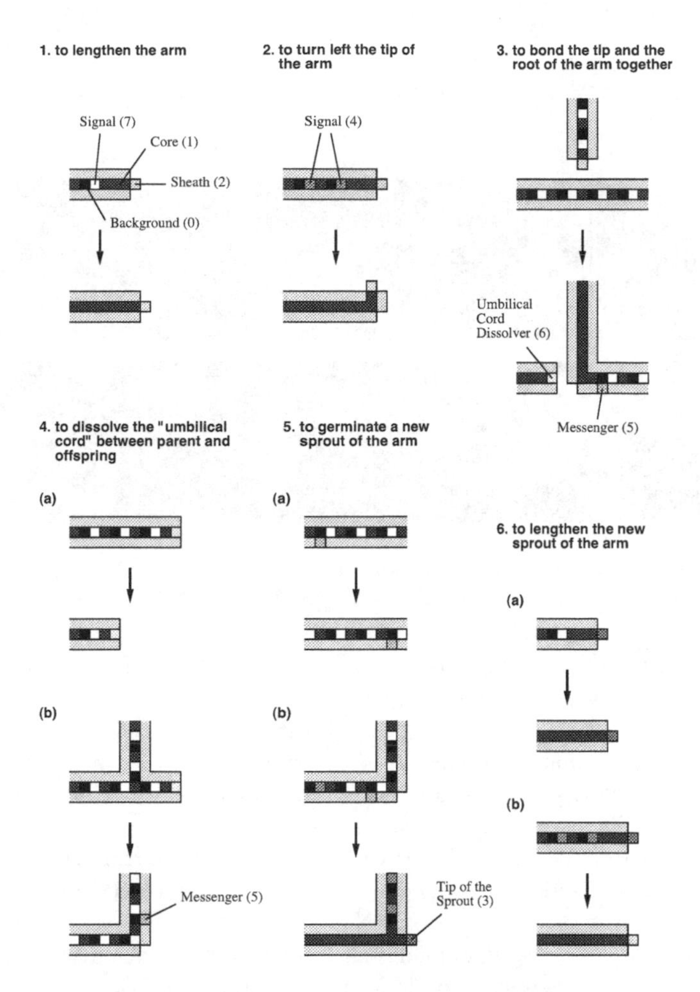

Figure 3: Behaviors of six functional phases of the self-reproductive process of the *evoloop*. **3.** When the tip of the arm collide with the middle of the arm, they will bond together, and both an umbilical cord dissolver '6' and a messenger '5' will emerge after several transient configurations. **4.** The umbilical cord dissolver '6' will travel along the umbilical cord against the signals' flow, dissolving structures of the cord (a), and it will turn into another messenger '5' when it arrives the T-junction of the parent loop (b). **5.** The messenger '5', having been generated in the aforementioned phases, will be trailed on the sheath by a signal '4' or '7' traveling along the tube (a), and it will begin to wait at the corner for a signal '4' arriving. When a signal '4' arrives there, it will germinate a new sprout there, while the messenger '5' will disappear (b). **6.** The growth of a new sprout will be stimulated by signal '7's (a), and the sprout will be changed into an ordinary arm by a couple of signal '4's (b).

evoloop is explicitly stimulated to grow by signal '7's contained in its body, the length of its umbilical cord is longer than that of the SR/SDSR loop. Thus, the colony of the *evoloop* looks a little sparse than that of the SR/SDSR loops. In the shown case, the loop contains thirteen signal '7's in its body. Hereafter the number of signal '7's in a loop will be used as a label of "species" of that loop.

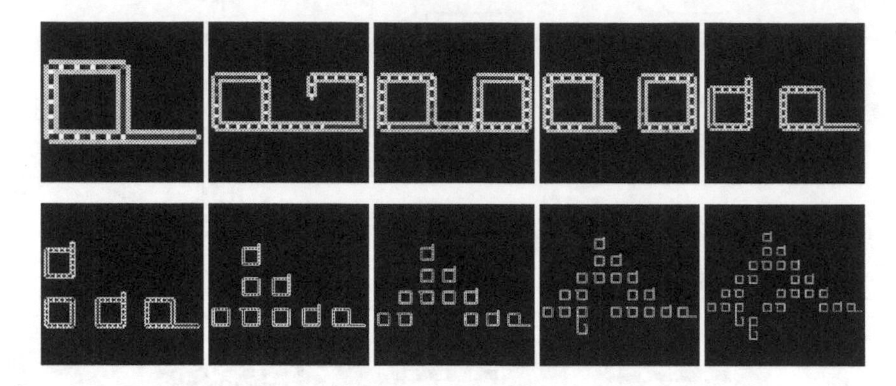

Figure 4: Self-reproduction of the *evoloop* of species 13 (i.e., a loop that has thirteen signal '7's). Each picture is scaled differently to the size of the colony.

It is remarkable that, owing to the adaptability enhanced above, some intriguing interactions of loops emerge in the *evoloop* world that have never occurred in the SDSR loop world. Fig. 5 shows, for example, a takeover of the arm happening between two *evoloops*. In this case, the right loop takes over the arm of the left loop, and consequently a small rectangular variant is produced between two loops. Due to the high adaptability of their self-reproductive mechanism, the parent loops as well as the produced variant can continue their self-reproductive activity after the accident. I expected such a direct interaction of phenotypes of *evoloops* to drive their evolution. Note again here that the state-transition rules of the *evoloop* have *no* explicit mechanism for evolution; they are merely composed of phases necessary for self-reproduction of loops.

Figure 5: Takeover of the arm caused by collision of two *evoloops*.

3.2 Modifying the initial structure

To examine evolvability of the *evoloop*, I carried out several preliminary experiments of breeding *evoloops* in finite spaces. The results indicated that the *evoloop* actually has some evolvability, because in some cases the loop evolved to that of larger species, and in other cases it generated some variants that lost their self-reproductive ability but became capable of reproducing smaller offsprings than themselves. However, self-reproductive smaller species could not emerge yet in these preliminary experiments.

One explanation for why evolution toward self-reproductive smaller species did not emerge in the preliminary experiments would be that the loop in that stage did not have a mature capability of injecting enough signals into its off-spring if the form of its arm was altered by some collisions with other structures during the self-reproductive process. This capability of signal injection may be affected by the order of signals in the loop. Based on this idea, I looked for new genotypes of the *evoloop* that would have a stronger self-reproductive ability than before by examining various genotypical patterns. This effort by trial and error fortunately resulted in discovering that some *evoloops* with slightly modified genotypes (shown in Fig. 6) have a stronger self-reproductive ability. The function of these genotypes is exactly the same as before, while only the order of signals differs. In the new genotypes, signal '4's are located near the front of a signal stream instead of its end. These genotypes seem convenient for a loop to inject more signal '7's into its offspring than before when some collision happens to itself. It must be noted that such genotypes were not viable without the new state-transition rules implemented in this study. In distinction from the old loop, these new loops with new genotypes are tentatively called 2-*evoloop*, 3-*evoloop*, and so on, by prefixing the number of signal '7's in front of signal '4's. According to this naming manner, the old loop should be called n-*evoloop*.

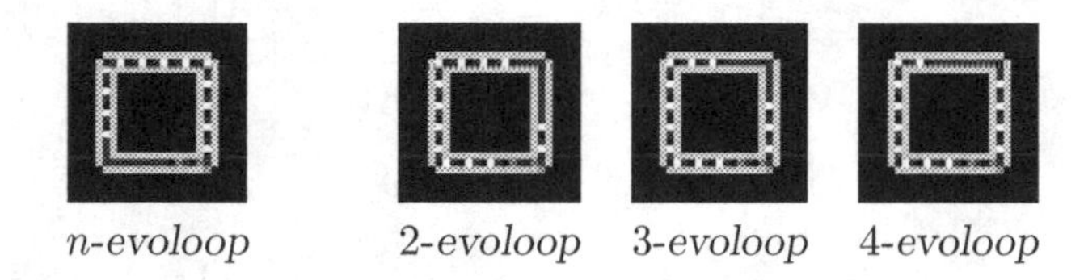

Figure 6: New genotypes of *evoloops* of species 13 that have a stronger self-reproductive ability. The right three loops have new genotypes of the strong self-reproductive ability in comparison with the original (leftmost one).

4 Results

I carried out full-scale experiments of breeding *evoloops* with new genotypes. The size of the space was decided, in consideration of both computing speed and feasibility of visualization, to be 199×199 to 201×201 sites. The loops of species 10 to 13 were selected to be ancestors as they were the largest species that rarely became extinct in the spaces of the aforementioned sizes. I examined 2-, 3-, and 4-*evoloops*. These experiments resulted, in almost all cases, in *evoloops* varying through direct interaction of phenotypes, the whole population gradually evolving toward smaller species, and finally the space filled with the smallest one.

A result using 2-*evoloops* of species 13 in a space of 200×200 sites is shown here for a typical example. In almost all other cases, behaviors of the whole system are qualitatively the same as this.

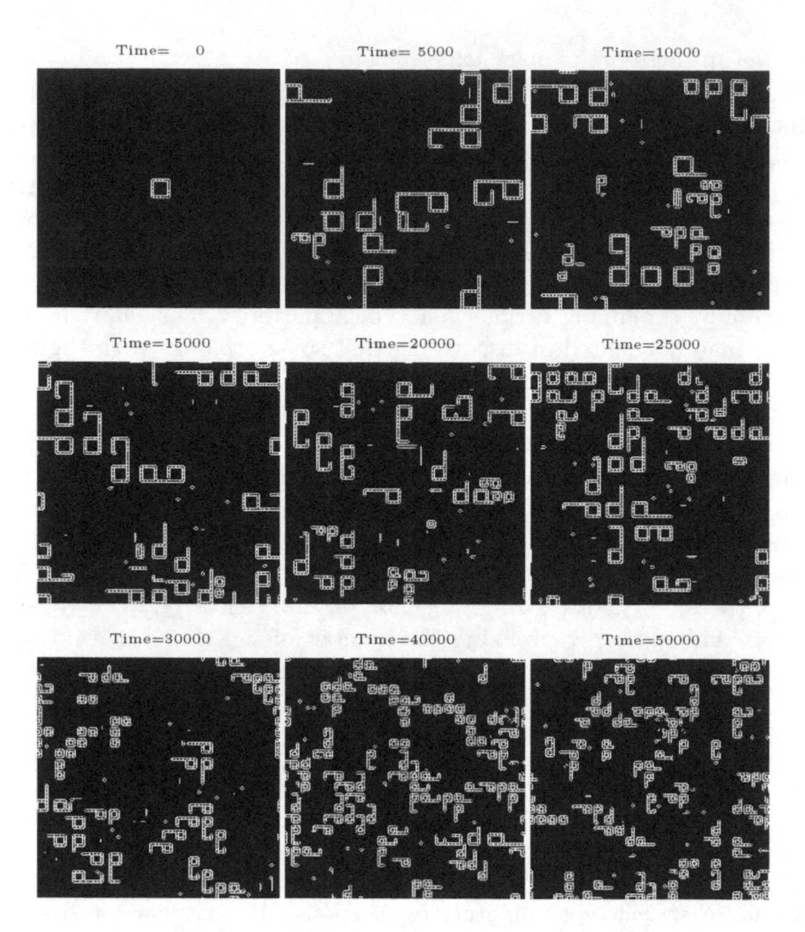

Figure 7: Temporal development of configuration of the evolutionary process of 2-*evoloops*. The ancestor is of species 13. The space is of 200×200 sites with periodic boundary conditions.

Fig. 7 shows temporal development of configuration in the evolutionary process in that case. At first an ancestral loop is set alone in the center of the space. When simulation begins, the ancestral loop soon proliferates all the space. Then, self-reproduction and structural dissolution of loops begin to happen frequently in the space, which produce various kinds of variants such as sterile loops, loops with two arms, loops not self-reproducing but reproducing smaller offsprings than itself, and so forth. A self-reproducing loop of smaller species also emerges by accident from this melee, and once it appears, it is naturally selected to proliferate in the space, due to its quicker self-reproductive ability. Such an evolutionary process develops in the space as time proceeds, and eventually, the whole space becomes filled with loops of species 4, which is the strongest species in this world.

A principal cause of evolution in this world is direct interaction of phenotypes such as a collision of two loops or a crash of a loop into a debris structure, which may change the length of their construction arms. It is quite characteristic of this evolutionary process that the variation in this world occurs first on the phenotype (not on the genotype) of the offspring being produced, consequently leading to alteration of the genotype. This manner of variation is in contrast to the idea of mutation we usually consider.

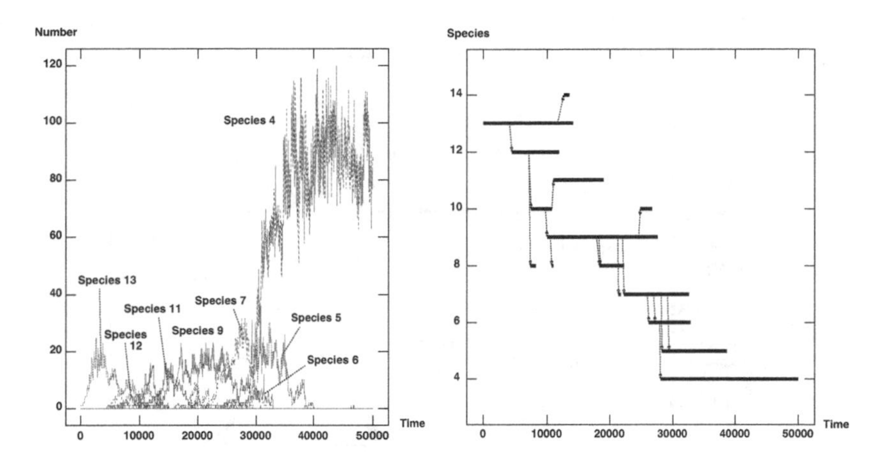

Figure 8: Temporal development of numbers of living *evoloops* (left) and their genealogy (right) in the shown case.

Fig. 8 shows temporal development of numbers of living loops and their genealogy in the aforementioned case. It is clearly observed from these graphs that various species of *evoloops* are produced in the course of evolution, and species 4 finally exterminates the other species. The genealogy (shown on the right) indicates that variation occuring in this world has some tendency to move toward smaller species, but it also leads to larger ones in relatively low probability. Anyway, the whole system seems to evolve toward the smallest species 4 approximately in proportion to elapsed time. In addition, it is found in this genealogy that larger species sometimes exterminate emergent smaller one that should, theoretically, have stronger power of self-reproduction. This indicates that selection in evolution of life can be affected to some extent by local, unpredictable conditions as well as by difference of fitness of competitive species.

Though the *evoloops* showed interesting evolutionary behaviors, we could not observe in their world either punctuated equilibrium of evolution or symbiosis of different species, which had been reported in other evolutionary systems[7, 9]. A main reason for this is that the *evoloops* have no ability to interact with each other in a functional way so that they cannot build complex relations by altering mutual fitness landscapes. In other words, the fitness landscape of *evoloops* is fixed throughout the run, where they merely adapt to a physical environment—a static space composed of a fixed number of finite sites.

5 Discussion

We may derive from this study some insights on the evolvability of artificial evolutionary systems. Fig. 9 shows a rough analogy between the development of digital organisms made by computer programs[1, 5, 7, 9] and the development of self-reproducing loops on CA including the *evoloop*. Behaviors of these artificial systems are classified here into three categolies: self-reproductive, competitive, and evolvable; the last category is further divided into two: adaptive to physical environment and adaptive to other individuals.

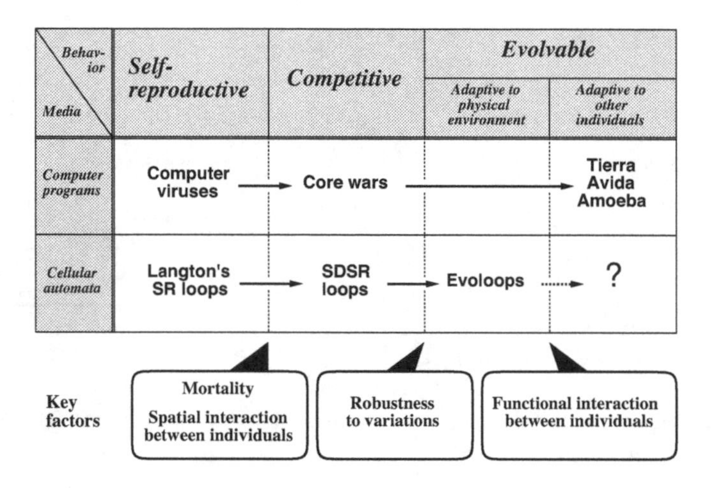

Figure 9: Analogy between the development of digital organisms made by computer programs and the development of self-reproducing loops on CA.

It would be possible to extract some factors common to both kinds of artificial systems in the same class of this analogy. The key factors to create competitive systems is obviously both mortality of individuals and spatial interaction between them. They are not sufficient, however, in advancing artificial systems to the evolvable class. The most important factor common to evolvable systems would be robustness of organisms to variations. For example, the famous evolutionary system *Tierra*[9] met with success by making both its instruction set and its addressing mode quite robust to genetic operation such as mutation and recombination of program codes. This factor also forms the main difference between the SDSR loop and the *evoloop*. Of course, as noted above, the *evoloop* is currently not in the same class as *Tierra*: It adapts only to a physical environment because it lacks a fourth key factor—functional interaction between individuals which causes emergence of diversity of digital organisms in *Tierra*.

If such a truly complex evolutionary system could be constructed on a simple deterministic CA space as a successor to the *evoloop*, it would be expected to have some characteristics inherited from the *evoloop*, as follows: (1) Evolution of life in such a system would be realized without any stochastic operations such as random mutation of genotypes. (2) There would be no need of a central operating system to maintain/evaluate information about the activities of all living individuals, since any particular information about the individual organism is maintained on the configuration of the CA space where it resides. (3) Such a system would be intrinsically very suitable for massively parallel processing, since all behaviors of the system emerge from only local computations between neighboring sites. These imply that, in the near future, we would be able to create extraordinary large-scale evolutionary systems on a fine-grained superparallel machine environment by using extremely simple algorithms, which would greatly advance our knowledge of both natural and artificial life.

Finally, from a biological viewpoint, the results obtained in this study can be regarded as a unique example of evolution in which variation occuring on phenotypes by their direct interaction consequently leads to variation of genotypes. This kind of evolutionary process emerging in the *evoloop* world would bear a close resemblance to the beginning of evolution of primitive life of small complexity, which might have actually occurred in the anscestral world. In such a world, organisms must have evolved not only by genetic mutation but also by interaction with the external environment, including other organisms.

For details of the *evoloop* refer to the other literature[11, 13], some of which can be retrieved from `http://necsi.org/postdocs/sayama/sdsr/`. This site carries several simulator software packages and color movies of acting loops too, which would be helpful for readers in understanding the behaviors introduced in this article.

Acknowledgments

I am indebted to the following for valuable advice and kind support: Chris Langton, Yoshio Oyanagi, Toshihisa Takagi, Wayne Dawson, Yaneer Bar-Yam, and Mari Sayama. I also thank an anonymous reviewer for helpful comments.

Bibliography

[1] ADAMI, C., and C. T. BROWN, "Evolutionary learning in the 2D artificial life system "Avida"", *Artificial Life IV: Proceedings of the Fourth International Workshop on the Synthesis and Simulation of Living Systems* (Cambridge, Massachusetts,) (R. BROOKS AND P. MAES eds.), MIT Press (July 1994), 377–381.

[2] BYL, J., "Self-reproduction in small cellular automata", *Physica D* **34** (1989), 295–299.

[3] CHOU, H. H., and J. A. REGGIA, "Emergence of self-replicating structures in a cellular automata space", *Physica D* **110** (1997), 252–276.

[4] CHOU, H. H., and J. A. REGGIA, "Problem solving during artificial selection of self-replicating loops", *Physica D* **115** (1998), 293–312.

[5] DEWDNEY, A. K., "In the game called Core War hostile programs engage in a battle of bits", *Scientific American* **250**, 5 (1984), 15–19.

[6] LANGTON, C. G., "Self-reproduction in cellular automata", *Physica D* **10** (1984), 135–144.

[7] PARGELLIS, A. N., "The evolution of self-replicating computer organisms", *Physica D* **98** (1996), 111–127.

[8] PERRIER, J. Y., M. SIPPER, and J. ZAHND, "Toward a viable, self-reproducing universal computer", *Physica D* **97** (1996), 335–352.

[9] RAY, T. S., "An approach to the synthesis of life", *Artificial Life II: Proceedings of the Workshop on Artificial Life* (Santa Fe, New Mexico,) (C. G. LANGTON, C. TAYLOR, J. D. FARMER, AND S. RASMUSSEN eds.), vol. X of *SFI Studies in the Sciences of Complexity*, Addison-Wesley (February 1990), 371–408.

[10] REGGIA, J. A., S. L. ARMENTROUT, H. H. CHOU, and Y. PENG, "Simple systems that exhibit self-directed replication", *Science* **259** (1993), 1282–1287.

[11] SAYAMA, H., *Constructing evolutionary systems on a simple deterministic cellular automata space*, Ph.D. dissertation Department of Information Science, Graduate School of Science, University of Tokyo (December 1998).

[12] SAYAMA, H., "Introduction of structural dissolution into Langton's self-reproducing loop", *Artificial Life VI: Proceedings of the Sixth International Conference on Artificial Life* (Los Angeles, California,) (C. ADAMI, R. K. BELEW, H. KITANO, AND C. E. TAYLOR eds.), MIT Press (June 1998), 114–122.

[13] SAYAMA, H., "A new structurally dissolvable self-reproducing loop evolving in a simple cellular automata space", *Artificial Life* **5**, 4 (1999), 343–365.

[14] SIPPER, M., "Non-uniform cellular automata: Evolution in rule space and formation of complex structures", *Artificial Life IV: Proceedings of the Fourth International Workshop on the Synthesis and Simulation of Living Systems* (Cambridge, Massachusetts,) (R. BROOKS AND P. MAES eds.), MIT Press (July 1994), 394–399.

[15] TEMPESTI, G., "A new self-reproducing cellular automaton capable of construction and computation", *Advances in Artificial Life: Proceedings of the Third European Conference on Artificial Life (ECAL'95)* (Granada, Spain,) (F. MORÁN, A. MORENO, J. J. MERELO, AND P. CHACÓN eds.), vol. 929, Lecture Notes in Artificial Intelligence of *Lecture Notes in Computer Science*, Springer-Verlag (June 1995), 555–563.

Regulation of Cell Cycle and Gene Activity Patterns by Cell Shape: Evidence for Attractors in Real Regulatory Networks and the Selective Mode of Cellular Control

Sui Huang and Donald E. Ingber
Departments of Surgery and Pathology, Children's Hospital
Harvard Medical School, Boston
huang_su@a1.tch.harvard.edu

1 Background

The central riddle of an integrative biology is the genome-to-phenome mapping problem, i.e. the functional relationship between the genes (the parts) and the organism (the whole). Despite formidable success in molecular biology which has unraveled countless gene regulatory pathways, a wide gap between the microscopic world of biochemical information, encoded in the genes and proteins, and the macroscopic, physical world, in which the cells, organs and organism exist, remains to be bridged. Of particular interest for biomedical science is the generation and maintenance of the intricate tissue architecture in higher organisms which underlies all physiologic function. How do cells sense "higher level" needs, that is, macroscopic mechanics and physiological requirements of the tissue, in order to regulate their genes and proteins, which in turn control their behavior in forming the essential tissue structure ? We start with a brief overview (section 1) on some fundamental concepts that has been helpful in addressing the different hierarchical levels of description in living systems. They form the framework leading to the predictions described in section 1.5., which will be followed by presentation and discussion of recent experimental data (section 2).

1. 1. Reaction-diffusion models

A paradigm for the structural aspect of the relationship between the parts and the whole is the reaction-diffusion mechanism, proposed by Alan Turing [1952] to explain the formation of biological patterns, such as segmentation or spiral waves. In this classical model of morphogenesis, long-range order is explained by short-range interactions described by appropriate spatio-temporal dynamics, in which diffusing molecules (morphogens) undergo chemical reactions which involve autocatalytic and inhibitory processes [see also Murray 1989]. Under certain conditions, such chemical systems can lead, via spontaneous symmetry-breaking, to simple geometric patterns, the Turing structures: Chemistry determines Geometry.

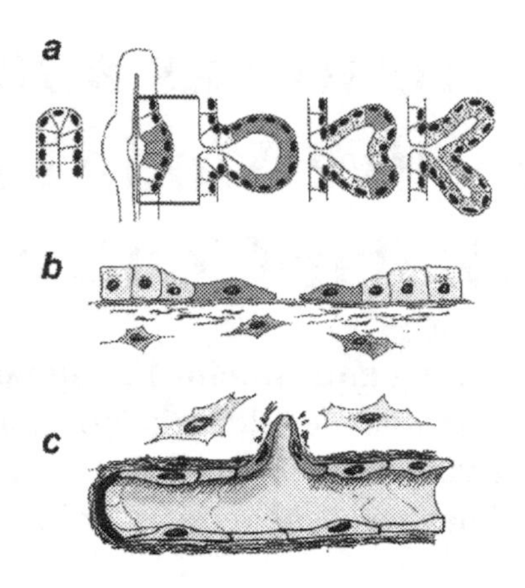

Figure 1. Development and maintenance of tissue architecture requires highly localized regulation of cell proliferation. Examples of tissue architecture remodeling where individual cells have to divide (dark cells), while their neighbors have to remain quiescent for generating the histologic patterns. Cells have to sense their physical environment (free space, geometry, forces) and adjust their proliferation.
a: Morphogenesis of glandular structure; *b:* Epithelial wound healing; *c:* Angiogenesis.

The intricate tissue architecture in higher organisms (Fig. 1) consists of patterns jointly generated by the cells and the extracellular matrix (ECM) that they produce [Adams 1993; Ingber 1989]. Based on the reaction-diffusion paradigm, an analogue, mechano-chemical model has been proposed by Murray [1989], where cells are the basic elements of transport and "reaction": the cells are subjected to convective transport by the ECM, random dispersal (analogue to diffusion), haptotaxis, and cell division (analogue to Turing's reaction term for molecules). In these models the basic element which 'reacts and diffuses' are molecules or cells, whose size S is negligible compared to a characteristic length L ("wavelength") of the patterns generated. However, unlike in the reaction-diffusion models inspired by chemical systems, for patterns of real tissues the characteristic length L of the convoluted morphological patterns is in the same range as the size S of the building blocks, the cells. This creates the highly complex patterns that cannot be explained by the Turing formalism. Moreover, in the reaction-diffusion model for cells, the growth behavior of a cell (the "reaction term" in the Turing equation) is often modeled with a logistic growth function, which should capture the phenomenon of density-arrest for a cell population. This type of regulation, however, reflects the long-range behavior of cells as a homologous population, and would not explain local spatial differences in proliferation, where one cell has to divide, while its immediate neighbor stays quiescent. Such a cellular behavior, regulated at the level of single cells according to the physical need of the local tissue micro-environment, is essential for generating the complex histologic tissue patterns in higher organisms (Fig. 1). Therefore, cells are not only regulated by soluble factors as classical biology assumes, but information from the physical structure of the local environment must also feed back onto the biochemistry of individual cells to regulate cellular behavior: Geometry must also control Chemistry.

1.2 Collective variables and control parameter

For the functional aspect of the part-whole relationship, the analogue to the geometric structure as a higher level pattern is the coordinated "genetic program" that governs physiological processes, such as cell differentiation, regeneration or an immune response. These processes represent ordered, emergent features which, similarly to the geometric patterns, arise through the interaction of the parts and cannot be studied as properties of the isolated parts. However, unlike geometric patterns which emerge from a small number of defined reagents and reactions, these abstract, ordered features are difficult to characterize because the relevant degrees of freedom are usually not known. One way to address such systems is to identify "phase transitions", that is, the macroscopic changes between qualitatively discrete, stable states, and then to analyze their dynamics as the collective variable(s) which depend on a relevant control parameter. This provides a handle to describe in precise terms the collective, ordered behavior and its relation to the possible underlying elements. In cell biology, we postulate that functional cellular states such as differentiation (or quiescence) and proliferation (or growth), represent the collective behavior of the genes and that the "switch" between these states can be viewed as phase transitions. How are these transitions regulated, i.e. what is the control parameter ?

1.3. Selectivity vs. instructivity in the regulation of cellular function

Although theorists of developmental biology now agree that regulation of cellular function is a selection of pre-existing (latent) behavioral modes of the cell [Wolpert 1994], practitioners of molecular and cell biology still implicitly regard soluble, diffusible factors that bind with high specificity on cell surface receptors as the primary regulator of cellular function. They assume that these messenger molecules (e.g. hormones, growth factors) instruct the cell with the information that is encoded in their three-dimensional structure how the cell has to behave, i.e. which "genetic program" it has to activate. This information is "read" by the cell via the cell surface receptor with a structure complementary to the messenger molecule. According to this instructivity view, the protein "nerve growth factor" (NGF), for instance, would encode the information about the set of genes to be activated to cause the development of a neural cell. However, numerous experimental evidence at the molecular level, including the use of chimeric receptors, have lent credence to the concept of selectivity. Cellular states, such as 'quiescence' and 'growth' can be seen as the pre-existing behavioral modes of the selectivity model, and thus, transition between them would be a mere selection process that could be regulated by a "non-specific" control parameter which does not need to contain the specific instruction for the genetic system of the cell. Altering the value of the control parameter would be able to cause the cell to switch between the pre-existing modes of behavior. Thus, according to the selectivity model the functional state of a cell is "self-organized" - there is no external instruction needed, only a defined trigger. This is reminiscent of the control of emergent features in physical systems by a non-specific parameter like the temperature in the Rayleigh-Benard instability [Normand 1977]. The question in biology is: how are the latent behavioral modes, as postulated by the selectivity view, established by the molecular constituents of the cell ?

1.4. Genetic networks and attractor states

In his studies of random boolean networks, in which binary genes are connected by regulatory interactions, Stuart Kauffman [1993] has shown that for a class of such networks, ordered collective behavior with respect to gene activity profiles spontaneously emerges. The network settles down into a few small attractors (in the state space defined by all the possible gene activity patterns of the genome). We propose now that such attractors correspond to cellular states, such as the growth or quiescence state studied here. The dynamics of such networks predict the existence of multiple states (gene activity patterns represented by attractors), which are qualitatively discrete and inherently stable, but can transit into each other with some efficiency upon an appropriate perturbation of the gene activity patterns. With such pre-existing alternative states being a natural consequence of the wiring diagram of the genomic regulatory network, the genetic network model provides an underlying mechanism for the formation of the latent, pre-existing behavioral modes postulated by the selectivity model.

1.5. Predictions for cellular behavior

The selectivity model requires and the idea of genetic networks predicts that the intracellular regulatory network of genes self-organizes into multiple, alternative cellular states and that therefore, a non-specific, physical control parameter would be able to evoke the same cellular responses known to be induced by specific biochemical signals which contain the regulatory information in their structure. This property of cellular regulation, together with the requirement for growth control at the level of individual cells by the tissue architecture, led us to postulate that cell shape, imposed by the physical environment, is such a non-specific parameter that regulates cellular states. Previous work by many groups has shown that cell shape can control cell growth. However, it wasn't clear to what extent instructive influences from protein ligands in the ECM played a role [Folkman 1978; Assoian 1997]. Further, it is not known how cell shape could interfere with the molecular mechanism of the cell cycle. Here we present some of our experimental findings showing the ability of cell shape *per se*, as a non-specific parameter, to regulate the switch between cell growth and quiescence and that it does so by harnessing the biochemical machinery of the cell in a coordinated manner.

2. Results and Discussion

2.1. Role of cell shape in growth control

We used several techniques to vary the shape of quiescent capillary endothelial cells attached to a planar surface, and measured their entry into the cell cycle. The experimental details were recently published [Chen 1997; Huang 1998]. In particular, the novel technique of micro-patterning cell-adhesive surfaces in a geometrically defined manner allowed us to precisely control the extent of cell spreading (i.e. "cell shape", measured as projected area) - independent from specific "instructive" influences arising from ECM proteins or growth factors in the culture medium. Thus, we could vary the cell shape as a control parameter, but keep biochemical culture conditions constant. We found that the probability of the quiescent cells to enter the cell cycle and progress to the DNA synthesis (= S) phase, decreased with the reduction of the projected cell area, while the period of the cell cycle was not affected. The latter supports the notion of discrete cellular states and phase transitions ("switches") between them. These results show that in addition to the classical soluble mitogens, cell shape is a critical (rate-limiting) factor in controlling the switch between growth and quiescence and acts independently of specific, biochemical, i.e. instructive, inputs. The significance of this finding is two-fold: first, biologically, it shows that not only chemically encoded information, such as hormones and growth factors, but also non-instructive physical cues, like the availability of space in the local environment of the cell, can be sensed by the cell and translated into a behavioral program. This could represent an important mechanism in morphogenesis. Second, from the perspective of complex dynamic systems, the results indicate that changes of cellular behavior can be treated as phase transitions which take place when the value of a single, non-specific, control parameter is continuously altered: qualitative change of behavior does not necessarily require qualitative information (e.g. in the form of a receptor ligand).

2.2. Analysis of gene activity pattern during shape-dependent growth regulation

(a) Experimental findings

For the regulation of cellular functions by specific molecules, the intracellular biochemical machinery involved in the switch from quiescence to proliferation (growth) and in cell cycle progression are pretty well understood [Sherr 1996]. But what is the molecular nature of the switch that is triggered by the changes in cell shape ? To study the relationship between cell shape and the biochemistry of the cell cycle, we analyzed the expression and activation of cell cycle genes and their proteins in cells that were either cell-cycle-arrested or allowed to proliferate by controlling the cell shape parameter. The genes studied here have an established role in the cell cycle regulation (Fig. 2, see legend). The results were reported recently in another context [Huang 1998]. In brief, cells that were kept small (non-spread) as described in 2.1. or by pharmacological disruption of the cytoskeleton (which also prevents spreading),

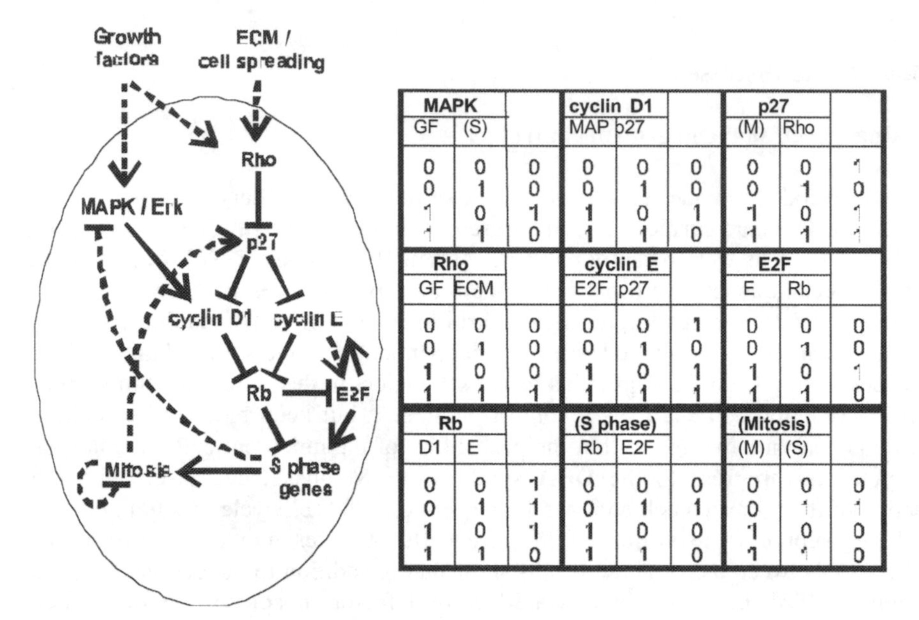

MAPK			cyclin D1			p27		
GF	(S)		MAP	p27		(M)	Rho	
0	0	0	0	0	0	0	0	1
0	1	0	0	1	0	0	1	0
1	0	1	1	0	1	1	0	1
1	1	0	1	1	0	1	1	1
Rho			**cyclin E**			**E2F**		
GF	ECM		E2F	p27		E	Rb	
0	0	0	0	0	1	0	0	0
0	1	0	0	1	0	0	1	0
1	0	0	1	0	1	1	0	1
1	1	1	1	1	1	1	1	0
Rb			**(S phase)**			**(Mitosis)**		
D1	E		Rb	E2F		(M)	(S)	
0	0	1	0	0	0	0	0	0
0	1	1	0	1	1	0	1	1
1	0	1	1	0	0	1	0	0
1	1	0	1	1	0	1	1	0

Figure 2. Real regulatory network of genes/proteins controlling the cell cycle and representation as boolean network of N=9 elements / k=2 inputs per element with 2 environmental inputs.

LEFT PANEL: Regulatory interactions between N=9 growth-related genes. Pointed arrow heads denote stimulation, barbed ends indicate inhibition. Depicted interactions are experimentally established relationships, except for induction of Rho by cell spreading (circumstantial evidence). Dashed lines indicate indirect functional regulation in real cells, implemented in the network simulation to obtain the necessary feedback loops. For simplicity, every gene was allowed to have only two inputs (k=2), corresponding to the best characterized upstream regulators. *RIGHT PANEL:* Look-up table for the boolean functions of the genes of the network with their two input ("upstream") genes (first and second column), and the output state (last column) according to the boolean function. For the full name of the genes and their biochemical functions see ref. in [Huang 1998; Sherr 1996]. (S) = S-phase: represents group of genes involved in DNA synthesis and whose transcription typically is inhibited by Rb and stimulated by E2F. (M) Mitosis: was used in the simulation to close the circle and represents in reality an event downstream of S phase which involves a whole cascade of effector genes. The self-limiting property of mitosis was implemented by a negative feed-back loop.

failed to induce the cell cycle-promoting protein cyclin D. Instead, they accumulated the cell cycle inhibitor protein p27. Further, unlike the large, spread cells, the small cells failed to induce the transcription of cyclin E and E2F-1, genes that are crucial for the entry into S phase. This profile of gene activation is known from classical cell cycle-arrest achieved by withdrawal of serum or growth factors. It leads, as observed here for the small cells, to the failure of inactivating the major gate-keeper of the cell cycle, the Rb protein, which in turn results in the exit from the cell cycle prior to S phase entry.

These findings demonstrate that cell shape regulates the cell cycle by impinging on the same specific molecular players that have been shown to mediate cell cycle control in response to soluble regulator molecules. There is a remarkable similarity between the gene activity profile of cells arrested by controlling cell shape and that in cells conventionally arrested by drugs which block biochemical growth signals or by growth factor withdrawal. This convergence to a common gene activity profile in response to completely different regulatory influences points to the existence of a coherent, stable pattern and suggests that such patterns correspond to attractors in the regulatory network's state space.

(b) Comparison with a simulated network

To illustrate this idea, we simulated the dynamics of a small network as a (non-random) boolean network which consisted of cell cycle genes/proteins, and whose wiring diagram was based on the biochemical or functional interactions between them (Fig. 2). All the connections have been experimentally proven in cell cycle research, with the exception that we had to assume that cell shape affects the signaling protein Rho (for which there is circumstantial evidence). We then compared the gene activity status of the attractor states of the simulated network with the observed activity status in our experimental cells. As shown in Fig. 3, in the presence of growth factors and ECM (i.e., cell spreading) the theoretical network exhibits a single limit-cycle attractor (state cycle) corresponding to the proliferation state (i.e., when the cell cycle is active - "growth attractor" in Fig. 3, right). The gene activity patterns of the individual states which form the state cycle as cells pass through the cell cycle was very similar to the activity profiles of genes that we observed. (Some minor oscillatory deviations in the simulation might be due to the boolean idealization). In the simulation, this attractor was the only attractor of the state space of the network. Upon removal of the growth factors ("serum-starvation"), or the ECM (preventing cell spreading), the limit cycle attractor "collapsed" to fixed-point attractors, which are unstable in the presence of growth factors or ECM, repectively ("Quiescence attractors" in Fig. 3, left). These attractors correspond to the quiescent states, and again, their gene activity patterns are consistent with observations in real cells. The attractor that forms in the absence of soluble mitogens (quiescence attractor I) and of cell spreading (quiescence attractor II) are very similar to each other, with the only difference being the signal transduction proteins MAPK and Rho - this reflects the difference of the external signals. In the simulation it is obvious that the shape-dependent signal will elicit a concerted response leading to the transition between attractors in the network since we had to explicitly feed in the shape signal via Rho as an input of a boolean function. But in reality, the molecular pathways for shape dependent control remain obscure. Still, the fact that in our cells, shape change regulated patterns of gene activation are almost identical to that of the corresponding simulated situation and to the ones under the regulation by specific soluble mitogens is remarkable and demonstrates that the non-specific "shape signal" is translated by the cell into coherent gene activity patterns that represent attractors.

Taken together, our experimental findings support the existence of latent, stable states in real cell regulation. These in turn are best explained by attractor states of a regulatory network which can relatively easily be "reached" by the cell when they are

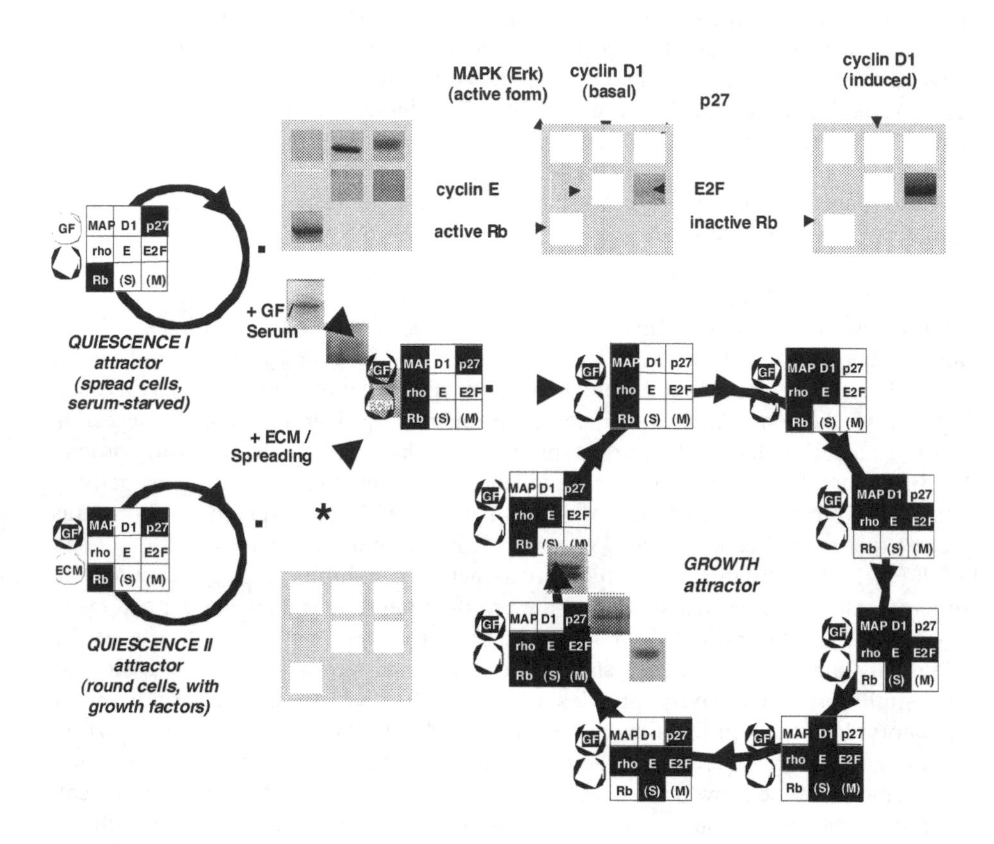

Figure 3. Simulation and experimental monitoring of gene activity patterns representing the attractors of the regulatory network and real cellular states.

Nine relevant growth-related genes (or group of genes, as explained in the previous figure 2) were chosen to form a regulatory boolean "subnetwork" of the size N=9 genes, with two external inputs. The 3x3 grids represent the network state with the activity pattern of the nine genes in the nine fields. Black fields indicate expressed or activated genes. The circles on the left of each grid indicate the status of the external inputs to the network: presence or absence of serum (with the growth factors, GF) and extracellular matrix, ECM (equal to cell spreading). Thick arrows denote transitions between network states upon spontaneous updating or perturbation by the external inputs. For four network states results from experimental monitoring of some of the genes by reverse transcription polymerase chain reaction (for transcriptional activation, in the case of cyclin E and E2F) or Western blot analysis (for translational or post-translational activation, for all the other genes) are shown in the corresponding fields in the gray inserts. Inactivation of Rb leads to the occurrence of a doublet band, reflecting hyperphosphorylation. (S) = group of S-phase specific genes, (M) = mitosis. MAP = MAPK/Erk; E, D1 = cyclin E and D1, repectively. Note that in real cells the transition (*) between the quiescence II attractor and the growth attractor is regulated by a "non-specific" control parameter, cell shape.

"tipped" into the corresponding basin of attraction. This alleviates the requirements for specific regulatory signals and allows physical parameters to elicit ("select") the same biological responses as those known to be induced by specific chemical messenger molecules. The compression of the complicated regulatory network of the genome into simple behavioral rules that can be exploited by higher order structures exposed to the laws of classical mechanics relieves the constraints imposed by the need for precise molecular signals at the microscopic level. This may facilitate the evolution of increasingly complex organisms adapted to the macroscopic world.

3. Conclusion

Information in biology is generally thought to reside in the DNA sequence and the encoded protein sequence, and to be retrieved and passed on by the three-dimensional structure of proteins when they can engage in highly specific intermolecular interactions. Here we study the interface between the microscopic world of bio-information and the macroscopic, physical world of geometry and mechanics by showing that cell shape can regulate gene activity patterns in the same way as do specific molecules. We provide experimental evidence to support the ideas of selectivity in cell regulation and show that genetic networks with attractors representing cellular states is a valid concept in a real cellular regulatory system. More work needs to be done to refine the theory of regulatory (boolean) networks, in particular, by implementing multi-threshold functions, non-synchronous updating etc., in order to bring the model closer to reality.

The mechanism by which cell shape, a physical control parameter devoid of a molecular specificity, is translated into distinct biochemical reactions remains to be established. Preliminary data indicate an important role for the actin cytoskeleton and the generation of isometric cellular tension in this process, supporting the concept that the cytoskeleton responds as a global, tensionally integrated, "tensegrity" system to the mechano-chemical signals emanating from the cell's matrix environment [Ingber 1999]. The flow of information from the macroscopic, physical world to the microscopic world of genetic information is an important novel aspect of the problem of genome-phenome mapping, and elucidating its molecular mechanism will be crucial in our attempt to understand the relationship between the parts (the genes) and the whole (the organism).

References

Adams, J.C., & Watt, F.M., 1993, Regulation of development and differentiation by the extracellular matrix", *Development,* **117**, 1183-1198.

Assoian, R.K., & Zhu, X., 1997, Cell anchorage and the cytoskeleton as partners in growth factor dependent cell cycle progression, *Curr. Op. Cell. Biol.,* **9**, 93-98.

Folkman, J., & Moscona, A., 1978, Role of cell shape in growth control, *Nature* **273** 345-349.

Chen, C.S., Mrksich, M., Huang, S., Whitesides, G.M., & Ingber, D.E., 1997, Geometric control of cell life and death, *Science,* **276**, 1425-1428.

Huang, S., Chen, S.C., & Ingber, D.E., 1997, Control of cyclin D1, p27(Kip1), and cell cycle progression in human capillary endothelial cells by cell shape and cytoskeletal tension, *Mol. Biol. Cell.,* **9**, 3179-3193.

Ingber, D.E., Folkman, J., 1989, Mechanochemical switching between growth and differentiation during fibroblast growth factor-stimulated angiogenesis in vitro: role of extracellular matrix, *J. Cell. Biol.,***109**, 317-30.

Ingber, D.E., 1998, The architecture of life, *Scientific American*, **278** (January), 48-57

Kauffman, S.A, 1993, *The origins of order*. Oxford University Press

Murray, J.D., 1993, *Mathematical biology*. Second Edition, Springer-Verlag.

Normand, C., Pomeau, Y., & Velarde, M.G., 1977, Convective instability: a physicist's approach, *Rev. Modern. Phys*.**, 49**, 581-624.

Sherr, C.J., 1996, Cancer Cell Cycles, *Science,* **274**, 1672-1677.

Turing, A.M., 1952, The chemical basis of morphogenesis, *Phil. Trans. Roy. Soc. Lond.*, **B237,** 37-72.

Wolpert, L., 1994, Do we understand development ? *Science,* **266**, 571.

A Model System and Individual-Based Simulation for Developing Statistical Techniques and Hypotheses for Population Dynamics in fragmented Populations

Jay Bancroft
University of Connecticut
Ecology and Evolutionary Biology U-43
Storrs, CT 06269
email:jsbancr@udel.edu

1 Background

We need to minimize the negative impacts of humans on biodiversity, and an obvious way to do that is to understand how our activities cause extinction (Erlich et al. 1981, Raven 1990, Soule 1991, and Wilson 1989). Human infrastructure fragments and degrades habitat, and in turn, subdivides and reduces previously continuous populations. In order to understand the dynamics of small populations, it is crucial to use individual-based models that allow for local patch heterogeneity and individual uniqueness. I used a combination of a model animal and computer simulation to determine the mechanisms that effect a population's response to resource fragmentation with the sawtoothed grain beetle, *Oryzaephilus suranemensis* (L.). Laboratory experiments were used to calibrate density dependent functions for oviposition, mortality, and dispersal. These empirically determined functions form the basis of an individual-based simulation model (IBS). Time-series census data was taken on populations subjected to resource limitation and fragmentation. The population time-series data for the metapopulations was used to test the predictions of the simulation. By calibrating the model with short-term empirical trials and testing the simulation using long-term, observed population dynamics, greater confidence is placed on model predictions. Finally, the simulation was used to quantify the most influential parameters on beetle abundance. This work aims to link theory to conservation biology using controlled experiments that can not be performed with endangered species or in landscape-sized systems.

In order to prevent extinction, we need to create a theoretical framework to rapidly access the dynamics of populations and identify endangered populations while we can still do something to help them. Tilman (1994) proposed the extinction debt hypothesis that predicts conservation efforts might be futile because of the lag time between habitat loss and species extinction. Many endangered populations are distributed as patches. A metapopulation is a set of discrete breeding populations of organisms between which some organisms may disperse (Hanski et al. 1997). Research on metapopulations, especially according to this broad definition, has exploded in recent years. Simulation experiments have predicted population persistence based on dispersal, patch size, patch destruction rate, and inter-patch distance (Adler 1994, Bascompte et al. 1996, Burkey 1989, Fahrig et al. 1988, Haydon et al. 1997, Keitt et al. 1995). Burkey (1997) has recently performed experiments with microorganisms that indicate habitat connections do not increase population persistence times. These results will be compared with my experimental persistence distribution, and differences between the two model systems will demonstrate the problems of scaling. Replication and speed make small-scale experiments a powerful method for testing the relative effect of various mathematical representations. Furthermore, mechanistic functions are more useful than phenomenological models for prediction across scales because parameters retain their biological meaning (Conroy 1995). Working with a small scale allows a much-needed empirical test of the theory of spatially structured populations (Doak et al. 1992).

Spatially explicit models have flourished in part because of advances in computational tools, but more importantly, through the development of a much clearer understanding of the complex processes of spatial heterogeneity (Fahrig 1997, Harrison 1991, Hassell et al. 1995, Kareiva 1990, Tscharntke 1991,

Turchin 1991, Turner et al. 1995, Wiens 1997). Within a patch details of life-history and behavior govern population carrying capacity. Wissel et al. (1994), predicted population persistence time: $T_K \approx \ln K$ for $2r <$ _^{2} where r is the intrinsic growth rate. T_K is strongly related to carrying capacity (K) and environmental noise (_^{2}). With my controlled system, the main environmental variation is manifest through resource replenishment. Demographic variation of parameters is also included in the IBS (Fig. 1).

2 System Description

O. suranemensis is generally encountered in grain silos. It is a worldwide pest of stored grain and can and rapidly increase in numbers if unchecked. It has an egg, larval, pupal, and adult stage, and the larva molts between 4 instars. Development lasts about 4, 14, and 4 days for the egg, larval, and pupal stages, respectively.

A number of studies of *O. suranemensis* have characterized reproduction and mortality at various temperatures, humidities, and in various grains (Back 1926, Howe 1956, Prus 1966). Stage specific mortality rates were recently reported (Collins et al. 1989). The important factors for the population dynamics are shown in figure 1, and each factor is represented by an algebraic function in the model. The parameter values from the literature and from my experiments are the basis of the key functions used in the simulation described below.

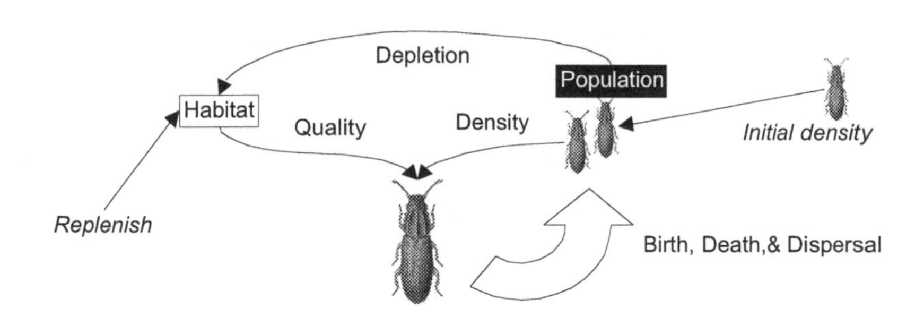

Figure 1. Key components

2.1 Description of Model

The IBS is based on a hierarchy of subprocesses that control the individual objects of the system (Figure 2). At 1-hour time steps, each patch is called to update its quality and call all of the beetles therein to update their state. When an individual is created, it is described by state variables particular to that individual. An individual uses its state variables in conjunction with shared parameters and instructions to guide its behavior. All individuals use the same template of program code. The model predicts population numbers of each life stages for an initial set of patches and their initial beetle population numbers.

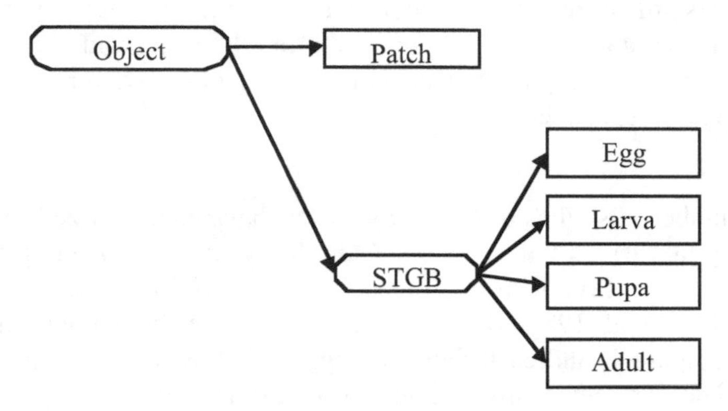

Figure 2. Object hierarchy

Input files hold parameters that describe initial populations, habitat structure, and natural history. The habitat consists of connected patches of various sizes. Each patch is described with a quality and a pointer list of simulated beetles, which reside there. Functions for mortality, reproduction, and dispersal are fitted to observed data that is described below.

3 Experiments

All experiments were performed in an environmental chamber (32±1°C and 65±10%RH). The primary diet was 95% wheat flour and 5% brewers' yeast by weight. Standard film vials were used as patches and are connected with Tygon™ or latex tubes that allow beetles to pass to the neighboring vials.

3.1 Parameter calibration experiments

Cohort density dependent immature survival. One experiment was performed to examine the effects of density dependence in cohorts. Small larvae (5, 10, 20, 40, and 80) were placed in vials with equal amounts of diet. The eclosion rate was almost 90% for single larvae and decreased to 60% for high-density cohorts. The fitted line in figure 4 is the basis for density dependent larval mortality in the simulation. Note that competition for limited food and cannibalism could not be differentiated.

Cohort density dependent immature survival. Another experiment showed a weak but significant effect of diet on the percent of eclosing pupae. In this experiment the same number of larvae were put into varying amounts of diet. This indicates that simple scramble competition is acting and larval cannibalism is not very important in this system.

Adult cannibalism dependent immature survival. Another experiment examined the effect of adults on the mortality of developing larvae (Figure 6). Ten larvae were placed in vials with 0.25g of diet. Each week 10 adults from the stock culture were added to a portion of the vials. There was a significant effect of adult presence on emergence during weeks 3 and 4. This is when pupation occurs and strongly suggests that adults kill the relatively defenseless pupae. Judging by the intact carcasses, the adults are not gaining much nutrition, but they are reducing future competitors for the limited food resources.

Density dependent adult dispersal. These ongoing trials will yield the daily rate of dispersal to a receiving vial based on the number of adults in the source vial. Five, 10, 20, or 40 adults are placed in a source vial with 0.25g of medium. Each day the receiving vials are checked for immigrants; if any are there it is noted and the trial is ended.

3.2 Time series

Sugar diet does not allow reproduction, and the wheat flour with yeast diet is preferred. Every two weeks, the larvae, pupae, adults, and dead adults were counted and the diet was replaced. The dead adults were kept for subsequent sexing to reconstruct sex ratios in the vials.

Time series, arrangement and fragmentation. I have recorded a time series of 32 censuses with two, four and eight vials per metapopulation ring (Figure 3). Some of this data was used to calibrate the model's oviposition, adult mortality, and dispersal. For example, the average number of beetles that died was divided by the total adults giving a 22% mortality rate per 2 weeks. Even simpler, dispersal was identified by an observed colonization of an empty vial. The rest of the time series data was held out for testing the simulation. All trials had a total of one gram of each diet type that was divided into the number of vials in the trial. For example, a two-vial ring had one gram of diet in each vial. A four-vial ring had one-half gram of diet in each vial: two with wheat-yeast and two with sugar.

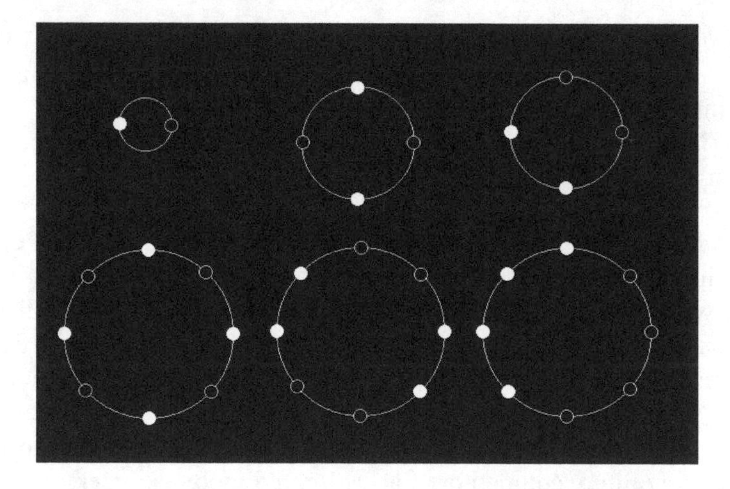

Figure 3. Structure of metapopulations

Fitting a model to the time series will test the hypothesis that threshold exists below which populations decline (Bascompte et al. 1996) Experimental replicates enable the creation of a confidence interval for this threshold. The resulting time series are also being used to enhance the model. The experimental data will be analyzed to determine spatio-temporal autocorrelations for all stages of the populations (Crone 1997, Dennis and Taper 1994, Turchin 1991). An example of the partial autocorrelation function (PACF) is shown in figure 4. I note that within vial regulation was occurring because total abundance in 2 vial series was significantly smaller than the 4 and 8 vial series. This showed that larvae should be represented by a first order autoregressive process. Finally, the average vial populations are shown in figure 5.

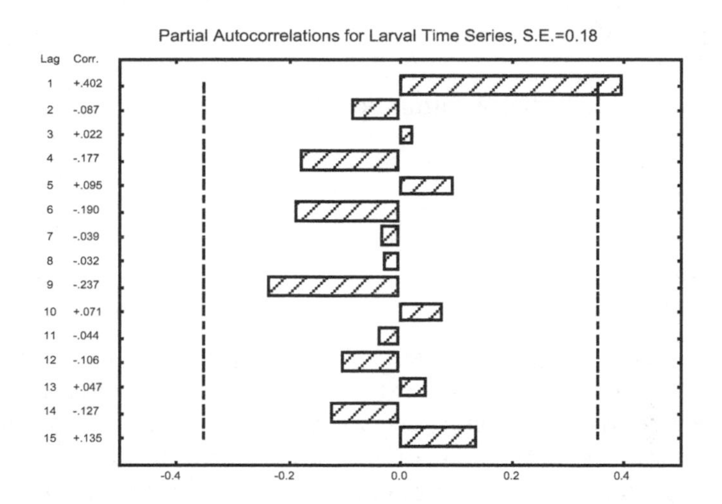

Figure 4. Partial autocorrelation for 1 time series

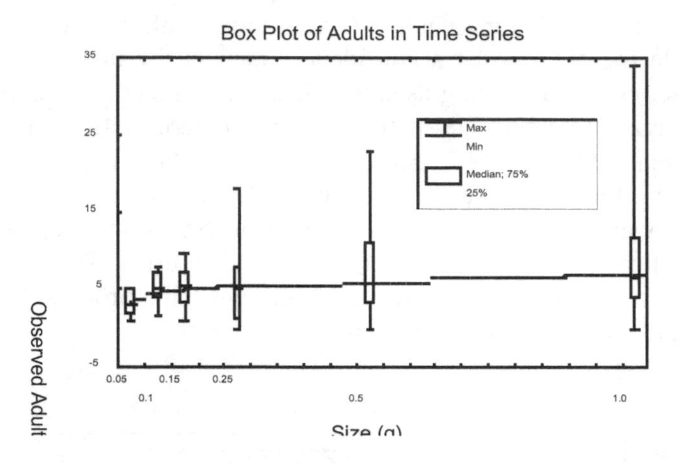

Figure 5. Average population by food availability

4 Model Testing

I have developed a parameter-optimizing program that changes input files, invokes the simulation, and evaluates the fit with observations. The simplex optimizer currently compares moments of the simulation output with test data and alters parameters to obtain the least squares fit. The observed mean, variance, and autocorrelation of the population stages are the response variables (Turchin et al. 1992, Kendall et al. 1998). The optimization process also directs effort where the population is most sensitive to the form of the theoretical function.

Elasticity analysis has determined the sensitivity of the IBS predictions to input parameters (Fig 6). To do this, a parameter value was increased by 5% of its standard value and then the simulation was run 50 times. Then the same parameter was decreased 5% and simulations were run again. By comparing the resultant populations with the output using the standard parameter value, sensitivity is determined for the parameter. This was done for all parameters in the model and the significant parameters are shown. Pupal and adult mortality, and immature development time are the key parameters.

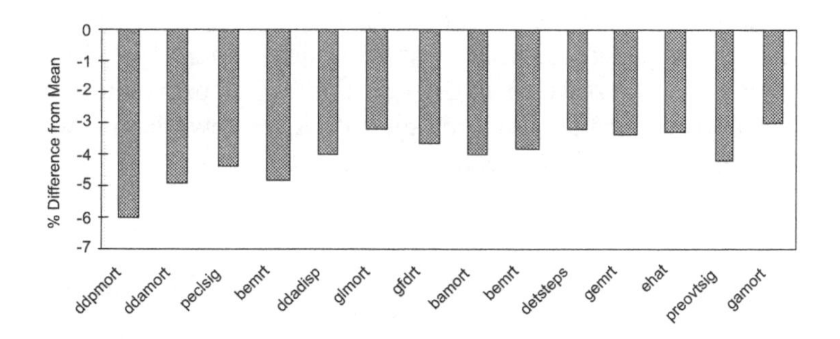

Figure 6. Elasticity analysis ranking. Parameters not be in strict magnitude order because of effects on variance.

5 General Significance

By working with real data, I aim to link theory to conservation biology. Much of the current debate in population conservation involves choosing larger continuous versus smaller fragmented reserves, but the realistic choices in nature reserve design are trade-offs between size, inter-reserve distance, and habitat types. The trade-off for these three depends on the specific sites and species in question. Policy and management are currently driven by keystone species that may be indicators of the entire ecosystem's health or by the umbrella concept where charismatic species may protect many other species in the same habitat. Many of these species are sensitive to environmental quality and are in conflict with human interests. Development of this IBS quantifies the effects of behaviors that influence spatio-temporal dynamics in a patch system and identifies the indices for use in the monitoring programs. My IBS is designed to scale-up for work with landscape systems and in the future will utilize geographic information system maps. This will be the best way to analyze population dynamics for a proposed park system, and this applies to any species whether they are endemic or invasive (Turner et al. 1991). For invading species, the analysis is the same, but the action taken with respect to the most sensitive environmental parameter is the opposite. As well as invasion, this work will be useful for management of areas with natural disturbance because the pattern of succession is a balance of local extinction and colonization. Conservation theory will benefit from empirical tests and a mechanistic IBS framework for prediction and hypothesis building. This research will yield quantitative results that will support or refute the importance of corridors, dispersal, patch size, and patch quality that are currently debated conservation issues (Beier 1992, Johnson and Wiens 1992). Finally, a preserve manager interested in a specific organism may improve their adaptive management strategy with a framework into which they may incorporate knowledge of the specific organism.

References

Adler, F. R. and B. Nuernberger. (1994). Persistence in Patchy Irregular Landscapes. Theor. Popul. Biol. 45: 41-75.

Back, E. A. (1926). Biology of the sawtoothed grain beetle, Oryzaephilus suranemensis Linne. J Agr Res 33: 435-452.

Bascompte, J. and R. V. Sole. (1996). Habitat fragmentation and extinction thresholds in spatially explicit models. J. Ani. Ecol. 65: 465-473

Beier, P. S. L. (1992). A checklist for evaluating impacts to wildlife movement corridors. Wildl. Soc. Bull. 20: 434-440.

Burkey, T. V. (1989). Extinction in nature reserves: the effect of fragmentation and the importance of migration between reserve fragments. Oikos 55: 75-81.

Burkey, T. V. (1997). Metapopulation extinction in fragmented landscapes: using bacteria and protozoa communities as model ecosystems. Am. Nat. 150(5): 568-591

Collins, P. J., J. C. Mulder, and D. Wilson. (1989). Variation in life history parameters of Oryzaephilus suranemensis (L) (Coleoptera:Silvanidae). J Stored Prod. Res. 25(4): 193-199.

Conroy, M. J., Y. Cohen, et al. (1995). Parameter estimation, reliability, and model improvement for spatially explicit models of animal populations. Ecological Applications 5: 17-19.

Crone, E. E. (1997). Delayed density dependence and the stability of interacting populations and subpopulations. Theor. Pop. Biol. 51: 67-76.

Dennis B. and Taper M. (1994). Density dependence in time series observations of natural populations: estimating and testing. Ecol. Monogr. 64:205-24

Erlich, P. R. and A. H. Erlich. (1981). Extinction: The causes and consequences of the disappearance of species. Random House, New York.

Fahrig, L. (1997). Simulation methods for developing general landscape-level hypotheses of single-species dynamics. Quantitative methods in landscape ecology. M. G. Turner and R. H. Gardner. New York, Springer-Verlag. 82: 417-442.

Fahrig, L. and. J. Paloheimo. (1988). Determinants of Local Population Size in Patchy Habitats. Theor. Pop. Biol. 34: 194-213.

Hanski, I. and D. Simberloff (1997). The metapopulation approach, its history, conceptual domain, and applications in conservation. *Metapopulation biology: ecology, genetics, and evolution*. I. A. Hanski and M. E. Gilpin. San Diego, Academic Press, Inc.

Harrison S. 1991. Local extinction in a metapopulation context: an empirical evaluation. Biol. J. Linn. Soc. 42:73-88.

Hassell, M. P., O. Miramontes, et al. (1995). Appropriate formulations for dispersal in spatially structured models: comments on Bascompte & Solé. Journal of Animal Ecology 64: 662-664.

Haydon, D. and H. Steen (1997). The effects of large- and small-scale random events on the synchrony of metapopulation dynamics: a theoretical analysis. Proc. R. Soc. Lond. B 264: 1375-1381.

Howe, R. W. (1956). The Biology of the Two Common Storage Species of Oryzaephilus (Coleoptera: Cucujidae). Ann Appl Biol 44: 341-55.

Johnson, A. R., J. A. Wiens, et al. (1992). Animal movements and population dynamics in heterogeneous landscapes. Landscape Ecology 7: 63-75.

Kareiva, P. (1990). Population dynamics in spatially complex environments: theory and data. Phil. Trans. Royal So. of London, ser. B: 53-68.

Keitt, T. H. and A. R. Johnson (1995). Spatial heterogeneity and anomalous kinetics: emergent patterns in difusion-limited predator-prey interaction. Theor. Pop. Biol. 172: 127-139.

Kendall, B., C. J. Briggs, W. W. Murdoch, P. Turchin, S. P. Ellner, E. McCauley, R. M. Nisbet and S. N. Wood. 1998. Understanding complex population dynamics: a synthetic approach. Submitted to Ecology.

Levin, S. A. (1994). Patchiness in marine and terrestrial systems: from individuals to populations. Phil. Trans. Royal So. of London, ser. B 343: 99-103.

Prus, T. (1966). Emigrational ability and surface numbers of adult beetles in 12 strains of Tribolium confusum and T. castaneum Herbst (Coleoptera, tenebrionidae). Ekologia Polska ser A 14(29): 547-588.

Raven, P. H. 1990. The politics of preserving biodiversity. Bioscience 40:769-774.

Soule, M. E. 1991. Conservation: Tactics for a constant crisis. Science 253:744-50.

Tscharntke, T. (1992). Fragmentation of *Phragmites* habitats, minimum viable population size, habitat suitability, and local extinction of moths, midges, flies, aphids, and birds. Conserv. Biol. 6: 530-535.

Turchin, P. (1991). Translating foraging movements in heterogeneous environments into the spatial distribution of foragers. Ecology 72: 1253-1266.

Turchin, P., and A. Taylor (1992). Complex dynamics in ecological time series. Ecology 73: 289-305.

Turner, S. J., R. V. O'Neill. (1991). Pattern and scale: statistics for landscape ecology. Quantitative methods in landscape ecology. M. G. Turner and R. H. Gardner. New York, Springer-Verlag.

Turner, M. G., G. J. Arthaud, R. T. Engstrom, S. J. Heil, J. Liu, S. Loeb and K. McKelvey. (1995). Usefulness of spatially explicit population models in land management. Ecol. Appl. 51: 12-16.

Wiens, J. A. (1997). Metapopulation dynamics and landscape ecology. Metapopulation biology: ecology, genetics, and evolution. I. A. Hanski and M. E. Gilpin. San Diego, Academic Press, Inc.

Wilson, E. O. (1989). Threats to biodiversity. Sci. Am. 261:108-117.

Wissel, C., and S. Zaschke. (1994). Stochastic birth and death processes describing minimum viable populations. Ecol. Model. 75: 193-201.

A New Paradigm for Evolutionary Psychology: Animal Social Systems as Complex Adaptive Systems

Jeffrey C. Schank
Department of Psychology
University of California, Davis
jcschank@ucdavis.edu

1. Animals Social Systems and Evolutionary Psychology

Social organization is found in a wide variety of animal species ranging from social insects such as ants, wasps, and even caterpillars to cowbirds, rats, primates and humans. Different types of social organization exist at different developmental stages, for example, parent-offspring relationships exist for only brief periods in an animal's life history, and are not static during these development stages [Alberts & Gubernick, 1983]—even largely non-social animals go through stages of sociality during their life history.

A basic question for the study of animal social systems (broadly construed as including human social systems) concerns the question of how these systems emerge. If we go back to the ethologist, Niko Tinbergen [1963], we find the question of emergence broken down into four basic questions. *First,* what are the proximate causes for social behaviors and organizations (e.g., hormonal, neurophysiological). *Second,* what are the adaptive functions ("survival value" in Tinbergen's terms) of social behaviors and organizations. *Third,* how do social behaviors and organizations develop? *Fourth,* how are the social behaviors and organizations of different species phylogenetically related? These four questions are by no means mutually exclusive, they are interrelated and serve to alert us to the central issues in understanding animal social systems. Moreover, because it is arguably the individual, which is our starting point for the study of animal social systems [Schank, 2001], these four questions constitute the starting point for a comprehensive *evolutionary psychology* of social behavior and organization.

The current paradigm for evolutionary psychology [Cosmides & Tooby, 1987; Buss, 1995], adopts a strong adaptationist approach [Gould & Lewontin, 1979] to the evolution of mental functions by viewing the brains of animals as consisting of neural circuits "designed" by natural selection to perform domain specific (i.e., relative to an ecological niche) functions that promote survival and reproduction. Evolutionary psychology, in this view, is a direct descendent of the neo-Darwinian approach of 1970's sociobiology [Wilson, 1975; Tivers, 1971] and the gene-oriented interpretations of evolution developed by Williams [1966] and Dawkins [1976]. What is new in the current evolutionary psychology paradigm is the focus on the evolution of domain specific neural circuits. In this view, Tinbergen's four questions have a straightforward neo-Darwinian interpretation. Causal questions concern the identification all the evolved neural circuits that constitute a brain as well as the elucidation of their neurophysiological mechanisms. Adaptive questions concern the role of these circuits in the survival and reproduction of an organism. Also, the logic of adaptationist thinking may help to identify what circuits have evolved in brains, and thus guide the search for the neuroanatomical and neurophysiological basis of these circuits. Development concerns the ontogeny of neural circuits just as with any phenotypic character of an organism. Finally, from a phylogenetic perspective, the task of evolutionary psychology is to trace the phylogeny of neural circuits and thereby the evolution of brains and cognition.

The strong adaptationist and selfish gene perspectives, which form the foundation of current evolutionary psychology, face the same problems that these perspectives face in evolutionary biology [Gould & Lewontin, 1979] as well as what I will call the *information overload* problem for genomes. If organisms and their brains are bundles of adaptations, then this information must be encoded in an organism's genome as a "blue print" or "program." The number of genes (in the classical sense of one gene, one protein) appears to be much less than expected; *C. elegans* has 19,000 genes and humans appear to have around 30,000 [International Human Genome Sequencing Consortium, 2001]. How are domain specific neural circuits of humans encoded in about half again as many genes as *C. elegans*? Theoretically, Kauffman [1993] has also argued, using computer simulations of gene-control networks, that complexity catastrophes are inevitable when a large number of genes must be maintained to preserve a large number of adaptations. Wimsatt & Schank [1988] have further argued that only the most essential genes can be maintained against even small mutational forces. On both empirical and theoretical grounds, it would appear that the information overload problem for evolving genomes casts serious doubt on the view that brains are bundles of domain-specific circuits designed by natural selection.

If evolving domain specific neural circuits poses a genome over-load problem for vertebrates such as birds and mammals, how did social behavior, organization, and cognition evolve? That domain specific rules have evolved is not in question here. Animals use a number of domain specific heuristic rules for solving problems [Todd & Gigerenzer, 2000] and for learning [Timberlake, 1994]. Rather, the fundamental problem for a viable evolutionary psychology concerns how these cognitive mechanisms can evolve and develop without being directly encoded as "blue prints" or "programs" in genomes?

In this paper, I will argue that a more viable evolutionary psychology must start by abandoning the neo-Darwinistic paradigm of 1970's sociobiology reborn in current evolutionary psychology [Cosmides & Tooby, 1987; Buss, 1995], and replace it with a more classical Darwinian perspective, which incorporates self-organization [Kauffman, 1993; Bar-Yam, 1997] and multiple levels of selective processes in evolution [Campbell, 1997], and multiple mechanisms for the transmission of adaptive behaviors and traits [Wimsatt, 1999]. The recognition that the phenomena of evolutionary psychology are far more complex than previously thought, creates new obstacles. However, ideas, methods, and technologies that are developing under the general rubric of *complex adaptive systems*, which have precisely the same general characteristics as just described [Holland, 1995], may allow us to hurdle these obstacles. Perhaps, the two most controversial ideas—as applied to evolutionary thinking—are that *self-organization* is critical for the functional organization of social systems and that there may be *non-genetic mechanisms* for the transmission of adaptive behavior.

2. Self-Organization

Even the most ardent adpatationists have recognized that not every character or behavior of an organism has adaptive function. It is generally agreed that some traits and behaviors are due to developmental constraints, allometry, and genetic correlations among traits such as pleitropy [Gould & Lewontin, 1979]. Kauffman [1993], however, has gone further by arguing that many traits and behaviors that have biological function emerge by near statistical necessity from the *self-organizing* interactions of the parts of organisms (e.g., gene control networks).

Similarly, *self-organization* is an emerging perspective on the social behavior and organization in animals [e.g., Bonabeau, 1998]. In this view, social behavior and organization emerges as a consequence of the *local interactions* among individuals, where the term *local,* implies that self-organized social structures and behavior do not emerge within hierarchical systems in which the flow of information and control is down a hierarchy. In strictly hierarchical systems, organization is determined by the flow of information from an individual (or group of individuals) at the top of the hierarchy to those at lower-levels [Bar-Yam, 1997]. The fundamental problem of hierarchical organization is that all the information for constructing the hierarchy is localized in the individual(s) at the top of the hierarchy [Bar-Yam, 1997]. If the behavior of the individuals constructing the hierarchy from the top is genetically determined, then we are again faced with the genome over-load problem. For example, ants moving along a trail generate a self-organizing path emerging from ants locally interacting by depositing pheromones (chemical signals) as they meander. There is no need for a controller ant at the top of a control hierarchy to direct the activity of other ants, nor must any ant store information about the social organization required to perform the functions of the colony [Pereira & Gordon, 2001]. The claim, however, that self-organization is typical of animal social organizations is not to deny that hierarchical relationships exist in animal social organizations, they do. Indeed, dominance hierarchies and cast systems are consistent with and can emerge from self-organizing interactions [Hemelrijk, 1999].

If organization emerges from interactions of individuals and other components of their environment, then these interactions can be described by rules (probabilistic) for the types of interactions that take place. Perhaps the simplest kind of rule-describable behavior in animals is the *taxis*—an orientation response and movement towards an object or along a gradient [Fraenkel & Gunn, 1961]. For example, ants display a *chemotaxis* towards the trail pheromones laid down by other ants. Norway rat pups exhibit *thermotaxis* from cold to warm and they also exhibit *topotaxes* toward other objects in their environment. Surprisingly complex social organizations—such as we see in social insects—emerge from relatively simple orientation responses (*taxes*) of individuals. More complex behavioral rules are found in animal social systems, however, *taxes* are the simplest rules from which complex social behavior and organization emerge.

3. Non-Genetic Modes of the Transmission of Adaptive Behaviors

The discussion in the last section may suggest that while complex social organizations emerge from simple rules of individual behavior, these rules are nevertheless directly encoded in an organism's genome. This is far too simple, however, especially for social vertebrates such as birds and mammals. Their behavioral rules can be influenced by development, learning, and can be culturally transmitted.

3.1. Cowbird Cultures

Cowbird (*Molothrus ater*) males must successfully communicate vocally to mate with females of their own species and females must recognize male song [see, e.g., King & West, 1977, 1983, 1990; King, West, & Freeberg, 1996]. Cowbirds, however, are unusual in that they parasitize the nests of other birds by laying their eggs in the nest of other species. The young cowbirds are then raised by other species in absence of adult cowbirds. The obvious problem is how do the young cowbirds that emerge from the nest acquire their species-typical behaviors? One answer is that these species-typical behaviors emerge from a closed genetic system of development, i.e., stimuli from adult cowbirds are not required during development in a foreign nest because all the information required for species-typical behavior is genetically programmed during development.

Research by King, West and co-workers [King & West, 1977, 1983, 1990; King, West & Freeberg, 1996] has revealed male and female cowbirds do not emerge from a foreign avian nest with a species-typical set of behaviors genetically preprogrammed. Young males do not produce coherent song after leaving the nest, but instead they produce sequences of highly variable sounds that adult cowbird females do not recognize as adult male songs. King, West, and co-workers have discovered in field and laboratory studies that cowbird song is both socially learned and culturally transmitted [King, West, & Freeberg, 1996].

Female cowbirds, though they do not sing, and thus cannot serve as a model for male song, can nonetheless act as social tutors. Young males generate a wide variety of sequences of sounds. When a female hears a sequence of sounds she prefers, she responds with head and wing movements that young males attend to closely [King, West, & Freeberg, 1996]. In this way, young males gradually learn the repetoir of adult songs. It should not be surprising that systems requiring social learning at some stage of development can evolve cultures. Cowbirds transmit cultural information about male-song repetoirs, female preferences and responses. Birds from a particular cowbird culture mate and aggregate preferentially with other birds from the same culture [King, West, & Freeberg, 1996].

The critical point for the perspective presented here is that a cowbird culture is socially learned and transmitted by interactions with adult and juvenile birds, and thereby cowbird enculturated rules evolve. This is not to say that any bird can become a cowbird. A cowbird must have a cowbird genome, but having a cowbird genome is not sufficient for transmitting all the information required to be a cowbird.

3.2. Parent-Offspring Relationships

Cowbirds provide an example of very low-level parent-offspring relations. A female cowbird chooses a nest, lays her eggs and flies away playing no further role in raising of her young. This is atypical for other avian and mammalian species. Even subtle differences in parental care early in development can have behavioral consequences later in development and across generations.

Mice (*Mus musculus*) are widely used in biomedical research, and, in particular inbred strains of genetically identical animals (i.e., inbred strains are derived through many generations of brother-sister matings). Ressler [1963, 1966] conducted a fascinating series of experiments that investigated the role of parental handling on visual exploration and learning in adult offspring. He used two strains of inbred mice: BALB/c (with white fur) and C57BL/10 (with black fur). Under identical laboratory conditions, the white strain (BALB/c) are more visually exploratory (in test environments) as adults than the black strain. BALB/c mice also handle their young more in laboratory tests (i.e., when pups are removed from the nest, BALB/c mothers are more likely to retrieve them). In cross fostering experiments, however, Ressler [1963] found that C57BL/10 mice were more exploratory if raise by a BALB/c mother than if raise by a mother of the C57BL/10 strain. Moreover, C57BL/10 mice were more visually exploratory and their rate of learning was enhanced if raised by grandparents (regardless of strain) that had been raised by BALB/c mice [Ressler, 1966]. Thus, not only did Resseler's experiments demonstrate the importance of developmental influences (e.g., parental care) on adult behavior in strains of genetically identical mice in identical environments, but also that these influences could be transmitted across generations without genetic change—though the mechanism of transmission remains unknown.

3.3. The Evolution of Social Behavior

From the perspective of an individual in a social system, there are costs and benefits (in terms of fitness) to participating in a social system. For example, in social insects with sterile casts (and in some social mammals in which females or males forgo or delay reproduction), members of these casts forgo reproduction and perform behaviors (e.g., foraging for food), that benefit the colony as a whole. Behavior such as this is often called *altruistic* and the problem has been how to explain its evolution.

Hamilton [1964] provided an analysis of altruistic behaviors that yielded a rule for when altruistic behaviors should evolve:

$$rB - C > 0 \tag{1}$$

where r is the degree of genetic relatedness between an individual and the recipient(s) of altruistic behavior, B is the sum of the fitness benefits to the recipient(s), and C is the sum of the fitness costs to the individual providing the benefit. If the coefficient of genetic relatedness r is sufficiently high so that (1) is true, then altruistic behaviors can evolve that incur severe costs to the individuals performing the behaviors.

There has been considerable controversy surrounding the logical status of this rule (1), but in a recent set of papers, Frank [1997 and references therein] has clarified the logical and causal status of (1), revealing diversity of possible mechanisms for the evolution of social organizations in animals. Frank distinguishes between direct selection *for* (i.e., the phenotypes themselves) and the transmission *of* predictors of phenotypes (e.g., genetic relatedness by descent). On the direct selection of phenotypes interpretation of (1), the coefficient r, depends only on phenotypic similarities. From a self-organizing perspective on the emergence of social behavior it should be added that phenotypic similarity of the rules themselves is not necessary, only that their *consequences* for social behavior and organization are similar. Phenotypic similarities in this broader sense can also have many sources such as a shared genotype (as a result of descent), but they can also be due to common developmental niches, common cultures, or even heritable similarities with other species [Frank, 1997].

In animal social organizations, whether the social organization of an ant colony or the parent-offspring relations between a mother mouse and her pups, the set of behavioral rules must be phenotypically similar within and across generations in order for adaptive organization or structure (e.g., the complex structure of a termite mound or wasp nest) to emerge from local interactions. In a self-organizing system as the similarity, r, between rules describing the interaction of individuals change, so do the benefits B. Thus, r and B are likely to be correlated in evolutionary time and together promote the rapid evolution of a set of behavioral rules by whatever mechanism of transmission.

A related point is that animals have the potential to realize a variety of behavioral rules, which is the set of rules an animal is *capable* of expressing at any given life stage. The expression of some subset of these rules depends on the organism's ecological and social context. For example, worker insects in sterile casts display a set of rules for maintaining the colony nest and promoting the reproduction of the queen (or queens), and yet they nevertheless have the potential to exhibit the behavioral rules of a queen. Whether or not an individual displays worker or queen rules depends developmentally on a type of social lottery at a critical stages in the life history of the colony [Darlington, 1981]. In Darlington's view [Darlington, 1981], even for a species in which there is a strict sterile caste system for allocating behavioral tasks, it is the entire set of behavior rules for an individual that evolves in the context of the colony. Thus, more adequate theories for the evolution of social behavior should consider the set of possible behavioral rules that an animal can express rather than analyzing only expressed rules at a given life-history stage.

Figure. 1. An illustration of the process of discovering rules of individual behavior with individual-based modeling. The discovery of rules always involves the parallel comparison of data from empirical social systems with model systems. Darwinian algorithms are use to find the best fit rules, where "fitness functions" for the models with respect to the data, are computed for both individual behavior, organizational structure, and dynamics [Schank & Alberts, 2000]. Once rules are discovered they have a variety of scientific functions including, testable predictions, the use of robotics as experimental probes, and the discovery of neural correlates and neural network models of the discovered rules.

The upshot is that we need to develop a paradigm for quantifying the rules of individual behavior in social and group contexts that encompass not merely genetic mechanisms of evolution and transmission, but other mechanisms including social learning, cultural transmission, and development. By developing such a paradigm we will be better able to quantify how these rules are learned, developed, evolved, and transmitted.

4. Developing a New Paradigm

Psychologists are interested in how social organization emerges from individual behavior and how individual behavior is shaped (in a very broad sense of this term, not limited to its meaning in traditional learning theories) by social organization. In the past, psychologists out of methodological necessity—which at times has become dogma—have been forced to study the behavior individuals in relative isolation (e.g., learning has traditionally been studied in isolation with devises designed to accommodate a species-typical response repetoir [Timberlake, 1994]). Such research has been valuable in revealing general principles of learning and behavior, but it does not reveal how rules are shaped socially. The new era of computational modeling, complexity theory, and robotics, which falls under the general rubric of *complex adaptive systems* [Holland, 1995; Caporael, 2000], free us from being held hostage to overly simplistic genetic reductionism (see Wimsatt [1980] for a discussion of the potential biases in reductionistic research).

The remainder of this paper will focus on some of the elements of a new paradigm for evolutionary psychology, which aims to integrate ideas and methods of complex adaptive systems for discovering and quantifying rules of individual behavior in social, developmental, and evolutionary contexts. To do so I will describe and point to some of the tools that can facilitate the realization of this new paradigm including: individual-based models, Darwinian algorithms, techniques for observing and analyzing the behavior of individuals in group or social contexts, and animates for experimentally probing and manipulating animal social

4.1. Individual-Based Modeling

With the explosive development of affordable computational power, computationally intensive computer simulations can now be performed on desktop computers. Individual-based modeling (including agent-based modeling and cellular automata), is the most rapidly growing style of modeling in ecology, artificial life, anthropology, and computational investigations of learning and evolution [Judson, 1994; Todd, 1996; Schank, 2001]. The central feature of this style of modeling is the representation of *individuals* and rules for their behavioral interactions in computer simulations.

A cutting-edge problem in the study of complex dynamical systems is how to infer the rules of individual behavioral interactions from the complex dynamics of the whole system. In mathematical systems such as cellular automata, there is no general solution to this problem of backward inference, because identical dynamics can be generated by different rules. Empirical systems, however, can be experimentally manipulated and viewed at multiple levels of organization. By iteratively comparing simulation and empirical experiments, it may be possible to identify specific rules and their neurological and physiological correlates (**Fig. 1**, Schank, 2001). If we can develop general techniques for inferring rules from social behavior, then (i) we can investigate the evolution of sets of rules, (ii) use these rules to experimentally probe and manipulate social systems, (iii) discover neural correlates and generate neural network models and other physiological models of these rules, and (iv) use these rules in empirically based "what if" modeling of the evolution and ecology of social systems **Fig 1**.

4.1.1. Rules as Conditional Probabilities

A completely general representation of behavioral rules in animals can be achieved with conditional probabilities (whether these rules are simple taxes or more complex rules conditionalizing sequences of behavior on sensory inputs and internal states). The general form of these rules is

$$P(B_h|x_1,\ldots,x_n,i_1,\ldots,i_m) \quad (2)$$

where B_h is a behavior expressed by an organism such as moving to contact another animal, grooming, or feeding; $x_1,\ldots,x_n$, are state variables such as sensorimotor activity or inactivity, hunger, hormone levels, or stress; $i_1,\ldots,i_m$ are input variables such as which animals are in contact with an animal, ambient temperature, or phase of the light dark cycle. Using individual-based models, the goal is to discover and thereby quantify probabilistic relationships between behavior and these variables. The next section outlines an example that illustrates how the simplest behavioral responses, *taxes*, can be represented as rules of individual behavior and applied to self-organizing group behaviors in developing Norway rat (*Rattus norvigicus*) pups.

4.1.2. An Example: Aggregation among Rat Pups

Born blind, deaf, and with limited sensorimotor capabilities, rat pups can still move about in their neonatal environment, forming huddles, moving towards a mother and attaching to a nipple [Alberts & Gubernick, 1983]. From a sensorimotor perspective, we can operationally define when pups are active, A, and inactive I [Schank & Alberts, 1997, 2000]. The probabilities of these activity states can be modeled with conditional probabilities as described above (2). In the simplest case, for pups 7-days or younger, we can assume that the probability of activity or inactivity of a pup depends only on its prior activity or inactivity [Schank & Alberts, 1997, 2000].

If a pup is active or inactive at time t, there are two conditional probabilities that determine whether or not a model pup is active at time $t{-}\Delta t$, $P(A_t \mid A_{t-\Delta t})$ and $P(A_t \mid I_{t-\Delta t})$. Similarly, there are two conditional probabilities that determine whether or not a pup is inactive at time t, $P(I_t \mid I_{t-\Delta t})$ and $P(I_t \mid A_{t-\Delta t})$. Because $P(A_t \mid A_{t-\Delta t}) + P(I_t \mid A_{t-\Delta t}) = 1$, $P(I_t \mid I_{t-\Delta t}) + P(A_t \mid I_{t-\Delta t}) = 1$, the conditional probabilities of activity and inactivity are completely determined by: $P(A_t \mid I_{t-\Delta t})$ = the probability of activity at time $t{-}\Delta t$ given inactivity at time t, and $P(I_t \mid A_{t-\Delta t})$ = the probability of inactivity at time $t{-}\Delta t$ given activity at time t. From these two conditional probabilities we can calculate the expected frequency of activity [Schank & Alberts, 1997, 2000].

By 10 days of age, although rat pups are still blind and deaf, sensory inputs from their littermates play an important role in their activity [Schank & Alberts, 2000]. We can represent this change again in terms of conditional probabilities

$$P(I_t \mid A_{t-\Delta t}, N_A, N_I) \quad \text{and} \quad P(A_t \mid I_{t-\Delta t}, N_A, N_I) \qquad (3)$$

that include, N_A, the number of active pups, and N_I, the number of inactive pups, which are in contact with a given pup.

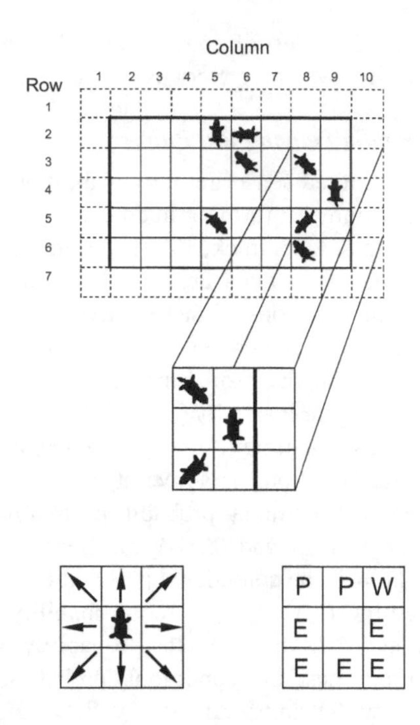

Figure 2. Top, a simple cellular-automata-like model of rat pups aggregating in an arena [Schank & Alberts, 1997, 2000]. Left, a focal pup is in the center and its view is outlined. Right, a focal pup's preference for pups (P), walls (W), empty cells (E). The rules for assigning preferences to cells are (E) cells not in a pup's view are or empty cells not next to a pup are empty; (W) cells in a pup's view and occupied by a wall; (P) Cells in a pup's view occupied by another pup or empty cells in a pup's view that are next to another cell in the pup's view which contains a pup. The first step in calculating the probability that a pup will attempt to move to any of the eight adjacent cells (given it is empty) is to sum up the preferences in all eight adjacent cells and divide each cell by the sum. During the movement of pups within a time interval, cells may become empty or occupied either allowing or blocking of movement of other pups into that cell [Schank and Alberts, 1997].

Taxes, orientation responses of organisms, can also be represented as conditional probabilities. The conditional probability that a pup will assume orientation, O_k, towards an object is given by

$$P(O_k \mid X_t, O_j, i_1,\ldots,i_m) \quad (4)$$

where X_t is the current activity state of a pup, O_j is the previous orientation, and $i_1,\ldots,i_m$ are inputs from objects (in this case other pups and walls). If $X_t = I_t$, then a pup is assumed not to reorient, and so $P(O_k = O_j \mid I_t, O_j, i_1,\ldots,i_m) = 1$. If $X_t = A_t$, then a pup can reorient, and so $P(O_k \neq O_j \mid A_t, O_j, i_1,\ldots,i_m) \geq 0$. In the individual-based models described in detail elsewhere [Schank & Alberts, 1997, 2000], both of these probabilities can only be calculated completely in the context of computer simulation.

4.1.3. Surrogate Experiments

With a set of parameterized rules for how individuals interact with each other (e.g., probabilistic rules for when to be active or what kind of object to move towards). Simulation experiments (i.e., *Monte Carlo* experiments) can then be run repeatedly, with the same parameter values to discover the limiting behavior for a set of social rules (**Fig. 1**). Data generated both virtually and empirically are often complex and highly interdependent, but using Monte Carlo approaches to data analysis, it is possible to do statistical tests of the goodness of fit of individual-based models to empirical data and to generate confidence intervals [Schank & Alberts, 2000]. If we can evolve models that behave like the social systems we are interested in, we can then conduct simulation experiments in which factors are introduced or removed that may not be practicable to do with empirical systems. For example, in studies on sensorimotor aggregation in rat pups [Schank & Alberts, 1997, 2000], coupled activity *and* preferences for other pups both facilitate aggregation, but together, surrogate experiments revealed that preferences and coupled activity interact synergistically to facilitate aggregation.

4.1.4. Darwinian Algorithms

A key innovation introduced into individual-based modeling is the use of *Darwinian algorithms* [Schank & Alberts, 2000; Schank, 2001]. At each step of the Darwinian algorithmic process (see **Fig. 1**), a large number of simulated experiments are run, in which model organisms can interact, socially learn, or culturally transmit information that may alter their behavioral rules. Data on observations of these simulated experiments are then compared with data from empirical experiments to determine how well the model (with a specific set of individual rules) statistically predicts the empirical data. This provides a measure of the "fitness" of the model to the data. Darwinian algorithms can be implemented using, for example, simulated annealing, genetic algorithms, or other evolutionary algorithms [e.g., Mitchell, 1998]. At each step of the process, rules are evaluated for their predictive power and changed ("mutated") rules are introduced into the model system (**Fig. 1**). When a criterion is reached for the "best" rules (e.g., failure to find a better set after a specific number of trials or some predictive criterion is achieved), we then have a predicted set of rules inferred from the dynamics of actual social behavior.[1]

Using this approach it has been possible to discover the rules underlying the developmental emergence of a component of sociality in rats [Schank & Alberts, 2000]. At 7-days of age, individual pups form aggregations, but whether a pup is active at any time is independent of other pups' activity [Schank & Alberts, 1997]. By day 10, however, the picture changes dramatically [Schank & Alberts, 2000]. Ten-day-olds are affected by the activity state of the pups with which they are in

---

1 V. Anne Smith is currently applying this kind of approach to aggregation and social learning in cowbirds. Rules of individual behavior are, however, more complex than simple taxes and correspondingly require more sophisticated models. In this case, she is using classifier systems to represent rules and genetic algorithms to evolve rules for cowbird aggregation and social learning of birdsong.

contact, and "spreading activation" and "spreading quiescence" can be observed in aggregates of pups [Schank & Alberts, 2000].

4.2. Extending the Paradigm

To fully develop a complex adaptive systems paradigm for evolutionary psychology, new empirical methods and technologies will have continue to developed. Two such methods and technologies that should facilitate the integrative use of individual-based modeling with experimental research include (i) methods and technologies for the collection of social behavior in context and (ii) the use robots to experimentally probe and test predictions in actual animal social systems.

4.2.1 Observing and Analyzing Data for Complex Adaptive System Studies

On the empirical side of the study of social organization, there is a bottleneck in the acquisition and analysis of animal of social behavior. Traditionally, an ethologist or psychologist would compile an ethogram (i.e. a list of the types of behaviors an observer categorizes, e.g., social grooming [Altmann, 1974]). Then either by taking notes or using a computer the occurrence, the time and/or duration of these behaviors is recorded.

The use of individual-based models for discovering probabilistic rules of behavior, however, requires more precise and fine-grained observations of behavior. For example, to collect sufficient data to estimate the conditional probability that a rat pup is active given that it is next to N_A active littermates and N_I inactive littermates, required assessing contact for 8 pups per litter over 15 minutes every 5 seconds. Depending on the accuracy desired and the number of parameters, large numbers of litters are required [Schank & Alberts, 2000]. Moreover, for other types of behavior (for example female wing flicking in cowbirds), behavioral observations may have to be assessed with even smaller time intervals.

The long-term success of this new paradigm will depend as much on the development of computer aided techniques for acquiring precisely timed behavioral data as it will on the development of computer modeling and analysis techniques. To this end, it will be critical to continue to develop computer aided image tracking and analysis software/hardware systems (e.g., Noldus, http://www.noldus.com/, has developed EthoVision, a program which can track up to 16 animals) as well as tracking systems that use other modalities. For example, there has been continued improvements in miniaturizing and tracking individuals with radio transmitters that can transmit not only information on location but also information on physiological states (Wolfram, 1999).

Even with the development of better technologies for collecting dynamic data within social contexts, there still remain issues of bias and error in data collection methods. The issue of error and bias is especially important if we wish to use the wealth of data available from previous studies. Fortunately, it is possible to model aspects of observer biases and error in the collection of data (**Fig. 1**). Although, it may not be possible to ever know exactly what errors and biases occurred in a given set of data, it is possible to determine ranges of errors and biases that may make a difference. We can thereby assess the likely robustness of models and statistical results derived from these data.

4.2.2. Robots

Individual-based modeling and Darwinian algorithms may facilitate the discovery of rules for individual social interactions from the dynamics of social behavior, but we must still be able to experimentally test these rules, and test them in ways that do not violate basic principles for studying complex systems. Thus, we must be able to gain greater experimental control over natural systems without resorting to reductionistic methodologies [Wimsatt, 1980]. A promising technology for achieving these goals is the by engineering of behavior-based robots [Mataric, 1998; Dean, 1998].[2]

The behavior-based approach reduces the complexity of building *reactive* robots by specifying simple sensorimotor rules that connect sensors to actuators. This kind of approach is ideal for the paradigm outlined here for two reasons. First, the design of robots that can probe social systems can initially start out very simple, but evolve into more complex designs as a consequence of learning more about the social behavior of the animals we study.[3] Breaking the sensorimotor behavior of robots into simpler rules should facilitate the evolution of robots as probes and tools for testing predictions by allowing the addition, change, and integration new rules. (More complex behavior and recognition may be possible by first integrating human recognition and control capabilities.) Second, this approach is *representational* [Webb, B., 2000] in the sense that either simple sensorimotor rules or integrated systems of rules of robotic behavior can be map onto the actual rules (i.e. the rules for social interactions discovered with individual-based modeling can be programmed into robotic probes).

5. Conclusions

The time is ripe for a Kuhnian [Kuhn, 1962] style revolution in evolutionary psychology and the study of animal social systems. Current evolutionary psychology faces severe problems especially in its assumptions of genetically evolved and domain specific neural circuits for behavior and cognition [see Schank & Wimsatt [2001] for issues concerning modularity and adaptations], which create information over-load problems for the evolution of genomes. The complex adaptive behaviors we observe in animals systems are due in no small part to self-organizing interactions at multiple levels of organization [Schank, 2001]. To study behavior in this way, however, requires a paradigm shift in conception, methodology, and technology.

We can discover and quantify rules of individual behavior in group and social contexts by analyzing the complex relationships between data and models. Specifically, we can do this by integrating individual-based modeling, new techniques

---

[2] There is a long history in animal behavior of using inanimate, non-reactive objects to simulate real animals (e.g., warm fury tubes to simulate rats). Thus, we already know that robotic animals will work as experimental probes.

[3] For example, mother-offspring antisnake behavior [Coss, 1991] could be initially be probed with robotic snakes that coil, extend and retract.

for collecting behavioral data, Darwinian algorithms for discovering behavioral rules from the data, and robots for probing and testing hypotheses in intact animal systems. We may thereby be better able to discover mechanisms of social learning and cultural transmission in group contexts. This newly emerging paradigm [also see Caporael, 2000] will have important implications for the study of animal behavior, but it will also have important implications for the study of human organizations, social structures, and their cultural evolution.

References

Alberts, J. R., 1978, Huddling by rat pups: multisensory control of contact behavior, *Journal of Comparative and Physiological Psychology*, **2**, 220–230.

Alberts, J. R. & Gubernick, D., 1983, *Reciprocity and resource exchange: a symbiotic model of parent–offspring relations, in Symbiosis in Parent-Offspring Interactions* edited by L. A. Rosenblum & H. Moltz, Plenum Press, 7–44.

Altmann, J., 1974, Observational study of behavior: sampling methods, *Behaviour,* **48**, 227–265.

Bar-Yam, Y., 1997, *Dynamics of Complex Systems*. Addison-Wesley.

Bonabeau, E., 1998, Social Insect Colonies as Complex Adaptive Systems, *Ecosystem,* **1**, 437–443.

Buss D. M., 1995, Evolutionary psychology: A new paradigm for psychological science, *Psychol. Inq*., **6**, 1–30.

Campbell D. T., 1997, From evolutionary episte-mology via selection theory to a sociology of scientific validity. *Evol. Cogn*. **3**, 5–38.

Caporael, L. R., 2000, Evolutionary Psychology: Toward a Unifying Theory and a Hybrid Science, *Annu. Rev. Psychol*., **52**, 607–28

Cosmides L., Tooby J., 1987, From evolution to behavior: evolutionary psychology as themissing link. In *The Latest on the Best: Essays on Evolution and Optimality*, ed. J. Dupr´e, pp. 277–306, MIT Press (Cambridge).

Coss, R. G., 1991, Context and animal behavior III: The relationship of early development and evolutionary persistence of ground squirrel antisnake behavior. *Ecol. Psychol*. **3**, 277–315.

Dawkins R. 1976, *The Selfish Gene*, Oxford Univ. Press (New York).

Dean, J., 1998, Animats and what they can tell us. *Trends in Cognitive Sciences*, **2**, 60-67.

Darlington, P. J., 1981, *Evolution for Naturalists: The Simple Principles and Complex Reality*.: John Wiley & Sons (New York).

Frank, S. A., 1997, The Price equation, Fisher's Fundamental theorem, kin selection, and causal analysis, *Evolution,* **51**, 1712–1729.

Fraenkel, G. S. & Gunn, D. L., 1961, *The Orientation of Animals*, Dover (New York).

Gould S. J., & Lewontin R. C. 1979, The spandrels of San Marco and the Panglossian paradigm. *Proc. R. Soc. London Ser. B,* **205**, 581–98.

Hamilton, W. D. 1964, The genetical evolution of social behiour, I, II, *Journal of Theoretical Biology*, **51**, 1–16, 17–52.

Pereira, H. M. & Gordon, D. M. , 2001, A trade-off in task allocation between sensitivity to the environment and response time, *Journal of Theoretical Biology,* **208**, 165–184.

Hemelrijk, C. K., 1999, Towards the integration of social dominance and spatial structure. *Animal Behaviour*, **59**, 1035–1048.

Holland J. 1995, *Hidden order: how adaptation builds complex-ity,* Addison-Wesley (Reading).

International Human Genome Sequencing Consortium, 2001, Initial sequencing and analysis of the human genome, *Nature,* **409**, 860–921.

Judson, O. P., 1994, The rise of the individual-based model in ecology. *TREE,* **9**, 9–14.

Kauffman SA. 1993, *The Origins of Order*. Oxford Univiversity Press (NewYork).

Kuhn, T., 1962. *The Structure of Scientific Revolutions*. University of Chicago Press (Chicago).

King, A. P., & West, M. J., 1977, Species identification in the North American cowbird: Appropriate responses to abnormal song. *Science*, **195**, 1002–1004.

King, A. P., & West, M. J., 1983, Epigenesis of cowbird song: A joint endeavor of males and females. *Nature*, **305**, 704–706.

King, A. P., & West, M. J., 1990, Variation in species-typical behavior: A contemporary theme for comparative psychology. In *Contemporary Issues in Comparative Psychology*, edited by D. A. Dewsbury, Sinauer (Sunderland, Massachusetts), pp. 331–339

King, A. P., West, M. J., & Freeberg, T. M., 1996. Social experience affects the process and outcome of vocal ontogeny in two populations of cowbirds. *Journal of Comparative Psychology*, **110**, 276–285.

Mataric, M. J., 1998, Behavior-based robotics as a tool for synthesis of artificial behavior and analysis of natural behavior. *Trends in Cognitive Sciences*, **2**, 82–87.

Mitchell, M., 1998, *An Introduction to Genetic Algorithms*, The MIT Press (Cambridge).

Ressler, R. H., 1963, Genotype-correlated parental influences in two strains of mice. *Journal of Comparative and Physiological Psychology*, **56**, 882–886.

Ressler, R. H., 1966, Inherited environmental influences on the operatn behavior of mice. *Journal of Comparative and Physiological Psychology*, 61, 264–267.

Schank, J. C., 2001, Beyond reductionism: refocusing on the individual with individual-based modeling. *Complexity*, **6** (in press).

Schank, J. C. & Alberts, J. R., 2000, The developmental emergence of coupled activity as cooperative aggregation in rat pups. *Proceedings of the Royal Society of London,* **267**, 2307—2315.

Schank, J.C. & Alberts, J.R., 1997. Self-organized huddles of rat pups modeled by simple rules of individual behavior. *Journal of Theoretical Biology*, **189**, 11–25.

Schank, J. C. & Wimsatt, W. C., 2001, Evolvability: adaptation and modularity. In *Thinking About Evolution*, Rama S. Singh (Ed.), Cambridge University Press (Cambridge), pp. 322-335.

Timberlake, W. (1994) Behavior systems, associationism, and Pavlovian conditioning. *Psychonomic Bulletin & Review*, **1**, 405–420.

Todd, P. M., 1996, The causes and effects of evolutionary simulation in the behavioral sciences, in *Adaptive Individuals in Evolving Populations*, SFI Studies in the Sciences of Complexity, Vol. XXVI, edited by R. K. Belew and M. Mitchell, Addison-Wesley, pp. 211–24.

Todd, P. M., and Gigerenzer, G. (2000). Simple heuristics that make us smart. Behavioral and Brain Sciences, 23(5), 727-742.

Trivers R. L. 1971, The evolution of reciprocalaltruism. *Q. Rev. Biol.*, **46**, 35–57.

Webb, B., 2000, What does robotics offer animal behaviour? *Animal Behaviour*, **60**, 545–558.

Wilson E. O., 1975, *Sociobiology*, Harvard Univ. Press (Cambridge).

Wimsatt, W. C., 1980 Reductionistic research strategies and their biases in the units of selection controversy, in *Scientific Discovery*, Vol. II: Case Studies, edited by T. Nickles, Reidel (Dordrecht) 213–59.

Wimsatt W. C. 1999, Generativity, entrenchment,evolution and innateness. In *Biology Meets Psychology: Constraints, Connectionsand Conjectures*, ed. V. Hardcastle, MIT Press (Cambridge).

Wimsatt, W. C. & Schank, J. C., 1988, Two constraints on the evolution of complex adaptations and the means for their avoidance. In *Progress in Evolution*, M. Nitecki, (ed.), Univ. of Chicago Press (Chicago) 231—273.

Wolfram, K., 1999, Telemetry in insects: The "intact animal approach", *Theory in Biosciences*,.**118**, 29–53.

Comparing causal factors in the diversification of species

C. C. Maley[1]
Fred Hutchinson Cancer Research Center

What have been the most important factors in the diversification of life? A configuration (individual-based) model (MoD) was constructed to examine the origins and maintenance of species diversity through time. The model represents a species as a connected component in a graph of the potential mating relationships between organisms. This allows us to detect speciation events and track species diversity over time, and so test many "untestable hypotheses" for the causes of diversification. The results suggest that much of the emphasis placed on the evolutionary innovations, in resource utilization and predation interactions, in order to explain the diversification of a group, is flawed at best. Furthermore, habitat heterogeneity had little impact on species diversity. Instead, the model points to the importance of geographical isolation as a primary causal factor for diversification, along with the evolution of specialization and sexual selection, in the form of positive assortative mating. The influence of assortative mating is particularly interesting because it is opaque to the methods of paleobiology.

1 Introduction

Evolutionary biology is characterized by a relatively solid understanding of microevolutionary phenomena, but little understanding of macroevolutionary phenomena. One of the most important open questions of macroevolution is what have been the most important causal factors that have driven the diversification of species.

For clarity of discussion, we will consider a species to be a reproductively

[1]This work was supported in part by NDSEG grant DAAH04-95-1-0557, the generosity of the MIT AI Lab and the Harvard Herbarium. I would like to thank Rod Brooks, Michael Donoghue, Hal Caswell, Dave Cliff, Bob Berwick, and Miller Maley for their guidance and support.

isolated gene pool. This is the dominant understanding of species in biology and is due in large part to the work of Ernst Mayr[16]. The question then becomes what factors tend to split gene pools apart and preserve them from extinction?

Most of the hypotheses that attempt to answer this question are both hard to test experimentally, and difficult to resolve in the fossil record. Thus, a model of diversification (MoD) has been developed that instantiates enough of the components of biological life to test and compare hypotheses for the diversification of life within a common framework.

2 Hypotheses

The hypotheses for the diversification of life fall into two broad categories: abiotic and biotic explanations[1].

2.1 Abiotic causes of diversification

Hypotheses for the diversification of species that depend on physical characteristics of the environment inevitably make reference to topology and/or climate. The conventional wisdom in theoretical biology holds that new species arise primarily from allopatric speciation. That is, populations of a species are separated geographically (topologically in the broad sense) and slowly diverge. This genetic divergence can emerge out of random genetic drift or differential selection between the two populations, as long as there are few enough migrants between the populations to avoid genetic homogeneity. Yet, if two subpopulations experience different climates, or more generally, different selective pressures, it is reasonable to expect them to diverge and perhaps eventually become reproductively incompatible. These are the two basic abiotic hypotheses. Experiments on laboratory organisms show little effect of geographical separation on hybrid fertility between members of the subpopulations. However, such experiments do demonstrate that when allopatry (or any other manipulation that prevents viable hybridization between subpopulations) is paired with divergent selection, reproductive isolation begins to evolve[18].

Cracraft[4] has proposed that the fact of environmental complexity may not be as important in the evolution of diversity as is change in the environment. He does not require the change to be in any one particular direction, as long as the environment is dynamic. Vermeij[24] also cites changes in climate and topography as being an important engine of speciation. Perhaps these changes have repeatedly isolated and then remixed populations, providing many opportunities for the separation of gene pools. A simpler variation on this idea is Benton's suggestion[1] that perhaps the physical environment has become more complex and fragmented over time. This should tend to fragment the populations and so boost speciation.

2.2 Biotic causes of diversification

Biotic causes for diversification have been more popular in the literature as compared to abiotic causes. Generally, biologists and paleobiologists have engaged in a search for the evolution of some "key innovation" that catalyzed the diversification of the group of interest[5, 10, 20]. Of course, the set of characters that have been proposed as key innovations have been biased by the set of characters that can be observed in the fossil record or inferred from phylogenies[9, 11].

Raup[17] observed a decline in the extinction rate of families over time. This should also lead to the observation of increasing diversity over time. Van Valen[23] argues that the observed decreases in extinction rates may be due to the decrease in the overlap between species' niches and so a reduction of extinction through competitive exclusion. Niche overlap can be reduced in two ways. Species may reduce the "size" of their niches through specialization, or they may find and shift to the use of new, unexploited resources. Benton[1] emphasizes both of these factors, specialization and innovation, as the most important causes of diversification. In fact, an innovation that helps to exploit a new set of under-utilized resources might lead to an increase in diversity regardless of any change in extinction rates.

There is another realm of biotic causes of diversification that has only recently received much attention. Since species are determined by reproduction, and reproduction involves a complex constellation of both behaviors and physical traits, its stands to reason that behavioral changes should have had an impact on diversification. Unfortunately, it is all but impossible to recover behavioral innovations from fossil data, except to the extent that the morphology of organisms evolved to support those behaviors. Todd[22, 21] has shown in simulations that sexual selection can quickly fracture populations into mutually exclusive mating groups. Other models[14, 6] have shown that positive assortative mating can evolve under divergent selection on some other trait even when the populations are mixed (sympatric). All this suggests that positive assortative mating may have played an important role in the diversification of life.

3 Model

We approach model design for theoretical biology through a focus on hypothesis testing. All models are simplifications. The difficulty of modeling lies in the decisions of what to include and what to exclude from the model. We endeavor to include only those dynamics which are necessary to test the hypothesis. More cannot be easily justified. The hypotheses for the diversification of life require an implementation of the following phenomena: organisms, reproduction, heredity, and mutation. This much is necessary to implement microevolution. Furthermore competition is required to model extinction through competitive exclusion. Hypotheses about niches, or at least resource utilization, dictate the inclusion of discriminate predation and the phenomenon of specialization. In addition, geographical barriers, migration, and habitats all feature prominently in hypotheses of diversification. Finally, we are interested in species diversity. Thus, we must have some representation of species as collections of organisms. As we have adopted the reproductive concept of species[16] we are obliged to implement sexual reproduction and reproductive barriers. Furthermore, if one assumes that offspring are generally not reproductively isolated from their parents, then speciation, and of course extinction, depend the phenomenon of death. Some parents must die in order to fragment gene pools.

MoD is a configuration[3] model (a.k.a., individual-based or agent-based model) at the level of the organism. It bears a strong similarity to Echo[8, 12] as well as a number of other models in Artificial Life[13, 19, 7]. Each organism is described by 4 different bit-pattern "chromosomes" (prey template, predation resistance, generalism, and reproduction chromosomes) where each bit represents a heritable character. The first three "ecological" chromosomes structure the predator-prey interactions between the organisms. If the prey template chromosome of one organism matches the bit pattern of the predation resistance chromosome of another organism, the first organism may prey upon the second. This interaction is further modified by the generalism chromosome in the predator. The bits set to 1 in the generalism chromosome flag which loci in the prey template chromosome are ignored when checking the match between predator and prey. The combination of the prey template and generalism chromosomes is called the search template of the organisms. An organism with all 1's in its generalism chromosome would be a complete generalist. While an organism with all 0's would be an extreme specialist, only able to digest organisms with one particular predation resistance genotype. The autotrophs (a characteristic encoded by a non-mutating flag in each organism) match their search template against a bit pattern in the environment that represents the characteristics of the habitat in their patch. For the particular details of the predation algorithm as well as the model in general, see [15]. Competitive exclusion arises from the fact that the autotrophs that better match their environment, and the heterotrophs that better match the prey population of their patch will be more likely to gather food and avoid starvation in any give time step than their more poorly adapted competitors.

The fourth and last chromosome is the reproductive chromosome. This encodes all of the characters that influence reproductive isolation between organisms, such as mating time, mating behaviors, reproductive organ morphology, gamete morphology, etc. Reproductive barriers are modeled by the rule that organisms can only mate with each other if their reproductive chromosomes differ in at most 1 locus.

Given this implementation of reproductive barriers, we can unambiguously identify the gene pools or species amongst the organisms. If we consider a graph where each extant reproductive genotype is a node and there is an edge between the genotypes that can mate (are within hamming distance 1 of each other), then a gene pool or species is just a connected component in this graph. Over the generations, genes can flow within these connected components but not between them. A speciation event occurs when all of the organisms bearing a particular reproductive genotype die, and thereby disconnect one of the previously connected components[2]. While all of the ecological chromosomes consist of 32 bits, the reproductive chromosome is represented by 64 bits. We doubled its size so as to reduce the chance that two species would coalesce into one[3].

The organisms are distributed on a 4 by 4 grid of patches, with each patch able to hold 2048 organisms. All ecological interactions occur between organisms within the same patch. Every organism is given a chance to eat in each time step of the model. If an organism fails to find a prey organism in its patch, it dies of starvation. If an organism survives for three consecutive turns, consuming three prey organisms, it is given the chance to find a mate in its patch and reproduce. The chromosomes of the offspring are constructed through two-point crossover of the parental chromosomes in conjunction with a Poisson mutation rate of on average 1 bit flip for every 10 new genomes. The only migration between patches happened when a new organism was born. New organisms take a random walk of 0-2 steps to locate their new patch. This random walk is modified by a "migratory barrier" parameter associated with each patch representing geographical obstacles to migration. The migratory barrier of a patch is a real number between 0 and 1 which encoded the probability that an organism would be prevented from entering the patch on a step of its random walk. If the new organism tries to settle in a patch that is already full[4] it has a 50% chance of displacing (and killing) one of the resident organisms of its type (autotroph or heterotroph).

4 Testing the model

The model was subjected to four tests in order to determine if its ecological and evolutionary dynamics were notably unrealistic as compared with accepted theory. We tested to see if the model demonstrated predator-prey population oscillations, trophic cascades, competitive exclusion, and adaptation.

Stable predator-prey oscillations do not appear at all three trophic levels unless the implementation of predation is stochastic and there is some representation of competitive interference between predators. This may be modeled as a spatial distributions within a patch such that not all predators may pursue a particular prey organism. We chose to implement this restriction by the

[2]There are several strange consequences of this implementation of reproductive barriers. The mating relationship is non-transitive. Thus, while two organisms may be members of the same species, they may not be able to produce fertile offspring. However, if two organisms can produce fertile offspring, they must be members of the same species. In addition, a speciation event, the severing of a connected component of the reproductive graph through the removal of a node, may often lead to the creation of more than one daughter species. In other words, speciation events are not necessarily just bifurcations.

[3]The coalescence of species in the history of life is thought to have been a rare event, but little data exists which bears upon this issue.

[4]Plants are not allowed to fill more than half of a patch (1024 organisms) so that they do not compete with the animals for space.

following equation.

$$Pr[\text{predation}] = ce^{-dn-m} \tag{1.1}$$

Where c is the *prey-location* parameter, d is the *predation-distribution* parameter, n is the number of times that the prey organism has escaped predation this time step (simulating competitive interference of the predators), and m is the number of bits that mismatch between the predator's search template and the prey's predation resistance characters. When the *predation-distribution* is 0, there is effectively no spatial structure. With this elaboration, predator-prey oscillations are stable at all trophic levels over a wide range of parameter settings.

The model also exhibits a trophic cascade[2]. Trophic cascades are observed in the population levels of a species when its predator and then its predator's predator are introduced to the ecosystem. In the absence of predation, a species should expand until its population reaches the carrying capacity of its environment. When a predator is introduced to the ecosystem, the prey species population generally declines to a level below the carrying capacity of the environment. Finally, when the predator's predator is added to the ecosystem, the predator's population is reduced, thus allowing some recovery in population levels of the initial prey species. This cascade effect is readily observable in MoD.

Competitive exclusion also occurs in the model. The model was seeded with two plant species, one slightly better adapted to the habitat (at 1 locus). The model was tested with the inferior competitor's search template matching the habitat bit pattern at 1, 8, 16, 24, 28 and 31 loci. In each case, the superior competitor's search template matched the habitat in one additional locus. The model was seeded with equal population sizes of each species. A competitive exclusion event was recorded if one species expanded to 90% of the total population in less than 2000 time steps. The model was run 100 times for each of the 6 competitions. In all cases, the superior competitor excluded the inferior one.

The previous experiments were all carried out in the absence of evolution. The mutation rate was set to 0. In order to test that simple adaptation occurs in Mod, the model was initialized with an autotroph that matched its habitat in 24 out of 32 loci. This is the expected proportion of matches if both the prey template and the generalism loci were evolving randomly. Adaptation should appear as a increase in the degree of match between the plant and its environment above the level for random mutation. The model was run 32 times for mutation rates of $0.02, 0.11, 0.155, 0.2, 1.1$ and 2 bits flipped on average per genome. In all cases the plants evolved adaptations to their environment as determined by a one-tailed t-test relative to neutral evolution ($p < 0.001$).

5 Experiments and results

MoD was used to test ten different hypotheses for the diversification of life. In all cases, the model was initialized with three genetically homogeneous species (plant, herbivore, and carnivore) and then run for 5000 time steps. The number

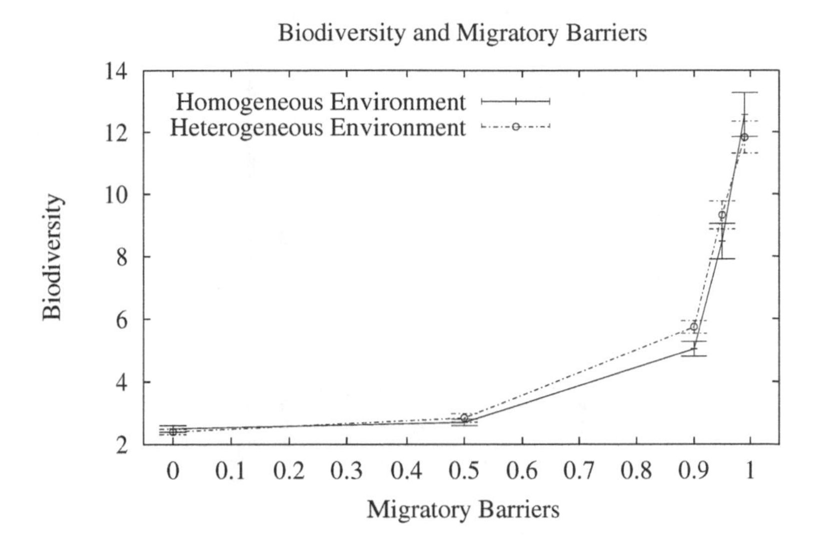

Figure 1: The effect of migratory barriers and habitat heterogeneity on biodiversity. The means of 50 runs in each condition are shown along with standard error bars. Migratory barriers have a strong impact on diversity only when those barriers are high. Habitat heterogeneity appears to slightly increase diversity but the effect is not statistically significant.

of surviving species was recorded at the end of that time. The model was run 50 times under each parameter setting. In most cases, the baseline for comparison was the result of running the model with no migratory barriers and the same habitat pattern in every patch (a homogeneous environment). Averaged over 50 runs, the baseline condition results in only 2.5 surviving species. Furthermore, most experiments were also run under "heterogeneous environmental conditions." This meant 0.95 migratory barriers and the random flipping of 8 bits in each habitat pattern (resulting in an expected average of 10.3 bits different between any two patches).

The first two experiments examined the effect of geographical isolation on diversity and the interaction of geographical isolation with habitat heterogeneity. The model was run under $0, 0.5, 0.9, 0.95$ and 0.99 migratory barriers, results shown in Fig. 1. The third experiment examined the effects of habitat heterogeneity by flipping 8 bits in every patch before running the model. The following experiments on abiotic factors tested Cracraft's[4] and Vermeij's[24] theories that change in the abiotic environment has been an important cause of diversification. In the fourth experiment, the habitat bit patterns of the patches were independently modified by flipping a bit, selected randomly, once every 156 times steps in each patch. This meant that by the end of 5000 time steps, the habitat in each patch would have changed 32 times. The fifth experiment varied the migratory barriers of all the patches over time so that they followed

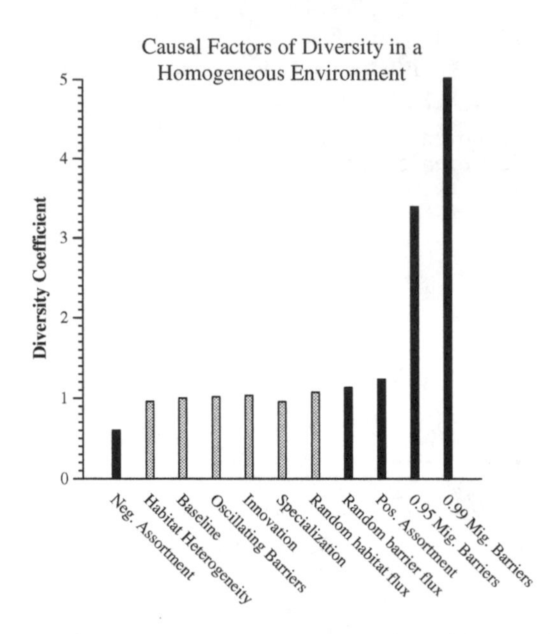

Figure 2: A summary of manipulations to species diversity. The chart shows the manipulations in the context of a homogeneous environment with no migratory barriers. Each effect has been normalized by its relevant baseline. The diversity coefficient indicates the magnitude of an effect relative to the baseline. The light grey bars indicate that the manipulation had no statistically significant effect. Black bars indicate a significant effect, with $p_i < 0.05$.

a sine wave with a period of 500 time steps, a maximum value of 1.0 and a minimum of 0. In this way the subpopulations were repeatedly isolated and mixed. The sixth experiment involved allowing the migratory barriers of the patches to individually and independently follow a random walk between 0 and 1 with increasing or decreasing steps of 0.05 every time step.

The seventh experiment examined the effect of allowing evolutionary innovations with the potential to allow species to capitalize on unexploited resources in their environment. A restricted innovation condition was modeled by preventing mutations in 10 loci of the generalism chromosome (set to 0), 10 loci of the prey template chromosome (set to 0), and 10 loci of the predation resistance chromosome (set to 0). This was then compared to evolution when mutations, and thus adaptations, were allowed in all 32 loci of those chromosomes. Similarly, the eighth experiment examined the effects of the evolution of specialism by fixing 10 bits of the generalism chromosome (set to 1). In this restricted specialism condition, all organisms were forced to be generalists in those 10 loci. This was compared to the case where species could evolve specialism in all 32 loci of their generalism chromosome. Finally, the ninth and tenth experiments examined the effects of assortative mating on diversification. Both positive and negative assortative mating were examined. Assortative mating was modeled by allowing

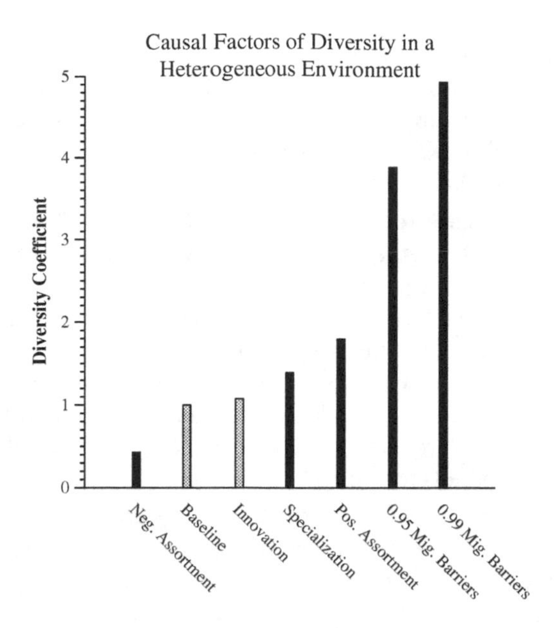

Figure 3: A summary of manipulations to species diversity in the context of a heterogeneous environment and 0.95 migratory barriers. Here the baseline is a heterogeneous environment with 0.95 barriers. In the case of the manipulations to the barriers, the relevant baseline is a heterogeneous habitat with no migratory barriers. Again black indicates a statistically significant effect with $p < 0.05$.

an organism to consider up to 8 potential mates in its patch. Each of these potential mates were still required to differ from the organism in no more than one locus of their reproductive chromosome. But in addition to this, the organism compared its predation resistance chromosome characters—a proxy for observable phenotypic characters—to those of its mates and either selected the most similar (positive assortment) or most dissimilar of the candidates for mating. The results off all these experiments are summarized in Fig. 2 and Fig. 3.

6 Discussion

The first surprise was that the increased capacity for innovations in resource utilization had no significant effect on diversification. This was the dynamic that Benton[1] thought best explained the increasing diversification of life. The same experiment was run with as many as 28 of the 32 loci fixed, restricting adaptive space dramatically. Yet, even this extreme restriction, with other parameters set to maximize diversity, does not significantly lower diversity levels. This results suggest that we should question one of the basic assumptions of the key innovation perspective. Why should expansion in adaptive space stimulate diversification? There is no direct relationship between adaptive space and speciation. Speciation depends on reproductive barriers. We should certainly expect that key innovations that modify characters involved in reproduction, and perhaps even geographical isolation, should have a strong impact on diversification. But it is unlikely that the availability of new resources in and of themselves should stimulate diversification.

There are many caveats to the model. First of all, the model is extremely simple and abstract. At the base of the food web there is only one resource, energy. Second, the spatial, population, and temporal scales are far below the scales at which most Paleobiologists work. With only 16 patches, a maximum of 32K organisms, and 5000 time steps (roughly 2000 generations) the model represents evolution and ecosystems on a regional and millenial scale at best. Third, we have not modeled morphological diversity. Fourth, all of the diversity dynamics of the model have taken place on a binary hypercube (the space of possible reproductive genotypes in the model).

These caveats may also help to explain the difference between Rice and Hostert's[18] findings that habitat heterogeneity had a strong effect on reproductive isolation and the model's result that habitat heterogeneity had little effect on diversification. MoD's representation of a habitat is so simple that it only directly affects the autotrophs. Yet, if one looks closely at Rice and Hostert's review, we see that divergent selection was only effective at producing reproductive isolation when the subpopulations were physically isolated, the hybrids were removed, or the divergent selection was particularly intense (e.g., 95% of the organisms were killed each generation). In the later case, reproductive isolation arises because the reproductive characters "hitchhike" along with the characters that are being intensely selected. Selection is too intense and quick for recombination to allow the reproductive traits to evolve independently of the selected traits. Simple disruptive selection on an arbitrary character does not tend to produce reproductive isolation. Most of the experimental cases where divergent selection leads to reproductive isolation are not represented in the experiments on habitat heterogeneity in MoD. The one exception is the case of divergent selection in allopatry (physical isolation). In MoD there was one case, when the migratory barriers where 0.9, that habitat heterogeneity had a significant effect on diversity levels ($p < 0.05$).

Under conditions of habitat heterogeneity and 0.95 migratory barriers, specialization does have a significant impact on diversity. When species are allowed to evolve specialism at all loci of their generalism chromosome, the diversity jumped from 6.7 to 9.32. However, specialization only became important in the presence of migratory barriers and divergent selection. Heterogeneous habitats with migratory barriers effectively creates more (geographically protected) niches upon which species might specialize.

MoD exhibits a striking influence of geographical barriers on diversification. Simply increasing the migratory barriers to high levels led to a dramatic increase in the number of species. This matches the dominant theory of speciation. Most speciation is allopatric, where populations are physically separated and slowly diverge to the point of reproductive isolation. Note that Fig. 1 shows that the relationship between migratory barriers and diversity is non-linear. This is why the experiment with randomly fluctuating barriers produces significantly more species than its baseline where the all the migratory barriers were held constant at 0.5.

Of particular theoretical interest is the effects of positive assortment. Such behavioral characteristics are difficult to recover from fossil data and so have gone largely untested as factors in speciation. Positive assortative mating significantly increases diversity in MoD under both homogeneous environmental conditions with no migratory barriers, and under heterogeneous environments with 0.95 barriers. Positive assortment acts as a proxy for geographical separation. The populations tend to fracture into the clusters of similar phenotypes (whatever characteristics organisms are using to select their mates) with little mating between these subpopulations. The subpopulations may then diverge to the point of reproductive isolation.

The results of the model must be predicated on acceptance of the simplifications in the abstractions and their implementations in the simulations. The predictions of the model should be tested with laboratory experiments, perhaps with *Drosophila* or *Saccharomyces*, as well as comparative phylogenetics.

Bibliography

[1] BENTON, M. J., "The causes of the diversification of life", *Major Evolutionary Radiations* (P. D. TAYLOR and G. P. LARWOOD eds.), Clarendon Press (1990) 409–430.

[2] BOOTH, G., "Gecko: A continuous 2-d world for ecological modeling", *Artificial Life* **3** (1997), 147–164.

[3] CASWELL, H., and A. M. JOHN, "From the individual to the population in demographic models", *Individual-Based Models and Approaches in Ecology* (D. DEANGELIS and L. GROSS eds.), Chapman and Hill (1992) 36–61.

[4] CRACRAFT, J., "Biological diversification and its causes", *Annals of the Missouri Botanical Garden* **72** (1985), 794–822.

[5] CRACRAFT, J., "The origin of evolutionary novelties: Pattern and process at different hierarchical levels", *Evolutionary Innovations* (M. H. NITECKI ed.), The University of Chicago Press (1990) 21–44.

[6] FIALKOWSKI, K. R., "Symptric speciation: A simulation model of imperfect assortative mating", *Journal of Theoretical Biology* **157** (1992), 9–30.

[7] HERRÁIZ, C. I., J. J. MERELO, S. OLMEDA, and A. PRIETO, "Biodiversity through sexual selection", *Artificial Life V* (C. G. LANGTON and K. SHIMOHARA eds.), MIT Press (1997) 308–318.

[8] HOLLAND, J. H., *Adaptation in Natural and Artificial Systems*, MIT Press (1992).

[9] HUNTER, J. P., "Key innovations and the ecology of macroevolution", *Trends in Ecology and Evolution* **13**, 1 (1998), 31–36.

[10] JABLONSKI, D., and D. J. BOTTJER, "The ecology of evolutionary innovation: The fossil record", *Evolutionary Innovations* (M. H. NITECKI ed.), The University of Chicago Press (1990) 253–288.

[11] JENSEN, J. S., "Plausibility and testability: Assessing the consequences of evolutionary innovation.", *Evolutionary Innovations* (M. H. NITECKI ed.), The University of Chicago Press (1990) 171–190.

[12] JONES, T., P. T. HRABER, and S. FORREST, "The ecology of echo", *Artificial Life* **3** (1997), 165–190.

[13] LINDGREN, K., and M. G. NORDAHL, "Artificial food webs.", *Artificial Life III* (C. G. LANGTON ed.), Addison-Wesley (1994) 73–104.

[14] LIOU, L. L., and T. D. PRICE, "Speciation by reinforcement of premating isolation", *Evolution* **48** (1994), 1451–1459.

[15] MALEY, C. C., *The Evolution of Biodiversity: A Simulation Approach*, PhD thesis Massachusetts Institute of Technology (1998).

[16] MAYR, E., *The growth of biological thought : diversity, evolution, and inheritance*, Belknap Press (1982).

[17] RAUP, D. M., and J. J. SEPKOSKI, JR., "Mass extinctions in the marine fossil record", *Science* **215** (1982), 1501–1503.

[18] RICE, W. R., and E. E. HOSTERT, "Laboratory experiments on speciation: What have we learned in 40 years", *Evolution* **47** (1993), 1637–1653.

[19] SARUWATARI, T., Y. TOQUENAGA, and T. HOSHINO, "Biodiversity through sexual selection", *Artificial Life IV* (R. A. BROOKS and P. MAES eds.), MIT Press (1994) 424–429.

[20] SCHLUTER, D., "Ecological causes of adaptive radiation", *The American Naturalist* **148** (1996), S40–S64.

[21] TODD, P. M., and G. F. MILLER, "On the sympatric origin of species: Mercurial mating in the quicksilver model", *Proceedings of the Fourth International Conference on Genetic Algorithms* (R. K. BELEW and L. B. BOOKER eds.), Morgan Kaufmann (1991) 547–554.

[22] TODD, P. M., and G. F. MILLER, "Biodiversity through sexual selection", *Artificial Life V* (C. G. LANGTON and K. SHIMOHARA eds.), MIT Press (1997) 289–299.

[23] VAN VALEN, L. M., "A resetting of Phanerozoic community evolution", *Nature, London* **307** (1984), 50–52.

[24] VERMEIJ, G. J., *Evolution and Escalation*, Princeton University Press (1987).

An Experimentally-Derived Map of Community Assembly Space

Craig R. Zimmermann, Tadashi Fukami, and James A. Drake
Complex Systems Group
Department of Ecology and Evolutionary Biology
University of Tennessee
Knoxville, TN 37996

1. Introduction

Ecological communities – assemblages of species occcurring together in space and time – do not emerge *de nova* as intact entities. Rather, they are assembled structures that derive their organization from a dynamic interplay of processes acting across a wide range of spatiotemporal scales. Regularities of pattern among some systems suggest the influence of common determinants of community organization. Investigation into the processes of *community assembly* comprises a broad research program that seeks the general principles or *assembly rules* by which species both accumulate in time and become integrated into functional organizations. One avenue of investigation in this area has explored how variation in the history of species invasion can influence such formation. Consisting of both empirical and theoretical studies, this research observes how manipulations of the sequence and timing of controlled invasions affects community structure and dynamics (Rummel and Roughgarden 1985; Wilbur and Alford 1985; Robinson and Dickerson 1986, 1987; Drake 1990, 1991; Drake et al. 1993; Lawler 1993; Post and Pimm 1993; Law and Morton 1996; Law 1999).

Here, we present an experimentally-derived community assembly map for a laboratory-constructed freshwater ecosystem that represents a synthesis of a suite of experiments conducted in our lab to explore various facets of the assembly process. We present our findings within a dynamical systems framework that addresses community assembly as a process of self-organization. We feels that such a framework offers a promising approach for an improved understanding of ecological patterns and processes. While we are not alone in advocating this position (Odum 1988; Kay 1991; Jorgensen

1998; Levin 1998; Ulanowicz 1998), our hope is that a wider exposure of these concepts to both ecologists and system scientists alike will foster future cross-disciplinary dialogue.

2. Community Assembly and Self-Organization

We begin by adopting a general model of community assembly that borrows concepts and terminology from dynamical systems theory. Foremost, we view community assembly as a process of *self-organization*. Self-organization - arising through the *local* action and interaction of system components - can bootstrap a system from an initially random state into one with a high level of *global* organization (Prigogine 1980). Inherent non-linearities, threshold effects, and feedback loops can result in cause and effect that need be neither proportionally expressed nor strictly top-down or bottom-up. Sensitive-dependence on initial conditions is also common such that prediction of the resulting structures can be difficult despite detailed descriptions of system components. As these processes are typically temporally irreversible, historical influences can play a powerful role.

We visualize the process of community self-organization as occurring in an *n*-dimensional state space - which we refer to as "assembly space" - wherein dimensions are comprised of all biotic and abiotic state variables relevant to the ecosystem(s) under consideration (Drake et al. 1999). Any given point within this space represents a unique community state and any series of subsequent states represents a trajectory of community development. Clearly, not all states are possible. Physical, chemical, and other abiotic constraints define fundamental limits within which allowable community configurations can occur. By considering the nature of multiple community trajectories as they arise under various conditions, one can begin to discern the underlying topology that influences the formation of alternate community states. Regions of high asymptotic convergence suggest the presence of *attractors* - stable or quasi-stable configurations - toward which the system is drawn when falling within the range of an attractor's influence. Regions of meandering trajectories, on the other hand, suggest a reduced influence of attractors and concomitant increase in variation due to stochastic processes. Finally "holes" - regions void of trajectories - may be present that represent untenable compositional configurations, sensu "forbidden species combinations" (Diamond 1975, c.f. Gavrilets 1999).

Assembly processes define the attractor landscape through which community development proceeds. Such attractors can govern not only the current community configuration, but also establish the suite of plausible states through which future development can proceed. Depending on context, the introduction of novel species can initiate any number of changes that alter the attractor landscape and transform community structure. The ways in which any given invasion can elicit such reorganization are numerous. These include, but are not limited to: a) addition of a new trophic level, b) competitive-exclusion of subordinate species, c) mediation of competitive interactions, d) facilitation/inhibition of species colonization through mediation of environmental conditions or resource availability (Connell and Slatyer 1977), e) allelopathic and other antagonistic interactions, and f) formation of symbiotic relationships. Note that these are not mutually exclusive. More than any single mechanism may be operating and synergetic interactions among them can result in complex higher-order mechanisms. Of course, not all invasions necessarily cause such reorganization. Some invading species, for example, may never reach an abundance at which any significant influence is achieved. Such species may, therefore, be regarded as peripheral to the essential core of community structure. In other cases, the invading species may possess functional characteristics that are too similar to other extant species to induce any significant qualitative change in community organization. The invader, in this case, is simply not "novel" enough.

3. A Map of Community Assembly Space

The assembly space map presented here represents a synthesis and reanalysis of a series of allied freshwater microcosm experiments that address various questions of community assembly. All experiments were conducted under controlled laboratory settings. Growth conditions - i.e. light, temperature, media concentration - were established at levels for optimal growth and starting conditions were standardized across replicates to eliminate initial environmental conditions as a source of variation in the phenomenon observed. Any exhibited community divergence, therefore, occurred as a result of species interaction and/or biological-mediation of the abiotic environment that arose as a function of assembly history.

We present the map in a flow-chart style that shows various assembly routes and their associated final community states. See Figure 1 for map schematic. Routes of assembly encorporate points of bifurcation that arise either: 1) between experiments due to fundamental differences in their respective designs, or 2) within experiments due to the influences of community assembly processes. Final community states are presented as rank abundance hierarchies by producer and consumer species. We now discuss a number of points of interest that address phenomenon that arose under these various assembly routes. The numbered points correspond to parenthesed numbers on the map proper.

1. Species Pool. Communities were assembled out of green algae and freshwater invertebrates. The specific species used and their associated map abbreviations follow. Green Algae: *Ankistrodesmus falcatus*, ANK; *Chlamydomonas reinhardtii*, CHL; *Euglena gracilis*, EUG; *Scenedesmus quardricauda*, SCE; *Selenastrum bibrium*, SEL. Invertebrates: *Cyclops vernalis*, CYC; *Cypris* sp., CYP; *Hyalla azteca*, HYA; *Daphnia magna*, MAG; *Moina macrocarpa*, MOI; *Paramecium multimicronucleatum*, PAR; *Pleuroxus truncates*, PLE; *Daphnia pulex*, PUL; *Simocephalus vetulus,* SIM.

2. Isolated Patch vs. Interconnected Patch Landscapes. The experiments fell into two broad categories: a.) isolated community patches wherein interaction among patches via species migration was excluded (Drake, 1991), and b) interconnected patch landscapes wherein dispersal between patches was allowed (Drake et al. 1993). See Figure 2 for a schematic of the landscape patch array used. Each landscape contained a single *primary*

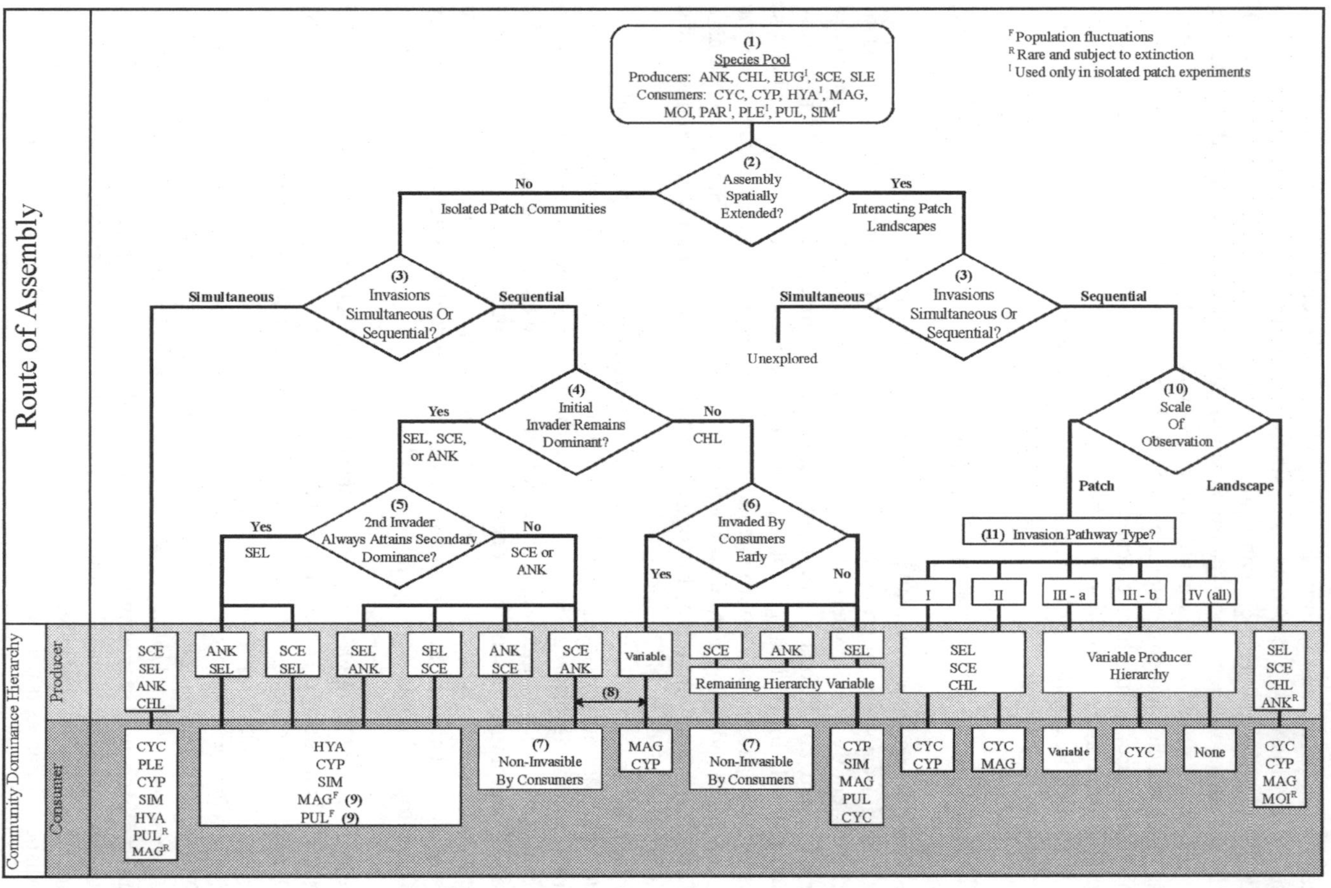

Figure 1. Flow-chart schematic of community assembly space showing various routes of assembly and their associated community states. Final states are delineated by rank abundance for producer and consumer species.

source patch into which each new invasion was first introduced. Beginning with the second primary invasion, aliquot transfers were made in a "stepping stone" fashion from the primary source patch to progressively more distant patches along predetermined pathways. See Figure 2 for the direction of spread. Patch labels indicate the terminal patch in each pathway analyzed. Note that some pathways contained one (Type III-b and IV-b) or two (Type IV-c) patches that received transfers from two intermediate source patches. As the source patch for any given transfer accumulated species, the potential for transfers containing multiple species increased. The combination of multiple species transfers, alternate routes of invasion, and sampling effects (see #11) created conditions under which variation in community states could arise.

3. Simultaneous vs. Sequential Invasion. Communities constructed by the simultaneous invasion of all species consistently produced a replicable dominance hierarchy that arose through differentials in producer growth rate, relative competitive ability, and grazing vulnerability. Communities constructed by sequential invasions, on the other hand, added a temporal element to the assembly space that allowed for considerable variation in the production of community structure. Regions of the assembly space visited by particular invasion sequences were unavailable to communities otherwise assembled. Dominance hierarchies, in these cases, were readily altered. Ten sequence permutations were tested with the isolated patch experiment. A single invasion sequence - one that consistently converged on the same community configuration in isolated patch experiments - was used in the landscape experiment. Thus any variation in the communities produced was an outcome of the interaction among landscape patches. See Point #11. Note that a simultaneous invasion treatment - one in which all invasions into the primary source patch are made concurrently - has not been explored in the landscape experiment.

4. Priority Effects of Initial Producer Invaders. Producer species that were the initial invaders into a patch encountered a nutrient-rich, competitor-free environment. Three of

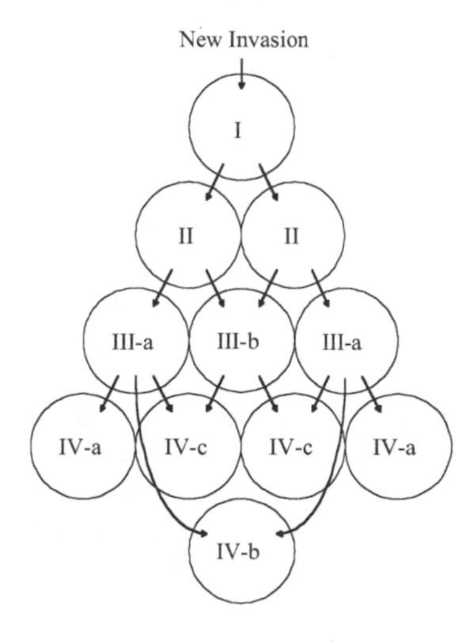

Figure 2. Schematic of patch landscape array. Arrows indicate the direction of transfers and patches labels ccorrespond to invasion pathway type. Patch I is the primary source patch into which all new species were first introduced.

the four producer species exhibited *priority effects* under such conditions in that they maintained numerical dominance throughout the experimental period. Only *Chlamydomonas* lost its numerical advantage when subsequently invaded by a competing producer species and eventually became extinct in all sequences explored.

5. Secondary Producer Ranking. The producer *Selenastrum* always attained secondary dominance when introduced as the second invader. The deterministic nature of these trajectories suggests the presence of a strong attractor. When either *Ankistrodesmus* or *Scenedesmus* was the second invader, however, the remaining dominance hierarchy remained variable. These trajectories appear to be indeterministic.

6. Invasion Timing of Consumers. Early invasion of consumers produced indeterministic assembly trajectories resulting in variation in the expressed community state. Late invasions, on the other hand, consistently produced communities with the same dominance hierarchy.

7. Invasion-Resistance to Consumers. Regions of the assembly space wherein either *Ankistrodesmus* or *Scenedesmus* was the dominant producer were found to resist invasion by all consumers. The specific mechanisms involved are unknown in most cases.

8. Crossing Trajectories. Some trajectories that were divergent in their early development were found to later temporarily converge in terms of their species rank abundance. Their later redivergence was apparently caused by a difference in algal sinking speed among the communities. Where such sinking was rapid, food availability to consumers was greatly reduced and communities resisted invasion by consumers. Consumers readily invaded communities, however, where the phytoplankton remained largely suspended. Thus, community invasibility cannot be simply predicted by rank abundance, but is dependent on events triggered early in their respective histories.

9. Fluctuations in Population Dynamics. History-induced variation was manifested not only in the species composition, but also in the dynamical behavior expressed by constituent populations. Some trajectories traversed regions of the assembly space wherein some species exhibit sustained - and often highly variable - fluctuations in abundance. Whether these represent genuine chaotic dynamics is unknown.

10. Scale of Observation. Replicate landscapes exhibited high similarity in terms of species composition when integrated at the landscape scale, despite considerable variation expressed among the constituent patches. This contrast underlies the need for selection of appropriate observational scale(s) when addressing ecological systems.

11. Species Dispersion and Invasion Pathways. The nature of invasion events in the landscape experiment differed from those of the isolated patch experiments in three key ways. First, transfers had the potential to contain more than a single species. Second, species could have multiple opportunities to colonize a given patch as long as it remained present in the source patch. Finally, given the relatively small transfer volume used, sampling bias effects could occur in the transfer process such that 1) the relative abundance among competing species could be changed, and 2) rare species could be excluded. As consumer species were a relatively small component of the primary source patch, this last effect was particularly influential in reducing consumer abundance. Consumers declined rapidly with distance from the primary source patch and were

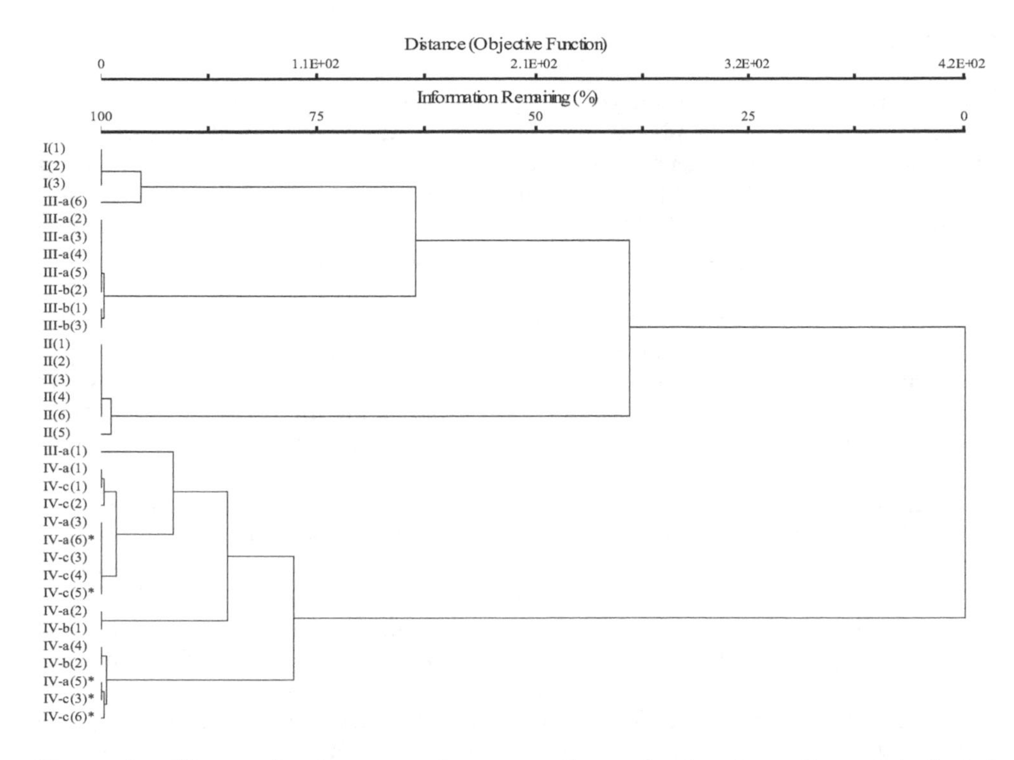

Figure 3. Cluster dendrogram of community rank abundance by patch. Labels correspond to invasion pathway type with numbers in parentheses indicating replicate number. Asterisks denote patches contaminated by premature invasion of *Selenastrum*. See Figure 2 for a landscape schematic with pathway identification.

completely absent at the landscape edges. Given these species can play an influential role in mediating competition between producers, their diminishing role resulted in considerable variation in producer composition exhibited in outer landscape patches. Cluster analysis of community rank abundance illustrates this trend of increasing variation with distance from the primary source patch. See Figure 3 for a cluster dendrogram. While replicate patches of types I, II, and III (a, b) were tightly clustered according to the type, patches of type IV (a, b, c) exhibited considerable variation. Note that contamination, in the form of premature invasion of *Selenastrum*, occurred in some of outer patches of one of the three replicate landscapes. Whereas *Selenastrum* lost its dominance in many of the equivalent patches of the unaffected replicates, the altered history caused by the contamination resulted in sustained dominance by *Selenastrum* and reduced compositional variability among contaminated patches.

4. Conclusions

Ecological communities are complex, self-organized entities whose extent of organization variably depends on environmental factors and local species pools. Historical effects can, consequently, have a strong influence of community organization. Drawing from results obtained from series of laboratory-constructed ecosystems, we have shown how manipulating invasion history can produce a rich diversity of community states. By constructing communities through the ordered addition of the same species, however, alternate community states were found to arise from different sequence permutations. Some of the sequences tested behaved deterministically in that a particular sequence repeatedly converged to a single community state configuration. Such convergence suggests the presence of strong attractors. Other sequences, however, behaved in a seemingly indeterministic fashion wherein multiple states arose from the same invasion history. It is possible that such communities are traversing a region of assembly space where the relative influence of stochastic processes is amplified. It is also possible, however, that self-organization is being governed by multiple attractors that emerge from sensitivity to parameters other than invasion sequence alone. Allowing for dispersion of species between communities was also found to contribute significantly to the production of variation in community states. Multiple species invasions, barriers to dispersal, and stochastic effects combined to generate unique invasion histories associated with each landscape patch. Despite considerable variation among individual patches, however, remarkable similarity in species composition was displayed among replicate landscapes when constituent patchs were integrated to the landscape scale. This suggests the presence of global constraints that may define regions of plausible community states.

The complex nature of ecological systems makes them difficult subjects of study (Fukami et al. 1999). We believe, however, that complex systems approach offers a powerful framework from which progress towards a robust theory of ecosystem development and behavior can be made. Microcosm studies like ours offer the potential of systems of managable complexity in which tractable questions can be posed. Clearly, the results presented here represent a small fraction of the possible questions on community assembly that can be addressed. Currently, we are exploring a broad range of new questions that address 1) the interaction of basal resource availability with invasion history, 2) the interaction role of phenotypic plasticity with invasion history, and 3) the contexts in which invasion historical effects are most influential.

References

Connell, J. H. and Slatyer, R. O. 1977. Mechanisms of succession in natural communities and their role in community stability and organization. *Am. Nat.* **111:** 1119-1144.

Drake, J. A. 1990. The mechanics of community assembly and succession. *J. Theor. Biol.* **147:** 213-233.

Drake, J. A. 1991. Community assembly mechanics and the structure of an experimental species ensemble. *Am. Nat.* **137:** 1-26.

Drake, J. A., Flum, T. E., Witteman, G. J., Voskull, T., Hoylman, A. M., Creson, C., Kenney, D. A., Huxel, G.R., LaRue, C .S., and Duncan, J. R. 1993. The construction and assembly of an ecological landscape. *J. Anim. Ecol.* **62:** 117-130.

Drake, J. A., Zimmerman, C. R., Purucker, T., and Rojo, C. 1999. On the nature of the assembly trajectory. In: *Ecological Assembly Rules - Perspectives, Advances, Retreats* (E. Weir and P. Keddy eds.), Cambridge University Press, Cambridge, MS.

Diamond, J. M. 1975. Assembly of species communities. In: *Ecology and Evolution of Communities* (M. L. Cody and J. L. Diamond eds.), Belknap, Cambridge, MS.

Fukami, T., Zimmermann, C. R., Russell, G. J., and Drake, J. A. 1999. Self-organized criticality in ecology and evolution. *Trends. Ecol. Evol.* **14(8):** 321.

Gavrilets, S. 1999. A dynamical theory of speciation on holey adaptive landscapes. *Am. Nat.* **154 (1):** 1-22..

Jorgensen, S. E. 1998. Ecosystems as self-organizing critical systems. *Ecol. Model.* **111:** 261-268.

Kay, J J. 1991. A non-equilibrium thermodynamic framework for discussing ecosystem integrity. *Environ. Mgt.* **15(4):** 483-495.

Law, R. and Morton, R. D. 1996. Permanence and the assembly of ecological communities. *Ecology* **77:** 762- 775.

Law, R. 1999. Theoretical aspects of community assembly. *Advanced ecological theory : principles and applications* (J. McGlade ed.) Blackwell Science, Malden, MA.

Lawler, S. P. 1993. Direct and indirect effects in microcosm communities of protists. *Oecologia* **93:** 184-190.

Levin, S. A. 1998. Ecosystems and the biosphere as complex adaptive systems. *Ecosystems* **1(5):** 31-36.

Odum, H. T. 1988. Self-organization, transformity, and information. *Science* **242:** 1132-1139.

Post, W. M. and Pimm, S. L. Community assembly and food web stability. *Math. Biosciences* **64:** 169-192.

Prigogine, I. 1980. *From Being to Becoming: Time and Complexity in the Physical Sciences.* W. H. Freeman and Co., San Francisco, CA.

Robinson, J. V. and Dickerson, J. E. 1986. The controlled assembly of microcosmic communities: the selective extinction hypothesis. *Oecologia* **71:** 12-17.

Robinson, J. V. and Dickerson, J. E. 1987. Does invasion sequence affect community structure? *Ecology* **68:** 587-595.

Rummer, J. D. and Roughgarden, J. 1985. A theory of faunal buildup for competitive communities. *Evolution* **39:** 1009-1033

Ulanowicz, R. E. *Ecology, the ascendent perspective.* Columbia University Press, New York, NY.

Wilbur, H. and Alford, R. A. 1985. Priority effects in experimental pond communities: responses of *Hyla* to *Bufo* and *Rana*. *Ecology* **66:** 1106-1114.

TRANSIMS for transportation planning

Kai Nagel
Los Alamos National Laboratory and Santa Fe Institute,
kai@santafe.edu

Richard L. Beckman
Los Alamos National Laboratory, rjb@lanl.gov

Christopher L. Barrett
Los Alamos National Laboratory, barrett@tsasa.lanl.gov

The TRANSIMS (TRansportation ANalysis and SIMulation System) project at Los Alamos National Laboratory attempts to model all aspects of human behavior related to transportation in one consistent simulation framework. The key to the TRANSIMS design is a radically microscopic simulation of the travelers. In consequence, as the first step, a synthetic population is created from demographic data. Next, another module of TRANSIMS generates synthetic activities (such as sleeping, eating, working, shopping, etc.) and activity locations for each synthetic individual. Each individual in the simulation then plans her transportation, making modal and routing choices. Finally, the plans of each individual are executed in a detailed transportation micro-simulation. Relaxation iterations are necessary to achieve consistence between the modules. Since this means running all modules, especially the micro-simulation, many times, this makes the already considerable computational challenges even harder. Results from micro-simulation runs can be used to extract data, for example for stake-holder analysis or for emissions calculations. In this way, different transportation alternatives can be compared in detail.[1]

---

[1]To appear in: Proceedings of the "International Conference on Complex Systems, Nashua, NH, 1999"

1 Introduction

Nobody likes traffic jams. For this and other reasons, most developed economies set aside a significant part of resources to reduce transportation bottlenecks. However, it has become increasingly clear that expansion of capacity may be limited, for example by safety concerns, by concerns about emissions and pollution, by restrictions to open space availability, etc. This means that detailed quantitative techniques would be useful, for example to analyze trade-offs between options.

In addition, it is now clear that increasing capacity increases traffic, often leading to the same bottlenecks as before. This means that transportation is *not* a simple infrastructure problem, where to find a good or optimal solution to move a given demand, but a problem where individual people's values and preferences play an enormously important role. In consequence, any quantitative technique needs to be able to represent aspects of human decision-making.

This leads, maybe naturally, to the idea to approach the problem with an *agent-based* micro-simulation. Each traveler is an agent; agents make plans about what to do during a day; in order to get from one activity location to another, agents can, for example, walk, use bicycles, drive cars, or use busses. This is exactly the idea of the TRANSIMS (TRansportation Analysis and SIMulation System) project: to represent each individual agent in a metropolitan region and to simulate all aspects of his/her decision-making as long as it is related to transportation.

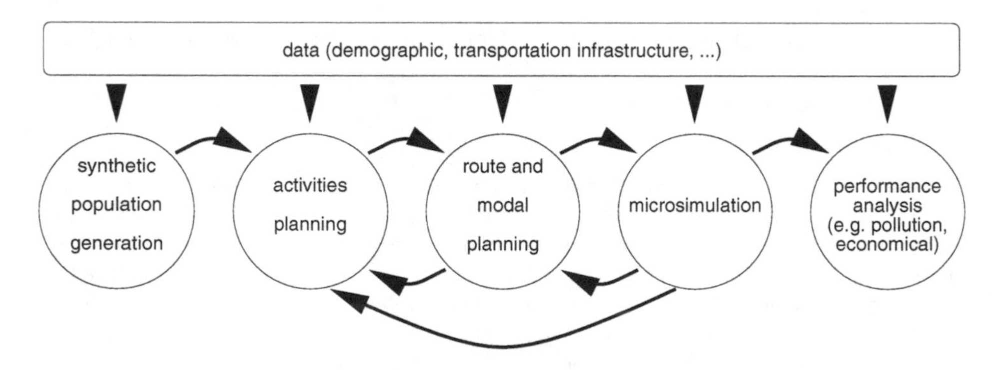

Figure 1: Modules of an agent-based simulation approach to transportation.

In principle, this sounds like a straightforward simulation approach (see Fig. 1): Derive synthetic individuals from demographic data and locate them on the network; use the demographic information together with land use information to derive activities (working, sleeping, eating, shopping, etc.) and activities locations for them; and let them decide about mode and routing of their transportation. So far, all these are *plans*, i.e. intentions of the simulated individuals. These plans can then all be fed into a realistic transportation micro-simulation, which can be used as the basis for analysis, such as emissions calculations.

However, this approach encounters challenges from several directions:

- **Size of the problem:** Metropolitan regions typically consist of several millions of travelers. Executing a second-by-second transportation micro-simulation on a problem this size within reasonable computing time is only possible with the use of advanced statistical and computational techniques.

- **Behavioral foundation:** We are far from understanding human behavior. For that reason alone, we are unable to predict the behavior of *individual* travelers. However, there seems to be a realistic chance that the *macroscopic* (emergent) behavior that is generated by thousands or ten-thousands of interacting individuals is considerably more robust than the behavior of an individual agent. This would be similar to Statistical Physics, where the trajectory of a single particle is unpredictable, yet, useful macroscopic properties of gases such as equations of state can still be derived.

- **Consistency problem:** The approach outlined above is not as straightforward as it sounds because the plans depend on *expectations* about traffic conditions during execution. For example, if a person expects congestion, he or she may make different plans than when no congestion is expected. Yet, congestion occurs only when plans *interact* during their simultaneous execution. – This problem is not unknown in traditional economic theory and is traditionally overcome by the assumption of rational agents. Both with and without the assumption of rationality, this problem of consistency between plans and micro-simulation makes the computational challenge even bigger.

- **Robustness:** Any approach to a problem needs to have reproducibility of the results under a wide enough range of changes, or otherwise the results are useless for practical purposes.

In the remainder of this paper, we want to give information on how the basics of such a transportation simulation can look like. Because of our own experiences, we will focus on the TRANSIMS project; however, there are other projects in the area (e.g. [5]).

2 Traditional approach

Before writing about TRANSIMS, let us introduce the approach that is currently in use in many transportation planning organizations. It is called the "four step process" because it consists of the following four steps (e.g. [10]): (i) **Trip generation** – determine sources and sinks of traffic. Note that this only counts departures and arrivals, but not where traffic comes from or goes to. (ii) **Trip distribution** – connect sources and sinks via traffic streams. The result of this procedure is sometimes called origin-destination (OD-) matrix. (iii) **Modal split** – determine which fraction of travel is done by the respective modes of transportation (walk, transit, car, etc.) (iv) **Traffic assignment** – assign traffic streams on the network, i.e. determine routes for the streams.

This method has several shortcomings, some of which we want to mention here. The most important *fundamental* shortcoming is probably that the mathematical foundation of the traffic assignment is only valid for time-independent problems. A consequence of this is that time-dependent dynamics, such as time-dependent queue-buildup from bottlenecks, or the sequential "chaining" of trips, cannot be represented easily. Probably as a consequence of the resulting difficulty to represent time-dependent processes in detail, practical implementations of the four step process decouple the modules from each other, with for example the consequence that congestion does not reduce the overall number of trips, or that analysis by sub-populations is not possible because there is no information on demographic characteristics of travelers available in step (iv). Also, the representation of the driving dynamics is too unrealistic to for example calculate emissions.

In spite of these shortcomings, the four-step process has the advantage that it has more than 30 years of history, during which the theoretical foundations have been well developed, and much practical experience has been gained. It is therefore understandably a difficult and slow process to build similar confidence in a new technology based on agent-based simulation.

3 Synthetic population generation module

The idea behind the TRANSIMS approach is that it simulates *individual* travelers. However, the typically available demographic input data is aggregated over geographical zones. The synthetic population generation module [2] of TRANSIMS generates instances of stochastic populations. These stochastic populations have the property that, if one would take a census off them, they would result in the same demographics as the original input data. The synthetic population can include information about household membership, home location, income, vehicle availability, etc. A functional implementation of such a module is currently available as part of TRANSIMS.

4 Activities planning module

The geographical and demographic information of the synthetic individuals is then used to generate activities for them. This can for example be done via complicated activities models with hundreds of free parameters obtained from the estimation of surveys (e.g. [4]), or via simplified "toy" models for research purposes (e.g. [12]). Such models should include activity *location* because that is where the travel demand is derived from.

5 Modal and route planning module

Activities at different locations need to be connected by transportation. Synthetic travelers need to select their modes (walk, drive, transit, etc.), and their routes. Again, the methods can range from simple, Dijkstra-type fastest-path algorithms [6] to sophisticated formal language constrained path problems [1].

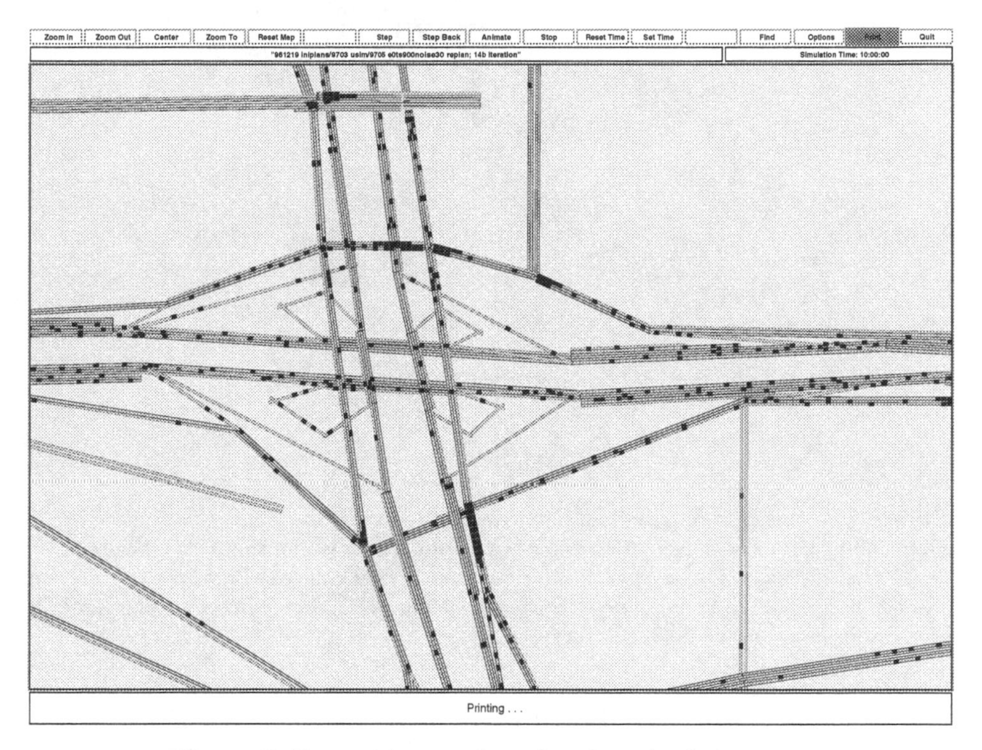

Figure 2: Zoomed-in snapshot of a micro-simulation run.

6 Micro-simulation module

All agents' plans are executed simultaneously in a transportation micro-simulation. This is the only place where interactions between agents are actually computed, although they could be *anticipated* in the other modules. Micro-simulations, too, can reach from relatively simple queue models [11] to arbitrarily realistic implementations; the currently maybe most realistic implementation that still fulfills the computing requirements is based on a cellular automata technique for car following and lane changing, enhanced by additional rules for elements such as signals, unprotected turns, turn pockets and weaving lanes, etc. [8].

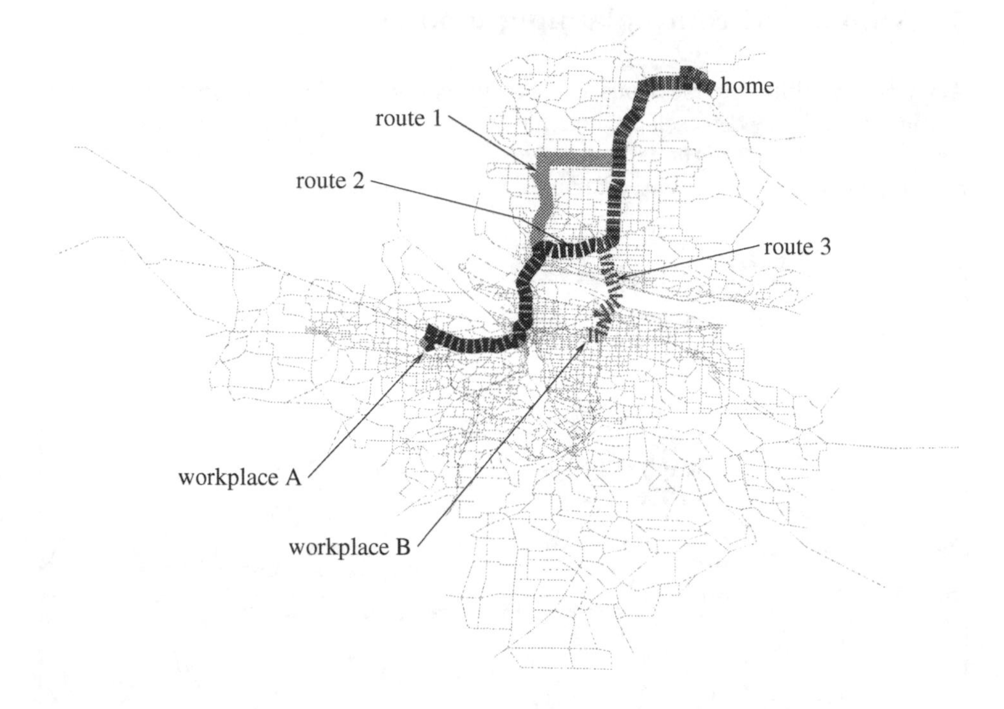

Figure 3: Example of re-planning. The traveler first drives to workplace A via route 1. Eventually, he/she decides to switch to a supposedly better route 2. Again later, he/she decides that this results still in too much driving time, and moves to a different workplace B closer to home. The street network shows Portland/Oregon. The example is taken from an actual feedback series.

7 Feedback and re-planning

As stated in the introduction, for recurrent situations such as rush-hour traffic, the conditions expected during plans-making need to be consistent with conditions during the execution of the plans. If they are not consistent, people are likely to change their plans. For example, if congestion turns out to be much worse than expected when the plans were made, people may drop less important destinations from their trip. That means that a situation where expected conditions are not consistent with conditions during execution are transient. Since we are currently interested in "average" (i.e. non-transient) results, such inconsistency is undesirable.

Yet, when the plans are made for the first time, no information about the resulting dynamics is available. This is true both for models and for reality. In our simulations, we currently solve this problem by a relaxation method, i.e. by iterating between the plans-making modules and the micro-simulation until expectations during plans making and conditions during plans execution are consistent ([7, 9], see Fig. 3). Such an approach does not seem implausible in behavioral terms. However, it is worth noticing that conventional economics solves the same problem by assuming complete rationality and in consequence assumes that agents will be able to find non-transient plans without ever executing them.

8 Analysis

Once expected and encountered conditions are consistent, we can assume that we are simulating non-transient scenarios. Such scenarios can be analyzed in detail in any way one wishes. Note that, because of the nature of the microscopic approach, all microscopic variables are still accessible at this stage. For example, information can be separated out by income of the travelers, by trip purpose, by age, etc. Since one can run the simulations for different alternative scenarios regarding the transportation infrastructure, it is straightforward to for example consider the effects of certain infrastructure alternative on certain sub-populations [3]. Also, since driving is –at least in principle– realistically modeled, vehicular emissions can be calculated [13]. All this is perhaps not surprising for people used to thinking microscopically about their problems, but it is a huge step for transportation which traditionally does not have access to these quantities as output from the four step process.

9 Summary

An agent-based simulation approach to transportation is possible with current technology, although it is still a significant computational challenge, especially when one strives for a realistic representation of the outside world. It is possible to break up the problem into modules for "population generation", "activities plans generation", "transportation plans generation", "micro-simulation, and "analysis". The computational challenge stems from the necessity to simulate realistic second-by-second driving of metropolitan regions with many millions of travelers. It is made even harder by the fact that the micro-simulation needs to be run several times so that the plans from the plans generation modules can relax towards "consistency" with the micro-simulation.

Acknowledgment – This work was performed at Los Alamos National Laboratory, which is operated by the University of California for the U.S. Department of Energy under contract W-7405-ENG-36 (LA-UR 98-5628).

Bibliography

[1] BARRETT, C. L., R. JACOB, and M. V. MARATHE, "Formal language constrained path problems", Los Alamos Unclassified Report (LA-UR) *98-1739*, see http://transims.tsasa.lanl.gov (1997).

[2] BECKMAN, R. J., K. A. BAGGERLY, and M. D. MCKAY, "Creating synthetic base-line populations", *Transportion Research Part A – Policy and Practice* **30**, 6 (1996), 415–429.

[3] BECKMAN ET AL, R.J., "TRANSIMS–Release 1.0 – The Dallas-Fort Worth case study", Los Alamos Unclassified Report (LA-UR) *97-4502*, see http://transims.tsasa.lanl.gov (1997).

[4] BOWMAN, J. L., *The day activity schedule approach to travel demand analysis*, PhD thesis Massachusetts Institute of Technology Boston, MA (1998).

[5] Forschungsverbund für Verkehr und Umwelt (FVU) NRW. See `www.zpr.uni-koeln.de/Forschungsverbund-Verkehr-NRW`.

[6] JACOB, R. R., M. V. MARATHE, and K. NAGEL, "A computational study of routing algorithms for realistic transportation networks", *ACM Journal of Experimental Algorithms* (in press), see http://www.santafe.edu/~kai/papers.

[7] NAGEL, K., and C.L. BARRETT, "Using microsimulation feedback for trip adaptation for realistic traffic in Dallas", *International Journal of Modern Physics C* **8**, 3 (1997), 505–526.

[8] NAGEL, K., P. STRETZ, M. PIECK, S. LECKEY, R. DONNELLY, and C. L. BARRETT, "TRANSIMS traffic flow characteristics", Los Alamos Unclassified Report (LA-UR) *97-3530*, see http://www.santafe.edu/~kai/papers (1997), earlier version: Transportation Research Board Annual Meeting paper 981332.

[9] RICKERT, M., *Traffic simulation on distributed memory computers*, PhD thesis University of Cologne Cologne, Germany (1998), see http://www.zpr.uni-koeln.de/~mr/dissertation.

[10] SHEFFI, Y., *Urban transportation networks: Equilibrium analysis with mathematical programming methods*, Prentice-Hall Englewood Cliffs, NJ, USA (1985).

[11] SIMON, P. M., and K. NAGEL, "Simple queueing model applied to the city of Portland", *International Journal of Modern Physics C* (in press).

[12] WAGNER, P., and K. NAGEL, "Microscopic modeling of travel demand: Approaching the home-to-work problem", Annual Meeting Paper *99 09 19*, Transportation Research Board, Washington, D.C. (1999), see http://www.santafe.edu/~kai/papers/.

[13] WILLIAMS, M. D., G. THAYER, and L. L. SMITH, "A comparison of emissions estimated in the transims approach with those estimated from continuous speeds and accelerations", Los Alamos Unclassified Report (LA-UR) *98-3162*, see http://transims.tsasa.lanl.gov (1998).

Mapping Multi-Disciplinary Design Processes

Cirrus Shakeri & David C. Brown
Mechanical Engineering Department
& Computer Science Department,
WPI, Worcester, MA 01609, USA.
cirrus@wpi.edu & dcb@cs.wpi.edu

This paper describes research in progress whose goal is to synthesize design methodologies for rapid product development in multi-disciplinary design situations. The potential outcome is superior design methodologies that facilitate integration and collaboration between different disciplines, conduct design tasks concurrently, and apply to a wide range of design problems, thus reducing costs and time-to-market.

The approach is based on simulating the design process using a multi-agent system that mimics the behavior of the design team. The multi-agent system activates the pieces of design knowledge when they become applicable. The use of knowledge by agents is recorded by tracing the steps that the agents have taken during a design project. Many traces are generated by solving a large number of design projects that differ in their requirements. A set of design methodologies is constructed by using clustering and abstraction techniques to generalize the traces generated. These methodologies then can be used to guide design teams through future design projects.

1. Introduction

The general scope of this research is the multi-disciplinary design of engineered systems. This work extends the concept of analysis-by-simulation to the area of engineering design research. Analyzing the behavior of physical systems in engineering applications by computer simulation using mathematical models has been a powerful tool in engineering, reducing costs and time in comparison to physical prototyping and experimentation.

In this work the same concept is applied to the design *process* instead of the design product. A computational model in the form of a knowledge-based multi-agent system is built that simulates the design process. By running the simulation under different conditions, and examining the performance, detailed understanding of the design process is gained [12]. As for simulations of physical systems, the computational model of the design process is a simplified one in which the design activities that are usually carried out by humans are performed by software agents in a slightly simplified manner. We have developed these ideas using the multi-disciplinary domain of robot arm design [11].

The current practices of multi-disciplinary design are based on ad-hoc strategies for handling the complexities that multiple points-of-view bring to the design process. These techniques solve the problem of complexity at the expense of giving up the potential advantages of diversity. The common methodologies for multi-disciplinary design are based on compromising between different disciplines rather than collaborating between them. These methodologies do not use a systematic, holistic approach to the problem of multi-disciplinary design and thus these approaches to multi-disciplinary design are not as efficient and effective as they could be.

The most common strategy to overcome the complexities of multi-disciplinary design is sequential design, in which different disciplines take part in the design process sequentially. In sequential design, information sharing between different disciplines is limited to the interfaces between disciplines [8]. As a result, conflicts between disciplines are not discovered until they are very expensive to resolve, because their resolution may need to destroy the partial designs generated by the previous discipline.

"In sequential design, a tentative design synthesis is developed by one designer, often acknowledged as the *lead discipline* designer, which address some of the key performance specifications and constraints for the artifact" [8]. Having a lead discipline that makes some of the key decisions reduces the number of conflicts. The other disciplines conform to the decisions made by the lead discipline. However, that may prevent them from producing their best solutions. In a lead-discipline approach a single point-of-view dominates the decision making process and therefore constraints from that discipline are favored. This produces a lower quality design product and increases the number of iterations required to reach an answer.

1. Simulation of the Design Process

We have proposed a new approach to the problem of producing better design methodologies for multi-disciplinary design based on the integration of different disciplines. The discipline-sequential approach, while poor, is quite simple. Integration tends to make the design process more complicated. To overcome this complexity, a computer system is developed based on a multi-agent systems paradigm in order to automate the simulation of the design process.

The system simulates examples of multi-disciplinary design processes while applying integration principles to the problem. These include common design knowledge representation schemes and common communication mechanisms; design knowledge sharing among participants; cooperative problem-solving strategies among participants; simultaneous design process where possible; and comprehensive mechanisms for conflict discovery and resolution.

The large chunks of discipline-specific knowledge are broken into small pieces and are represented in the system by agents. Agent activation is triggered in an opportunistic manner and is unaffected by discipline boundaries. Agents might participate sequentially or in parallel.

The traces of the agent activations (i.e., knowledge use) during the course of the design process are recorded. The recorded traces consist of orderly patterns of different design tasks that have led to the solution. Some candidate design methodologies are extracted by generalizing the patterns using clustering and inductive learning techniques. Some of these candidates will be reinforced by solving more examples and accepted as design methodologies for that particular class of problems.

A design methodology is a scheme for organizing reasoning steps and domain knowledge to construct a solution. It provides both a conceptual framework for organizing design knowledge and a strategy for applying that knowledge [13]. A design methodology can provide the knowledge for decomposing the problem into sub-problems, synthesizing partial designs, evaluating and then combining them into more complete partial designs, ordering design tasks by considering proposals from all participants, and discovering and resolving conflicts. Figure 1 shows one of the methodologies generated by the system for robot design.

IF

- constraints on deflection and the gain are both tight, and
- requirements on workload is rather "easy";
- workspace is of type "small-M";

THEN

- **choose** the location of the base of the robot: "left or below of the workspace length"
- **choose** the material: "steel stainless AISI 302 annealed"
- **select** the shape of the cross section of the link: "hollow round"
- **choose** the structural safety factor: "3"
- do the design and proceed to the next step

- **choose** the link 2 to link 1 length ratio: "0.5"
- do the design and proceed to the next step

- **pick** the configuration of the arm: "left-handed"
- **select** the ratio of section dimension to min required: "4"—if it fails select "3"
- do the design and proceed to the next step

- **find** the accessible region: use Equation 2-4
- **find** the deflection of the tip: use Equation 2-14
- **choose** the type of controller: "PD"
- do the design and finish the process.

FIGURE 1. A Methodology Discovered

2. Related Research

Recently there has been increasing recognition that multi-disciplinary design is important. A large amount of very good research has been focused on Multi-disciplinary Design Optimization (MDO) [14]. MDO tries to produce an effective product by recognizing and using appropriate combinations of parameters to be controlled and optimized by the designer. A key part of the process is the use of decompositions (see the discussion of decomposition in [14]) [3] [9].

In multi-disciplinary design problems the values of design parameters may determine what design method will be employed, as methods may have applicability conditions. As different design methods may introduce different dependencies, dependency chains, and potentially problem decompositions, can be dynamically determined. Hence the sequencing of design tasks can also be dynamically determined. Some approaches provide some user interaction to help determine what task sequence will be used [5] [4] [17]. However, while Multi-disciplinary Design problems often require the user's investigation of design trade-offs, for each problem and related set of requirements, there are a number of common design task sequences that are used. Such sequences form the basis of a design methodology for that problem or class of problems.

Our work requires that the results of the discovery process be well integrated and concurrent. Fine-grained tasks are needed, as opposed to the large grained tasks used by some research [4] [16]. Our agent-based approach can accommodate qualitative, experiential, and heuristic knowledge. Lander [6] provides a detailed review, while other work on multi-agent Systems in Concurrent Engineering is reported in [1].

The use of Machine Learning methods in support of design has been well documented [2]. While the use of Case-Based Reasoning (CBR) and inductively formed user (i.e., designer) models is becoming familiar, our method for generalizing design traces is not. Depending on what is included in the design traces, and its representation, we can take advantage of work on clustering and on induced finite-state transition networks, inductive learning for state-space search, or flexible macro-operators [7, pp. 258, 304, 348].

2. The System and the Results

A knowledge-based model of design is adopted in order to implement the proposed strategies for integration. To implement the proposed model a knowledge-based design tool based on a multi-agent architecture is developed that simulates the design process. The multi-agent paradigm intuitively captures the concept of deep, modular expertise that is at the heart of knowledge-based design [6].

An agent as a self-contained problem solving system capable of autonomous, reactive, pro-active, social behavior is a powerful abstraction tool for managing the complexity of software systems [15]. A multi-agent system is composed of multiple interacting agents, where each agent is a coarse-grained computational system. Agents are used as an abstraction tool for conceptualizing, designing, and implementing the knowledge-based design approach.

A Java-based computer program called RD (Robot Designer) has been implemented for parametric design of a two degrees of freedom (2-DOF) planar robot arm. We used RD to solve a set of 960 design projects. Figure 2 shows how many projects followed a specific trace. The promising results is that the distribution of the traces is quite scattered—that is, many projects followed similar traces. The total number of possible traces is the product of the number of design approaches of all the designer agents. For the experiments shown in Figure 2 the total number of possible traces is 2304. Among all 2304 possible traces only 84 were followed to generate successful designs, i.e., less than 4%.

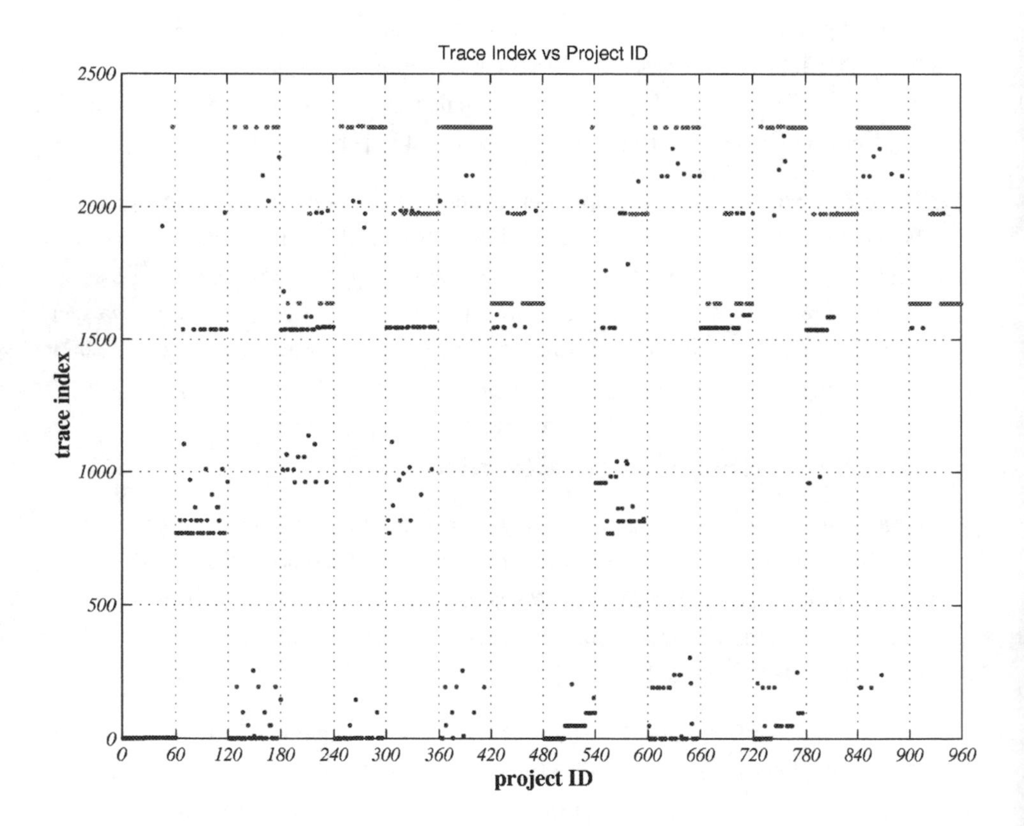

FIGURE 2. Distribution of Traces versus Projects

The low percentage of successful traces indicates that for each group of projects that followed a particular trace there is a unique combination of approaches leading to successful designs, hence there is a high chance that if similar projects follow the same trace they will succeed in generating a successful designs. As a result, the path followed by those projects can lead us to formulating a design methodology for the projects that followed that trace as well as projects that are similar to those projects.

The traces in the set of successful traces that are close enough can be clustered together to form a generalized trace. A generalized trace covers all the projects that followed each of the traces incorporated in the generalized trace. Design methodologies are formulated by extracting the correlation between a generalized trace and the design projects that followed that trace. The sample design methodology that is shown in Figure 1 is the English translation of the correlation between design projects and the corresponding traces.

2. Importance of this Research

Using system-developed methodologies allows effective and efficient practices to be used from the start of a project instead of being learned from experience. These new methodologies are radically different from the sequential, discipline-based ones. They also reduce time-to-market and save resources. To be able to compete in today's global market, companies need continuous improvements in product quality and improvements to the performance of their design and manufacturing processes. Integration reduces the number of failures and backtracking by facilitating information sharing, thus saving resources and reducing design time. Integration also provides collaboration between different participants that, as a result, enhances the quality of the design.

Agent-based systems allow the incorporation of new technologies systematically and quickly through the addition or deletion of agents. Thus new knowledge can be added, and old knowledge removed rapidly. Running the system with the new set of agents will result in new traces and thus new and different methodologies. In addition, design processes can be biased toward more environmentally friendly products, as the alternative methods that are built into each agent are tried in a preferential order, and as each method tends to contribute differently towards the final properties of the design.

The research attacks the problem of integration in multi-disciplinary design. The number of specialists is increasing, while the number of generalists, capable of doing system integration, is decreasing. Also the knowledge burden on the designer keeps increasing due to more materials and more options [10]. Thus it is becoming harder to develop methodologies for the integration of multiple disciplines in design. An increasingly specialized technological environment tends to force designers to concentrate on some disciplines more than others. This research allows designers to see the whole design problem.

Computers have mostly been used to support the manipulation and analysis of design product information. This work focuses on the design process, an aspect that has not benefited from computers very much. It applies computers to new areas of engineering design by incorporating new software methods. Simulation of design processes based on a multi-agent paradigm is a new area of research that has a high potential for practical as well as theoretical impact on the design of products. This research has produced convincing, preliminary results. The use of multi-agent systems technology is growing rapidly with the development of Java-based systems and agent access across the world-wide web.

The research incorporates judgement and experience. "System integration, many consider, is an ill-structured problem... No specific rules have to be followed when doing integration... Experienced designers deal with system integration using judgement and experience. Knowledge-based programming technology offers a methodology to tackle these ill-structured integration and design problems" [13].

This work benefits from inter-disciplinary contribution from the state-of-the-art in both Artificial Intelligence and Engineering Design. According to NSF's report on Research Opportunities in Engineering Design [10], "research areas that will have greatest impact on engineering design over the next 10 years are: Collaborative Design Tools and Techniques, Perspective Models/Methods, System Integration Infrastructure/Tools, and Design Information Support Systems". This work covers all of these areas of research and hence is expected to have a strong impact.

2. Conclusions

The potential applications of this research are in multi-disciplinary design situations, such as those that occur throughout the automotive industries, where large gains can be achieved with integrated methodologies. In addition, current methodologies can be analyzed for flaws and bottlenecks, and necessary refinements made. New methodologies can be customized so that they are biased toward specific objectives such as manufacturability or being environmentally friendly. By applying this approach the response time for the incorporation of new technologies in design processes will be reduced. Methodologies can be refined as soon as a change occurs in the market or in the organization of the company.

This research has proven that the following hypothesis is true: *Computers can provide us with better ways of doing design by discovering superior design methodologies that integrate different points-of-view of multiple disciplines in the design process.* It has been shown that it is possible to use the computers to simulate the design process. We can then analyze the results of the simulation to synthesize design methodologies that have superior features. The approach that we have proposed has been developed based on parametric design problems. Applicability of the approach to other types of problems needs to be investigated.

2. References

[1] BROWN, D. C., S. E. LANDER & C. J. PETRIE, "The Application of Multi-agent Systems to Concurrent Engineering", Editorial: *Concurrent Engineering: Research and Applications*, Technomic Publ., V.4, N.1, (March 1996), pp. 2-5.

[2] DUFFY, A. H. B., "The 'What' and 'How' of Learning in Design", *IEEE Expert: Intelligent Systems & Their Applications*, Vol. 12, No. 3, (May-June 1997), pp. 71-76

[3] GEBALA, D. G. and S. D. EPPINGER, "Methods for Analyzing Design Procedures", *Proceeding of ASME Design Theory and Methodology Conference: DTM'91*, ASME DE- Vol. 31, (1991), pp. 227-233.

[4] HALE, M., J. I. CRAIG, F. MISTREE, D. P. SCHRAGE, "DREAMS and IMAGE: A Model and Computer Implementation for Concurrent, Life-Cycle Design of Complex Systems", *Concurrent Engineering: Research and Applications*, Vol. 4, No. 2, (June 1996), pp. 171- 186.

[5] KROO, I. and M. TAKAI, "Aircraft Design Optimization Using a Quasi-Procedural Method and Expert System", *Multidisciplinary Design and Optimization Symposium*, (Nov. 1990).

[6] LANDER, S., "Issues in Multi-agent Design Systems", *IEEE Expert: Intelligent Systems and their Applications*, (March-April 1997), Vol. 12, No. 2, pp. 18-26.

[7] LANGLEY, P., *Elements of Machine Learning*, Morgan Kaufmann, Inc., (1996).

[8] LEVITT, R. E., Y. JIN, and C. L. DYM, "Knowledge-Based Support for Management of Concurrent, Multidisciplinary Design", *AI EDAM*, Academic Press Ltd., Vol. 5, No. 2, (1991), pp. 77-95.

[9] LIU, J., and D. C. BROWN, "Generating Design Decomposition Knowledge for Parametric Design Problems", *AID-94: Artificial Intelligence in Design'94*, (Eds.) Gero & Sudweeks, Kluwer Academic Publ., (1994), pp. 661-678.

[10] NATIONAL SCIENCE FOUNDATION, *Research Opportunities in Engineering Design*, NSF Workshop Final Report, (April 1996).

[11] RIVIN, E. I., *Mechanical Design of Robots*, McGraw-Hill Book Company, (1988).

[12] SHAKERI, C., D. C. BROWN, and M. N. NOORI, "Discovering Methodologies for Integrated Product Design", *Proceedings of 1998 AI & Manufacturing Workshop: State of the Art and State of Practice*, (Aug. 31-Sept. 2, 1998), Albuquerque, NM, pp. 169-176.

[13] SOBOLEWSKI, M., "Multiagent Knowledge-Based Environment for Concurrent Engineering Applications", *Concurrent Engineering: Research and Applications*, Technomic Publ., V.4, N.1, (March 1996), pp. 89-97.

[14] SOBIESZCZANSKI-SOBIESKI, J. and R. T. HAFTKA, "Multidisciplinary Aerospace Design Optimization: Survey of Recent Developments", *34th AIAA Aerospace Sciences Meeting and Exhibit*, Reno, Nevada, AIAA Paper No. 96-0711, (January 15-18, 1996), pp. 32.

[15] WOOLDRIDGE, M., "Agent-based software engineering", *IEEE Transactions on Software Engineering*, (1997), 144(1): 26-37.

[16] WOYAK, S., B. MALONE, and A. MYKLEBUST, "An Architecture for Creating Engineering Applications: The Dynamic Integration System", *Proceeding of ASME Computers in Engineering Conf.*, Boston, MA, (September 1995).

[17] WUJEK, B. A., J. E. RENAUD, S. M. BATILL, and J. B. BROCKMAN, "Concurrent Subspace Optimization Using Design Variable Sharing in a Distributed Computing Environment", *Concurrent Engineering: Research and Applications*, Vol. 4, No. 4, (December 1996), pp. 361-377.

What Can We Learn From Learning Curves?

Gottfried J. Mayer-Kress[1]
Karl M. Newell[1]
Yeou-Teh Liu[2]
[1] Department of Kinesiology, Penn State University
University Park, PA 16801
[2] Taipei Physical Education College, Taipei, Taiwan

One of the universal characteristic features of complex adaptive systems is their capability to learn in the sense of exhibiting persistent change of behavior in response to changing external conditions This applies to natural as well as artificial complex systems across different levels of complexity. There exists a vast amount of literature that deals with specifics of how learning takes place and how it can be accelerated or made more robust to external perturbations. Since the early years of systematic studies of motor learning a number of different types of functional forms have been proposed to describe the shape of these learning curves. In many publications this description was based purely on curve fitting without much theoretical foundation. Others are dealing with models for learning in an artificial intelligence context (for instance the "chunking model" of Newell and Rosenbloom) as a theoretical basis for general learning behavior. The recent interest in a dynamical systems approach to motor behavior provided the interpretation of learning as transient approach of state space trajectories to attractors that correspond to the behavioral pattern that needs to be learned. This is closely related to hill climbing methods in neural network learning. The corresponding learning curves therefor would be expected to show an exponential characteristics. Other interpretations associated learning with non-equilibrium phase-transitions which would imply power-law behavior with critical exponents. In this report we describe conditions for motor learning for which we would expect exponential or power law behavior. We fit both learning curve representations to empirical data-sets from motor tasks. We present simulations that demonstrate how observed learning curves can be reproduced by a number of different methods that can lead to testable hypotheses about the learning mechanism.

1 Introduction

A common feature of all complex system is that in one way or another they show a form of adaptation that we can call learning in a broad sense of the word. In the following we present a number of historical examples of human motor learning that suggests a power law behavior of performance as a function of practice time. There have been theoretical explanation attempt based on "chunking" of information that is learned algorithmically[5]. We want to re-examine the generic form of learning curves in the framework of non-linear complex systems where we would expect both exponential[9] as well as power-law regularities in learning curves, depending on the dynamical state the system is in.

2 Functional Classes of Learning Curves

2.1 Exponential Functions And Powerlaws

Already in the beginning of this century it was observed that there are strong regularities in the general shape of learning curves in different areas of human motor learning[15]. Without much theoretical foundation a number of different functions have been proposed to fit the data[14]. It turns out, however, that all these functions fall into two categories[7]:

1. Exponential functions: They correspond to a constant learning rates and one fixed time-scale. In this case the performance function $p(n)$ changes with the number of practice sessions as:

$$p(n) = A + Be^{g(n+n_0)} \tag{1.1}$$

where A is a constant corresponding to the asymptotic performance parameter, B indicates the difference to the initial performance, g indicates the magnitude of the learning rate (generally g is negative) and therefore its inverse defines an intrinsic time-scale of the system. The constant n_0 allows taking into account for a shift in the start time (in the following we assume $n_0 = 0$). If we plot the logarithm of the performance function we obtain a linear graph with slope g :

$$\log(p(n) - A) = g(n + n_0) + \log(B) \tag{1.2}$$

2. Powerlaw functions: The learning rate is decreasing and there is no single time scale ("fractal scaling", self similarity)[13]. The general form of the performance function $p(n)$ is given by:

$$p(n) = A + B(n + n_0)^a \tag{1.3}$$

The parameters have analogous meanings as in the exponential case 1. The main difference is that a does not play the role of a time-scale as g did but is an example of a "scaling exponent". For certain classes of complex systems scaling exponents play a central role and are "universal" in the sense that their value does not depend on system details. The exponent a can be estimated from the slope of the performance function (minus its asymptotic value) in a double logarithmic representation:

$$\log(p(n) - A) = a \log(n + n_0) + \log(B) \tag{1.4}$$

Theoretical models based on an algorithmic information processing approach used a "chunking" hypothesis which lead to a powerlaw prediction for the learning curve[5]. The experimental findings indicate that the corresponding exponents are individual and task specific. On the other hand, from a complex systems viewpoint one can argue that learning corresponds to non-equilibrium phase transitions[1, 2] and therefore one would expect that the power-law (scaling) exponent characterize universality classes. To our knowledge there has been not much empirical work that tried to classify learning curves according to their scaling exponents. On the other hand in the context of dynamical systems motor learning has been interpreted as basically transient behavior en route to an attractor[11, 12, 10, 3] and in that case one would predict an exponential shape of the learning curve.

2.2 Mirror Tracing

In his classic experiment Snoddy asked his subjects to trace a circuit of a 12-edge, star shaped path with one fourth inch width[8]. The direct vision of the tracing instrument and the hands were obstructed by a screen so that only the indirect mirror image of the tracing device and hands was available to the participants. The instruction to the participants was to "move around as fast as possible and avoid making contact". Each trial consisted of completing one circuit, and the ratio of 1000 over the sum of tracing time (T) and number of contact made (E) within each trail [$1000/(T + E)$] was used as the trial score and performance parameter. Fig. 1 shows the experimental data scanned from the original figure. They are from 100 participants practicing 20 trials per day, 4 days' performance. The performance parameter is plotted as a function of "practice time" which is given by the number of trials a subject has performed that task. In Fig. 1 we have plotted the results in two different representations:

1. Log-Linear: We plot the logarithm of the performance parameter as a function of the practice time. Linear segments correspond to exponential domains of the learning curve. (Fig. 1a)

2. Log-Log: We plot the logarithm of the performance parameter as a function of the logarithm of practice time. Linear segments correspond to powerlaw domains of the learning curve. (Fig. 1b)

We can see the "warm-up decrement" at the beginning of each day's data segment. Also the power-law seems to fit the data slightly better. That is also confirmed by calculating regression functions but the difference is not very dramatic. Therefore we want to study in the following the problem of discriminating the two alternatives under the common constraints of small data sets that are contaminated by noise.

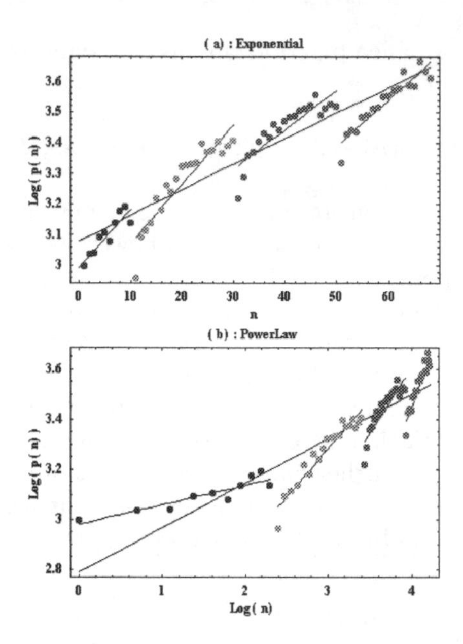

Figure 1: Two different representation of Snoddy's mirror tracing experiments[8]. Each data point represents the median performance of 100 participants practicing 1 circuit of mirror tracing task. The entire practice consisted of groups of 20 circuits, the circuits being separated by 1 minute interval and the groups by 24 hours interval. The first 10 circuits of the first group were not shown on the figure. In (a) we fit an exponential to each group and to the whole data set, in (b) we fit a power law function.

3 Approximating Power Laws by Exponential Functions

In the following example we approximate the power law $P_7(n) = n^{-0.7}$ (red line in the log-log plot of Fig. 2) with the sum of two exponential functions with only real exponents g (see (1.1)). The two amplitudes and exponents in Fig. 1 are $B_1 = 0.3961$, $g_1 = -1.36$ (blue) and $B_2 = 0.6038$, $g_2 = -0.161$ (yellow). We can see that the sum of the two exponentials (green) approximates the power law fairly well over a certain range of practice times.

In the example above we assumed that all solutions with different time scales are active at the same time. If we look at the decay rates of a process that evolves according to a power law then we observe that those decay rates also decay. For example, for the power law $x(n) = n^{-0.7}$ we can fit different exponents in consecutive time units.

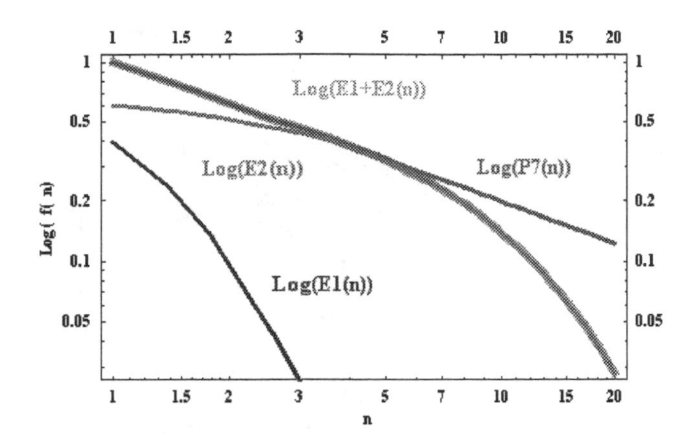

Figure 2: Approxiation of a power law function $P_7(n) = n^{-0.7}$ (red) as the sum of two exponential functions $E(n) = E_1(n) + E_2(n)$, where $E_1(n) = 0.3961e^{-1.36n}$, $E_2(n) = 0.6038e^{-0.161n}$. We observe a reasonably good approximation over seven trials.

4 Time Scales and Power Laws

We have discussed that in the context of learning a fixed time scale can be expressed by a fixed learning rate that then leads to an exponential learning curve. This means that the distance of the function that is measured by the learning curve to its asymptotic value decreases at a constant rate. If the learning curve is a power law then the rate is not constant but rather decreases continuously.

As an example, let us consider a learning curve that measures errors E_n of consecutive trials and is characterized by a power law with exponent: $E_n = En^g$. Here n represents trial number (time) and g the rates at which the errors are assumed to decrease with trial number. For each trial number we can approximate the power law by an exponential function by estimating the learning rate $R(n)$ at that specific trial number n. Under the assumption that we are far away enough from the target so that we have a continuous improvement of the performance (i.e. $E_n > E_{n+1}$, $E_n - E_{n+1}$ small compared to E_n) this can be done by dividing the difference in consecutive errors ($E_n - E_{n+1}$, performance increase due to learning) by the value of the error E_n at that trial.

If we do this for consecutive trials then we observe a systematic decrease not only of the error but also of the rate at which the error changes, the learning rate $R(n)$. In the situation of the power law learning curve ($E_n = En^g$) as shown in Fig. 3 the learning rate $R(n)$ decreases as:

$$R(n) = \frac{E_n - E_{n+1}}{E_n} = \frac{n^g - (n+1)^g}{n^g} \tag{1.5}$$

If we interprete the inverse of a rate as a system timescale then the decreasing rate in (1.5) corresponds to a stretching of time scales. Fig. 3 shows the change in learning rate as a function of practice time for our example function $p_7(n)$:

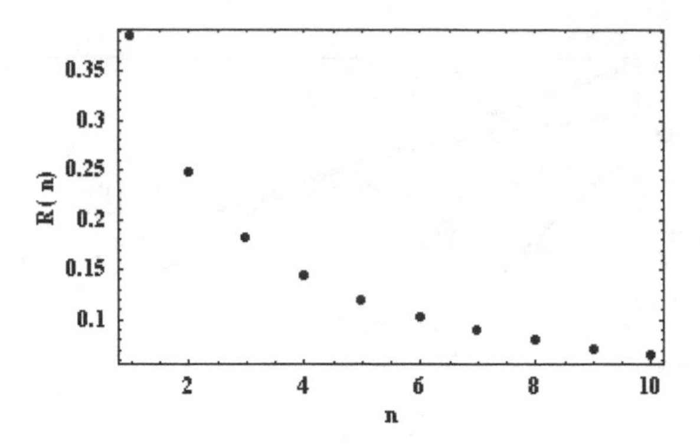

Figure 3: Decreasing local learning rate $R(n)$ for power law $E_n = n^{-0.7}$.

The above technique of taking the pair-wise trial rate of change provides a direct evaluation of the local time scales and a new way to discriminate between exponential and power law behavior. In the above example, if the learning data were exponential then the plot $R(n)$ would have shown a horizontal function as a reflection of change from trial to trial that is proportional to the level of performance. If the data were that of a power law then a $R(n)$ would be a decreasing function. We use this method later to assess the function of learning for some published data and show that it has some advantages over using the percent of variance accounted for in curve fitting to determine the appropriate function of change.

5 Power Laws From Concatenated Exponentials

In this example we show that one can approximate a power law by concatenated exponentials. This is a feature that will arise when different learning rates dominate in different phases of learning. Since we have a given (local) learning rate $R(n)$ at any point of the learning curve, we can approximate a power law by a sequence of processes with fixed but decreasing rates. For instance we can interpolate an exponential function $h_{mn}(t)$ between values of the learning curve E_m and E_n for times t between m and n. In Fig. 4 we illustrate that process with the example of the power-law $E_n = En^g$ from above with $g = -0.7$ (green) and exponential interpolation between consecutive trials, i.e. $n - m = 1$ (red) and $n - m = 2$ (blue) for trials $n, m < 10$.

$$\eta_{m,n}(t) = E_m e^{\log(\frac{E}{E_m})\frac{t-m}{n-m}} \tag{1.6}$$

This simple simulation confirms that a sequence of processes governed by exponential laws of decreasing exponents could approximate a learning curve that can

be best fitted by a power law. Besides overall quality of global fit we therefore also need to consider non-random modulation of the learning curve that provides additional evidence for the presence of exponential processes if those deviations are convex downward over specific ranges of time scales.

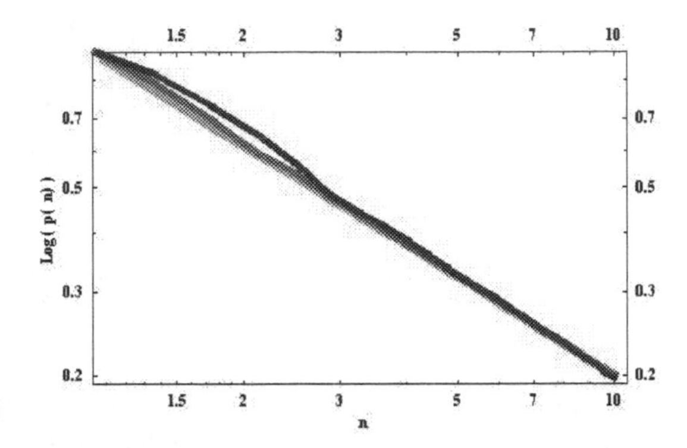

Figure 4: Log-Log plot of power law $E_n = En^g$ with $g = -0.7$ (red) and exponential interpolation between consecutive trials (red) and next to consecutive trials (blue).

6 Power Laws and Superpositions of Exponentials

Although the class of exponential functions with real exponents does not constitute a basis set in function space (as opposed to, for example, trigonometric functions as Fourier components) they nevertheless can approximate power-law functions over a finite range with limited precision. To illustrate this case let us assume for the moment that we have N participants with exponential learning curves with fixed rates $g_i > 0$ centered at a mean rate g and with a Gaussian distribution of width D_g.

This means that for each participant "i" we have a learning curve: $E_{i,n} = E_i e^{g(i)n}$. If we average across participants then we observe an approximate averaged learning curve $E_n = Ee^{gn}$. For large values of D_g, however, the observed averaged data can have a better fit with a power law than with an exponential. In Fig. 5 we illustrate such an example. In summary, this simulation shows that averaging learning data across individuals that exhibit an exponential function with different exponents can lead to a power law for the collective function of change. To address the hypothesis of exponential functions being more likely to emerge in simple tasks, we have reanalyzed data from a study that was set-up to examine the time evolutionary features of learning a single biomechanical freedom timing task[6, 4].

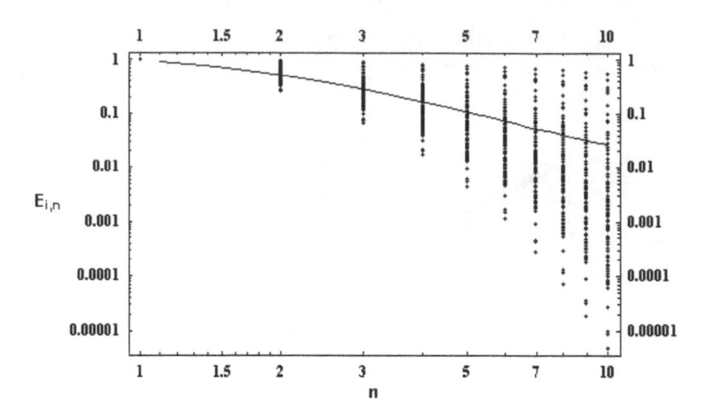

Figure 5: Log-Log plot of distribution of simulated exponential learning curves $E_i(n)$ (dots) with average rate $g = 0.7$ and variance $s^2 = 0.25$. The average $E(n)$ was calculated over $N = 1000$ simulated participants with 10 trials each.

7 Time Scales and Motor Tasks

Based on what we know from complex systems we are in a position to predict when we would expect which functional form of learning curves: Simple tasks that do not involve qualitative changes in movement coordination but simply improve timing or accuracy should have a tendency towards exponential relaxation. More complex tasks including bifurcations would be expected to display power laws.

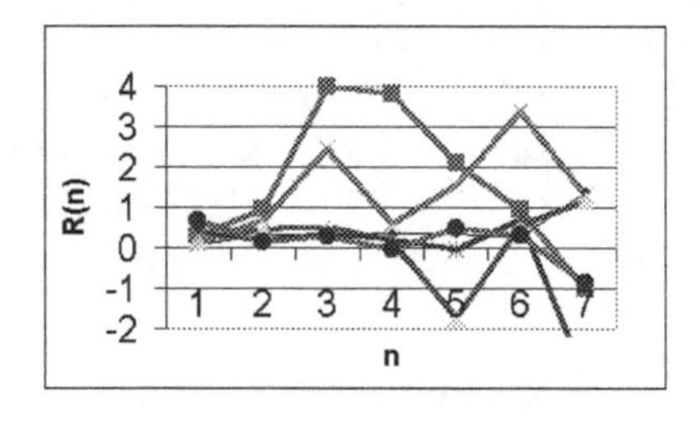

Figure 6: Observed learning rates from a single bio-mechanical freedom timing task[4]. Target time is 125ms for 5-20 deg. flexion.

In Fig. 6 we present local learning rates from six different subjects for a elbow flexion timing task[4]. The selected data show significant performance improvement over the first seven trials. The rates are initially positive and do not show a systematic decrease with n. The negative values indicate the influence of stochastic variations. The approximately constant learning rate is consistent with an exponential form of the corresponding learning curve.

Bibliography

[1] Haken, H., 1983, Synergetics: An introduction (3rd Ed.), Springer-Verlag (Berlin).

[2] Haken, H., Kelso, J. A. S., & Bunz, H., 1985, A theoretical model of phase transitions in human hand movements, Biological Cybernetics, 51, 347-356.

[3] Kelso, J. A. S., 1995, Dynamic Patterns, MIT press (Cambridge).

[4] Liu, Y.-T., Mayer-Kress, G., & Newell, K. M ., 1998, A piecewise linear, stochastic map model for the sequential trial strategy of discrete timing tasks, Manuscript under review.

[5] Newell, A., & Rosenbloom, P. S., 1981, Mechanisms of skill acquisition and the law of practice. In Cognitive skills and their acquisition, edited by J. R. Anderson, Erlbaum (Hillsdale), 1.

[6] Newell, K. M., Liu, Y.T., & Mayer-Kress, G., 1997, The sequential structure of movement outcome in learning a discrete timing task, Journal of Motor Behavior, 29, 366-382.

[7] Newell, K. M., Mayer-Kress, G., & Liu, Y.-T., 1998, Time scales in motor learning and development, Manuscript under review.

[8] Snoddy, G. S., 1926, Learning and stability, J. Applied Psychology, 10, 1-36.

[9] Shaw, R. E., & Alley, T. R., 1985, How to draw learning curves: Their use and justification. In Issues in the ecological study of learning, edited by T. D. Johnston & A. T. Pietrewicz, Erlbaum (Hillsdale), 275.

[10] Schoner, G. (1989). Learning and recall in a dynamic theory of coordination patterns. Biological Cybernetics, 62, 39-54.

[11] Schoner, G., & Kelso, J. A. S. (1988a). A synergetic theory of environmentally-specified and learned patterns of movement coordination. I. Relative phase dynamics. Biological Cybernetics, 58, 71-80.

[12] Schoner, G., & Kelso, J. A. S. (1988b). A synergetic theory of environmentally-specified and learned patterns of movement coordination. II. Component oscillator dynamics. Biological Cybernetics, 58, 81-89.

[13] Schroeder, M., 1991, Fractals, chaos, power laws: Minutes from an infinite paradise, Freeman (New York).

[14] Thorndike, E. L., 1927, The law of effect, American J. Psychology, 39, 212-222.

[15] Thurstone, L. L., 1919, Psychological Monographs, XXVI, Whole No. 114.

Time Series Analysis of Ecosystem Variables with Complexity Measures

Holger Lange
Ecological Modeling Group
Bayreuth Institute of Terrestrial Ecosystem Research (BITÖK)
University of Bayreuth
D-95440 Bayreuth, Germany
holger.lange@bitoek.uni-bayreuth.de
http://www.bitoek.uni-bayreuth.de/Mitarbeiter/000125/EN.html

1.1 Science and Management of Ecosystems

Natural forested watersheds are prototypes of ecosystems. They have a relatively well-defined boundary, closed cycles for fluxes of elements, contain a collection of species of all kinds of spatial sizes (microbes to trees), and are of important human concern as timber source and drinking water supply. They have been the focus of scientific investigation since many years now, and it seems to be a rather trivial statement that they exhibit rather complex behavior. Considering the amount of scientific knowledge gained, the intricate nature of food webs, transport mechanisms for matter and energy, growth and succession patterns, or responses to natural hazards, it almost occurs like a miracle that they are nevertheless manageable from a practitioner's point of view. Forestry has been working since centuries and has come up with simple empirical rules which allow timber yield predictions for a full rotation period (e.g. 80 years). Drinking water supply works properly in many cases. Science has not yet come up with an explanation for the existence or effectiveness of these empirical rules. Some of these rules are devaluated nowadays due to anthropogenic disturbances, e.g. in the context of forest decline symptoms. However, thus far science failed as well to give conclusive hints why these managing rules do not work properly. There is an obvious conflict between the managing and the scientific approach to such ecosystems.

The reason behind this conflict is in our opinion the importance of the unique historical development of each given system (Hauhs and Lange 1996). They are rendered manageable by an effective *biological reset*, e.g. by planting operations, successive thinning etc., and thus a reduction or "amputation" of historical influence. On the other hand, a de novo (re)construction of such systems in the laboratory has not been possible; no scientific substitute for practical experiences has been found. It may well be that the type of knowledge residing in experienced practitioners is not only nonscientific in nature, but to a certain extent nonverbal also. An approach relying heavily on visualization tools to explicate the inferential behavior of foresters is currently under investigation and discussed in (Lange et al. 1998a).

1.2. Problems with process-oriented models

The convincing success of deterministic models in physics and other "hard" sciences has let ecosystem researchers to follow similar lines, especially in the context of transport simulation. Here, appropriate simplifications of the Navier-Stokes equation have been developed for the description of water movement in soils and other porous media, and the movement of solutes with convection-diffusion models (Bear and Verruijt 1990).

Direct modeling of observed transport patterns is virtually never possible. The available field data are rare and of limited quality; the experimental constraints on parameter functions describing soil hydraulic and hydrochemical properties are almost never very restrictive. Mismatches between model reconstructions and measurements are easily attributable to spatial heterogeneity. Therefore, inverse modeling is applied, fitting the transport properties of the system at an assumed level of heterogeneity and a given scale (Lange et al. 1996). In many cases, this procedure leads to successful data reproductions; however, the fitted properties are far from being unique or very near so; the danger of overparametrization is notorious. In addition, the information contained in output variables is too limited to justify an approach requiring such a high amount of assumptions on the interior. This could be demonstrated with a series of tracer experiments for a well-controlled small catchment (Lange et al. 1996).

We conclude that if the focus is on characterizing the relationship between input and output of an ecosystem, an explicit modeling assuming lots of internal and unobserved processes should be avoided. The mapping performed by the system should be characterized rather by direct data analysis.

2. Time Series Investigation Methods

The selection of appropriate measures to quantify ecosystem input and output is primarily dictated by the length and quality (w.r.t. gaps, outliers etc.) of the available data. Some of the monitored data, especially related to hydrology, are taken since decades on a daily basis. Others are irregularly and sparsely sampled, e.g. monthly runoff chemical composition. Secondary, the type of information seeked for determines the choice as well. Some of the methods quantify short-term dynamics, others long-term correlations, and yet another group the relationship between two or more data sets. We therefore suggest to use a variety of methods, and to start with standard techniques like periodograms, or auto- and cross-correlation functions. However, experience shows that in our case, some nonlinear methods are more powerful or give important additional information. We give three examples for them in the following.

2.1 Hurst analysis

The phenomenon of persistence (extension of periods with systematic deviations from the long-term mean), first considered in the context of Nile floodings (Hurst 1951) and often present in hydrological data sets, is quantified by calculating the Hurst exponent H using the rescaled range or R/S statistics (Mandelbrot and Wallis 1969). This is a prototypical example of a long-term nonlinear measure.

Given a time series X of length n and a time scale k, we plot the quantity

$$q(n,k) = R(n,k) / S(n,k) \tag{1}$$

versus k for various values of n (e.g. 20 different realizations), and the expected persistence behavior $q \propto k^{H}$ is fitted to the results. Here, R denotes the *total range statistics* : Building partial sums from the original series X,

$$Y_n = \sum_{i=1}^{n} X(t_i) \tag{2}$$

the deviation from a linear increase (as for a constant or essentially stationary series) is calculated as

$$D(n,i,k) = Y_{n+i} - Y_n - \frac{i}{k}(Y_{n+k} - Y_n) \tag{3}$$

and its range is given by

$$R(n,k) = \max_{0 \le i \le k} D(n,i,k) - \min_{0 \le i \le k} D(n,i,k) \tag{4}$$

In order to obtain a dimensionless test quantity, R is divided by the standard deviation of the series at timescale k :

$$S(n,k) = \sqrt{\frac{1}{k} \sum_{i=n+1}^{n+k} (X(t_i) - \bar{X}(n,k))^2} \tag{5}$$

thus arriving at eq. (1). If Hurst scaling is found, the expected H values are $0 \le H \le 1$; $H \le 0.5$ means antipersistent behavior, which is seldom observed in experimental data, H=0.5 is ordinary Brownian motion, and H ≥ 0.5 is fractional Brownian motion with increasing persistence strength as H approaches 1.

A possible caveat is that the fixation of the range of time scales where Hurst scaling is assumed is a somehow subjective manner. In some cases, multiple scaling regimes are observed, indicating a finite memory contained in the data, or broken scaling in general. Then it is more a matter of taste whether one attributes Hurst behavior to the data at all.

2.2 Complexity measures

Short-term fluctuations can be investigated through information-theoretic measures. They allow to quantify the information content, or the randomness, of data sources on one hand, and their complexity on the other. By the latter we intuitively mean a measure which vanishes or is very small for constant or periodic sequences, but also for completely random data, as both types are easy to describe. For time series which are not amenable to such an easy description involving only a few parameters, the complexity measures should show high values.

We will discuss here only one randomness and one complexity measure appropriate for our purposes. Their calculation is performed by first constructing a symbol string from the data X via partitioning:

$$\Sigma: X \rightarrow \mathbb{A},\ x_i \rightarrow a_j \quad \mathbb{A} = \{a_j\},\ j=0,...,A-1 \tag{2}$$

where $\mathbb{A}$ is the alphabet for the symbol sequence. In the examples presented, we have chosen a binary alphabet, and the partitioning was performed in a static manner: the values of the observed quantities were cut at the median of their distribution, and all values below (above) the median were assigned the symbol 0 (1).

Then, one defines a word length L to group symbols together (in this article, L=4). The relative word frequencies $p_{L,i}$ for each possible word i (i runs from 1 to A^L) and conditional (or transition) probabilities $p_{L,ij}$ (i.e. the probability to find word j immediately after word i) are calculated. These are the ingredients to calculate the *generalized Shannon entropy*

$$H_L = -\sum_{i=1}^{A^L} p_{L,i} \log_2 p_{L,i} \tag{3}$$

and our measure for randomness, the *mean information gain*

$$MIG_L = H_L - H_{L-1} \tag{4}$$

Finally, the fluctuation complexity (Bates and Shephard 1993) is given by (index L suppressed for simplicity)

$$\sigma_{fc}^2 = \sum_{i,j} p_{ij} (\log \frac{p_i}{p_j})^2 \qquad (5)$$

These two quantities have the desired features, as can be demonstrated e.g. for binary Bernoulli sequences (Lange 1999). Whereas MIG is nonlinearly proportional to randomness, being more sensitive to structural changes in the region of low randomness, FC exhibits a maximum and vanishes for constant as well as completely random sequences. Thus, FC is close to our intuition of what a "true" complexity measure should be.

2.3. Recurrence quantification analysis

Possible instationarities, trends, extreme periods, periodicities and many other (nonlinear) features of the time series are visualized and quantified by this technique (Zbilut et al. 1998a). It essentially calculates distances between time series values at all possible different times and stores and displays them in a matrix. The properties of this recurrence matrix are the basis for several derived quantities like the local recurrence rate, the degree of determinism, the Shannon entropy of line segments, or an approximation to the highest Ljapunov exponent present in the system. The method does not presuppose stationarity or large amounts of data and is not very restrictive when data points are not equidistant.

After normalization, one also can compare two different time series for the same observation period, which is the subject of cross-recurrence quantification (Zbilut et al. 1998b). One application for this rather new technique will be given.

3. Data sets and results

In this short overview, examples from long-term monitoring data on hydrology are shown only. Precipitation and runoff from natural catchments have been measured daily for decades at a reasonable large number of places. The quality of the data is generally high, with the possible exception of data from places where a substantial amount of precipitation occurs as snowfall.

Fig. 1 shows a Hurst analysis for one of our investigation sites, located in the Fichtelgebirge, NE Bavaria, Germany. The typical result is that the output from the system is much more persistent than the input. This long-term correlational structure is imprinted by the system and not present in other driving variables. It is also a universal feature of runoff from catchments of drastically different sizes (Pelletier and Turcotte 1997). The analysis also shows that rainfall is definitely different from a pure noise process.

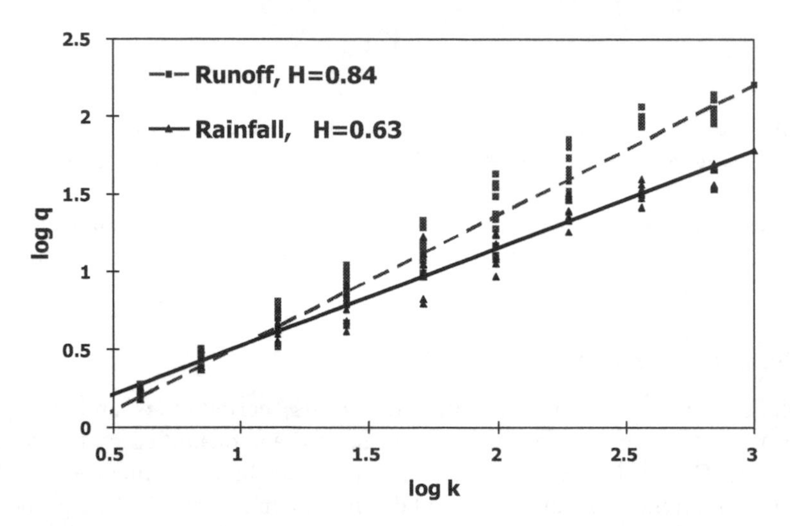

Figure 1: Hurst analysis for a small (4.2 km^2) forested catchment in NE Bavaria. Runoff is more persistent than rainfall, which is different from pure noise. The scale k is in days.

An example of a complexity analysis is shown in Fig. 2. Here, 30 catchments from different climatic regions are compared, using the randomness and complexity measures discussed above as abscissas and ordinates, respectively. The Bernoulli process provides a limiting curve: apart from possible finite size effects, all experimental data sets should lead to points below it (Lange 1999). Our observation is that for most data from natural systems, the deviation from the maximum complexity at given randomness (mean information gain) is small. On the other hand, artificial or highly controlled systems usually have much larger distance from the limit. This is demonstrated here: whereas runoff from the natural catchments operates at high complexities and intermediate randomness, urban (channeled) systems have much more randomness, indicating that biological uptake (dominantly evapotranspiration) smoothes signals. It is thus feasible to think of ecosystems as information filters (Hauhs and Lange 1996).

Precipitation is characterized by extremely high randomness and consequently low complexity. The difference in randomness between input and output from the same system serves as measure for its efficiency, whereas the distance to the limiting curve is considered as an indicator for the « naturalness » of the system.

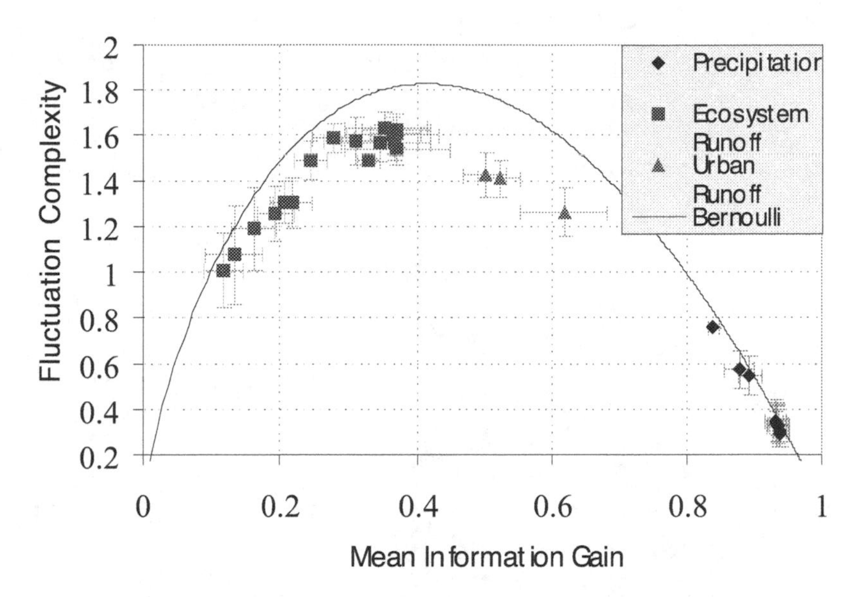

Figure 2: Complexity diagram for daily data from 30 different catchments. Natural runoff is significantly different from urban one ; rainfall is highly random. The symbols and error bars denote the mean and standard deviation for windows of length 4 years each.

An example of recurrence analysis is given in Fig. 3. The data are runoff values from watershed 2 of the Hubbard Brook Experimental Forest (White Mountains, New Hampshire), the time span is 9 years (1962-1970). Both axes of this plot are time axes. A dark pixel at position (i,j) of this plot indicates that the time series values at position i and j are similar in magnitude (according to a threshold parameter which has to be fixed); otherwise the pixel is left white.

At the beginning, a clear periodic pattern can be recognized (chess-board like structure of black and white rectangles), which is to be expected since tree water uptake during the vegetation period induces seasonality. The dark rectangles correspond to dry periods in summer (or low flow conditions).

After four years, however, the periodic pattern is suddenly lost, obviously a dramatic change in the underlying dynamics. This is the consequence of a drastic experimental manipulation : a clear-cut of the vegetation, followed by pesticide treatment for the subsequent 3 years (Likens and Bormann 1995). Whereas hydrochemically, clear indications of the clear-cut were found, the hydrological response was not detected by linear methods. This demonstrates the power of recurrence quantification.

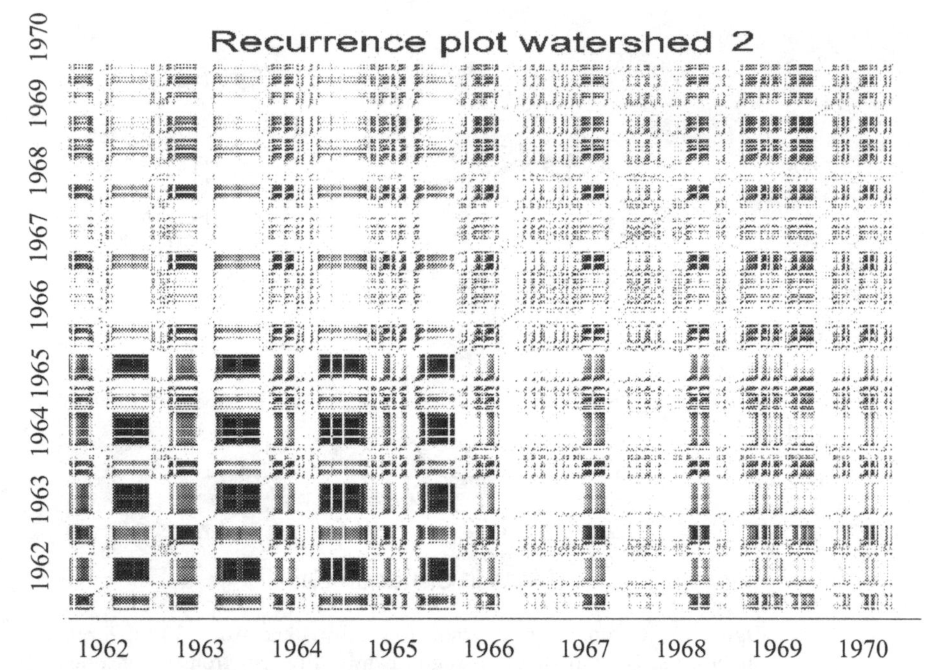

Figure 3: Recurrence plot for the Hubbard Brook watershed 2 for 9 years (1962-1970). The periodic pattern visible for the first four years vanishes suddenly at the end of 1965 : the watershed was clear-cut in that year and prevented from regrowth (by pesticides) for three more years.

4. Conclusions

It could be demonstrated that new (generally nonlinear) techniques of time series analysis can provide new information about the detailed structure of data from natural and anthropogenic ecosystems, which is not available by standard tools. The results indicate that extended memory effects are present in the systems and that biological activity induces them (cf. recurrence plots). As the history of system development seems to be crucial, an approach to model the system as a whole (biotic as well as abiotic parts) as a state-determined dynamical system is not expected to work.

Acknowledgements: This work was funded by the German Ministry of Education and Research (BMBF) under grant No. PT BEO 51-0339476B.

References

Bates, J. E., & Shephard, H. K., 1993, Measuring complexity using information fluctuation. *Physics Letters A* **172**, 416-425.

Bear, J., & Verruijt, A., 1990, *Modeling groundwater flow and pollution*. Reidel (Dordrecht).

Hauhs, M., & Lange, H., 1996. Ecosystem dynamics viewed from an endoperspective. *The Science of the Total Environment* **183**, 125-136.

Hurst, H.E., 1951, Long-Term Storage Capacity of Reservoirs. *Transactions of the American Society of Civil Engineering* **116**, 770-799.

Lange, H., Lischeid, G., Hoch, R., & Hauhs, M., 1996, Water Flow paths and Residence Times in a Small Headwater Catchment at Gårdsjön (Sweden) during Steady State Stormflow Conditions. *Water Resources Research* **32**, 1689-1698.

Lange, H., Thies, B., Kastner-Maresch, A., Dörwald, W., Kim, J.T., & Hauhs, M., 1998a, Investigating forest growth model results on evolutionary time scales. In: Adami, C., Belew, R.K., .Kitano, H., & Taylor, C.E. (eds.), *Artificial Life VI* , pp. 418-422. MIT Press (Cambridge, Mass.).

Lange, H., Newig, J., & Wolf, F., 1998b: Comparison of complexity measures for time series from ecosystem research. *Bayreuther Forum Ökologie* **52**, 99-116.

Lange, H., 1999, Are Ecosystems Dynamical Systems? *International Journal of Computing Anticipatory Systems* **3**, 169-186.

Likens, G.E., & Bormann, H., 1995, *Biogeochemistry of a Forested Ecosystem*, Springer-Verlag (New York).

Mandelbrot, B.B., & Wallis, J.R., 1969, Robustness of the rescaled range R/S and the measurement of noncyclic long run statistical dependence. *Water Resources Research* **5**, 967-988.

Pelletier, J.D., & Turcotte, D.L., 1997, Long-range persistence in climatological and hydrological time series: analysis, modeling and application to drought hazard assessment. *Journal of Hydrology* **203**, 198-208.

Zbilut, J.P., Giuliani, A., & Webber Jr., C.L., 1998a, Recurrence quantification analysis and principal components in the detection of short complex signals. *Physics Letters A* **237**, 131-135.

Zbilut, J.P., Giuliani, A., & Webber, C.L., 1998b, Detecting deterministic signals in exceptionally noisy environments using cross-recurrence quantification. *Physics Letters A* **246**, 122-128.

The Dynamics of Minority Competition

Robert Savit, Radu Manuca, Yi Li, Rick Riolo
Center for Study of Complex Systems and Physics Department,
University of Michigan, Ann Arbor
savit@umich.edu

1 Introduction and Motivation

Complex adaptive systems of agents with co-adaptive or co-evolving strategies are systems in which agents compete for some resource, or set of resources, and by their competition alter their environment. In such systems agents are typically endowed with heterogeneous strategies, beliefs and behaviors. The collective action of these agents in general alters their environment, and so, as time goes by, agents must reevaluate and possibly change their strategies to more effectively compete in the altered environment. Markets, ecologies, political structures and competition among businesses are all examples of such systems. Indeed, it is difficult to imagine an interesting system in the biological or social sciences which does not encompass elements of co-evolving competitive systems.

One very important element in many such competitive systems, is the drive for agents to be in the minority, or to be different from their competitors. There are many situations in which one is rewarded for being in the minority. Examples of systems that are, at least in part, driven by minority-membership dynamics include vehicular traffic on roads, in which each agent would prefer to be on an uncongested road [2], packet traffic in networks, in which each packet will travel faster through lesser used routers [3], and ecologies of animals looking for food, in which individuals do best when they can find areas with few competitors [1]. Finally, a trader will often be rewarded by being on the short side of a sale. For example, if there are more buyers than sellers, demand will tend to drive the price up, and the sellers (who are in the minority) will be rewarded by "selling high." In this paper we will examine some of the dynamics associated with the drive to be in the minority by studying a very simple class of models. While the models are very simple, and far from a realistic representation of any particular biological or social system, it will not be difficult to see the relevance of the dynamics that are incorporated in the model.

The model we will study is a repeated game of N players, choosing (independently) according to some set of strategies to join one of two groups. At each time step only members of the minority group are rewarded. As explained in more detail below, this system has a most remarkable behavior [4],[5]. The system manifests a phase change from a phase which is in a certain sense, efficient, but maladaptive, to a phase which is, in a certain sense, inefficient, but in which the system as a whole utilizes its resources effectively. As one increases the size of the strategy space available to the agents, one moves from the former to the latter phase. The dynamics whereby this transition takes place are subtle and fascinating, and are the result of an emergent coordination in the decisions taken by the agents. This coordination is mediated only by aggregate, publicly available information, and does not result from any detailed knowledge of the behavior of one agent by any other. The system also has some intriguing properties reminiscent of a spin-glass in statistical mechanics.

2 The Model

The simple model of competition we discuss here consists of N agents playing a game [8][1]. The rules of the game are as follows: At each time step of the game, each of the N agents playing the game joins one of two groups, labeled 0 or 1. Each agent that is in the minority group at that time step is awarded a point, while each agent belonging to the majority group gets nothing. An agent chooses which group to join at a given time step based on the prediction of a strategy. (In general, different agents use different strategies, as we shall explain below.) A strategy of memory m is a table of $2m$ rows and 2 columns. The left column contains all $2m$ combinations of m 0's and 1's, and each entry in the right column is a 0 or a 1. To use a strategy of memory m, an agent observes which were the minority groups during the last m time steps of the game and finds that entry in the left column of the strategy. The corresponding entry in the right column contains that strategy's prediction of which group (0 or 1) will be the minority group during the current time step. Thus, a strategy uses information from the historical record of which group was the minority group as a function of time, which is the information publicly available to all the agents.

In each of the games discussed here, all strategies used by all the agents have the same value of m. At the beginning of the game each agent is randomly assigned s (generally greater than one) of the 2^{2^m} possible strategies of memory m,

[1] A related model is described in [9].

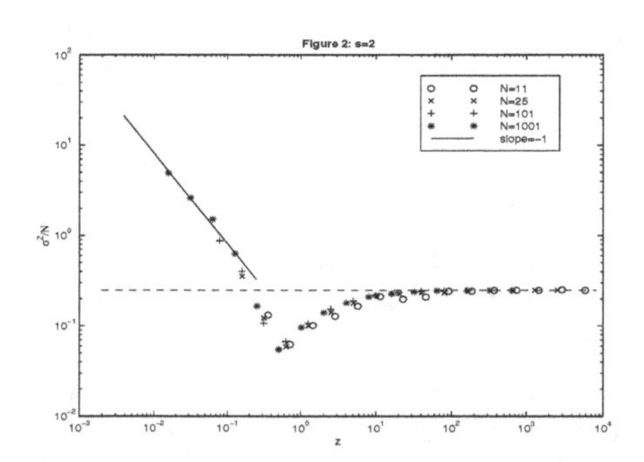

Figure 1: σ^2/N as a function of $z \equiv 2^m/N$ for $s = 2$ and for various values of N and m, on a log-log plot.

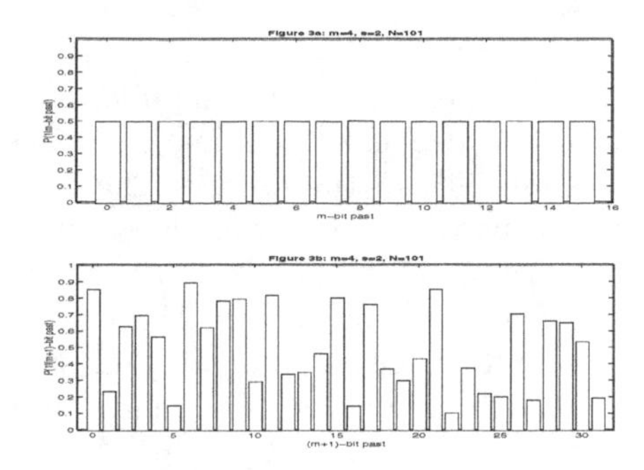

Figure 2: **a)** A histogram of the conditional probabilities $P(1|u_k)$ with $k = m = 4$. There are 16 bins corresponding to the 16 combinations of four 0's and 1's. The bin numbers, when written in binary, yield the strings U_k. **b)** A histogram of the conditional probabilities $P(1|u_k)$ with $k = 5$ (32 bins) for the game played with $m = 4$.

with replacement. For his current play the agent chooses from among his strategies the one that would have had the best performance over the history of the game up to that time. Ties among an agent's strategies can be decided either by a coin toss, or by choosing to play the strategy that was most recently played[2]. Following each round of decisions, the cumulative performance of each of the agent's strategies is updated by comparing each strategy's latest prediction with the current minority group. Because the agents each have more than one strategy, the game is adaptive in that agents can choose to play different strategies at different moments of the game in response to changes in their environment; that is, in response to new entries in the time series of minority groups as the game proceeds. Although this system is adaptive, the versions we analyze here are not, strictly speaking, evolutionary. The strategies do not evolve during the game, and the agents play with the same s strategies they were assigned at the beginning of the game. Evolutionary mechanisms can be simply incorporated into this model, and are discussed in [10].

3 Basic Results

In Fig. 1, we present a graph which summarizes the basic system-wide behavior of the game described above. The results presented here are for the case of $s = 2$ strategies per agent. As a measure of overall systems efficacy (i.e., the

[2]The later is default for our models, but the system-wide behavior does not generally depend on which method is used.

ability of the system to generate points), we consider the standard deviation, σ, of the time series of the number of agents belonging to group 1 as a function of time. Because the game is symmetric, the mean of this time series will be close to 50% of the agents. The larger σ, the standard deviation, the smaller, typically, will be the minority groups, and so the less effective the system is at generating resources. (Note this is not a zero-sum game.) In this figure we plot σ^2/N as a function of $z \equiv 2^m/N$ on a log-log scale for various N and m. We see first that all the data fall on a universal curve. Let m_c be the value of m at which σ^2 has its minimum. The minimum of this curve is near $2^{m_c}/N \equiv z_c \equiv 0.5$, and separates two different phases[3]. The value of σ^2/N which is approached for large z is the value that σ^2/N would have if the agents made their choices randomly and with equal probability. We call this the random choice game (RCG). For $z < z_c$ the system is in a phase in which over-all system efficacy is poor. In this phase agents exhibit a kind of herding behavior and on average do worse than they would even in the RCG. For $z > z_c$ the system is in a phase in which agents are able to coordinate their choices to achieve a good utilization of resources. However, the agents' ability to coordinate their choices decreases as z increases beyond z_c. The slope for $z < z_c$ approaches -1 for small z, while the slope for $z > z_c$ approaches zero for large z. The most important take home lesson from this graph is that *maximum system wide efficacy is achieved at the point of a phase transition when the dimension of the strategy space from which the agents draw their strategies is on the order of the number of agents playing the game.*

4 Information Content and the Nature of the Phase Structure

The system clearly behaves in a qualitatively different way for small and large m. To further study the dynamics in these two regions, we consider the time series of minority groups, ($\equiv G$), the data publicly available to the agents. We want to study the information content of strings of consecutive elements of this time series of various lengths (including strings of length m) for different values of m and N. Since the strategies use only the information contained in the most recent m-time steps of G, it is natural to ask how much information is accessible to the strategies. To do this, we consider the conditional probability $P(1|u_k)$. This is the conditional probability to have a 1 immediately following some specific string, u_k, of k elements of G. For example, $P(1|0100)$ is the probability that 1 will be the minority group at some time, given that minority groups for the four previous times were 0,1,0 and 0, in that order. Recall that in a game played with memory m, the strategies use only the information encoded in strings of length m to make their choices. In Fig. 2, we plot $P(1|u_k)$ for G, the time series of minority groups generated by a game with $m = 4$, $N = 101$ and

[3]Since these experiments involve only integer values of m, this value of z_c should be considered only an estimate; this and additional comments about the thermodynamic limit are presented in [5].

$s = 2$. This value of m and N puts the system in the low m_c phase. Fig. 2a shows the histogram for $k = m = 4$ and Fig. 2b shows the histogram for $k = 5$. Note that the histogram is flat at 0.5 in Fig. 2a, but, remarkably, is not flat in Fig. 2b. Thus, any agent using strategies with memory (less than or) equal to 4, will find that those strings of minority groups contain no predictive information about which group will be the minority at the next time step. But recall that this time-series was itself generated by agents playing strategies with $m = 4$. Therefore, in this sense, the market is efficient (at least with respect to the strategies[4]) and no strategy playing with memory (less than or) equal to 4 can, over the long run, accumulate more points than would be accumulated randomly[5]. We call this phase informationally efficient with respect to the strategies. But note also that the time series of minority groups is not a random (IID) string. There is information in this string, as indicated by the fact that the histogram in Fig. 2b is not flat. However, that information is not available to the strategies used by the agents in playing the $m = 4$ game which collectively generated that string in the first place. The histogram for $k = m$ is flat for $m < m_c$ (although it does begin to show some small $O(T^{1/2})$ fluctuations for m just below m_c). It turns out [5] that the dynamics in this phase generate a maladaptive kind of herding or faddish behavior among the agents. Approximately half the time, a very large (of order N) fraction of the agents all choose to join the same group. This gives rise to a very large value of σ, and is also responsible for the flat conditional probability histograms, such as that shown in Fig. 2b.

We can repeat this analysis for $m \geq 6$ ($N = 101$, $s = 2$). For this value of m and N the system is close to the transition, in the high m phase. For this range of m, the corresponding histogram for $k = m$ is not flat, as we see in Fig. 3 for the $m = 6$ game. In this case, there is significant information available to the strategies of memory m and the "market" is not informationally efficient with respect to the strategies[6]. Indeed, some individual agents using their strategies do accumulate significantly more points than they would by simply making random guesses. However, for $m < 5$, no agent ever achieves more than 50% of the possible available points.

---

[4]Actually, the structure in this phase is a little more subtle. The system is informationally efficient with respect to the strategies, but is not efficient with respect to the agents. In fact, there are strong correlations between the minority groups over long time periods of order 2^m time steps, which results in the very poor performance of the system in this regime. Note that these long time correlations are not to be confused with the non-IID conditional probabilities (which indicate information over time scales of order $m + 1$) illustrated in Fig. 2b. Despite these subtleties this phase does partake of the usual idea that efficiency implies an absence of usable information.

[5]In fact, asymptotically, for $m < m_c$, no strategy can be right more than 50% of the time. If the behavior for these values of m were truly random, there would be fluctuations of order $T^{1/2}$ in the number of times a strategy won over a time T. But in our case these Guassian fluctuations are absent, and the strategies are limited to being correct no more than 50% of the time. A corollary of this is that the historgram in Fig. 2bis extremely flat, and does not contain even the random fluctuations that would be present for an IID series.

[6]Note the particular shape of the histograms in Fig. 3 as well as Fig. 2b depend on the initial distribution of strategies, but in all cases with $m > m_c$ or with $m < m_c$ and $k > m$ the histograms due indicate the presence of significant predictive information

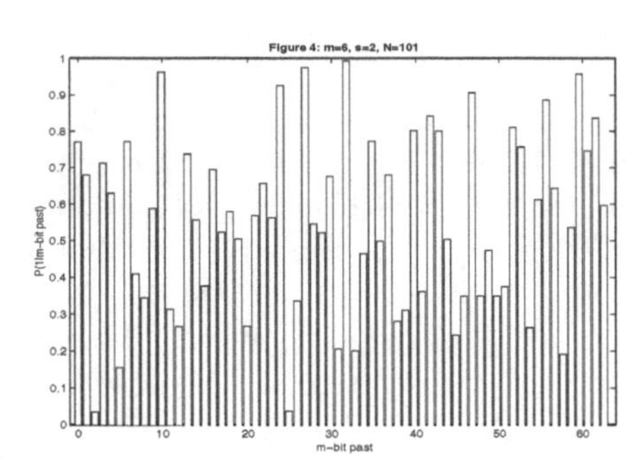

Figure 3: A histogram of the conditional probabilities $P(1|u_k)$ with $k = m = 6$ (64 bins).

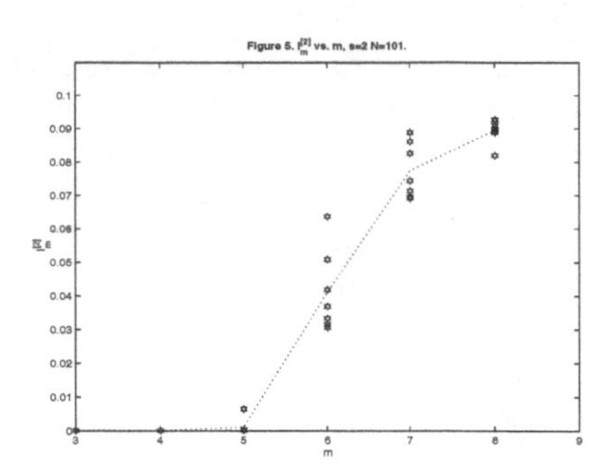

Figure 4: $I_m^{[2]}$, the square summed information available to the strategies in the time series of minority groups G, defined in the text, as a function of integer m for $N = 101$, $s = 2$. Values for 8 independent runs are plotted for each m.

If for $m < m_c$ the time series of minority groups contains no information available to the strategies, as m increases beyond m_c the amount of information available to the strategies in G increases. To see this, look at Fig. 4 in which we plot

$$I_m^{[2]} = \frac{1}{2^m} \sum_{u_m} I^2(u_m) = \frac{1}{2^m} \sum_{u_m} (P(1|u_m) - \frac{1}{2})^2$$

as a function of m. The more $P(1|u_m)$ differs from 1/2, the more information there is available to the strategies about what the minority group following u_m will be. $I_m^{[2]}$ is just a measure of that information averaged over all strings of length m. In Fig. 4 we see that $I_m^{[2]}$ increases monotonically with m for $m > m_c$. Thus, as m increases, the system becomes increasingly *inefficient* with respect to the strategies, in that increasing amounts of information are left in G. It may seem counter-intuitive that as m increases there is an increasing amount of information contained in G while at the same time overall system performance and individual wealth decrease. The explanation is that the time series of minority groups is *endogenous*. It is created by the choices of the agents themselves. The fact that more information is left in G, just reflects that fact that the agent's choices are less coordinated, so that those choices use less of the information inherent in the strategies of the agents themselves. For this system, uncoordinated choices means that each agent chooses randomly, independently and with equal probability between its two strategies at each time step[7]. Note, however, that this is not the same as the RCG in which each agent chooses 0 or 1 randomly, independently and with equal probability at each time step. In the latter case G would be an IID sequence with no information, (and the

[7]It is easy to show that as m increases, the information in G reflects, increasingly accurately, statistical features of the particular distribution of strategies amoung the agents [5].

system would be in a mixed-strategy Nash equilibrium) but in the former case G contains significant information about the structure of the agents' strategic preferences (and the system is not in a Nash equilibrium). The existence of this fixed strategic preference structure is the reason that the system becomes increasingly inefficient (so that there is more information in G) as m increases, while at the same time the agents' choices become as uncoordinated and as random as possible[8].

In addition, the wealth distribution and the nature of the strategies that are most successful are markedly different in the two different phases, and reflect, in a very interesting way, the different informational content of G. In the small strategy space, maladaptive phase in which the agents exhibit herding behavior, agents whose strategy sets are composed of very similar strategies do best, without regard to the nature of those strategies, or to the nature of the strategies of the other agents. In the large strategy space phase, agents whose strategies are maximally distant in behavior space from all other strategies do best. Thus, the value of a given strategy is dependent on details of that strategy in the context of all other strategies in the game.

Finally, and contrary, perhaps, to naive expectations, the best coordination among agents is obtained when the agents possess only moderate degrees of adaptivity. In the case in which there is only one strategy per agent, this structure disappears, and the game becomes trivial and periodic. On the other hand, as the number of strategies available to each agent increases beyond two, coordination becomes more difficult. The general phase structure still obtains, but the degree of coordination diminishes and the system generally performs more poorly at the transition and in the inefficient phase than it does when each agent has fewer strategies (but still more than one).

5 Summary and Discussion

In this paper we have presented an overview of the rich and complex behavior that very simple minority games display. The behavior seen in these simple games is quite remarkable, as is its apparent robustness. We have also studied the robustness of the general phase structure of these games to changes in the information set used by the agents' strategies. We have found that when other kinds of information are given to the agents—for example, more specific information about the number of agents in a group as a function of time—the general phase structure of the system remains the same [10].

Nevertheless, there are a number of effects that could alter the system's dynamical structure significantly. One feature of the games described in this paper is that all of the agents are looking at the same signals. That is, the strategy space is the same for all strategies. It may be that the existence of a set of common signals on which the strategies base their predictions allows the agents to coordinate their strategy choices in a relatively simple way. Another

[8]The behavior of more traditional measures of information are discussed in [5].

very important issue to address is the role of evolution in such games. Although the systems discussed in this paper were adaptive, they were not evolutionary, in the sense that the agents' strategies were fixed from the beginning of the game. The inclusion of evolution results in very interesting modifications in the results of these games, including marked improvement in overall system performance for some ranges of parameters. Remarkably, though, the general phase structure persists [11].

A natural question to ask about our results is their relation to Nash equilibria. The minority games have many Nash equilibria. There are clearly Nash equilibria in this system. However, it appears that the dynamics of this game do not typically send the system to a Nash equilibrium. This observation, of course, is just a special case of the general idea that dynamics in an adaptive system can prevent the system from achieving equilibria, and in particular, Nash equilibria. In our opinion, there is a need to seriously consider dynamics in the analysis of economic and other complex adaptive systems, and not rely solely on equilibrium analyses.

It is also interesting to consider again the nature of the strategies and their success in these games. In particular, consider the inefficient phase in which the strategies in play represent only a small sample of the total number of strategies in the strategy space. Here some strategies perform better at predicting minority groups than others. But *a priori* there is nothing special about any particular strategy. They are all just randomly generated strings of 0 and 1. Nevertheless, in the context of the other strategies represented in that particular game, certain strategies perform better than others. A necessary condition for the emergence of this ordering is the fact that the strategies in play represent only a small sample of the total strategies possible. Indeed, increasing the number of strategies in play, either by increasing N or s sufficiently, puts the system into a region in which no strategy performs better than any other. Thus, there is a sense in which there is an emergent meaning to specific *a priori* random strategies, which meaning emerges contextually in an environment representing a small random sample. We believe that a sparsely sampled configuration space may be a general condition for the emergence of meaning. The simplest example of this is the observation that it is always possible to find structure and patterns in a finite string of random numbers. As the length of the string grows, however, finite length patterns lose their significance. The kinds of meaning that emerge in different contexts, and the criteria for what constitutes a "small" random sample will certainly depend on the dynamics associated with the system. Nevertheless, we believe that a sparse sampling of the configuration space of the system may well be a necessary condition for the emergence of meaning[9].

This system also has some characteristics that are reminiscent of a spin-glass

[9]Of course this discussion can be put into the larger context of symmetry breaking. In that case, one can suggest a number of mechanisms that might break pre-existing symmetry and cause associations and meaning to emerge. Which of these mechanisms can be subsumed under the rubric of sparsely sampled space is a semantic question. But in the context of many social and biological systems, a sparsely sampled configuration space (interpreted more narrowly) may turn out to have been a precondition for the emergence of meaning.

in statistical mechanics. In a spin-glass, the interactions among particles are mediated by a set of fixed, randomly distributed pair-wise interactions, which may differ in sign, some being ferromagnetic (causing spins to align in the same direction) and others being anti-ferromagnetic (causing spins to align in opposite directions). In general the system may not be able to satisfy all these tendencies for a given distribution of interactions, while at the same time achieving a low energy. This is the phenomenon of frustration, and leads to a very rich behavior of the system. In our game, the strategies that are distributed to the agents typically represent a small random sample of all possible strategies. They are fixed for the duration of the game, and they endow agents with strategic preferences which, in general, cannot all be satisfied, while at the same time achieving an optimal utilization of resources. Much of the rich behavior of our game is intimately bound up with the frustration associated with highly constrained agent strategic preferences. It will be interesting to study the ways in which frustration effects change if evolutionary dynamics are introduced into the system, so that agents are allowed to alter their strategies in response to selective pressure. The general dynamic of frustration is likely to be a very important one in the study of adaptive competition and evolution in biological and social systems [13], and deserves more intensive study in that context.

Finally, it is clear that our work has implications for the study of a wide range of complex adaptive systems. The apparent generality of the phase structure suggests that this may be a persistent feature of a large number of specific systems, even when more realistic details are included. At the very least, this underlying phase structure is likely to be a base upon which the behavior of more complicated specific systems are built. In addition, we believe our work raises questions about the underlying epistemology of complex systems. There are currently few, if any, established general principles for the emergent behavior of complex systems. In fact, we do not even know what the proper questions are to ask about such systems. For example, we do not know, in general, what features of systems are specifically dependent on the details of the systems, and what are more generic. In our simple model, for instance, one might not have expected that σ in the efficient phase would have been so strongly dependent on the particular distribution of strategies to the agents. This absence of guiding principles, of course, has profound implications for the veracity of conclusions that can be drawn from more realistic models of specific systems. We believe that an epistemology for complex systems will emerge from the aggregation of insights garnered over time from a range of research in this area. The investigation of very simple prototypical systems, such as the one presented in this paper, has much to contribute to the evolution of an epistemology of complex systems.

Bibliography

[1] DeAngelis, D. and L. Gross (eds), *Individual-Based Models and Approaches in Ecology : Populations, Communities, and Ecosystems*, Chapman and Hall, New York (1992).

[2] NAGEL, K., RASMUSSEN, S., and C. BARRETT, "Network Traffic as a Self-Organized Critical Phenomena," in *Self-organization of Complex Structures: From Individual to Collective Dynamics*, (F. SCHWEITZER ed.), Gordon and Breach, London (1997), p579.

[3] HUBERMAN, B. and R. LUKOSE, *Science, 277*, 535 (1997).

[4] R. SAVIT, R. MANUCA and R. RIOLO, "Adaptive Competition, Market Efficiency and Phase Transitions." *Phys. Rev. Lett. 82 (10)*, 2203-2206 (Mar 8 1999).

[5] R. MANUCA, Y. LI, R. RIOLO and R. SAVIT, Submitted for publication; "The Structure of Adaptive Competition in Minority Games." UM Center for the Study of Complex Systems preprint 98-11-001 (1998). Available at http://www.pscs.umich.edu/RESEARCH/pscs-tr.html.

[6] M. DE CARA, O. PLA, and F. GIUNEA, Los Alamos preprint archive, cond-mat/9811162 (1998).

[7] N. JOHNSON, M. HART and P. HUI, Los Alamos preprint archive cond-mat/981227 (1998).

[8] D. CHALLET and Y.-C. ZHANG, *Physica A, 246*, 407 (1997).

[9] N. JOHNSON, S. *etal*, *Physica A 256*, 230 (1998).

[10] LI, Y., MANUCA, R., RIOLO, R. and R. SAVIT, in preparation.

[11] LI, Y., MANUCA, R., RIOLO, R. and R. SAVIT, Forthcoming in *Physica A, 276*, pp 234-264 (I), and pp 265-283 (II). (Feb. 2000)

[12] ANDERSON, P, ARROW. K. and D. PINES (eds.), The Economy as an Evolving Complex System, Addison-Wesley, Reading MA (1988);

[13] AXELROD, R. *The Complexity of Cooperation*, Princeton University Press, Princeton NJ (1997).

Spatial Autocatalytic Dynamics: An Approach to Modeling Prebiotic Evolution

Dimitris Stassinopoulos, Silvano P. Colombano, Jason D. Lohn, and Gary L. Haith
NASA Ames Research Center, Computational Sciences Division.
Jeffrey Scargle
NASA Ames Research Center, Planetary Systems Branch Division.
Shoudan Liang
NASA Ames Research Center, SETI Institute.

This paper addresses the origin of robust and evolvable metabolic functions, and the conditions under which it took place. We propose that spatial considerations, traditionally ignored, are essential to answering these important questions in prebiotic evolution. Our probabilistic cellular automaton model, based on work on autocatalytic metabolisms by Eigen, Kauffman, and others, has biologically interesting dynamical behavior that is missed if spatial extension is ignored.

1 Introduction

One difficulty central to any theoretical approach to the study of *prebiotic evolution* is the origin of self-reproduction. Once a reproductive mechanism is in place, the process of mutation and selection of fitter variants acts as feedback that biases the system's parameters towards the direction of improved performance. But how is such a mechanism initiated? The difficulty is far from trivial and there is a significant body of research addressing this issue. The central challenge is to understand and describe primitive processes that may have been the precursors of the reproductive and metabolic mechanisms employed by present living systems.

One suggestion is that "the origin of life came about through the evolution of autocatalytic sets of polypeptides and/or single stranded RNA." [8, 10, 11, 9] The central premise is based on the so-called *autocatalytic set*, a set of chemicals in which each member is the product of at least one reaction catalyzed by at least one other member. Such a set allows — in principle at least — for the development of a fortuitous cycle or *catalytic closure*. This mutual support allows the autocatalytic set as a whole to sustain and proliferates at the expense of other species in the chemical soup that do not benefit from such mutual "altruism."

Despite their lack of detail, these models have shown the utility of theoretical approaches to the origin of life. Specifically, they demonstrate that some of the difficulties in making these models work are fundamental ones, and thus likely to be encountered in actual biochemical processes. Two such problems are: i) the problem of generating an evolving chemistry, or more precisely, a chemistry that does not reach fixed points or local minima, and ii) the related problem of understanding how to make such systems responsive to environmental changes.

We argue that the spatial aspect might be an essential ingredient in addressing both problems. Whereas in previous models of autocatalytic sets the underlying assumption was that the chemicals are *perfectly mixed*, here we will allow for *incomplete mixing*. In the context of our model this suggests that chemicals can only interact locally. Our paper is outlined as follows: In Section Two, we give some motivation for the consideration of space in our model, as well as the model specifics. In Section Three, we present some results that show that previous studies on the formation of spiral waves in hypercyclic catalytic sets [4] can be extended to include cross-catalytic reactions. Interestingly, the dynamics of the spatially extended case is qualitatively different from the one observed in the perfectly-mixed case — a situation with no analogue in the previous study based on hypercycles [4]. In Section Four we discuss the potential significance of spatial considerations in the context of the broader research regarding prebiotic evolution.

2 A model of spatial autocatalytic dynamics

Most of the previous work on autocatalytic sets was formulated and studied in terms of coupled ordinary differential equations. Such formalism implies well-stirred chemostat in which each chemical can interact with every other chemical in the system. Here, we relax the assumption of the perfectly-stirred container. More specifically, we introduce a cellular space in which abstract chemical species can be transported and interact with each other only locally.

The main hypothesis underlying our model is that *spatial organization arising from conditions of incomplete mixing can facilitate chemical organization.* This assertion is encouraged by similar notions of the role of incomplete mixing in related disciplines:

(1) In *population genetics* one school of thought holds that geographical separation speeds up speciation while mixing leads to evolutionary stagnation e.g., [3]. (2) Results in *Artificial Life* literature on modeling ecologies indicate that spatial geometry is conducive to the development of diversity and thus improves the adaptive capabilities of the ecology as a whole [1, 6]. (3) Research in *complex systems*, and *connectionist artificial intelligence* suggests that sparse connectivity (related to the notion of incomplete mixing) is an important ingredient for learning and adaptation [11, 2]. A suggestion arising from this diverse body of work is that sparse connectivity and spatial segregation (incomplete mixing) are important requirements for adaptability and self-organization whereas high-connectivity systems (perfect mixing) have a stronger tendency to move very fast to either an ordered or a chaotic steady state.

2.1 The Model

We base our model on a probabilistic *cellular automaton* (CA) [13, 15]. Cellular automata are discrete dynamical systems, both in time and in space. An attractive feature of CAs is that typically the update rule of each cellular state depends on a small number of neighboring states — usually its nearest neighbors. This feature enables massively parallel implementations of the CA dynamics.
More specifically, in our CA:

1. The dynamics takes place in a two-dimensional cellular grid.

2. Each cell can either be occupied by a chemical, I_s, denoted by a non-zero cellular state, s, or empty, denoted by a zero cellular state. We denote the total number of cellular states with M.

3. The entire space of cells is permeated by an invisible "ether" from which molecules may be created and into which they may decay. Specifically, the update of the cellular states involves three processes:

    - *Decay.* Occupied states are updated to zero with probability d (see Eq. 2).
    - *Replication.* This process requires an empty cell and one occupied cell in its neighborhood to serve as a replication template [8]. During replication an occupied cell makes a copy of itself, $I_s \stackrel{r_s}{\rightarrow} 2I_s$. I_s denotes the chemical species that undergoes replication and r_s the replication constant.
    - *Auto- and Cross-catalysis.* This process requires an empty cell and two occupied states in its vicinity, one being the template, the other being the catalyst. Figure 1 shows a special case in which the four nearest-neighbor cells can act as template cells and the eight nearest-neighbor cells can act as catalyst cells (for details see figure caption).

4. A *diffusive* process that transports chemicals throughout the system. For simplicity we use the algorithm of Toffoli and Margolus [13] which ensures conservation of chemicals during each diffusion step. To simulate a higher diffusion constant we simply allow more than one diffusion steps.

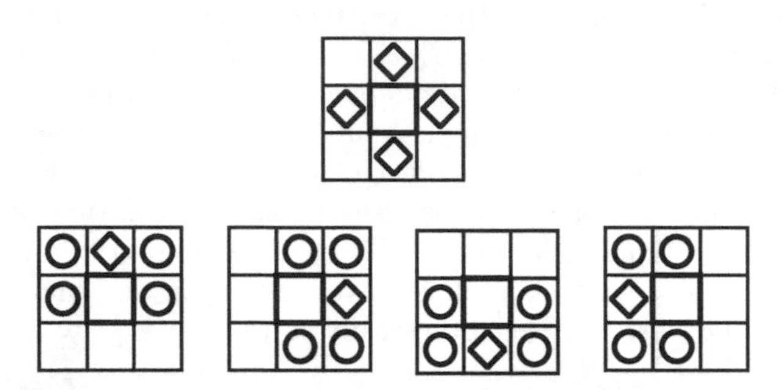

Figure 1: Example of CA rules required for updating the state of an empty cell (denoted here by the heavy-bordered-square in the center). The state of the empty cell, in the next time step, is determined by the product, $I_{s''}$, of reaction, $I_s + I_{s'} \overset{c_{ss's''}}{\rightarrow} I_s + I_{s''} + I_{s'}$ (Eq. (1)). Here, only the four nearest neighbors to the empty cell can serve as a template (diamond-occupied cells in top figure). Furthermore, for every given choice of a template cell, four eligible catalyst cells can be uniquely specified (circle-occupied cells in bottom figures).

Equations 1-3 provide a summary of the probability functions required to update the CA's state. Whenever possible we tried to adhere to the Boerlijst-Hogeweg implementation [4]. Whereas that work focuses on the spatial extension of the hypercycle [8], we are more interested in studying spatial dynamics in autocatalytic sets. Specifically, whereas [4] considered autocatalytic reactions,

$$I_s + I_{s'} \rightarrow 2I_s + I_{s'},$$

where I_s and $I_{s'}$ denote the chemical species serving as template and catalyst respectively, we consider a generalization where the outcome of the templating process, $I_{s''}$, can differ from the chemical species, I_s, acting as template,[1]

$$I_s + I_{s'} \overset{c_{ss's''}}{\rightarrow} I_s + I_{s''} + I_{s'}, \tag{1}$$

where $c_{ss's''}$ is the catalytic strength for the particular reaction. First we define the probability function $p(i)$, denoting the probability that an occupied site, i, remains occupied in the next time step:

$$p(i) = 1 - d, \text{if site i is occupied.} \tag{2}$$

[1]Although from a mathematical point of view the change seems small, it greatly broadens the space of allowed reactions and the resulting dynamics. Such a kind of generalized template replication is also experimentally realizable [12].

Next we define the probability functions $p_s(i), s = 0, \dots, M$, the probability that an empty site, i, will be occupied by chemical I_s, or remains empty, in the next time step:

$$p_s(i) = P_s(i) / \sum_{w=0}^{M} P_w(i), s \in \{0, \dots, M\}, \tag{3}$$

where, $P_0(i) = P_0$, a non-zero constant, allowing for the possibility that an empty site remains empty, and where P_s is defined as follows if $s = 1, \dots M$,

$$P_s(i) = \sum_{j \in O_i} \delta_{q(j)s} r_{q(j)} + \sum_{w=1}^{M} \sum_{(j,k) \in Q_i} \delta_{ws} c_{q(j)q(k)w}, \tag{4}$$

where we use $q(i)$ to symbolize the cellular state of cell i. O_i is the set of four nearest neighbors of cell i that can act as templates. For instance, in the example of Fig. 1, O_i consists of the four sites indicated by $\diamondsuit$ (top part of Fig. 1). Q_i is the set of all template-catalyst pairs of cells in the vicinity of cell i. In the example of Fig. 1, Q_i includes all possible $(\diamondsuit, \bigcirc)$ pairs (bottom part of Fig. 1).

3 Results

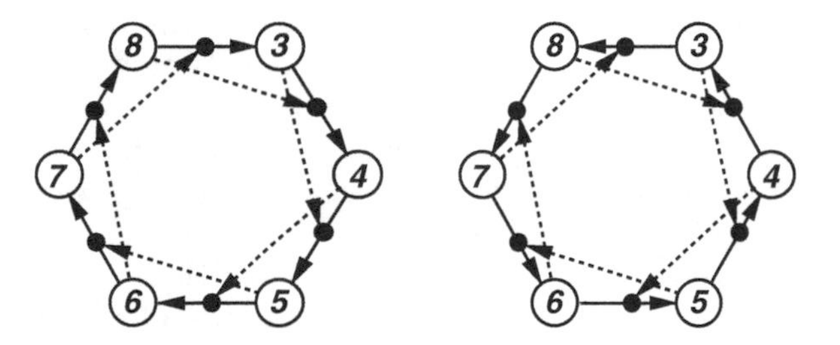

Figure 2: Chemical Reaction graphs (A) and (B).

Here we are concerned with aspects of the spatial dynamics that are relevant for the study of prebiotic evolution and *cannot* be captured by the perfectly-stirred-container approximation. In that regard a qualification is in order: a broad body of research in pattern formation [5], in general, and in reaction-diffusion systems, in particular, has produced a great variety of spatial dynamics. These studies have clearly demonstrated that the incorporation of space greatly broadens the range of dynamics that can be exhibited by spatially extended systems. However, this new level of complexity might be totally irrelevant to the study of prebiotic evolution if it so happens that the mechanisms of prebiotic processes can be described — at least qualitatively — with models based on the perfectly-stirred-container approximation. In such a case, consideration of imperfect mixing would only improve our understanding of the details rather than of the fundamental underlying mechanisms. Thus, for our approach to be

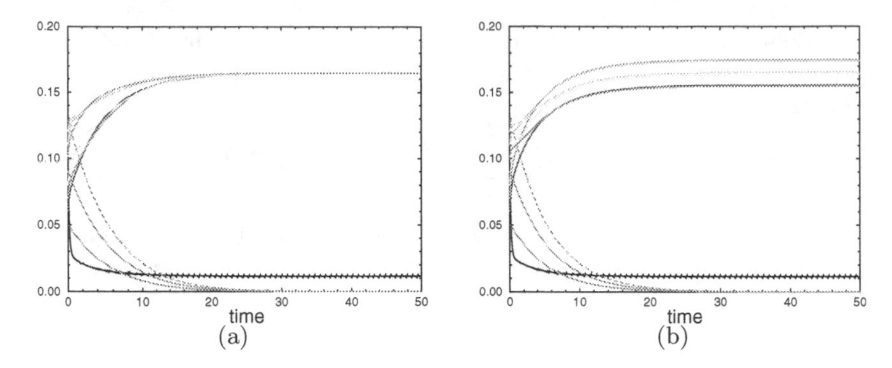

Figure 3: Perfectly mixed approximation — Evolution of chemical concentrations for two $M = 10$ chemistries (for color code see Fig. caption 5): (a) reaction graph (A); (b) reaction graph (B);

meaningful, it is imperative that it focus on phenomena that i) are essential for the understanding of prebiotic evolution and ii) cannot be studied with the simpler formalism of coupled ordinary differential equations.

We illustrate the point with a simple system of ten ($M = 10$) chemical species. We will consider catalytic reactions of the form, $I_s + I_{s'} \rightarrow I_s + I_{s''} + I_{s'}$ (Eq. 1). In particular, we will focus on two catalytic reaction sets, (A) and (B), graphically depicted in Fig 2. Solid arrows emanate from template chemicals and point to products, while dashed arrows emanate from chemicals that act as catalysts, and point to the reaction that they catalyze. The two graphs are identical except the orientation of the solid arrows. Because we consider a small fraction of all the possible reactions, $c_{ss's''}$, here, is a sparse matrix. For the results reported here we consider a space consisting of size 300×300 with periodic boundary conditions. Furthermore, we set the model's parameters as follows: $d = 0.2, P_0 = 11, r_s = r = 1, s = 1, \ldots, M$, and all non-zero elements of the catalytic matrix $c_{ss's''}$ to $c = 100$. We have chosen c to be one hundred times larger that r to address the fact that catalytic self-replication is expected to be much more efficient than spontaneous self-replication.

How does such difference in the choice of chemistry affect the resulting dynamics? First we studied the perfectly-mixed case. The results of our numerical simulations are summarized in Figure 3. We note that in both cases the chemical concentrations stabilize to constant values.[2] The situation changes drastically for the imperfectly-mixed case: whereas graph (A) gives rise to dynamics analogous to the ones described above, graph (B) generates a new kind of dynamics (Fig. 4). Now the system has a much higher concentration of empty sites. Furthermore the chemical concentrations follow persistent oscillations where all six chemicals participating in the catalytic set alternate in relative strength (Fig.

[2]More careful study of the dynamics reveals that in chemical graph (A) there is a stronger causal relationship between neighboring reactions (the template of one reaction is also the catalyst of the subsequent reaction) however these differences do not alter the qualitative nature of the overall dynamics.

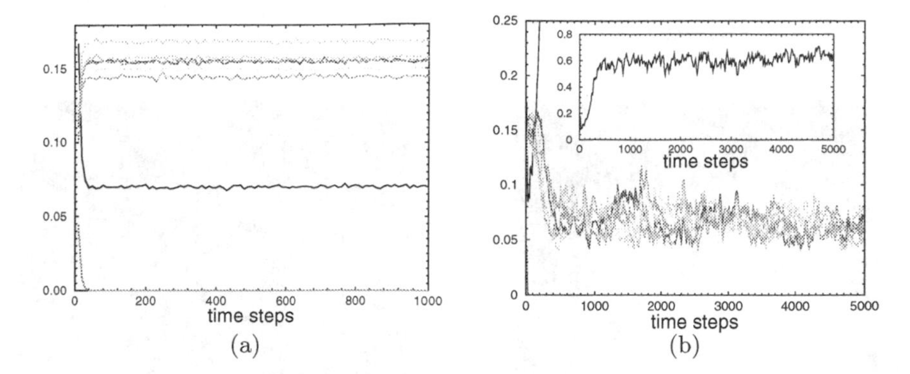

Figure 4: Imperfectly mixed approximation — Evolution of the chemical concentrations for two 10-chemicals chemistries (for color code see Fig. caption 5): (a) Chemical graph (A); (b) Chemical graph (B) (Inset - concentration of empty sites).

4b). Observation of snapshots of the cellular space (Figs. 5abc) reveals further differences: Case (A) is characterized by fairly homogeneous distribution of the six chemicals participating in the reaction set; very little structure is present in this configuration. Case (B), on the other hand, displays clear signs of spatial self-organization. Fig. 5b displays the state of the system after 1000 time steps; a spatial structure has already started to form. This organization finally leads to the formation of a pair of spiral waves (Fig. 5c). Note the six spiral arms — each color representing a different chemical species — emanating from the spiral center.[3] Further study of a sequence of snapshots of the CA dynamics has shown that the oscillatory pattern of Fig. 4b is directly related to the rotation of the spiral-arm formation. Indeed, the period of the spiral-arm rotation agrees with the period of the oscillation of the chemical concentrations. Finally, it should be noted that for species I_s to receive catalytic support from species I_{s-2} they have to overcome the barrier generated by species I_{s-1}. It is the diffusive process, present in the system, that ensures that some of the members of the I_s will eventually be catalyzed by I_{s-2}.

The particular reaction graph was chosen i) to demonstrate that spiral dynamics can be extended to autocatalytic sets, and ii) to provide a concrete example of a chemistry whose dynamics cannot be described by the fully-stirred approximation. Furthermore, it suggests that the mechanism underlying the dynamics described above might be extended in situations where the efficient catalytic reactions are relatively few and the ones that lend little or no catalytic support are numerous. In such a system the chemicals forming the autocatalytic set would have to overcome the barrier of a large number of irrelevant chemicals. It would be interesting to know under what conditions — if at all - a diffusive mechanism, similar to the one described above, might allow the members of the autocatalytic set to overcome this barrier.

[3]See ref. [4] for a detailed description of similar spiral dynamics observed in hypercyclic reaction sets.

4 Discussion

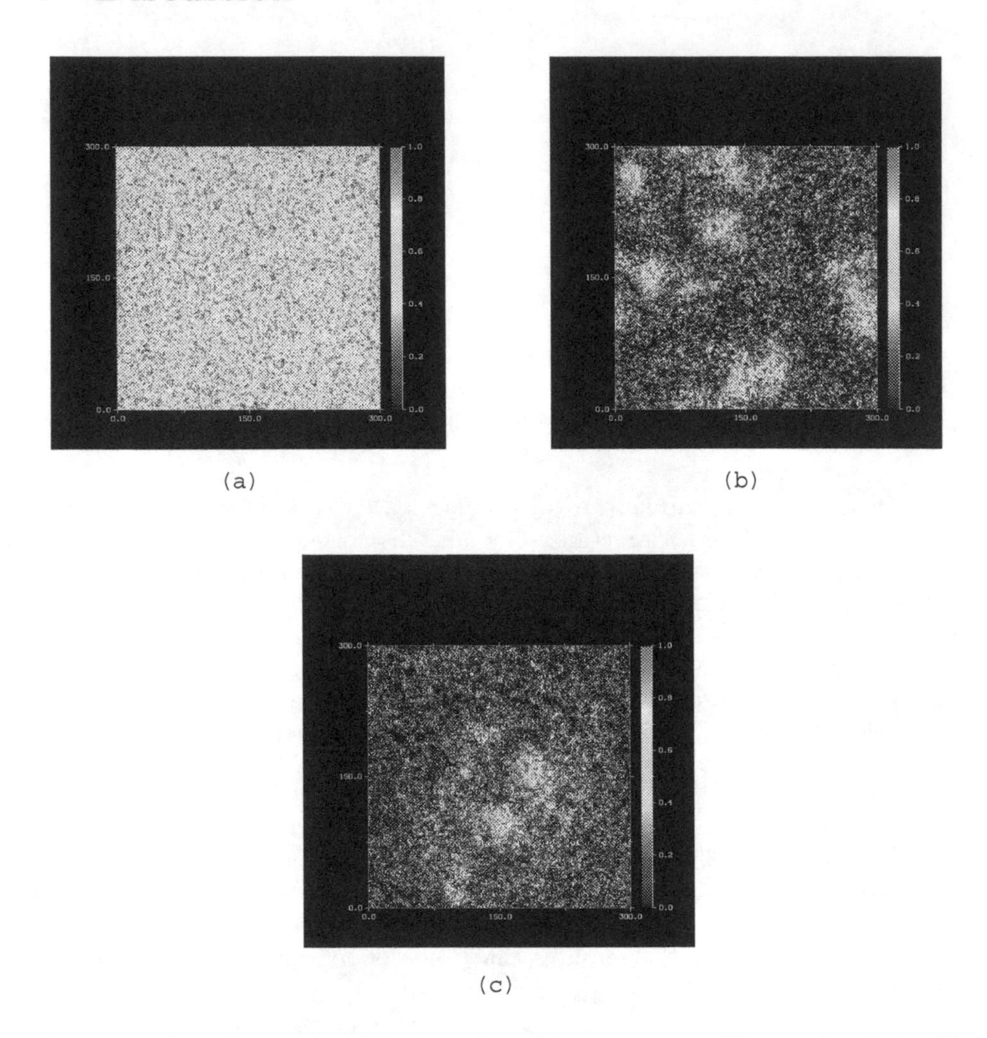

Figure 5: Snapshots of the CA dynamics; Colors represent different chemicals, (0) black; (3) blue; (4) cyan; (5) green; (6) light green; (7) yellow; (8) orange. (a) reaction graph (A) – system configuration after 1000 time steps; (b) reaction graph (B) – system configuration after 1000 time steps; (c) reaction graph (B) – system configuration after 5000 time steps;

Spiral waves have been previously observed, in the context of prebiotic evolution, in a CA-based model [4] that has added spatial inhomogeneity to the original model on hypercycles [8] by Eigen and co-workers. One aspect of the original work that generated excitement was that in order to proliferate, a hypercycle relies on altruism rather than selectionism. In other words, each member, I_s, of the hypercycle lends support to some other member, I_{s+1}. From a selectionist perspective this should lead to an increase of species I_{s+1} at the expense of species I_s. In the hypercycle this does not happen because it has been designed so that *every* member lends catalytic support to some other member of the system and at the same time *every* member receives catalytic support.[4] The work in ref. [4] has shown that the same type of mutual interdependence that makes the hypercycle function can be found in the spiral-wave organization found in the extended model. Although that work has shown that the element of space increases the stability of this mechanism (in the sense that spiral-waves seem to be more resilient to parasites) it leaves open the main question namely, "how did a hypercyclic organization come to be in the first place?" We feel that if it can be shown that spiral waves can emerge spontaneously under fairly generic initial conditions, their relevance for modeling prebiotic evolution will deserve serious consideration.

Our work is a first step toward that goal. By opting for a more general catalyzed template reaction (Eq. (1)) we have greatly broadened the space of possible reactions making the accidental occurrence of a hypercycle exceedingly unlikely. Based on a CA-model we have shown that spiral waves are indeed realizable by more general reaction sets. We argue that our choice of set of catalytic reactions is interesting because i) it sustains an organization of mutually dependent species only in the imperfectly-mixed approximation (in the well-stirred case it collapses to a fixed point), and ii) furthermore, the mechanism that generates spiral organizations for this reaction set might also work for more general choices of reaction sets. Ongoing work focuses on a more systematic study of this second point.

5 Acknowledgments

The authors like to thank Mike New, Andrew Pohorille, Karl Schweighofer, and Karim Shariff for stimulating discussions and comments. This work has been supported in part by NASA Program UPN 632-14-40. D.S. has also been supported by a NRC-NASA Research Associateship.

Bibliography

[1] C. Adami and C. T. Brown, "Evolutionary Learning in the 2D Artificial Life Systems," in *Proc. of Artificial Life IV, 1994,* eds. R. Brooks and P. Maes, MIT Press, Cambridge, p. 377, 1994.

[2] P. Bak, H. Flyvbjerg, and B. Lautrup, "Coevolution in a Rugged Fitness Landscape," *Phys. Rev. A* **46**, 6724 (1992); H. Flyvbjerg and B. Lautrup, "Evolution in a Rugged Fitness Landscape," *Phys. Rev. A* **46**, 6714 (1992).

[4] Indeed, it is known that hypercycles are vulnerable to *parasites* i.e., chemical species that receive support but do not lend support themselves. In the presence of a parasite a hypercycle simply dies out [14].

[3] L. L. Cavalli-Sforza and W. F. Bodmer, *The Genetics of Human Populations,* W. H. Freeman and Co., San Fransisco, pp. 423-430, 1971.

[4] Boerlijst, M., and P. Hogeweg, "Spiral wave structure in pre-biotic evolution: Hypercycles stable against parasites", *Physica D* **48** (1991), 17–28.

[5] Cross, M.C., and P.C. Hohenberg, "Pattern formation outside equilibrium", *Reviews of Modern Physics* **65** (1993), 851.

[6] T. S. Ray, "Selecting Naturally for Differentiation: Preliminary Evolutionary Results," *Complexity* **3**(5), 25-33, 1998.

[7] Edwards, S.F., and P.W. Anderson, "Theory of spin glasses", *J. Phys. F* **5** (1975), 956.

[8] Eigen M. and P. Schuster, *The Hypercycle,* Springer Verlag, 1979.

[9] Farmer, J. Doyne, Stuart A. Kauffman, and Norman H. Packard, "Autocatalytic replication Polymers", *Physica D* **22** (1986), 50.

[10] Kauffman, Stuart, "Autocatalytic sets of proteins", *J. Theoret. Bio.* **119** (1986), 1.

[11] Kauffman, Stuart, *The Origins of Order: Self-Organization and Selection in Evolution,* Oxford University Press (1993).

[12] Lee, D.H., K. Severin, Y. Yokobayashi, and M.R. Ghadiri, "Emergence of symbiosis in peptide self-replication through a hypercyclic network", *Nature* **390** (1997), 591–594.

[13] Toffoli, Tommaso, and Norman Margolus, *Cellular Automata Machines,* MIT Press, 1987.

[14] Smith, J. Maynard, "Hypercycles and The Origin of Life", *Nature* **280** (1979), 445.

[15] Wolfram, Stephen, *Theory and Applications of Cellular Automata,* World Scientific (1986).

Geophysiological modeling: a new approach at modeling the evolution of ecosystems

By **Francesco Santini** (1)(2), **Lodovico Galleni** (2), **Paola Cerrai** (2)(3)

(1) Department of Zoology, University of Toronto.
fsantini@zoo.utoronto.ca
(2) Centro Interdipartimentale per lo Studio dei Sistemi Complessi (CISSC), Universitá di Pisa. lgallcni@agr.unipi.it
(3) Dipartimento di Matematica, Universitá di Pisa. cerrai@dm.unipi.it

1. The theoretical framework.

In the recent years many of the inherent problems in the neo-darwinian view of evolution have been underlined by several authors. These include, among others, a lack of attention for biological hierarchies; for the importance of adaptive mutational systems and the epigenetic inheritance systems; for the developmental constrains caused by morphogenetic fields; for the relationships that exist between thermodynamics, information and evolution [Goodwin 1984, 1994; Wan Ho 1984; Eldredge 1985; Brooks and Wiley 1988; Jablonka et al. 1998].

Living organisms, of course, evolve within ecological associations (ecosystems, landscapes, biomes, the biosphere) that are constituted by a biological component and a physico-chemical component. Until recently the attention of evolutionary ecologists has been mainly directed towards the understanding of the effects that the interactions between the biological components of ecosystems have on the dynamics of populations and communities. Such interactions (that is, infra- and interspecific competition and predation) were supposed to be, by far, the main cause for the observed organization of ecosystems. This was due to the fact that only the eukaryotic component (but it would be better to say: multicellular component) of ecosystems was investigated. At these levels competition and predation do indeed have a main role in shaping the constitution and interactions of the community. However, when microbial communities, and especially the prokaryotic ones, are taken into consideration, a very different picture emerges [Margulis and Sagan 1991; Markos 1996]. At this point two additional factors have to be taken into account, in order to be able to understand correctly the functioning of ecological systems: cooperation among organisms and the interactions between the living and non living parts of the ecological systems.

We believe that in the search for a more comprehensive theory of evolution a much greater attention should be paid to the ways in which the physico-chemical environment interacts with the living organisms. This aspect is mainly based on the suggestion of taking into consideration the Biosphere as a whole evolving object, which was the very basis of Geobiology, the science of the Biosphere proposed by Teilhard de Chardin [Galleni 1995, 1996].

These concepts are at the core of the Gaia theory (formerly Gaia hypothesis), advanced by the British Scientist James Lovelock in the early seventies [Lovelock and Margulis 1974; Lovelock 1995a, 1995b]. The Gaia theory claims that it is not possible to talk about an independent evolution of the biological and chemico-physical components of the biosphere, because both parts are strongly linked and influence and direct the evolution of each other. Hence, according to Lovelock, we should think about a unique integrated system, that is Gaia, in which the self-regulation that leads to the homeostasis of certain parameters is a self-emergent property obtained by the interactions between the several parts of the system [Lovelock 1995a]. To try to better visualize his concept, Lovelock recurred to the metaphor of the "Superorganism", claiming that the Earth is, **as a** superorganism, able to maintain the homeostasis of some of its components, such as the salinity of the oceans and the atmospheric composition, being for example all the gasses of the Earth's atmosphere of biological origin [Lovelock and Margulis 1974]. Hence the name **Geophysiology** to describe the new science of Gaia; the science that studies ecological systems as if their components were co-evolving and influencing each other, and not just a mere sum of independently functioning parts.

2. Where are we now ?

Mathematical models have always been massively used by ecologists in an attempt to simulate the dynamics of biological populations and communities and hence predict the behavior of ecological systems. Until recently, however, only few attempts have been made to try to include also the physico-chemical environment and the interactions that exist between the living and non living part of the ecosystems in the mathematical modeling of temporal evolution [Watson and Lovelock 1983; Maddock 1991]. In our opinion, in order to be able to develop more realistic models, it is of paramount importance to take into consideration these interactions.

According to this line of thought, in our opinion [Benci and Galleni 1998] the factors that influence the evolution of ecosystems can be split in 3 groups:
1) X factors, which represent the living beings and the interactions among themselves.

2) Y factors, which represent the "non-living species", that is the environmental parameters which are subject to the influence of the biota. In reality, as first foreseen by the American Biologist Lotka and the Russian geochemist Vernadsky [Lovelock 1995a], X and Y factors are tightly bound and heavily influence the evolution of each other.
3) λ factors. These are environmental parameters that are out of the biota's control, such as, for example the increase in luminosity of the sun, or the tectonic plate drive. All these events are often not taken into consideration in the vast majority of attempts to simulate the evolution of ecosystems through modeling, despite the great importance that they have on evolutionary processes (e.g. mass extinctions caused by asteroids and comets). Interacting with the X and Y factors the λ factors are also constituents of natural selection.

3. The model applied to Ecosystem studies

For our study, we have decided to apply this approach to a recently discovered new kind of ecosystems: the hydrothermal vents. These vents are produced by the geological processes that drive plate tectonics and are responsible for the genesis of oceanic crust at spreading centers. The biological communities of these particular ecosystems do not depend on photosynthesis but on bacterial chemosynthesis, which obtains carbohydrates and other biological molecules from inorganic compounds, such as CO_2, exploiting as electron donors the reduced inorganic compounds present in the heated fluid. Thus the source of energy is not solar light but is geothermal energy produced by the Earth's interior. The main reduced compounds used in chemosynthetic reactions are dissolved sulphides [Jannasch and Mottl 1985; Tunnicliffe 1991]. Chemoautolithotrophic bacteria are the only primary producers in deep-sea hydrothermal ecosystems. They form the base of the trophic structure of the communities. Vent communities have a much bigger biomass than that of the classic hadal communities even if they show a more limited diversity. Chemoautotrophy and symbiosis are the reason for the presence of a great amount of biomass. Heterotrophic bacteria, protists and several species of invertebrates, many of which are in symbiosis with chemotrophic bacteria, make up the remaining part of these communities. The fauna of the inhabitants of the hydrothermal vents is endemic and taxonomically distinct from that of the normal deep-sea environment [Tunnicliffe 1991].
For the purposes of this study this kind of ecosystems presents many advantages, including a short life-time that makes it easy to test theoretical predictions.

4. Our Geophysiological model

Given that it is possible to identify among the several components of ecosystems both biotic and abiotic, phenomena of "self-limitation", "competition" and "predation", the mathematical model that we propose is a Lotka-Volterra variation [Murray 1989]. This kind of modeling in our opinion presents several advantages that have convinced us of its validity for our research. It was in fact created by Volterra to study the dynamics of biological populations, and independently by Lotka to model chemical reactions (and in our model we have both kinds of interactions). This is a standard modeling technique, and for the purposes of our study it gives acceptable results, even considering the degree of approximation of those.

In order to simplify the ecological interactions we have divided the biota into producers, called X_1, and consumers, called X_2. Producers include not only the free-living chemoautotrophic bacteria, but also the chemosynthetic bacteria that live in symbiosis with macroinvertebrates and their metazoan hosts in cases where these have no other mean of finding food if deprived of their associates. Consumers include both primary (grazers and filter feeders) and secondary consumers.

In our model we have also decided to take into account the main chemical compounds that perform the role of electron donors and acceptors. These two roles are performed mainly by hydrogen sulphide and molecular oxygen. Molecular oxygen is produced by the activity of living organisms outside the thermal ecosystem, while the hydrogen sulphide is mainly produced by venting, making the production of these compounds beyond the control of living beings inhabiting the hydrothermal vents. Oxygen and sulphide are indicated as λ_1 and λ_2 respectively. In our investigation of hydrothermal ecosystems we could not identify any Y factor that has a vital role in the functioning of these ecosystems. Being the chemical elements used as source of energy and as source of matter mainly of thermal origin or coming from non-thermal organisms (such as in the case of O_2), we decided to label them as λ factors. However Y factors are usually extremely important in any other kind of ecosystem investigated [Benci and Galleni 1995, 1998].

Due to needs of model manageability, we have decided to limit it to a set of 4 differential equations in 4 variables.

$$\begin{cases} \dfrac{dX1}{dt} = X1 \cdot (-C1 - X1 - X2 + \lambda 2) \\ \dfrac{dX2}{dt} = X2 \cdot (-C2 + X1 - X2) \\ \dfrac{d\lambda 1}{dt} = \lambda 1 \cdot \left(B1 \cdot (1 - \lambda 1) - A \cdot \lambda 2\right) \\ \dfrac{d\lambda 2}{dt} = \lambda 2 \cdot \left(B2 \cdot (1 - \lambda 2) - D \cdot X1 - C \cdot \lambda 1\right) \end{cases}$$

X_1, X_2, λ_1, λ_2 represent respectively the density of autotrophic organisms, heterotrophic organisms, electron acceptors and electron donors. dX_1/dt, dX_2/dt, $d\lambda_1/dt$ and $d\lambda_2/dt$ represent the variations through time of those densities. The coefficients that appear in the equations are all positive, and to make matters simple we have put the coefficients of self-limitation, competition and predation for X_1, X_2, λ_1, λ_2 equal to 1. A, B, C, D, C_1, and C_2 are coefficients. As in the Lotka-Volterra models, the advantages and the disadvantages that are obtained by every species/compound from the interactions with other components of the ecological system are supposed to be directly proportional to the density of the interacting species/compounds.

The first equation represents the variation through time of the producers. In the absence of λ_2 we would have an exponential decrease to zero, with a decrease coefficient C_1, to which is added the internal self-limitation (that is, competition) and predation. The advantage given by the presence of inorganic reduced compounds, which are necessary for chemosynthesis, is represented in the model by the expression $X_1\lambda_2$. The disadvantage, determined by the competition for resources among autotrophs and the predation from heterotrophs, is represented by $-X_1X_1$ and $-X_1X_2$ respectively.

The second equation, dX_2/dt, represents the evolution of the consumer's guild. Without autotrophs, represented by X_1, there would be an exponential decrease to zero with coefficient C_2, to which is added the effect of self-limitation. X_2 receives advantage in the form of predation from the presence of X_1, always with a coefficient considered to be 1. The advantage, here represented by the expression $+X_1X_2$, is still considered to be directly proportional to the density of the 2 populations.

The equation $d\lambda_1/dt$ represents the variation in time of molecular oxygen. This would tend to reach, through a sigmoid growth, a maximum concentration that corresponds to that of sea water, which we have considered to be 1. However this is altered by the presence of reduced compounds, which perform what might be considered to be the equivalent of a biological "predation" with coefficient A. The disadvantage caused by the reduced inorganic compounds for the molecular oxygen is given by the expression $-A_1\lambda_2$, and is still considered to be directly proportional to the densities of the interacting species, even if now chemical and not biological anymore. We have not considered in our equation the consumption of oxygen through respiration of the biota, given the fact that this quantity is extremely low, compared to the consumption caused by sulphides.

The last equation, $d\lambda_2/dt$, represents the variation through time of the hydrogen sulphide. This also depends on the tendency of this compound to reach its maximum concentration, which corresponds to the one of the thermal fluid, also in the sea water, and even this maximum concentration is considered to be 1 in the model. The consumption by chemotrophs and by oxygen, represented by $-D\lambda_2X_1$ and $-C_1\lambda_2$, keeps the concentration of hydrogen sulphide much lower.

By selecting the growth coefficients in accordance to the experimental evidences, it is now possible to simulate the evolution of these ecosystems and their associated biological communities.

5. Results

Given that it is not possible to graphically represent the evolution of the four components at the same time, we decided to split them and analyze the evolution of the biological and the chemical components of the ecosystem separately. Figure 1 and figure 2 show that the amount of sulphide and oxygen that is present in the hydrothermal ecosystems remains constant since the early stages of the life of a vent. Soon after the beginning of the emission of thermal fluids a stable state is reached, and this state is maintained for the whole life of the vent. At the end of the emission of thermal fluids (figure 3), the concentration of sulphides in the hydrothermal ecosystem tends to drop to zero, given that no new sulphides are emitted by the vent and all the sulphides already emitted are either consumed by the living organisms or oxidized by the molecular oxygen present in the sea water. Hence, given that sea water represents a continuous source of oxygen while the H_2S concentration goes to zero, the concentration of O_2 in the hydrothermal ecosystem approaches the same level present in the sea water, which in our model we have considered to be 1.

For the biological components, the model shows that in the early stages of the life of a vent, producers start to increase quite rapidly, while consumers are more slow to follow (figure 4). As time passes by, the population of producers tends to reach a climax stage more quickly than the population of consumers, which keeps growing, as illustrated by figure 5. When the hydrothermal emissions stop, primary producers tend to collapse very rapidly, while the population of consumers converges towards zero more slowly (this because they find preys to consume even within their own category), as can be seen from figure 6.

6. Conclusions

Our model shows that when there is a flow of thermal fluids rich in hydrogen sulphide, and presence of oxygen in the marine water, in the area of mixing of fluids rich biological communities flourish and start to modify their surrounding environment (e.g. the chemotrophs lower the concentration of sulphides in the water). In order to keep the biological communities alive, there has to be a constant supply of both oxygen and sulphides. The amount of these two elements that is present in the area surrounding a single vent is massively influenced not only by the interactions among the chemical species, but also by those among biological species. Furthermore, the fate of living organisms of thermal communities, is determined not only by the classic biological phenomena, such as competition and predation, but also by the interactions between biological and chemical species. When the flow of thermal fluids is reduced, available sulphides are rapidly oxidized and this event leads to the death of chemoautotrophs and the collapse of the thermal community. This illustrates the need to take into consideration also the chemico-physical environment and its interactions with the biota when modeling ecosystem functioning. Otherwise it would be very difficult (not to say impossible) to explain the disappearance of the hydrothermal communities if only the biological interactions are considered.

We hope that our model may illustrate that in order to be able to predict in a more realistic way the functioning of ecosystems, it is not possible to avoid considering the core idea of the Gaia theory. Evolution is not just a process that concerns living organisms. It is a phenomenon that should be investigated with a "Biospherocentric" perspective, keeping in mind that living organisms cannot be taken into consideration without also considering their surrounding environment, which they have a great influence in forming [Lowenstam, 1981; Lovelock, 1995a].

Figure 1: Interactions between H_2S and O_2 at beginning of hydrothermal emissions.

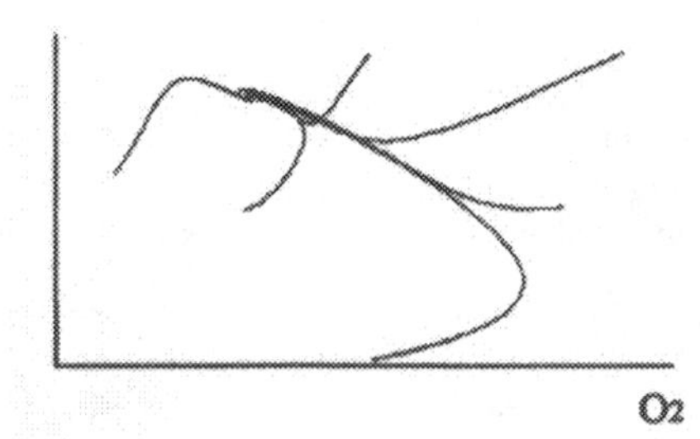

Figure 2: interactions between H_2S and O_2 at mature life stages of hydrothermal vent.

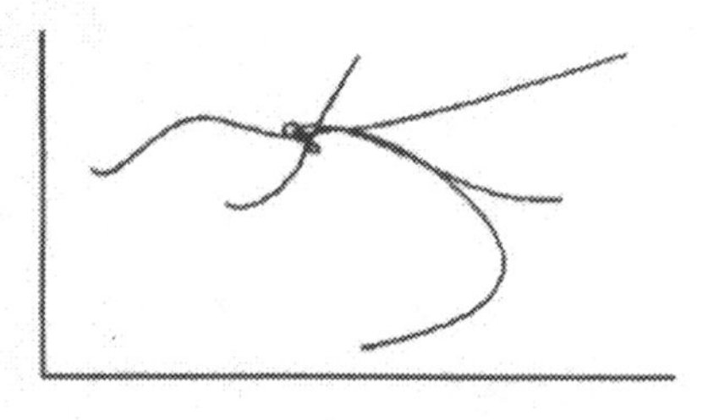

Figure 3: interactions between H_2S and O_2 at end of vent activity.

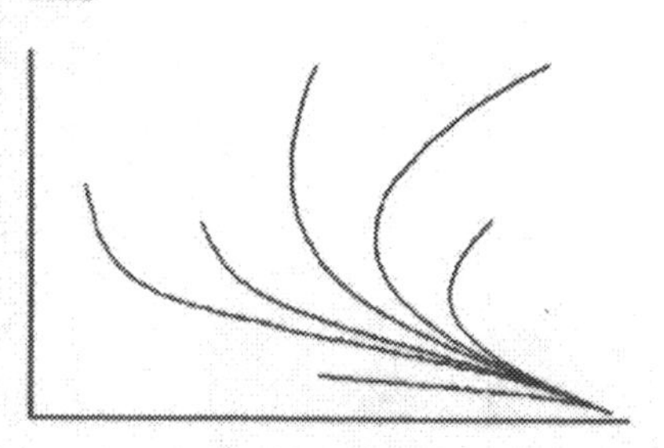

Figure 4: interactions between primary producers (X_1) and consumers (X_2) during early stages of vent life.

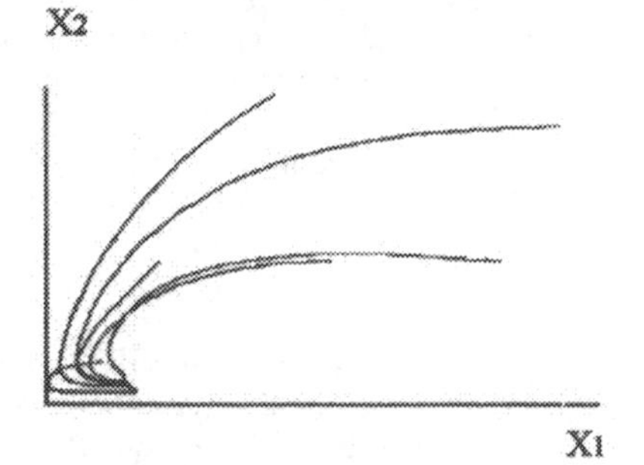

Figure 5: interactions between consumers and producers during mature stages of vent life.

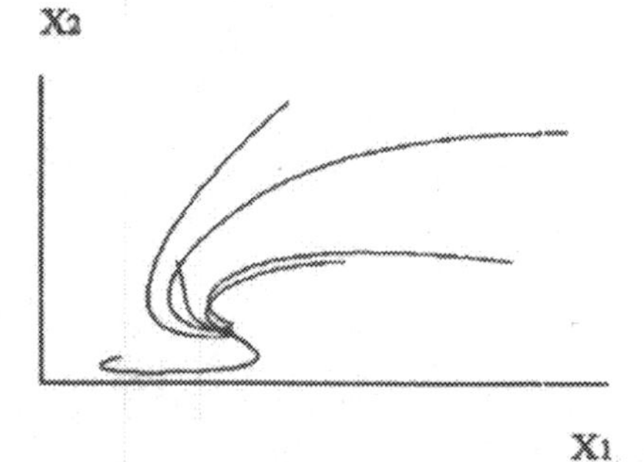

Figure 6: interactions between consumers and producers at the end of vent activity.

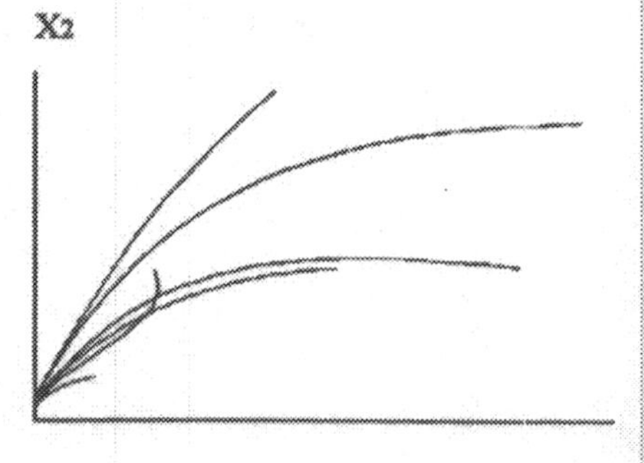

References

Benci, V., & Galleni, L., 1998, Stability and instability in evolution, *J.theor. Biol.*, **194**, 541.
Brooks, D.R., & Wiley, E.O., 1988, *Evolution as Entropy. Towards a Unified Theory of Biology*, University of Chicago Press (Chicago).
Eldredge, N., 1985, *Unfinished Synthesis*, Columbia University Press (New York).
Galleni, L., 1995, How does the teilhardian vision of evolution compare with contemporary theories, *Zygon*, **30**, 25.
Galleni, L., 1996, Levels of organization and evolution, From Teilhard de Chardin to Lovelock, *Studies in Science and Theology*, **4**, 109.
Goodwin, B., 1984, A relational or field theory of reproduction and its evolutionary implications, in *Beyond Neo-Darwinism*, edited by M.W. Ho & P. Saunders, Academic Press (London).
Goodwin, B., 1994, *How the leopard changed its spots,* Charles Scribner's sons (New York).
Ho, M.W., 1984. Environment and heredity in development and evolution. in *Beyond Neo-Darwinism*, edited by M.W. Ho and P. Saunders, Academic Press (London).
Jablonka, E., Lamb, M.J., & Avital, E., 1998, 'Lamarkian' mechanisms in darwinian evolution, *Trends Ecol. Evol.*, **13**, 206.
Jannasch, H.W., & Mottl, M.J., 1985, Geomicrobiology of Deep-Sea Hydrothermal Vents, *Science*, **229**, 717.
Lovelock, J.E., & Margulis, L., 1974, Atmospheric homeostasis by and for the Biosphere: the Gaia hypothesis, *Tellus*, **26**, 2.
Lovelock, J.E., 1995a, *The Ages of Gaia: a Biography of our Living Earth Updated and Revised*, Norton and Company (New York).
Lovelock, J., 1995b, *Gaia: a New Look at Life on Earth*, Oxford University Press (Oxford).
Lowenstam, H.A., 1981, Minerals Formed by Organisms, *Science*, **211**, 1126.
Maddock, L., 1991, Effects of simple environmental feedback on some population models, *Tellus*, **43B**, 331.
Margulis, L. & Sagan, D., 1991, *Microcosmos: four billion years of microbial evolution*, Touchstone Book (New York).
Markos, A., 1996, The Gaia theory: the Role of Microorganisms in Planetary Information Network, *J. theor. Biol.*, **176**, 175.
Murray, J.D., 1989, *Mathematical Biology*, Springer-Verlag (Berlin).
Tunnicliffe, V., 1991, The biology of hydrothermal vents: ecology and evolution, *Oceanogr. Mar. Biol. Annu. Rev*, **29**, 319.
Watson, A.J., & Lovelock, J.E., 1983, Biological homeostasis of the global environment: the parable of Daisyworld, *Tellus*, **35B**, 284.

Organizational Intelligence

Paul W. Morris
paulmorris@earthlink.net

1. Introduction

Business organizations are *intelligent* entities. An understanding of this provides new ways of interpreting events and patterns, and offers new principles useful in deciding actions.

Organizational intelligence is the intelligence achieved *collectively* by a group of people, by an organization. It is not the sum of all of the individuals' intelligences, but is the unique intelligence of the group as a whole. This intelligence is in no way given, or fixed; it can be influenced along several of its dimensions, better fitting it to specific needs. Particularly as a result of advances in information technologies, the competitive potential of such improvements is rising rapidly.

Several others have written about, or at least touched upon, the topic of collective intelligence in business organizations. [e.g., Rothschild 1990, Pinchot 1993, Kelly 1994, Shimbel 1975, Florida 1991] Approaches to the topic range from a conception of collective intelligence as the information possessed by an organization to recognition of the reality of a distinct collective intelligence, but without elaboration of its functioning. Most typical is a treatment in which it is considered as a simple sum of the parts, a matter of more fully engaging each individual's intelligence. In this approach, collective intelligence is something essentially pre-existent, to be exploited as fully as possible. In this white paper, the concept is proposed that this intelligence is a real, separate intelligence--dependent on individuals' intelligences but distinct from them; something to be understood, created, and optimized; not quite something to be managed, but certainly to be nurtured. (While this approach to understanding organizations is largely coincident with complex adaptive systems frameworks, the differences are productive for considering implications and application of the concepts for business organizations.)

2. Intelligence

Intelligent organizations? To begin to bring this concept into focus, it's helpful first to consider a simpler example: ants.

Individual ants are obviously rather limited in their problem-solving capabilities. Ant colonies, however, display impressive capabilities: locating and retrieving remote food in a complex environment; building, maintaining and defending the complicated nest structure; rapidly moving the entire community to a new location. How do such obviously simple creatures achieve such amazing results? Research into how ants solve problems [Franks 1989] provides intriguing answers. Ants are seen to employ several basic techniques to rapidly solve complex problems:

- interaction, or "mixing" with the environment, with the problem to be solved;
- experimentation;
- communication of success;
- reinforcement of success; and,
- a high degree of redundancy.

These techniques suffice to give ant colonies their impressive capabilities to evolve solutions to problems. We see an extreme case of simple individuals dealing successfully with a world that is complex beyond their capacities for comprehension or even perception. But can these capabilities correctly be described as intelligence?

2.1. What is Intelligence?

Dictionaries offer quite a diverse range of definitions of intelligence, addressing different facets of the concept. Many of the definitions are specific to the special case of human intelligence, rather than helping with a general definition. Research into definitions proposed by experts in the fields touching on the question does not simplify matters. It unearths even more diverse definitions, with the most persuasive being the longest and the vaguest. And again, most of the focus is on human intelligence rather than other forms. However, analysis of the various approaches to defining intelligence turns up several common elements:

- intelligence is a capacity, an ability; [Gove 1986, Sperber 1994, Morris 1975]
- intelligence involves perception, action, behavior; [Butterworth 1994, Calvin 1994]
- knowledge and learning are central to intelligence; [Gove 1986, Schank 1994]

From these components a general, working definition of intelligence can be synthesized:

Intelligence is the ability to interact adaptively with an environment.

A couple of key points implicit in this definition need to be emphasized. First of all, intelligence exists only where some form of interaction is taking place. The intelligent entity must be able somehow to perceive, and act in response to, what it encounters in its environment. We tend to think of intelligence as residing *between* perception and action, between stimulus and response; but intelligence cannot be found as something separated from capabilities for perceiving and acting. In fact, these capabilities must be seamlessly integrated with whatever *central* information processing faculty may exist. [e.g., Gilder 1989]

Secondly, in addition to interaction, intelligence requires a capacity for learning. Interacting *adaptively* means altering behaviors over time, based on experience, so that actions bring improving results (by whatever measure). The product of learning--knowledge--must be retained, and then must be changeable with further learning.

2.2. Specificity of Intelligence

A major source of confusion regarding intelligence is the illusion of *general* intelligence--seeing it as one sort of thing that can be measured on a linear scale. While useful in normal usage, the concepts of "higher" and "lower" intelligence are, on a deeper level, meaningless. [Gardner 1983, Mackintosh 1994] "Increasing" intelligence has meaning, but only with respect to a specific task or problem; it means driving relevant capabilities toward a better fit with the requirements of that particular problem, any measurement of which has no meaning for a different problem. So, essentially,

Intelligence = Fit

Relative intelligence reflects success in achieving ends (explicit or implicit), interacting adaptively with a particular environment or niche. Intelligence is very close in meaning to *adaptability,* but also factors in the time constraints of a particular problem.

3. Collective Intelligence

Consider again the ant colony: Its actions evidence intelligence, but *what* is it that is intelligent? We expect intelligence to be the property of a single entity and hesitate to ascribe it to a collection of creatures. Is the ant colony a 'thing' that can be intelligent? This is somewhat like asking which is the doll in a nested series of Russian dolls. Cell, ant, colony, species--meaningful units can be defined at several levels; and, there are arguments for, and problems with, considering any one of them as a stand-alone unit. At any level selected, units can be approached as unitary entities or as aggregations of sub-units. The ant colony is a collective organism, a "superorganism." [Seeley 1989] However, it still seems strange to say that it is intelligent. Doesn't the intelligence need somehow to be localized? How can it be distributed? A little more insight into the nature of intelligence helps to clear up this problem.

3.1. Intelligence is Collective

Most of our experience teaches that each of us is a separate entity, possessing a unitary intelligence--a single thing locked up inside our skulls. While this is a valid perspective, it is far from being complete or totally accurate. We are only beginning to learn how our brains work, but what we do know argues against a simple view of the brain functioning as a single thing.

> "The brain itself resulted from several hundreds of megayears of adapting to different environments, and now consists of hundreds of specialized sub-organs called brain

centers, each one linked by more or less specific connections to several others, to form a huge system of what clearly must be a myriad of mechanisms for synchronizing, managing, and arbitrating among different functions." [Minsky 1993; see also Minsky 1986 & Damasio 1992]

Of course, these individual components are woven together into an integrated whole, and their fates are completely intertwined (more obviously than is the case for ants). They are interdependent to an obvious extreme; but this should not obscure the fact that even the brain much more closely resembles a community than a unitary, machine-like unit operating 'logically.' What we are looking at, with a brain or an ant colony, is a *system* of intelligence. It appears that all forms of intelligence, on close inspection, are actually such collective systems of interacting, intercommunicating components. Intelligence is an emergent phenomenon in such systems--as seen in the ant colonies--something "that cannot be readily attributed to the individual behaviors of the components of that process." [Clancey 1994]

3.2. How Does It Work?

Human intelligence, artificial intelligence efforts, and ants: consideration of these reveals several common elements necessary for achieving intelligence:

- Multiple units, each having an independent ability to process information (generate a response to a stimulus)
- The capacity for gathering information, individually and/or jointly--the ability to perceive;
- Capabilities for acting on an environment;
- Capabilities for sending/receiving information;
- Communication channels, interconnections between the units;
- Redundancy of units and of problem solving methods/ experiments;
- A capacity for reinforcing/repeating successful actions;
- Compelling interdependence, the inadequacy of the individual unit, giving rise to:
- Common purpose, effective alignment of actions toward mutual benefit.

It appears that intelligence is achieved through a set of processes which, in fundamental terms, are the processes of evolution and natural selection. These processes seem to be sufficient for the implementation of intelligence, "regardless of the material which instantiates them." [Schull 1990]

4. Organizational Intelligence

Let's now explore these concepts further in the context of human organizations. What determines the intelligence an organization achieves? The above elements can be combined to construct a formula. It is not a rigorous mathematical equation, but is rather a *conceptual* formula showing the factors of organizational intelligence and how they interrelate. It provides a useful framework for exploring how organizational intelligence (OI) is achieved and how it can be enhanced:

OI = N x A x C x P

where: **OI** = the ***organizational intelligence*** achieved;
N = ***number*** of individuals interconnected to form the organization; (can also refer to sub-groupings)
A = ***ability,*** or capability, per individual;
C = effectiveness of ***communications;***
P = alignment and intensity of ***purpose.***

Value ranges for OI and its factors can be defined as follows:

OI: 0 = perfect stupidity: a complete lack of fit between capabilities and 'problem;'
1 = maximum intelligence: a perfect fit, an ideal capability to solve problem.
N: Can increase without limit; (but C and P--and, in some instances A--tend to drop exponentially with increasing N).
A: Has to do with perception, modeling and action, but is essentially about *knowledge*. *A* = the fractional (unique) knowledge possessed per individual.
0 = no knowledge, no information relevant to the problem
1 = complete knowledge
C: 0 = no communication
1 = ideal communication (number of interconnections and information flow)
P: 0 = no alignment
1 = complete alignment; 'life-or-death' motivation

Maximizing OI means *optimizing* its four factors. Some of the factors tend to work against each other, and a low value on one may offset higher values on the others. A balanced approach is required, optimizing for the best overall fit with requirements--minimizing the losses that an advance in one may create in another. Following are brief expansions of the factors' definitions.

N is the simplest of the factors: the number of individuals comprising the organization. It is the means by which sufficient knowledge is aggregated. All other things being equal (which they are not), increasing *N* enhances intelligence. As *N* increases, it tends to work against and dramatically drive down the other factors--especially *C* and *P*. (Of course there are cases where it works: markets offer examples of very large *N* yielding, for instance, unbeatable resource allocation capabilities.)

A is an obvious factor: the abilities of the individuals comprising the organization. The focus here is on skill rather than will (which comes later). At a fundamental level, *A* refers to capabilities for perception, processing (basically modeling of external reality), and action. Essentially, however, *A* is about *knowledge*. Aside from basic physical capabilities, humans' abilities derive from knowledge and, hence, are exceedingly diverse and plastic. Even our physical limitations are often overcome through extension or substitution by various technologies (all knowledge-based). Human intelligence confers great potential for acquiring and applying almost limitless varieties of knowledge. This makes *A* a highly variable and *somewhat* manageable factor. Counter-intuitively, optimization within the overall system is the issue, *not* simplistic maximization. More is not *always* better.

C--the effectiveness of **communications**--addresses the communication channels between individuals. *A* incorporates individuals' skills in using communication channels; *C* is about the channels themselves. These are the nerves of the (collective) organism, the means of interconnection between members of the organization. Considering how *C* interacts with the other factors, we can see that there tends to be an inverse correlation between *C* and *N:* the larger the organization, the more difficult it becomes to knit everyone effectively into a coherent unit. The 'value' for *C* is determined by the number, capacity and speed of communication channels. Again, however, more is not necessarily better. [Kelly 1994] The optimum interconnectivity (defined to be $C = 1$) depends on the capabilities of the individuals *(A)* to process the volume of information efficiently and productively.

P is the soft correlate of *C:* the alignment and intensity of **purpose** shared by the individuals comprising an organization. It is not about skills or technologies; it is about will. It reflects the alignment of individuals' objectives with the overall objectives of their organization, and the intensity of commitment to those objectives. Simply put, the more committed members of a group are to achieving common objectives (and the more committed they are to cooperative efforts to do so), the higher the value of *P* that is achieved. It is a factor recognized as fundamental and vital to an organization's success. [Pinchot 1993, Senge 1990, Katzenbach 1993, Bartlett 1994] As for *C,* there tends to be an inverse correlation between *P* and *N:* the larger and more complex the organization becomes, the more removed from the individual and abstract its purposes tend to become. With increasing size comes increasing potential, but also growing challenges in achieving high values for *P*.

5. Information Technology and Knowledge

So far, the focus has been mainly on the static picture, on the unchanging fundamentals. Consideration of current dynamics requires an analysis of the nature and impacts of information technologies (IT).

5.1. Impacts of IT

IT advances simultaneously alter organizations' environments and their intelligence potentials. New technologies find application in new 'tools;' new tools make possible new practices and products; these, in turn, lead to new companies, industries and markets--and to new structures of existing companies, industries and markets. These changes can be thought of as the transformation of existing economic landscapes, and the emergence of new ones. Business 'organisms' must adapt to this warping and reforming of existing fitness landscapes, and can colonize and search the new ones. [Todd 1993, Kelly 1994]

The potential dimensions of differentiation for business organizations are extended by IT advances. Considering these within the organizational intelligence context: Potentials for *C* are being drastically altered with the proliferation and rapid advance of communications technologies. Businesses can differentiate along this dimension or leverage it to achieve differentiation along another dimension, such as *N*. With the new technologies, larger values of *N* (at multiple levels) can be incorporated into effective systems of intelligence. And there are tremendous changes in potential for *A*, as it is essentially about knowledge. Rapidly expanding knowledge, coupled with new communications technologies, creates endless possibilities for innovation and differentiation. Clarity regarding the nature of *knowledge* helps to make this impact of IT more apparent.

5.2. What is Knowledge?

Information technologies are increasingly being used to handle information in its most complex and valuable form: encoded representations of knowledge. Witness the current focus on such topics as "the knowledge-based company" and "knowledge management," usually relying on advanced computer-based technologies. A question arises here: are these systems truly dealing with knowledge, or with more inert information? Knowledge is certainly much discussed, but seems to be only vaguely and inconsistently defined. As for intelligence, there is a nice smorgasbord of definitions. Clearly, knowledge is a complex concept having many facets, allowing for several meaningful but partial definitions; but some clarity can be achieved by considering the knowledge in our own brains.

First we have to move away from a common but misleading view of the brain as actually storing information in memory and operating logically. (This point developed in Damasio 1992, Edelman 1992, and Clancey 1991.) Consider how we solve a simple math problem. We 'know' that 2+2 = 4. When we solve this problem, it seems as if we simply 'retrieve' the answer from memory; but, in fact, we generate the response, "4," every time we give ourselves the input, "what is 2+2?" This input (stimulus) starts electrical impulses racing along neural circuits that end up generating "4" (response). It appears that all mental functioning works essentially this way, albeit on a staggeringly grand scale.

When we *learn* new things, when we add to or alter our knowledge, the neural circuitry of our brains physically changes: certain synaptic connections between neurons are strengthened or weakened, to varying degrees, or are eliminated entirely. [Shatz 1991] Learning is a process of altering the circuitry of the brain. It is not the storing of information in some passive matrix that is essentially unaffected by the experience. So, what is knowledge? A useful distinction between knowledge and information can be made with a concise working definition of knowledge based on our understanding of the brain:

Knowledge is the architecture of intelligence.

This is meant quite literally. Knowledge is the acquired design or physical layout of the brain's circuitry that makes possible the generation of specific responses to stimuli. It is the basis for perception, interpretation, choosing among possible responses, and execution of planned actions. As implied by the definition, knowledge and intelligence do not exist apart from one another. Knowledge is integral to intelligence while information is its feedstock and product.

5.3. Technologies of Mind

Let's use the above definition of knowledge to gain other insights into the impacts of IT. A key implication arises from the inseparability of knowledge and intelligence: If knowledge does not exist apart from intelligence, then information that is *not* acting as the architecture of intelligence is not knowledge. There is a continuum, of course, between obvious knowledge and inert information, but it is useful to segment a spectrum running from tacit knowledge, through the many forms of articulated knowledge, to information not incorporated in processing (which might be considered *potential* knowledge).

Figure 1 shows these categories of knowledge with an indicative scale of processing times. It also shows schematically the impacts of IT advances.

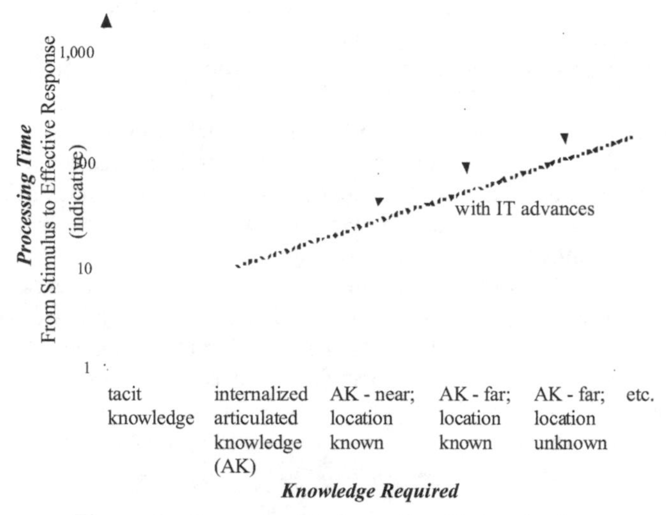

Figure 1. Impacts of Information Technologies

IT advances make possible the conversion of information into knowledge and effective integration of a much larger body of knowledge. Rapidly advancing technologies for storing, sharing, finding and retrieving representations of knowledge enlarge and accelerate the systems of cognition that can be achieved. These are dramatic changes in *A*, with commensurate changes in intelligence potential.

6. Implications for Strategic Management

So what does this mean in practice? Let's consider a few of the implications.

6.1. If Intelligence = Fit . . .

If intelligence = fit, then management of organizational capabilities means optimizing OI for a particular strategy, a chosen niche--optimizing across the OI factors to increase intelligence specific to a chosen 'problem.' The type and scale of organizational system to be nurtured is dictated by the relationship between capabilities and problem complexity.

The challenge is matching intelligence to specific *complexity*. The following analyses help to make needs and options more apparent.

6.2. Dimensions of Complexity

The complexity of the environment (nature of the 'problem') that organizations must deal with can be thought of as having two basic dimensions: the number of parameters that must be taken into account in decision-making, and the speed required in making decisions. The first is obvious, but its importance and difficulty are clearly functions of how much time is available for the task--which reveals the relevance of the second dimension.

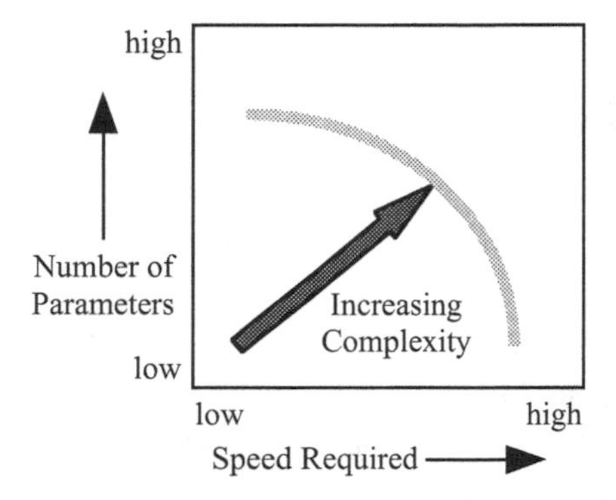

Figure 2. Effective complexity is a function of the number of parameters (span of knowledge) and speed requirements.

So, complexity comes in many varieties. The specific challenges, and the fundamental responses to those challenges, vary with the type of complexity facing the organization:

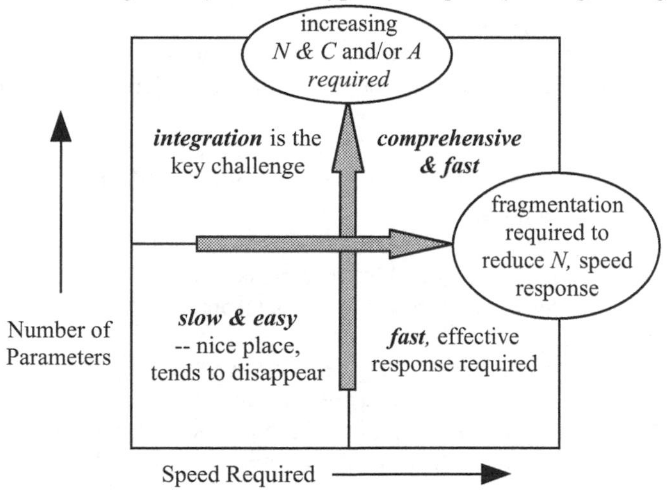

Figure 3. The nature of complexity defines the demands on organizations.

Of course, the appropriate organizational response will also vary. The first general response is to adapt to the demands of the organization's particular environment (current or targeted). Several structural responses are shown in Figure 4.

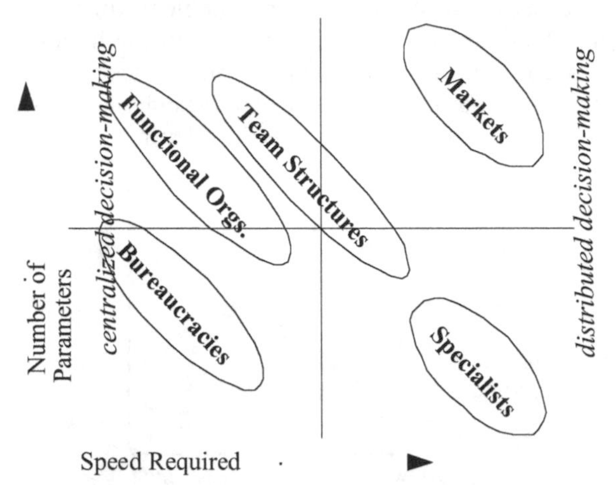

Figure 4. Organizational response.

Expanding knowledge is altering all business environments, shifting organizations to more complex environments and causing an automatic diversification over time. This helps to explain the general trends toward distributed decision-making and increasing specialization.

6.3. Errors of Level and Scale

Applying models of this type to business problems requires thinking broadly: at multiple levels and within, outside of, and across organizational boundaries. This can be seen in the following typical errors:

- **Applying model within constraints such as existing organizational boundaries;** this obscures many sound options, e.g.:
  - makes "*redundancy* of units . . . methods/ experiments" seem nonsensical, even when such increases in N may be required by growing complexity;
  - *externalization of redundancy* might be an appropriate response, but would not be considered;

    Example: consider the shift in much corporate R&D from in-house development to quick accessing of externally sourced technology. When growth in knowledge and blurring of industry boundaries necessitate larger N, such externalization can make affordable levels of redundancy that would otherwise be cost-prohibitive.

- **"One size fits all" prescriptions for maximum adaptability** (closely related to above). This prognosis is common in recent management literature. It may well be that at the larger system level highly distributed intelligence--high *N*--is increasingly required, but this is not the right response for *all* organizations. It is right at the overall *system* level, but this may be at the market/industry/joint venture level. Within that system there will be diverse opportunities for organizations to differentiate successfully--even by moving in the opposite direction.

6.4. Going Forward

With expanding knowledge and the resulting increases in environmental complexity, business organizations at all levels will be pushed simultaneously toward increased adaptability and narrower specialization. However, differentiation remains the first principle of business strategy. Within this overall migration of systems and structures, organizations will continue to exploit diverse types of organizational intelligence in pursuit of successful differentiation. The organizational intelligence model offers fundamental conceptual tools for considering just such forces and creative responses to them.

References

Bartlett, C.A., & Ghoshal, S., 1994, Changing the Role of Top Management: Beyond Strategy to Purpose, *Harvard Business Review*, **72** (79-88).

Butterworth, G., 1994, Infant Intelligence, in *What is Intelligence?*, edited by J. Khalfa, Cambridge University Press (Cambridge).

Calvin, W.H., 1994, The Emergence of Intelligence, *Scientific American*, **271** (100-107).

Clancey, W.J., 1991, "Israel Rosenfield, *The Invention of Memory: A New View of the Brain*," in *Contemplating Minds: A Forum for Artificial Intelligence*, edited by W.J. Clancey, S.W. Smoliar, & M.J. Stefik, The MIT Press (Cambridge, Mass.).

Clancey, W.J., Smoliar, S.W., & Stefik, M.J., 1994, Introduction to Part III (Architectures of Interaction), in *Contemplating Minds: A Forum for Artificial Intelligence,* The MIT Press (Cambridge Massachusetts).

Damasio, A.R., & Damasio, H., 1992, Brain and Language, *Scientific American*, **267** (89-95).

Dawkins, R., 1987, *The Blind Watchmaker*, W.W. Norton & Company, Inc. (New York).

Edelman, G.M., 1992, *Bright Air, Brilliant Fire: On the Matter of the Mind,* Basic Books (New York).

Florida, R., 1991, The New Industrial Revolution, in *Futures*, (559-576).

Franks, N.R., Deneubourg, J.L., & Goss, S., 1989, Collective Problem Solving, University of Bath, Research Proposal.

Gardner, H., 1983, *Frames of Mind*, Heinemann.

Gilder, G.F., 1989, *Microcosm: the quantum revolution in economics and technology*, Simon and Schuster (New York).

Gove, P.B. (ed.), 1986, *Webster's Third New International Dictionary,* Merriam-Webster Inc. (Springfield, Massachusetts).

Katzenbach, J.R., & Smith, D.K., 1993, *The Wisdom of Teams,* Harvard Bus. School Press (Boston).

Kelly, K., 1994, *Out of Control: the rise of neo-biological civilization*, Addison-Wesley.

Mackintosh, N, 1994, Intelligence in Evolution, in *What is Intelligence?,* edited by J. Khalfa, Cambridge University Press (Cambridge).

Minsky, M., 1986, *The Society of Mind*, Simon & Schuster (New York).

Minsky, M., 1993, "Allen Newell, *Unified Theories of Cognition*," in *Contemplating Minds: A Forum for Artificial Intelligence*, edited by W.J. Clancey, S.W. Smoliar, & M.J. Stefik, The MIT Press, (Cambridge, Massachusetts).

Morris, W. (ed.), 1975, *The American Heritage Dictionary of the English Language, New College Edition,* American Heritage Publishing Co., Inc. (New York).

Pinchot, G., & Pinchot, E., 1993, *The End of Bureaucracy & the Rise of the Intelligent Organization*, Berret-Koehler Publishers, Inc. (San Francisco).

Rothschild, M., 1990, *Bionomics: Economy as Ecosystem*, Henry Holt & Co., Inc. (New York).

Schank, R., & Birnbaum, L., 1994, Enhancing Intelligence, in *What is Intelligence?,* edited by J. Khalfa, Cambridge University Press (Cambridge).

Schull, J., 1990, Are Species Intelligent?, *Behavioral and Brain Sciences*, **13** (63-108).

Seeley, T.D., 1989, The Honey Bee Colony as a Superorganism, *American Scientist*, **77** (546-553).

Senge, P.M., 1990, The Leader's New Work: Building Learning Organizations, *Sloan Management Review*, (7-23).

Shatz, C.J., 1991, The Developing Brain, *Scientific American*, **267** (60-67).

Shimbel, A., 1975, Collective Intelligence, *General Systems*, **20** (205-208).

Sperber, D., 1994, Understanding Verbal Understanding, in *What is Intelligence?,* edited by J. Khalfa, Cambridge University Press (Cambridge).

Todd, P.M., 1993, "Stephanie Forrest, ed., *Emergent Computation: Self-Organizing, Collective, and Cooperative Phenomena in Natural and Artificial Computing Networks*," in *Contemplating Minds: A Forum for Artificial Intelligence,* edited by W.J. Clancey, S.W. Smoliar, & M.J. Stefik, The MIT Press (Cambridge, Mass.).

The Complex Evolution of International Orders and the Current International Transition

Robert M. Cutler
Institute of European and Russian Studies
Carleton University (Canada)
rmc@alum.mit.edu

1. Introduction

The history of modern international relations is composed of a succession of international orders. Each international order comprises a succession of international systems. The last international system of one international order is also, as a transitional phenomenon, the first international system of the next international order. The systems composing any order are all based upon the same norms, but they may manifest different structural configurations. These structures are animated by coordinative and collaborative tendencies [Stein 1982] that may or may not survive, contributing to the definition of the succeeding system. Notwithstanding March/Olsen's [1998] assertion of the "small-N" problem, inferences are possible about the logic of succession of international systems and orders. Categorical reasoning, using such structural concepts as unipolarity, bipolarity, and multipolarity [Kaplan 1966], is the vehicle for motivating generalizations that permit inferences on what we may expect in the twenty-first century.

2. A Comparative International Orders Approach

The Treaty of Westphalia effectively ended the Counter-Reformation, but it only codified changes in international law and practice that had been evolving for some time before [Krasner 1993]. Although the norms to which we usually refer as "Westphalian" were codified in 1648, there was no structurally unified state "system" even at that time. Rather, there were various geographically delimited systems, each pursuing its own dynamic in response to the interests and behavior of more or less local powers, but without reference to geographically more distant states in other regions which participated in separate systems. It is nevertheless possible to define a previous order, beginning from the early fifteenth-century Council of Constance, with which the analysis here presented is consistent [Figgis 1907]. However, length constraints on the present chapter do not permit its elaboration.

2.1. The "Crowned Society" Order

In ordinary language, the order inaugurated by the transitional system that lasted from the Treaty of Westphalia to the Treaty of Utrecht is frequently called the "Westphalian order." However, for present purposes it is more appropriate to refer to it as the Crowned Society Order, referring to its dominance by the royal sovereigns and the monarchist principle. This international order comprises three international systems.

The first Crowned Society system after Utrecht lasted until the launch of the democratic wars by the French Revolution in the early 1790s. (It is convenient to date the end of this system at 1794, the year of Robespierre's fall.) Collaboration was most evident through the Holy Alliance, implemented through family relations, economic ties, and a common domination of the ideological discourse. This was basically a unipolar, Franco-centric system. The successor system took shape with the definitive defeat of Napoleon and the convocation of the Congress of Vienna in 1815. The transitional period from 1794 to 1815 was characterized by coordination through the institutionalization of common norms of self-preservation in the Powers. Their numerous alliances against Napoleon were thus transformed afterwards.

The second Crowned Society system was the Concert of Europe, from 1815 until the mid-1850s. It broke up over the Crimean War, and specifically over the British decision to back Turkey against Russia. The resulting decline and chaos were resolved only with the unification of the German states, beginning in 1866. The second transition thus lasted from the mid-1850s until the mid- to late 1860s when it saw the emergence of the German State which would dominate the subsequent system.

The third Crowned Society system stretched from the late 1860s to the First World War. It oscillated between Bismarck-centered unipolarity and generalized multipolarity, finally breaking down into bipolarity after a mini-transition lasting from 1890 to 1894. This mini-transition separates "Bismarckian" from the "post-Bismarckian" moment within this single international system. The coordinative aspects of the Bismarckian moment of this system (from the late 1860s to 1890) repose in the policies of reversals of alliances in the search for the resolution of the security dilemma. Its collaborative aspects are in the collaboration between the industrialists (and, variously, sometimes the aristocracy) plus the military within each state; and in the collaboration, at least implicit, among states at the level of their military apparatuses and the headquarters of these.

The second (post-Bismarckian) moment lasted from 1894 until 1914. The coordinative aspects of the post-Bismarckian moment of this third, Crowned Society system are represented by collective security. These carried over after the end of the First World War, to animate the concert of the victorious Powers against the vanquished, in the League of Nations. The collaborative aspects of the post-Bismarckian moment of the third Crowned Society system (and its subsequent transition) are those of the Holy Alliance, under another guise: after the First World War, they were transformed into a coalition of industrialized republics to the exclusion of Germany and the Soviet Union.

2.2. The "International Society" Order: The Long or Short Twentieth Century?

The breakdown of the Crowned Society order into bipolarity in the two decades preceding the First World War prefigured the bipolarity of the "International Society" order. The latter began in the early 1920s, marked notably by the beginning of the end of the British Empire through the London Conference of 1925. It is unclear whether the present international transition, which began in 1989/1991, marks the end of the International Society order (the "Short Twentieth Century") or the transition to another international system within that order (the "Long Twentieth Century"). If the former, then we are entering a new order that will be characterized by a tension between unipolarity and multipolarity over time across its constituent systems; if the latter, then we are entering another mainly bipolar international system within the same order.

The first international system of the ("International Society") Twentieth Century order—whether Long or Short—is the Interwar System from the early 1920s to 1941. The coordinative aspects of the system are represented in the military coalition against the Axis powers. The collaborative aspects emerge in the creation of the U.N. on the basis of the League of Nations, plus an ideological collaboration on two sides.

The second international system of the ("International Society") Twentieth Century order—whether Long or Short—is the Cold War system, from 1946/47 to 1991. It is possible that the years 1974/75–1979/80 mark a mini-transition between the two moments of the Cold War system. (Their significance is that they mark the decline and fall of Soviet-American détente, from Angola to Afghanistan. The biennium 1973/74 also marks the oil embargo that irrevocably changed post-1945 international politics and economics.) If so, then the years 1946/47–1973/74 represent system's the Tight Bipolar moment, and the years 1979/80–1991 represent its Loose Bipolar moment.

Such a mini-transition, and the years following, are susceptible to two interpretations. They may introduce a new international order, as did years 1894–1914 after the mini-transition within the third international system of the Crowned Society order. If this is so, then just as unipolar/multipolar tension degenerated into bipolarity, we may suppose that Cold War bipolarity "degenerated" into what may be called Multilateral Interdependence towards the end of the twentieth century. In this case, the coordinative and/or collaborative aspects of Multilateral Independence are what will carry over into the next international order, which we may call the "World Society" order.

2.3. Regularities in System Transformation

Regularities are evident from the foregoing analytical review of the succession of international orders and international systems within those orders. The first Crowned Society system lasted 81 years and was followed by a transition of 21 years. The second Crowned Society system lasted 39 years and was followed by a transition of 12 years. The third Crowned Society system lasted about 48 years, divided into two moments of equal length by an interim mini-transition of four years, and was followed by a transition of 11 years. The first Twentieth Century system lasted 16 years and was followed by a transition of five years. The second Twentieth Century system lasted 45 years, divided into two moments of unequal length by an interim mini-transition lasting four years.

Two regularities may be observed. The first is that the length of an international transition is roughly one-quarter the length of the international system it succeeds. On this basis, it is possible to conclude that the present international transition, which started in 1991, will end during the first half of the first decade of the twenty-first century.

The second regularity is that the last international system of each international order is split into two "moments" by an interim mini-transition that is about one-quarter the length of the first moment. Of those two "moments," the second contains the seeds of the normative essence of the succeeding international order.

The further significance of these generalizations will become clear from a discussion of the differing analytical significance of the disintegration of the Soviet bloc in East Central Europe *versus* the abolition of the Soviet Union itself.

3. International Orders and Structural Transformation

3.1. Summary of the Historical Record

The Crowned Society order is characterized by the "classical" balance of power [Gulick 1955]. The principal characteristic of the Crowned Society order is thus the tension between unipolarity and multipolarity, and moreover the frequent shifting of alliances depending upon the exigencies of the moment. The breakdown of the tension between German unipolarity generalized multipolarity into bipolarity, after a mini-transition lasting from 1890 to 1894, heralded the end of this order.

The International Society order is characterized by bipolarity: between the status quo and revisionist powers under the Interwar System, before the Second World War; and between the two superpowers and their blocs under the Cold War system, after the Second

World War. However, as described above, it may be conceived that the Cold War system of the International Society order is, like the last international system of the Crowned Society order, divided into two moments by the second half of the 1970s.

If the years afterwards, up until the end of the Cold War system in 1991, are designated to be a separate and multilateral "moment" of that system—say, for example, the Multilateral Interdependence moment—then this represents the breakdown of Short Twentieth Century bipolarity and the transition to a new international order—let us call it the World Society order—that will be characterized by the tension between multipolarity and unipolarity. However, if the current international transition, which began in 1991, inaugurates only another bilateral system, then there is no new international order and only a Long Twentieth Century.

3.2. Inferences from the Historical Record

The Concert of Europe hid an ideological (normative) opposition between the republican and monarchical principles. This later animated, without determining in all its details, the principal structural basis for the geopolitical bipolarization that led to the First World War. If a similar pattern is followed today, then the contemporary postmodernization of the Enlightenment will become the basis for a system-wide ideological bipolarization that will in turn, after the middle of the twenty-first century, relegate the now-emerging multipolarity to secondary status as a characteristic of the system as a whole.

Some have argued, for example, that the United States seems to be adapting newly proposed international norms—e.g., "the law of humanitarian intervention in civil conflict"—to its own particular influence-projection interests. Chechnya and Tatarstan in Russia, and Tibet and Uighuristan (Xinjiang) in China, provide domestic political-control explanations why Russia and China oppose this new norm. To oppose a new normative basis is to seek to conserve the old one, which is expressed through the bipolarity of the Twentieth Century order. We would then have the interesting development, at first glance paradoxical, of the status quo power in the new system—the United States—becoming the innovator of norms for that international order.

So what about the emerging "World Society" international order, if there is one? By previous reasoning, its first international system would be characterized by a tension between multipolarity and unipolarity. The consensus of a wide variety of "long-cycle" and "world-systems" research in political science, all with different assumptions, that a system-wide struggle over the structure of the international system will occur—whether peaceful or not—around 2030–2050, supports this analysis [Denemark 1999]. The next section reveals that, under closer examination, a transformed bipolar structure may also be maintained, with at least one of the poles being an emerging international actor not yet evident on the world stage.

4. Towards a "World Society" Order?: 1989 *versus* 1991

The key to understanding the nature of the current international transition is to distinguish between what happened in 1989 (the disintegration of the Soviet bloc in East Central

Europe) and what happened in 1991 (the abolition of the Soviet Union). These separate events motivate mutually distinct interpretations of the nature of the current international transition. The 1989 events mark the end of what we may call a Short Twentieth Century. Under this interpretation, the current international transition is a transition to a normatively new international order that will manifest as a succession of international systems, all animated by the tension between unipolarity and multipolarity. The 1991 events, on the other hand, mark the end of current transition as a bridge to another bipolar system within an always bipolar Long Twentieth Century.

4.1. The Significance of 1989

If 1989 marks the end of the International Society order (as the Short Twentieth Century), then it means the end of bipolarity as an organizing principle of the international order. That in turn would mean that the bipolarity of the Cold War was not European in its origin. Rather, under this hypothesis, the division of Europe by the Cold War manifested something more fundamental, *viz*., the projection into Europe of the ideological and great-power confrontation between the U.S. and the Soviet Union also to be found in the developing world. This ideological conflict began to disappear even before 1989 with the doctrinal innovations introduced by Gorbachev.

From this perspective, we have today a unipolar, U.S.-centered transition to a multipolar system. If the unipolar transition initiated by Bismarck in 1866 is taken as a model, leading to the multipolar system that ended resolved into a bipolar conflict between opposing blocs before culminating in war in 1914, then we might explore the hypothesis that the multipolar system to which the current unipolar transition may lead, will likewise resolve into a bipolar conflict between blocs.

Under this interpretation, the unification of Germany under Bismarck appears as a nineteenth-century European-level analogue to today's global-level unification of an extended "Atlantic Civilization" (pan-Europe and North America) under NATO and the European Union as a joint framework. However, the analogy goes still deeper. Following Bismarck's passage from the scene in 1890, the international structure reverted to an "updated" Holy Alliance in the form of the Three Emperors' League. Analogously, we have today a reversion to an "updated" NATO through the adjoinment of the Partnership for Peace.

Moreover, the East European former members of the Warsaw Treaty Organization are now nearly all members of NATO, ready to become NATO members or *de facto* NATO protectorates (e.g., Albania and the successors to Yugoslavia). If the superstructure of the Holy Alliance was an ineffectual Quadruple Alliance, today an ineffectual OSCE stands as superstructure to the new NATO/EU composite framework. This reasoning motivates the inference that the current international transition is a transition to a new international order, and not just to a new system within an existing order. If that is the case, then historical logic compels the conclusion that this order should be characterized by a succession of international systems animated by the tension between unipolarity and multipolarity. Under such a conclusion, the unipolar Power is the United States.

Historical logic likewise compels the conclusion that this tension between unipolarity and multipolarity will ultimately resolve into a bipolar conflict between opposing blocs. One of these will be the composite NATO/EU bloc on a global geostrategic level.

4.2. The Significance of 1991

It is possible that the years 1979/80–1991 (representing the Loose Bipolar moment of the second International Society system) do not catalyze a fundamental change in the International Society order's bipolar dynamic, which is merely transformed. The sense in which a new bipolarity comes from 1991, arises from the contemporary turning of Russian public opinion against the West, and of its political elites towards Asia.

Here, too, there is an historical parallel. After losing Crimea in the mid-nineteenth century, Russia turned its attention to Poland, the Pacific coast (now meaning and including China as well as Japan) plus south and southwest Asia, and the Balkans. That is the same pattern as today, with Poland being replaced by Poland/Lithuania/Baltics. The orientations are regionalized but the patterns are the same. This reversion to an old pattern creates the basis for continued bipolarity in Europe, if we recall that geographers have always defined Europe as extending to the Urals. Current tendencies in Russian foreign policy therefore provide a foothold on the Continent for a geopolitical pole opposing Euro-Atlantic community on the global scale.

The present paper reveals that over recent centuries, the geometric mean of the lifetime of any international system is 40 years, and that the length of the international transition following the death of any international system is about one-quarter the length of that system's lifetime. This finding would project a new realignment to take place towards the middle of the twenty-first century [Denemark 1999].

5. Conclusion

Complexity science dictates the need for multi-scale analysis. It therefore offers the insight that both the 1989–based and the 1991–based interpretations may be simultaneously valid, because it posits that the realities represented by these interpretations are not coterminous but instead interact and intersect at multiple levels. Coemergent multiple realities are natural to complex systems.

Without necessarily endorsing Huntington's [1996] "clash of civilizations" thesis, it is possible to uphold the idea that one geopolitical/geocultural) bloc now consolidating itself is Euro-Atlantic. In historical perspective, this could even be called the Renaissance/Enlightenment bloc. The strongest candidate at present to play the role of counterweight is an Asian bloc. Rising Asian powers have already offered an "Asian model" of political and economic development. This "Asian model" may have an Islamic cultural-normative component, yet without Islamic scholars' preservation of ancient Greek texts in translation, there would have been no Renaissance. Indeed, a non-Islamic Asian (e.g., Chinese) model is also possible.

Such a now-emergent mesolevel Asiatic geocultural unification could be the catalyst, in the middle of the twenty-first century, for a system-wide restructuring that would eventually reduce the now-emerging multipolarity into a more bipolar framework. In fact, this unification need not even be political like Bismarck's of Germany: it may be transnational, particularly insofar as all these trends really cross-cut through individual societies. If the major blocs in the twenty-first century will be geocultural, then the centrality of Central and Southwest Asia becomes evident. The significance of Iran is especially heightened. Uzbekistan emerges as a potential a flash-point in the early twenty-first century, probably around 2020–2025, principally for demographic and economic reasons [Spruyt 1994]. That may be the spark that leads to the more general realignment—i.e., period of system transformation—that other work projects for 2030–2050. This work independently places that realignment to begin in the early 2040s and estimates its duration at 10 years, in conformance with the "one-quarter rule."

References

Denemark, R.A., 1999, World System History: From Traditional International Politics to the Study of Global Relations, in *International Studies Review*, Blackwell (Malden), **1**, 2.

Figgis, J.N., 1907, *Political Thought from Gerson to Grotius, 1414–1625*, Cambridge University Press (Cambridge).

Gulick, E.V., 1955, *Europe's Classical Balance of Power*, Cornell University Press (Ithaca).

Huntington, S.P., 1996, *The Clash of Civilizations*, Simon & Schuster (New York).

Kaplan, M.A., 1966. Some Problems of International Systems Research, in *International Political Communities: An Anthology*, Doubleday & Co. (New York).

Krasner, S., 1993, Westphalia and All That, in *Ideas and Foreign Policy*, edited by J. Goldstein and R.O. Keohane, Cornell University Press (Ithaca).

March, J.G., & Olsen, J.P., 1998, The Institutional Dynamics of International Orders, in *International Organization*, MIT Press (Cambridge), **52**, 4.

Spruyt, H., 1994, *The Sovereign State and its Competitors*, Princeton University Press (Princeton).

Stein, A., 1982, Coordination and Collaboration: Regimes in an Anarchic World, in *International Organization*, MIT Press (Cambridge), **36**, 2.

Cooperation, nonlinear dynamics and the levels of selection

Leticia Avilés
Department of Ecology and Evolutionary Biology
University of Arizona
Tucson, Az 85721
laviles@u.arizona.edu

1. Introduction

The history of life is the history of the association of lower level units into higher levels of organization [Buss 1987; Maynard Smith and Szathmáry 1995; Michod 1997; Frank 1997]. Genes became associated in chromosomes, prokaryotic cells in eukaryotic cells, single-celled organisms in multicellular organisms, and multicellular organisms into social groups. In understanding these evolutionary transitions two questions need to be addressed: What caused the lower level units to become associated into a higher level entity? And then, once the associations were under way, whether and, if so, under what circumstances was there a transition from selection acting primarily on the lower level units to selection acting primarily on the higher level entities? Using insights gained through the study of a particular group of social organisms —the cooperative spiders (see below)— I will explore these questions as they pertain primarily to the association of individuals into groups. The issues raised, however, should be relevant to other transitions involving conspecific associations.

In the following sections, I will first consider the possibility that intercolony selection may be an effective evolutionary force in the cooperative spiders. I will do so by discussing the evolution of their highly female-biased sex ratios within the context of a multilevel selection model. This model illustrates the interplay between population structure and group turnover in shifting selection from individuals to groups. I will then discuss empirical and theoretical evidence that cooperation may be responsible for giving rise to sufficiently cohesive groups on which selection may act. I will argue that the nonlinear effects of cooperation on individual fitness may have important consequences on the existence, persistence, and dynamics of social groups. Portions of the ideas discussed here have been presented in greater detail elsewhere [Avilés 1986, 1993, 1997, 1999; Avilés and Tufiño 1998], although outside of the common framework being considered here.

2. The non-territorial permanent social or cooperative spiders

The cooperative spiders are a phylogenetically diverse set of primarily tropical species. In the 17 known non-territorial permanent social or "cooperative" spiders, groups of related individuals occupy communal nests in which colony members cooperate in nest maintenance and repair, prey capture, feeding, and brood care [reviewed in Avilés 1997]. In a situation somewhat unusual for a majority of highly social organisms, the cooperative spider colonies constitute not only social groups, but also relatively closed and self-sustaining populations. This situation arises because spiders of both sexes —who are all potentially reproductive— remain within their natal nest to mate generation after generation [reviewed in Riechert and Roeloffs 1993; Avilés 1997]. Through this process of intracolonial mating, the colonies grow for a number of generations until they either proliferate or become extinct. Colonies that reach a relatively large size give rise to daughter colonies by budding, swarming, or the production of small propagules. There is apparently no mixing among colony lineages during this founding stage or at any other stage of a colony's life cycle [Riechert and Roeloffs 1993; Smith and Engel 1994; Smith and Hagen 1996; Avilés 1997]. The cooperative spiders, therefore, constitute an extreme form of a metapopulation system where separate colony lineages grow, proliferate, and become extinct without mixing with one another. As I will argue next, this population structure is an almost word by word match to even the most stringent conditions suggested by some models [e. g. Leigh 1983] to be necessary for interdemic selection to prevail over counteracting individual selection within demes (note that in the cooperative spiders a colony is also a deme).

3. Sex ratio and the levels of selection

The possibility that intercolony selection might be an effective evolutionary force in the cooperative spiders was first hinted by the highly female-biased sex ratios present in their colonies. These sex ratios, which are of the order of 4 to 40 females per male, have been confirmed in several instances to represent an overproduction of female embryos [Avilés and Maddison 1991; Rowell and Main 1992]. I have suggested that this situation, which is in clear violation of the evolutionary principle of an equal investment in males and females [Fisher 1930], may be the result of intercolony selection for faster rates of colony growth and proliferation [Avilés 1986, 1993].

I investigated this hypothesis by means of computer simulations [Avilés 1993] in which evolving diploid sexual organisms were represented by two binary strings (16 loci) coding for the proportion of males they produced among their progeny. This polygenic system of inheritance allowed for processes that mimicked recombination, meiosis, and the formation of zygotes. Also, as in a natural system, population subdivision and drift had the effect of increasing homozygosity at the various loci. The individuals mated panmictically within colonies, but the colonies were isolated from one another to different extents. As in the spiders, there was no general mixing of colony lineages at any stage in their life cycle. New colonies originated by the proliferation of existing colonies that had grown above a certain threshold size. A fixed fraction of randomly chosen colonies were removed (i. e. went "extinct") every generation to make room for newly-founded ones. The rate of migration among colonies (introduced here at 0-5% migrants per colony per generation), their size at foundation (2 - 10 adult inseminated females or half the individuals in the maternal colony), and their rate of turnover (10 - 25% colonies per generation) were varied as parameters of the simulations. The following conclusions were drawn from this modeling work [Avilés 1993]:

(i) Selection both among- and within-groups was involved in determining the overall equilibrium sex ratio. Within the colonies, the sex ratios became less biased every generation as the colonies aged. However, an overall bias was generated or maintained if, and only if, larger colonies were more likely to proliferate. The latter condition created an advantage for faster-growing, more female-biased colonies. In the absence of this group-level selective advantage, female-biased sex ratios did not evolve despite strong population subdivision and large rates of group turnover. The global equilibrium was, thus, dynamic and a compromise between the values favored by selection acting at the two levels.

(ii) More female-biased sex ratios evolved under lower migration rates, smaller propagule sizes, and higher rates of turnover. Propagule size and migration rate determined the amount of genetic variance present among and within groups and, thus, the *strength* of selection acting at each of these levels (see Fisher 1930; Slatkin and Wade 1978; Falconer 1981, Ch. 13; Slatkin 1981). Smaller values of the two parameters resulted in lower amounts of within and greater amounts of between-group variance. The group turnover rate, on the other hand, affected the *opportunity* for selection to act at each of the levels. Faster rates of group turnover reduced the time available for within-group selection to act while providing more frequent opportunities for between-group selection.

Unlike earlier models for the evolution of sex ratios in subdivided populations [reviewed in Antolin 1993], this model separated gene flow among existing groups from lineage mixing during the founding stage. In doing so, it showed that a period of general mixing of groups [as in the haystack or trait-group selection models, Bulmer and Taylor 1980; Wilson and Colwell 1981] is *unnecessary* for selection among more or less isolated lineages to counteract opposing individual selection within them. The model also highlighted the fact that the outcome of a multilevel selection process depends on two distinct and independent properties of the entities at each of the levels: the amount of genetic variance available among them and their relative rates of turnover. As already mentioned, these two properties determine, respectively, the strength of and opportunity for selection at each of the levels.

In a more general sense, these results illustrate properties that are common to multilevel selection processes, regardless of the type of traits or levels involved. The term "more-female biased sex ratio" in the conclusions laid out above can be substituted by "lower levels of virulence", "greater degree of altruism", "lower transposition or replication rate", "lesser extent of meiotic drive", etc. Likewise, the levels involved could be transposons within genomes, organelles within cells, cells within bodies, parasites within hosts, individuals within social groups, trait groups, or populations, or the latter within populations, metapopulations, or species. Finally, given the hierarchical organization of life, more than two levels will often be laid upon each other. In all cases, the outcome of the multilevel selection process will depend on how the particular idiosyncrasies of the various systems shape the relative amounts of genetic variance and the rates of turnover of the entities at each of the levels.

In our empirical case at hand, the question is whether the cooperative spiders fulfill the conditions of strong population subdivision and significant rates of group turnover that the simulations suggest would be necessary for their highly female-biased sex ratios to evolve. The short answer, when such information has been gathered, is "Yes, they do." Estimated rates of colony turnover range from 20 to close to 70% of the colonies per generation [reviewed in Avilés 1997]. These rates are more than enough to give the opportunity for intercolony selection to act. The spiders also exhibit extraordinarily low levels of within-group variance and higher levels of between-group variance, as demonstrated by allozyme studies in four different species [Riechert and Roeloffs 1993; Smith and Engel 1994; Smith and Hagen 1996]. Estimated values of Wright's fixation index —a measure of population subdivision [Wright 1951; Weir and Cockerham 1984]— in these four species are on the order of 0.8-0.9, with 1.0 being the maximum possible degree of population subdivision.

4. Individual fitness effects of cooperation and the origin and dynamics of social groups

Given solitary or periodic-social ancestral species [Krafft 1979; Buskirk 1981], the existence in the cooperative spiders of group-level selection capable of overriding counteracting individual selection within the groups implies, at least for the sex ratio trait, a switch in the primary focus of selection from the individual to the group. The next question is: What led to the formation of groups with an amount of cohesiveness and rate of turnover that made this transition possible? Empirical and theoretical work that I will describe next suggests that cooperation and its synergistic effects on individual fitness may be responsible both for the formation of groups and their emergence as units on which selection may act.

By directly measuring the average lifetime reproductive success of females in colonies of different sizes, Avilés and Tufiño [1998] found that individuals of a neotropical cooperative spider gained a significant fitness advantage by living in groups. Individuals in colonies with 60 - 100 spiders had more than twice the average lifetime reproductive success than solitary conspecifics in the same environment. This effect was primarily the result of a higher probability of offspring survival as the size of the colonies increased. The probability of female reproduction within the colonies, however, decreased with colony size. As a result, individual fitness was maximum at intermediate colony sizes. I would argue that a unimodal fitness function explains why groups form and why they remain cohesive while small to intermediate-sized, but disintegrate or bud-off new colonies once they have grown to a size at which the benefits of group-living no longer outweigh its costs [Avilés and Tufiño 1998]. A similar relationship between colony size and individual fitness has been found in other social organisms [e. g. Itô 1987; Raffa and Berryman 1987; Heinsohn 1992; Cash et al. 1993; Wiklund and Andersson 1994] and I expect it to be fairly general provided that all phases in the life cycle of individuals and groups are accounted for when individual fitness functions are estimated.

Besides cohesiveness, the occurrence of selection among the isolated breeding units requires a certain degree of turnover of those units. In a model to appear in Evol. Ecol. Res. [Avilés 1999], I show that the nonlinear effects of cooperation on individual fitness may not only be responsible for the existence of social groups, but may also contribute to their dynamical instability. This model —a difference equation representing the growth of a social group— contains positive and negative density-dependent factors representing, respectively, the individual fitness effects of cooperation and competition within the groups. One of the model's predictions is that group-living and cooperation may allow the colonization of environments or ecological niches in which solitary individuals would be unable to replace themselves. A second prediction involves the dynamics of the social groups beyond this absence-persistence boundary. In this region, group dynamics are expected to switch from point equilibria to cycles or chaos with increasing values of the cooperation parameter. This effect is the result of increased rates of growth due to cooperation, combined with negative density-dependence, scramble competition, and discrete generations [see May 1974 or a recent review by Hastings et al. 1993, for comparable models with negative-density dependence only]. A third prediction is the existence of a second region of deterministic extinction at large values of the three parameters of the model. This upper extinction region arises when extreme dynamical instability leads to crashes below an intermediate unstable equilibrium which is a novel feature of this model. The existence of this second extinction region suggests strong selection near the chaos-extinction boundary to prevent the evolution of extreme parameter values (or to modify growth patterns in violation of the assumptions of discrete generations or scramble competition). Because in the chaotic region the groups are expected to be strongly cohesive and subject to large rates of turnover, selection near the chaos-extinction boundary is expected to take the form of intercolony selection.

In support of the first prediction, there are numerous examples in which sociality has apparently allowed the colonization of environments where only groups can access or defend large but patchy resources or cope with high predation or competition rates [e.g. Heinsohn 1992; Jarvis et al. 1994; see also Emlen 1991]. A case in which the second prediction may be met are some cooperative spider species that, under more plentiful conditions, may have evolved towards the regions of intrinsic dynamical instability [Avilés 1997, 1999, in prep.]. There is empirical evidence that the colonies of these species meet the conditions of discrete generations, scramble competition, and density-dependence required by the above described model [e.g. Avilés 1986; Avilés and Tufiño 1998]. Evidence of the occurrence of colony size oscillations and of a reproductive output consistent with intrinsic dynamical instability in two neotropical cooperative spiders will be presented elsewhere [Avilés in prep.; Avilés and P. Salazar in prep.].

By causing individuals to become dependent on one another under harsh environmental conditions, cooperation may start to switch the primary focus of selection from the individual to the group. This may initiate a positive feedback loop (cooperation -> increasing interdependence -> intercolony selection -> greater cooperation -> dynamical instability -> greater colony turnover -> stronger intercolony selection) that in some cases may have led to the evolution of highly integrated and complex social systems. Once selection at the higher level takes over as the primary evolutionary force, we can expect the evolution of traits that suppress conflict among the lower level entities and give the groups superorganismic qualities. Cooperation, with its nonlinear effects on individual fitness and its dynamical effects on social groups, may thus create a new simplicity in which selection is concentrated on what could now be considered a higher level of organization.

References

Antolin, M. F., 1993, Genetics of biased sex ratios in subdivided populations: Models, assumptions, and evidence, in *Oxford Surveys in Evolutionary Biology*, edited by D. Futuyma & J. Antonovics, Oxford University Press (Oxford).

Avilés, L., 1986, Sex-ratio bias and possible group selection in the social spider *Anelosimus eximius*, *American Naturalist* **128**, 1.

Avilés, L., 1993, Interdemic selection and the sex ratio: A social spider perspective, *American Naturalist* **142**, 320.

Avilés, L., 1997, Causes and consequences of cooperation and permanent sociality in spiders, in *The Evolution of Social Behavior in Insects and Arachnids*, edited by J. Choe & B. Crespi, Cambridge University Press (Cambridge), 476.

Avilés, L. & Tufiño, P., 1998, Colony size and individual fitness in the social spider *Anelosimus eximius*, American Naturalist, **152**, 403.

Avilés, L., 1999, Cooperation and non-linear dynamics: An ecological perspective on the evolution of sociality, *Evolutionary Ecology Research* **1**: 459.

Bulmer, M. G. & Taylor, P. D., 1980, Sex ratio under the haystack model, *Journal of Theoretical Biology* **86**, 83.

Buskirk, R. E., 1981, Sociality in the Arachnida, in *Social Insects* edited by H. R. Hermann, Academic Press (New York), 281.

Buss, L., 1987, *The Evolution of Individuality*, Princeton University Press (Princeton, NJ.).

Cash, K., McKee, M. & Wrona, F., 1993, Short- and long-term consequences of grouping and group foraging in the free-living flatworm *Dugesia tigrina*. *Journal of Animal Ecology* **62**, 529.

Emlen, S. T., 1991, Cooperative breeding in birds and mammals, in *Behavioral Ecology An Evolutionary Approach* edited by J. R. Krebs & N. B. Davis, Blackwell Scientific Publications (Oxford), 305.

Falconer, D. S., 1981, *Introduction to Quantitative Genetics*, 2nd. ed., Longman (New York).

Fisher, R. A., 1930, *The genetical theory of natural selection*, Dover (New York).

Frank, S., 1997, Models of symbiosis, *American Naturalist*, **150**, S80.

Hastings, A., Hom, C. L.,Ellner, S., Turchin, P. & Godfrey, H.C.J., 1993, Chaos in ecology: is mother nature a strange attractor?, *Annual Review of Ecology and Systematics* **24**, 1.

Heinsohn, R. G., 1992, Cooperative enhancement of reproductive success in white-winged choughs, *Evolutionary Ecology*. **6**, 97.

Itô, Y., 1987, Role of pleometrosis in the evolution of eusociality in wasps, in *Animal Sociality: Theories and Facts* , edited by Y. Ito, J. L. Brown, & J. Kikkawa, Japan Sci. Soc. Press (Tokyo), 17.

Jarvis, J. U. M., O'Riain, M. J., Bennett, N. C. & Sherman, P. W., 1994, Mammalian eusociality: A family affair, *Trends in Ecology and Evolution*. **9**, 47.

Krafft, B., 1979, Organisation et évolution des sociétés d'araignées, *Psychology* **1**, 23.

Leigh, E. G., Jr., 1983, When does the good of the group override the advantage of the individual? *Proceedings of the National Academy of Sciences of the USA* **80**, 2985.

May, R. M., 1974, Biological populations with non-overlapping generations: stable points, stable cycles, and chaos. *Science* **186**, 645.

Maynard Smith, J. & Szathmáry, E., 1995, *The Major Transitions in Evolution*, W. H. Freeman (Oxford).

Michod, R., 1997, Evolution of the individual, *American Naturalist*, 150, S5.

Raffa, K. F., & Berryman, A. A., 1987, Interacting selective pressures in conifer-bark beetle systems: A basis for reciprocal adaptations? *American Naturalist*. **129**, 234.

Riechert, S. E. & Roeloffs, R. M., 1993, Evidence for and consequences of inbreeding in the cooperative spiders, in *The Natural History of Inbreeding and Outbreeding* edited by N. W. Thornhill, The University of Chicago Press (Chicago and London), 283.

Rowell, D. M. & Main, B. Y., 1992, Sex ratio in the social spider *Diaea socialis* (Araneae: Thomisidae), *Journal of Arachnology* **20**, 200.

Sober, E. & Wilson, D. S., 1998, *Unto Others The Evolution and Psychology of Unselfish Behavior*, Harvard University Press (Cambridge, Ma.).

Slatkin, M, 1981, Populational heritability, *Evolution* **35**, 859.

Slatkin, M. & Wade, M. H., 1978, Group selection on a quantitative character, *Proceedings of the National Academy of Sciences, USA* **75**, 3531.

Smith, D. R. & Engel, M. S., 1994, Population structure in an Indian cooperative spider, *Stegodyphus sarasinorum* Karsch (Eresidae), *Journal of Arachnology* **22**, 108.

Smith, D. R. & Hagen, R. H., 1996, Population structure and interdemic selection in the cooperative spider *Anelosimus eximius* (Araneae: Theridiidae), *Journal of Evolutionary Biology* **9**, 589.

Weir, B. S. & Cockerham, C. C., 1984, Estimating F-statistics for the analysis of population structure, *Evolution*, **38**, 1358.

Wiklund, C. G., & Andersson, M., 1994, Natural selection of colony size in a passerine bird, Journal of Animal Ecology, 63, 765.

Wilson, D. S. & Colwell, R. K., 1981, The evolution of sex ratio in structured demes, *Evolution*, **35**, 882.

Wright, S., 1951, The genetical structure of populations, *Annals of Eugenics* **15**, 323.

Examining the dynamics of a vector representing neurophysiological state during intensive care

André J. W. van der Kouwe, Richard C. Burgess

The Ohio State University, Columbus, Ohio and Division of Neurological Computing, The Cleveland Clinic Foundation, Cleveland, Ohio

Several monitors are used to monitor vital physiological variables in stroke patients in the neurointensive care unit. Decisions which ultimately affect the patient's outcome must be made on the basis of these and other data such as medical images. A continuous bedside neurological monitor could give a timely indication of changes in the patient's neurological condition. This research represents the development of such a monitoring system. The system combines several physiological and electroneurophysiological parameters into a neurophysiological state vector. Such a state vector representing physiological condition was first described by John Siegel et al. [5]. Goldberger demonstrated that a loss of heart rate variability over time corresponds with a state of pathology and in general that a loss of complexity of the dynamics of the associated physiological parameters typifies several diverse disease states [3]. In this work, the dynamics of a time-varying neurophysiological state vector representing the condition of stroke patients in the neurointensive care unit are examined. The neurophysiological state vector consists of parameters derived from the latencies and amplitudes of significant peaks in the brainstem auditory and somatosensory evoked potentials recorded at regular intervals together with spectral information derived from the electroencephalogram and parameters routinely monitored in the intensive care unit, including heart rate and oxygen saturation.

1 Introduction

When a stroke patient is transferred to the neurointensive care unit, various bedside monitors are used to keep track of the patient's condition. A comprehensive cardiac monitor tracks the condition of the circulatory system in terms of parameters such as heart rate, blood pressure and blood oxygen saturation. Renal output, respiration and the functioning of other systems are also routinely monitored. Regular neurological examinations give an idea of the neurological condition, but there is no single continuous monitor of brain function.

CT scans and MR images give an excellent idea of large structural changes in the brain. Although these imaging techniques provide a wealth of information, they reflect only structural changes at a particular instant in time, and there is some risk associated with moving a critically ill patient to the scanning division of the hospital.

Electrophysiological techniques, such as EEG and evoked potentials, are not typically employed on a continuous basis in the intensive care unit. These tests may be ordered if changes are noticed in the neurological examination results. The techniques are valuable in that they directly reflect the functioning of the central nervous system, are non-invasive and can be applied continuously.

The purpose of this work is to provide a prototype for a continuous monitor of neurophysiological function for use at the bedside in the intensive care unit. Such a monitor could potentially give an early indication of changes in a patient's neurological condition.

2 State space representations in physiology

In this work, the brain's condition is represented as a vector of time-varying scalar parameters related to various aspects of brain function. Together these parameters make up a neurophysiological state vector.

In general the state of an object may be described by a number of variables x_1 to x_N. The range of possible values that the variables may take describes the extent of the state space of the system. Although the state space may be higher-dimensional, it can usually only be represented graphically in two or three dimensions. Siegel et al. represented an 11-dimensional vector on a circular chart [5].

The trajectory that a system's state follows in state space relates to the complexity of the system. If the system's state is fixed and constant, or if it is entirely random, the effective complexity of the system is very low. Alternatively, the dynamics of the state may describe an attractor of some sort in state space. The state of a complex system may describe a simple limit cycle or a more complicated strange attractor in state space.

Claude Bernard, the father of modern physiology, wrote [1]:

> "...all of the vital mechanisms [in an organism], varied as they are, have only one object: that of preserving *constant* the conditions of

> life in the [internal environment]."

In the wake of the success of this view, Walter Cannon coined the term homeostasis to describe:

> "...the coordinated physiological process which maintains most of the *steady states* in the organism."

In terms of state space representations, the concept of homeostasis as it was originally defined implies that the dynamics of the physiological state of a healthy organism describe a fixed point in state space, and that the pathological states are unstable. In the last 25 years, it has been shown that this is not true.

Siegel and his colleagues showed in 1979 [5] that under pathological conditions, the physiological state may repeateably and stably occupy the same abnormal region of state space. For example, Siegel constructed a vector of 11 parameters, such as partial pressure of carbon dioxide, cardiac output and mean arterial pressure, and represented these radially on a circular chart. The normal average value of a parameter was represented as a point at unit distance from the circle's origin. A complete set of normal values would result in a unit circle if the points were joined. Any deviation from normal was represented by the displacement of a point from the unit circle, scaled in terms of the normal standard deviation of the parameter. Siegel found that under pathological conditions, different constellations were formed, but that certain constellations were persistent and common to particular pathologies. They were able to use these charts to characterize such states as normal stress after trauma and primary myocardial infarction.

Goldberger and others [3], [2] have shown that the dynamics of the physiological state in a healthy person describe a complicated attractor and not a fixed point in state space. For example the heart rate should necessarily vary in a fractal manner in a healthy individual. It has been reliably shown that a decrease in heart rate variability accompanies certain pathological conditions.

3 State space dynamics in electroencephalography

The electroencephalogram (EEG) is a record of the electrical activity of the brain. The dynamics of the EEG display the property that they lose complexity under unhealthy conditions. For example, it has been shown that the complexity of the EEG is higher in a normal awake subject than when the subject is asleep, comatose or in a state of epileptic seizure. This is if complexity is quantified in terms of the embedding dimension of the EEG whch is normally very high. Arguably it is not meaningful to specify the embedding dimension under these conditions. Under abnormal conditions, the embedding dimension has been reported to fall as low as 4 to 6 [4]. This relates to the underlying number of degrees of freedom of the mechanism which gives rise to the dynamics of the

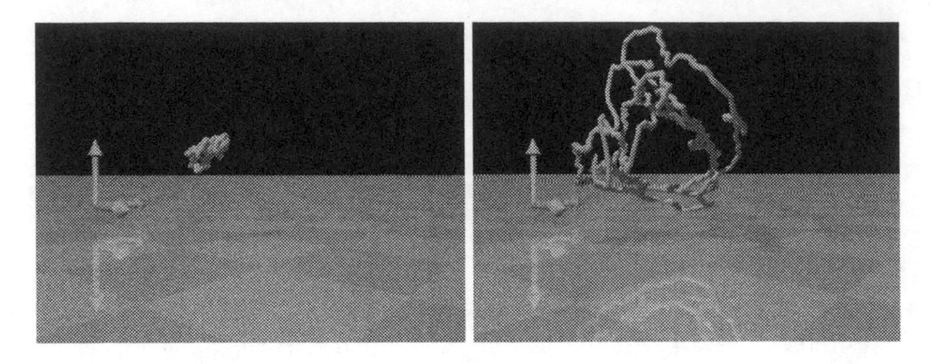

Figure 1: Phase portraits of EEG burst-suppression activity during induced coma.

observed variable. This corresponds to the dimensionality of the phase space needed to fully describe the variable's dynamics. For a single variable, the vector in phase space consists of that variable and the appropriate number of its derivatives.

Apart from embedding dimension, phase space portraits and Lyapunov exponents are used to study system dynamics. The Lyapunov exponents describe the rate at which two points which are initially close in phase space diverge in time. Positive Lyapunov exponents denote an unstable system and negative exponents denote a stable system. Under conditions of visual stimulation, it appears that the alpha activity generated in the occipital cortex becomes entrained and simpler, with an accompanying increase in negativity of the Lyapunov exponents [6]. This suggests that the system is stabilized by the entraining effect of the visual stimulation.

Figure 1 shows phase space portraits for the EEG exhibiting burst-suppression in a person in pentobarbital-induced coma. The left panel shows the phase portrait during 2 seconds of suppression and the right panel shows the phase portrait during 2 seconds of burst activity. The EEG amplitude is represented on the three axes with lags of 0, 0.1 and 0.2 seconds. Whilst activity is minimal during the period of suppression, the portrait from the burst period shows an attractor which suggests low-dimensional periodic activity. This is consistent with the observation that pathological conditions result in a decrease in complexity of the dynamics of functional measures such as heart rate and EEG. In further analogy with heart rate variability, it has been shown that under healthy conditions, the EEG alpha band energy must be variable.

4 Neurophysiological state

The neuromonitoring system developed for this study combines various parameters into a vector representing the state of the central nervous system. These parameters fall into two main groups: electrophysiological and other physiological parameters. The emphasis of the study is on continual electrophysiological monitoring, since this is not done routinely in the intensive care unit. To address in part the requirement for a comprehensive representation of the state of the central nervous system, it was decided to include other parameters in the neurophysiological state vector.

The parameters cover a range of functions of the brain. The electroencephalogram recorded from scalp electrodes reflects cortical activity. In the wakeful state, cortical electrical activity has a higher average frequency than during sleep or coma. In barbiturate-induced or severe coma, cortical electrical burst-suppression patterns are observed. This information is included in the state vector as the energy in the delta, alpha and beta energy bands at each electrode and the presence or absence of burst-suppression activity.

The more basic life-supporting functions of the brain are performed by its deeper structures - the midbrain and the brainstem. The function of these areas is evaluated by means of evoked potentials. In the somatosensory evoked potential (SEP), the response of the brain to electrical stimulation of a peripheral nerve is tested. For our purposes we stimulate at the median nerve at the wrist and measure the response of different areas of the brain along the sensory pathways. The latencies of particular peaks in the response are observed and included in the state vector as a means of assessing the integrity of the sensory pathways. The peaks originate at certain points along the sensory pathways, in the brainstem, diencephalon and cortex and are measured on the left and the right sides. In the brainstem auditory evoked potential (BAEP), the response of the brain to auditory stimulation is tested. The latencies and amplitudes of three particular peaks of interest (waves I, III and V) are recorded in the state vector, and these correspond to the time it takes the signal to get from the cochlea to the cochlear nerve, pons and inferior colliculus respectively. In brain-injured patients the absence of the SEP may indicate a poor prognosis but the bilateral absence of the SEP almost always indicates a bad outcome. The brainstem is a primitive part of the brain, responsible for many basic life-supporting functions such as regulating blood pressure, heart rate and breathing. Absent waves III and V in the BAEP may therefore indicate serious damage to the basic functioning of the brain. Of course, the BAEP only reflects damage to the brainstem along the auditory pathways, and this is one of the reasons that a comprehensive neurophysiological state vector should include some other parameters such as heart rate, blood pressure and breathing rate which indicate whether the brainstem is performing its regulatory function. Also included in the vector are the intracranial pressure and cerebral blood flow velocities obtained by transcranial doppler if they are available.

Figure 2 shows a stack of SEPs recorded from a patient who suffered a large hemispheric infarction. The recordings were made every 15 minutes over a period of 18 hours. The values in the neurophysiological state vector correspond to the peak latencies marked on these raw traces. The complete state vector consists of more than a hundred parameters, and the exact number depends on the data which are available from the patient. A meaningful subset which omits redundant parameters may be selected for representation. Such an 8-parameter

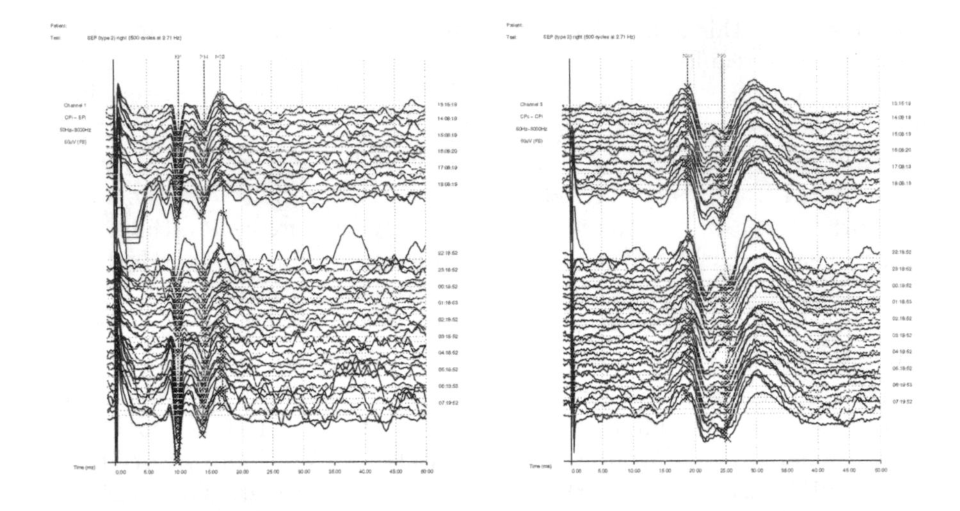

Figure 2: A stack of SEPs recorded in a patient with a large hemisphere infarction over a period of 18 hours. The traces on the left were recorded from CP-EPi and those on the right from CPc-CPi.

subset is shown in Figure 3. Some artifacts due to noise are evident in these traces. These data are from the same subject as Figure 2.

5 Regions of state space

Further work in this study is concerned with characterizing the dynamics of the neurophysiological state vector. Consistent with the discussion above, it is expected that the neurophysiological state vector should undergo a predictable pattern of changes as a a patient's condition changes. In the case of a patient for whom illness results in death, it is expected that cortical functioning should be lost first with an associated decrease in the average frequency and energy of the cortical EEG energy. Next the SEP is expected to exhibit shifts in latencies of the significant peaks, and finally the BAEP is expected to show latency shifts and loss of amplitude. A definite and predictable trajectory of the state through its space is anticipated. It is of course not possible to represent the entire space graphically, but some subspace could be represented. Figure 4 shows the 3-dimensional subspace of the state vector consisting of the wave I, III and V latencies of the BAEP. The values for a normal subject are located closer to the origin, within the boxed-off area, and the values for a patient who has suffered a posterior subarachnoid hemorrhage are shown beyond the boxed area. Work is currently underway to collect data to observe shifts from one region of the space to another. Such a shift occured in the patient of Figure 2, but the shift is not evident in the subspace diagram due to noise and the limited amount of data.

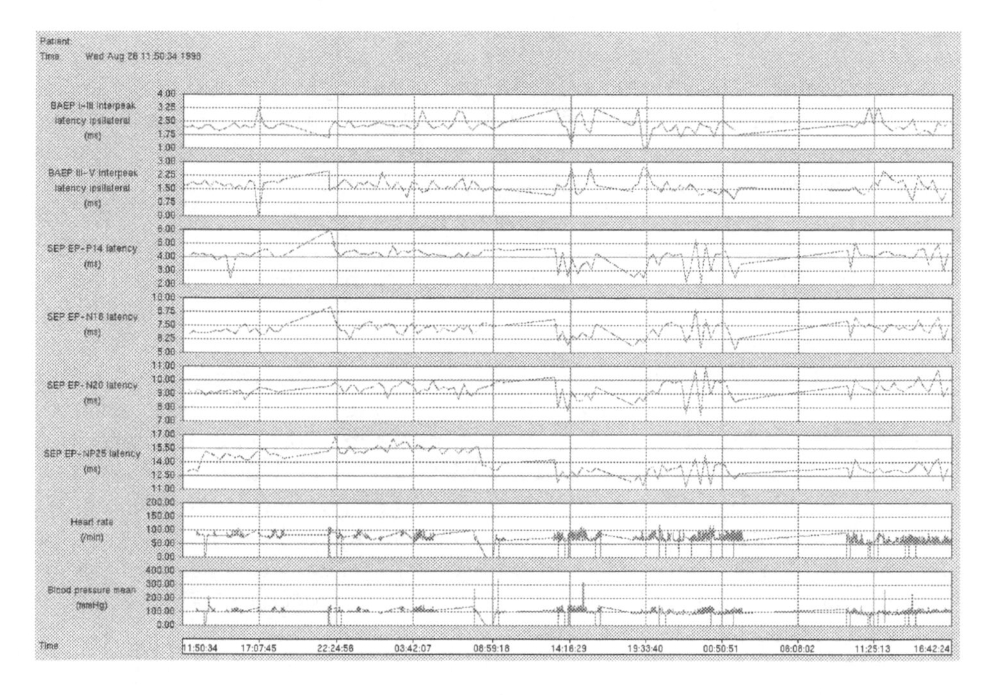

Figure 3: A subset of the parameters in the neurophysiological state vector, recorded in a patient with a large hemisphere infarction over a period of 18 hours. The parameters are BAEP wave I-III interpeak latency, BAEP wave III-V interpeak latency, SEP EP-P14 latency, SEP EP-N20 latency, SEP EP-P25 latency, heart rate and mean arterial blood pressure.

Figure 4: Diagram of subspace showing BAEP waves I, III and V latencies for a patient with and without subarachnoid hemorrhage (right and left clusters respectively).

6 Conclusion

Automatic analysis facilitates interpretation of electrophysiological data by clinicians unskilled in neurophysiology. With state space trajectory representations, it may be possible to have a computer identify normal and abnormal regions of the space and indicate the region of the space that the patient vector occupies. It may be possible by extrapolation to make a prediction of how the vector values may change in future with the purpose of triggering intervention if necessary. By combining the parameters, early subtle changes reflected across a number of parameters may be detected which would otherwise go unnoticed if the parameters were observed individually.

In ongoing work, neurointensive care patients are being monitored for an extended period during their stay in the intensive care unit. Patients are included in the study if they are suffering a large hemispheric or cerebellar infarction with edema or intracerebral hemorrhage with CT evidence of mass effect. It is expected that these patients may exhibit tissue shifts with associated changes in the BAEP and SEP results. The state space representation of the data will be examined to identify regions of the state space associated with different disorders or stages of illness. The dynamics will be examined to determine whether it is possible to identify trends with predictive value in the data.

Bibliography

[1] Timothy Buchman. Physiologic stability and physiologic state. *The Journal of Trauma: Injury, Infection, and Critical Care*, 41(4):599–605, 1996.

[2] Donald Coffey. Self-organization, complexity and chaos: The new biology for medicine. *Nature Medicine*, 4(8):882–885, Aug. 1998.

[3] Ary Goldberger. Fractal variability versus pathologic periodicity: Complexity loss and stereotypy in disease. *Perspectives in Biology and Medicine*, 40(4):543–561, 1997.

[4] J Roschke and J Aldenhoff. A non-linear approach to brain-function - deterministic chaos and sleep eeg. *Sleep*, 15(2):95–101, 1992.

[5] John Siegel, Frank Cerra, Bill Coleman, Ivo Giovannini, Mohan Shetye, John Border, and Rapier McMenamy. Physiological and metabolic correlations in human sepsis. *Surgery*, 86(2):163–193, Aug. 1979.

[6] André van der Kouwe and Pierre Cilliers. A comparison of linear and nonlinear techniques for distinguishing phase-dissimilar vep's. In *Proceedings of the 1995 IEEE Conference on Nonlinear Signal and Image Processing*, 1995.

Optimal Organizational Size in the Presence of Externalities

Bennett Levitan[1]
Bios Group, LP.
bennett.levitan@biosgroup.com
José Lobo
Graduate Field of Regional Science
Cornell University
jl25@cornell.edu
Richard Schuler
Department of Economics
Cornell University
res1@cornell.edu
Stuart Kauffman
Bios Group, LP.
stuart.kauffman@biosgroup.com

We consider a simple economic question: in a stochastic environment, how does group size relate to group performance in the presence of externalities? The search for improved organizational configurations is modeled as a random walk on a space of possible configurations whereby agents in a group periodically have the opportunity to accept or reject random changes in their characteristics. We have developed a modeling framework which allows the connections among the individual members constituting a group and the connections between groups (externalities) to be tuned independently. We present numerical results showing that for short search periods, large organizations with few externalities perform best; while for longer time horizons, the advantage

[1]Corresponding author

accrues to small organizations with few externalities. As the time horizon continues to lengthen, a modest number of externalities enhances the payoff of small organizations. Under all circumstances, organizations on the "edge of chaos" appear to perform best.

1 Introduction

A salient feature of economic behavior is the tendency of individual economic agents to cluster themselves into groups, organizations and firms. These economic groupings affect each other's performance in a myriad of ways not mediated by market mechanisms, giving rise to the phenomena of externalities. Indeed, a prominent answer to the question, "why are there firms?" is that firms exist primarily to internalize externalities (Coase 1937, Williamson 1985). Another salient, and widely documented, feature of economic life is the wide range of firm sizes found within and across industries (see, *e.g.*, Acs and Audretsch 1988, Schmalense 1989, Hannan and Carroll 1992, Hansen 1992, Davis, Haltiwanger and Schuh 1996, Sutton 1997). A fascinating aspect to this diversity is that there seems to be little pattern to optimal firm sizes. Although both of these two aspect of economic behavior, externalities and variation in firm size, have been the subject of much separate study, there is little understanding of how firm size relates to firm performance in the presence of externalities.

In this study we specifically consider: (1) How does optimal group size change as the extent of externalities among distinct groups varies?, and (2)How does optimal group size depend upon the length of the group's operational cycle? To this end, we devised a modeling framework which allows the connections among the individual agents constituting a group and the connections between groups (externalities) to be tuned independently. These externalities can have positive or negative effects on group performance. By simulating an economy constituted by groups constantly make conflicting decisions to improve their payoffs, we can characterize the effects of group size and externalities on group performance.

2 Modeling Framework

2.1 Agents and Groups

In our setting, economic activity is performed simultaneously by G *groups*, each composed of L individual *agents*. There are a total of N agents in the economy ($N = G \cdot L$) [Fig. 1]. Groups may correspond to work teams within a firm, divisions within a multi-unit firm, or firms within an economy, depending on how the model parameters are chosen and interpreted. Agents may correspond to individual workers within a work team, production processes within a plant or communities within a metropolitan economy.

Each agent is characterized by one of S discrete states. A *state* represents different choices of economic behavior undertaken by an agent (such as selecting a particular manufacturing technology or investment strategy). A *configuration*

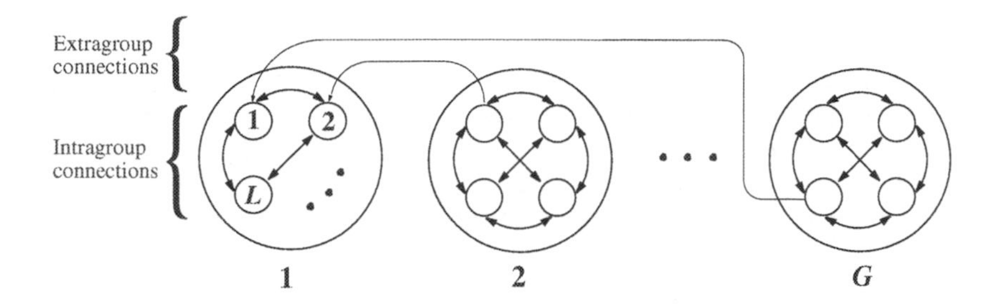

Figure 1: Depiction of connections in model for $L = 4, J = 2$. Agents are represented by small circles. Each large circle represents a group composed of L agents. Lines within a large circle indicate intragroup connections. Lines above the groups represent extragroup connections (*i.e.*, "externalities"). Only the extragroup connections to group 1 are shown.

denotes a specific assignment of states to every agent in the economy. Let Ω denote the space of all possible configurations. We define $a^c_{i,g}$ as the state of the i'th agent in group g in configuration c, $\vec{\omega}^c_g$ as the state of group g in configuration c, and $\vec{\Omega}^c \in \Omega$ as the state of the economy as a whole in configuration c. Notationally: $a^c_{i,g} \in \{1, \ldots, S\}$, $\vec{\omega}^c_g = [a^c_{1,g},\ a^c_{2,g}, \ldots a^c_{N,g}]$, and $\vec{\Omega}^c = [\vec{\omega}^c_1,\ \vec{\omega}^c_2, \ldots,\ \vec{\omega}_G]$, with $i = 1 \ldots L$ and $g = 1 \ldots G$. Since all groups are composed of L agents and each agent has S possible states, there are S^{GL} possible configurations for the economy (*i.e.*, $c \in \{0, \ldots, S^{GL}\}$).

Performance of a group may be measured using a variety of metrics such as time, costs or profits. We use the generic term *payoff* to refer to the property a group seeks to optimize. We define $\phi^c_{i,g}$ as the contribution of the i'th agent in group g in configuration c, ϕ^c_g as the payoff of group g in configuration c, and Φ^c as the payoff of the economy as a whole in configuration c. A group's payoff is the average of the contribution of its members, and the economy's payoff is the average of the payoffs of its constituent groups. Thus, $\phi^c_g = \sum_{i=1}^{L} \phi^c_{i,g}$ and $\Phi^c = \sum_{g=1}^{G} \phi^c_g$. All payoffs are in the range [0, 1].

The payoff of an agent depends on its own state and that of other agents, both within the same and other groups. We say agent $a_{i,g}$ is *connected* to agent $a_{j,k}$ if $\phi_{j,k}$ depends on $a_{i,g}$. Actions by members within a group (e.g., whistling while at work, deciding to use JAVA instead of C++ as a programming language) have direct impacts on the payoffs of other group members. Note that connections are not necessarily symmetric: one agent may be connected to a second without the second being connected to the first. By definition, every agent is connected to itself.

Many of our results hinge on the difference between *intragroup* connections and *extragroup* connections (or *externalities*). (An example of extragroup connections might be when a person in firm A, a competitor to firm B, pollutes the water supply or develops a better way to paint cars, hence altering B's payoffs.)

Let K be the number intragroup connections per agent, excluding the connection from itself, and J be the number of extragroup connections per group. Agent $a_{i,g}$ might be affected by K connections with agents within group g and by anywhere between 0 and J connections from agents in other groups. In our simulations the outside agents that affect a given group are chosen at random and are distributed as uniformly as possible.

Each agent is "maximally connected" within a group; that is, every agent is connected to every other agent in its group ($L = K+1$) [Fig. 1]. This assumption corresponds to an organization in which the choices of each individual affects the payoff of every other individual in the organization. An agent's payoff thus depends upon its own state and on those of between K and $K+J$ other agents. By assuming that essentially arbitrary interactions are possible among agents, we can assign payoff value to the $\phi^c_{i,g}$ at random; and they are drawn independently from the uniform [0,1] distribution in our simulations. While a group itself has only $S^{K+1} = S^L$ possible configurational arrangements, it has $S^{(K+1+J)}$ possible payoff values due to the external connections. In this manner a "table" of payoff values is constructed for each agent. The connections and these values define the entire economy in our model.

2.2 Group Search for Greater Payoffs

In our simulations the stochastic nature of economic environments is modeled as random changes in the agents' states. Every generation, a randomly-selected agent is given the opportunity to adopt a different state, also selected at random. Because each group is maximally connected, this agent's decision affects the payoffs of all other members of the group. The agent decides to accept the new state only if that change improves the payoff of its group, even if the payoff decreases to the agent in question. Due to externalities, there may be an effect on the payoffs of other groups. While attempting to optimize its own group's payoff, an agent's decisions will constantly alter the ability of other groups to optimize their payoffs. The resulting churning of agent states is responsible for all the interesting behavior in our model.

The group search process is as follows: (1) The economy starts a production run at $t = 0$ with initial configuration $\vec{\Omega}^t$ (using time to index configurations). (2) During production run t, a randomly chosen agent in group g is given the opportunity to adopt a new state. The economy's tentative new configuration is $\vec{\hat{\Omega}}^t$. (3) If payoff $\hat{\phi}^t_g > \phi^t_g$, the agent adopts the new configuration, so that $\vec{\Omega}^{t+1} = \vec{\hat{\Omega}}^t$; otherwise, the economy keeps the same economic configuration, such that $\vec{\Omega}^{t+1} = \vec{\Omega}^t$. (4) Increment$t$ and return to step 2. This search process is Markovian, since at the end of period t the group needs only to remember the configurations and payoffs for periods t and $t-1$ payoffs. The search can be therefore be regarded as a *random adaptive walk* in the space of economic configurations. This trial and error process for finding higher group payoffs imposes no decision cost since an agent is allowed to return to its previous state if no improvement is realized.

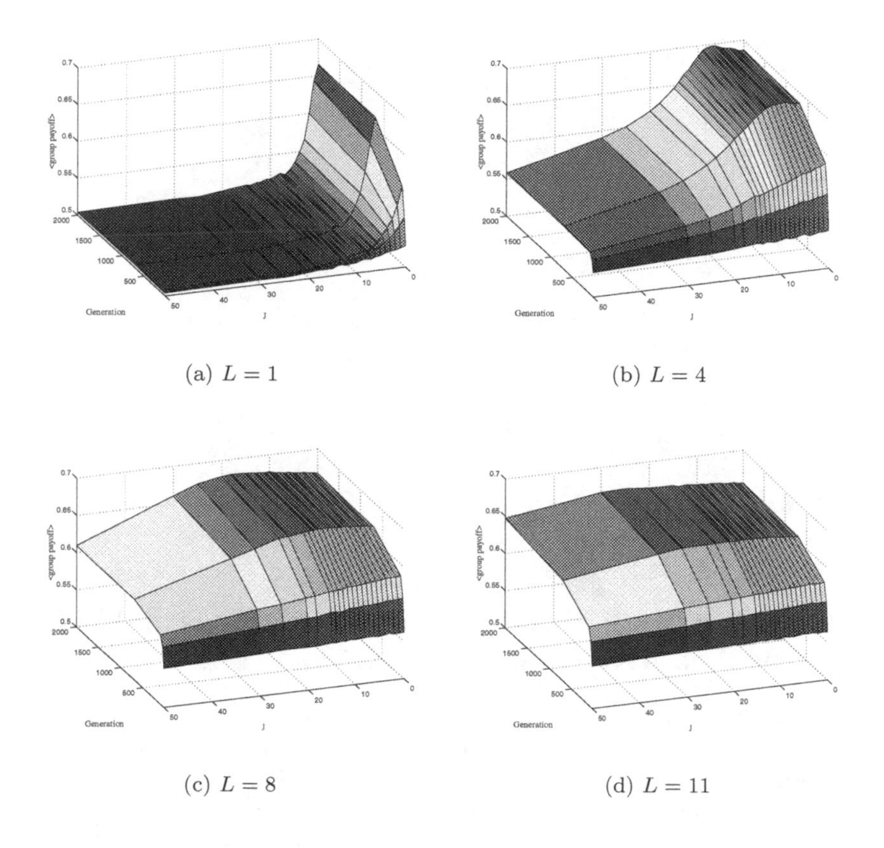

(a) $L = 1$ (b) $L = 4$

(c) $L = 8$ (d) $L = 11$

Figure 2: Average payoff vs. J and generation for $L = 1, 4, 8$ and 11.

3 Numerical Results

3.1 Characteristics of Simulated Searches

We have simulated the group search process for economies with $N = 100$ agents, $S = 2$ possible states per agent and group sizes L varying from 1 to 11. The number of groups $G = N/L$ varies from 9 to 100. We consider a large range of external connections, from 0 to 99. Simulations were run for 2000 generations, which generally proved sufficient to show steady state behavior of the dynamic economy. Ensemble averages were computed by compiling cumulative statistics for 500 different economies (sets of connections and payoff tables) with 2 simulations per economy. In all results below, "average group payoff," $\langle \Phi^c \rangle$, refers to the payoff of a group averaged over all groups and all simulations at the corresponding generation.

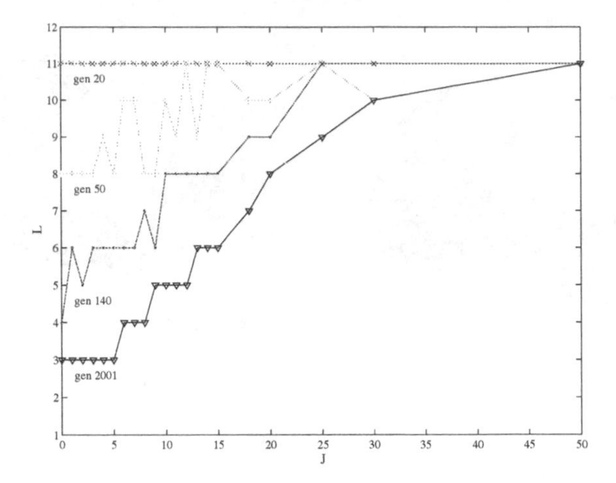

Figure 3: Optimal group size (L) corresponding to varying levels of externalities (J) after 20, 50, 140 and 2001 generations.

3.2 Optimal Group Size, Level of Externalities and Length of the Adaptive Random Walk

The interplay between group size and external effects is illustrated in figure 2, which shows the average group payoff per generation for fixed L and varying levels of external connections. For larger group size, group payoff is maximized by $J > 0$, and the larger the group size, the larger the level of external connections associated with the highest group payoff. As expected, higher payoff is achieved after longer searches (or, equivalently, longer adaptive walks). The advantage provided by a large group size is clearly an increasing function of externalities, suggesting that larger organizational size provides a buffer against the stochastic shocks transmitted by external connections. An alternative explanation is that in large organizations where operational changes are decided upon incrementally, the probability of getting stuck at a locally optimal configuration, or globally sub-optimal configuration, is high. In these cases, a larger number of external interactions enhances the likelihood of being dislodged from a sub-optimal position and of being able to continue the search for a global optimum.

Figure 3 shows optimal group size as a function of level of externalities for several generations. Larger groups achieve higher payoffs when adaptive walks are of brief. As the duration of search increases, smaller group size proves to be more beneficial. When the level of external connections is high, however, larger group-size is optimal regardless of the duration of the search. The advantage of a large organizational size is clearly a function of increasing levels of externalities. This is one of the surprising results from this work.

Figure 4(a, b) plots average group payoff as a function of group size (L) and magnitude of external connections (J) after 51 and 2001 generations. For short search walks, maximum payoff is associated with large groups and few externalities. But notably, as the length of the search increases, maximum payoff is associated with relatively small group size and few externalities. These results can be interpreted to mean that for organizations engaged in activities characterized by short turn-around periods or short production cycles, large organizational sizes may be preferred to small ones. The reason is that in short periods, the number of alternative configurations that can be sampled by each group is S^L and for the economy as a whole is S^{GL}. With J small, larger groups get more chances to find a better outcome – initially. As the duration of the walk increases, the chances of getting stuck at a local optimum increases with group size and overcomes the advantage of the larger number of alternatives.

3.3 Transition Between Ordered and Chaotic Regimes

Figures 4(c, d) show the average rate at which agents within the economy continue to choose new states, as a function of L and J for two durations of activity ($g = 51$ and 2001). The *flip rate* is the rate at which agents accept candidate states and measures the extent to which the search process continues. It was calculated as a running average over the 50 generations. It is easiest to understand these figures by focusing first on generation 2001. The behavior of the economy exhibits two broad regimes in the LJ plane, an "ordered" regime where sites stop flipping for low values of J together with high values of L, and a "chaotic" regime where sites keep flipping rapidly, seen for high values of J together with low values of L.

It is particularly interesting to compare the average payoffs to the flip rates at corresponding generations [Fig. 4]. Note that in figures 4(a, b) a curved ridge of high payoff runs from low L, low J values roughly diagonally to high L and J values. The ridge is a kind of watershed where payoff decreases down the two slopes flanking up to the ridge. Comparing this curved ridge of maximal payoff with the corresponding curves in the LJ plane in figures 4(c, d), shows that the ridge of maximum payoff in the LJ plane corresponds to a line of almost constant low flipping rates.

In the ordered regime each group finds a local or global optimum that is mutually consistent with the optima attained by the groups that affect its payoff and the groups whose payoffs it affects. Once these mutual optima are attained, motion of all groups stops. By contrast, in the chaotic regime, adaptive moves by one group alter the payoffs of other groups faster than those groups can find their own optimal payoffs. Given the restriction that each group must move incrementally (can change or flip only one state at a time), the ordered regime can be thought of as a local Nash equilibrium among all the groups, local in the sense that a stable optimum is a Nash equilibrium for each group with respect to single state changes.

The observation that optimal payoff is roughly associated with the boundary of a region with constant low flipping rate is deeply interesting. The model economy faces an extremely hard combinatorial optimization problem with a huge configuration space. Achieving optimal payoffs over time requires a bal-

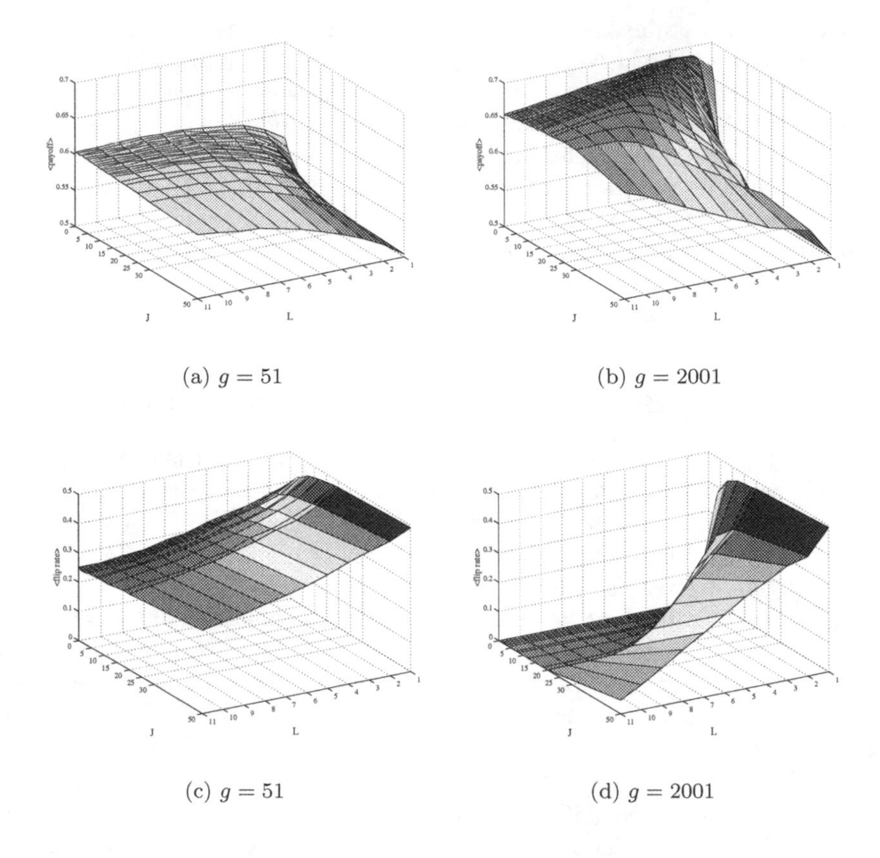

(a) $g = 51$ (b) $g = 2001$

(c) $g = 51$ (d) $g = 2001$

Figure 4: Average payoff (a, b) and flip rate (c, d) as a function of L and J for generations 51 and 2001.

ance between "exploration" and "exploitation" (March 1991) and is visually on the "edge of chaos." This balance appears to be optimally struck somewhere near the phase transition between the ordered regime and the chaotic regime, where persistent low levels of exploration (*i.e.*, flipping) continue. In much of the ordered regime (generally high L together with low J) the economy easily becomes trapped on locally sub-optimal configurations. By contrast, deep in the chaotic regime, (generally low L together with high J), few groups find or sustain locations of high payoff.

4 Discussion

We have presented a modeling framework which can be applied to the study of organizations at a variety of scales. We have shown how time horizon and the extent of externalities influence group payoff: For short search periods, larger organizations fare better since they can experiment with a larger number of alternatives. As the search period lengthens, this advantage does not keep pace with the ability of smaller groups to avoid local optima.

We also show that high levels of external connections enhance the payoff of a large group far more than they do for smaller groups, and in fact these externalities actually diminish the payoffs for very small groups since they groups are frequently knocked from their optima. For large groups that may be stuck at a local optimum, these external shocks more often improve performance. The magnitude of externalities is an exogenous variable in our model, but were the groups to represent firms; the level of external connections could then be seen as a parameter characterizing economic sectors. Whether an economic sector is dominated by small or large firms could be partly a result of which firm size is optimal given the extent to which firms in that sector affect each other.

The authors gratefully acknowledge research support provided by the Santa Fe Institute and Bios Group, LP.

Bibliography

[1] ACS, Zoltan, and David AUDRETSCH, "Innovation in Large and Small Firms: An Empirical Analysis", *Amer. Econ. Rev.*, **78** (1988), 678-690.

[2] COASE, R., "The Nature of the Firm", *Econometrica*, **4** (1937), 386-405.

[3] DAVIS, Steven, John HALTIWANGER and, Scott SCHUH, *Job Creation and Destruction*, The MIT Press (1996).

[4] HANNAN, Michael, and Glenn CARROLL, *Dynamics of Organizational Populations*, Oxford University Press (1992).

[5] HANSEN, John, "Innovation, Firm Size, and Firm Age", *Small Business Economics*, **1** (1992), 37-44.

[6] MARCH, James, "Exploration and Exploitation in Organizational Learning", *Organization Science*, **2** (1991), 71-87.

[7] SCHAMALENSEE, Richard, "Inter-industry Studies of Structure and Economic Performance", *Handbook of Industrial Organization* (Richard Schamalensee and Robert Willig eds.), North-Holland (1989), 951-1009.

[8] SUTTON, John, "Gilbrat's Legacy", *J. Econ. Lit.*, **35** (1997), 40-59.

[9] WILLIAMSON, Oliver, *The Economic Institutions of Capitalism*, The Free Press (1985).

Education in Complex Systems

Michael J. Jacobson
Allison~LoBue Group, LLC
mjjacobson@earthlink.net

Win Farrell
PricewaterhouseCoopers Consulting
winslow.k.farrell@us.pwcglobal.com

Kenneth Brecher
Department of Astronomy
Boston University
brecher@bu.edu

Jim Kaput
Department of Mathematics
University of Massachusetts-Dartmouth
JKaput@umassd.edu

Marshall Clemens
Idiagram
mclemens@world.std.com

Uri Wilensky
Center for Connected Learning and Computer Based Modeling
Tufts University
uriw@media.mit.edu

1. Overview

Session Chair: Michael J. Jacobson

This paper provides an overview of issues related to learning about complex systems and the application of complexity concepts and approaches that were considered as part of presentations at the Education in Complex Systems session given during the Second International Conference on Complex Systems. There are at least two main challenges related to the sciences of complex and dynamical systems for education and training broadly construed to encompass schools and the workplace. The challenge for science education is obvious. The scientific literacy of the next century may well require that students in school develop a deep understanding of new cross-disciplinary concepts and new ways of doing science that are related to complexity and complex systems. Unfortunately, given that students frequently have difficulty understanding the scientific knowledge from before the twentieth century, helping students learn ideas and concepts related to complex and dynamical systems may prove to be even more difficult. Further, there are important cognitive, learning, and curricular issues that to date have not been systematically considered.

The second major challenge relates to the application of scientific findings, intellectual perspectives, and modeling techniques based on complex and dynamical systems research to problems and issues in the "real world." Senior executives in business and industry, policy makers in government, and the general public may need to make important technological, economic, and social decisions for which knowledge about complex systems is relevant, yet about which they have had no formal training. Further, there is reason to believe that many complex systems ideas and concepts are not only cognitively demanding, but also counter intuitive, and thus may prove difficult for both students and adults in non-scientific and even scientific fields to learn. And lacking a solid understanding of these ideas, there is the danger that relevant scientific knowledge based on complex and dynamical systems research may not be appropriately applied to technical, economic, and social problems and issues. The presentations given as part of this session considered these, and other important issues, related to learning and applying knowledge from this emerging field of scientific inquiry.

2. Real Complexity For Real People

Presenter: Win Farrell

Win Farrell observed in his talk that there has been much intellectual debate recently about whether the sciences of complexity represent a metaphor for thinking about the way "the world works," or represent something more than a metaphor--a context for thinking about difficult and important problems. Several years have passed since "toy" complex systems models have been demonstrated to business executives who have real-world responsibilities. While instructive, these "toy" models have also served as a starting point to evaluate whether real complex systems models of the world, populated with real data from business data warehouses, can serve as predictive models that can be trusted.

Important features--reality, suitable predictive power, instruction, and context--are all brought into focus with the contemporary high-powered complexity models that are starting to be used by the business community. Although features such as these are the perceived advantages driving executives to resort to complexity models, the field's novelty raises important issues such as a model's testability and veracity. Consequently, the all-important trust that many executives have in complexity models is still below the level of trust in other modeling techniques that rely on more conventional "top down" systems thinking. Corporate executives' and managers' trust in these complexity models, just as they have in the traditional top-down judgment and decision making models to which they accustomed, however, is presumably the critical aspect to the success and permanence of complexity models in the real world. Farrell conjectured that in order for complexity models to flourish, the establishment of real-world users' comfort and trust in bottom-up, complexity-based models must be similar to that of the incumbent top-down models. Is it only a matter of time, or the revelation of some "moment of truth," when complexity models consistently outperform traditional top-down alternatives? Moreover, if this flash of insight is not achieved, is the metaphorical utility of complexity enough to sustain interest beyond the notion of complexity as "just another passing fad?" These questions remain to be answered.

3. Representations and Complex Systems

Presenter: Marshall Clemens

In the talk by Marshall Clemens, he asserted that the importance of the field of complex systems lay not just in the applicability of the complex systems concepts across the many scientific disciplines they transcend–as interesting and important as those applications may be–but in the fact that the field transcends these disciplines at all. Complexity's transcendent ability suggests that it might form a set of truly general concepts applicable across a wide range of domains, and thus should be of great value to all students no matter what field they may eventually specialize in. Complex systems theory has the potential to be the first truly viable candidate for a scientifically based "liberal arts" curriculum. Both the power and difficulty of such a knowledge backbone lies in its abstraction. Such high-level concepts are necessarily abstract and abstract concepts are often the most difficult to acquire and apply. Abstractions must be built, bottom up, from specific examples, grounded, top-down, in numerous instances, and fleshed-out in verbal, visual, tactile, and auditory representations. This is the challenge of teaching complexity: mediating between the concrete and abstract using all available modalities.

The unique nature of complexity science–its newness, transdisciplinary roots, and conceptual complexity–has made it difficult to synthesize into a concise yet complete educational package. Clemens argued that visual representations, in addition to verbal and mathematical ones, are an essential component in integrating and understanding complexity concepts. Following the account of human understanding pioneered by George Lakoff and Mark Johnson–that at its core, human knowledge is based largely on image-schema and their metaphorical extensions–Clemens proposed that image-schematic constructs are essential to grasping many complexity concepts. And, conversely, if one lacks the correct image construct, or an appropriate visual analogy, understanding will be partial or erroneous. In the task of integrating these concepts into a unified picture of complexity science, here again, but for a different set of reasons, visual-diagrammatic representations are essential. When the body of knowledge to be integrated becomes conceptually complex (i.e., when it exceeds the capacity of our working memory of 7+-2 concepts or relations) external representations become a practical necessity. Among external representations, diagrams–because of their unique abstractive, mnemonic, and relational affordances–are particularly well suited to the task of knowledge synthesis. Diagrams can explicitly show conceptual relationships hidden in verbal or mathematical arguments, and can help mediate between the concrete and abstract by linking general concepts to specific examples. The balance of Clemens' presentation discussed ways to design complex systems visual representations, such as shown in Figure 1, based on his recent work in this area.

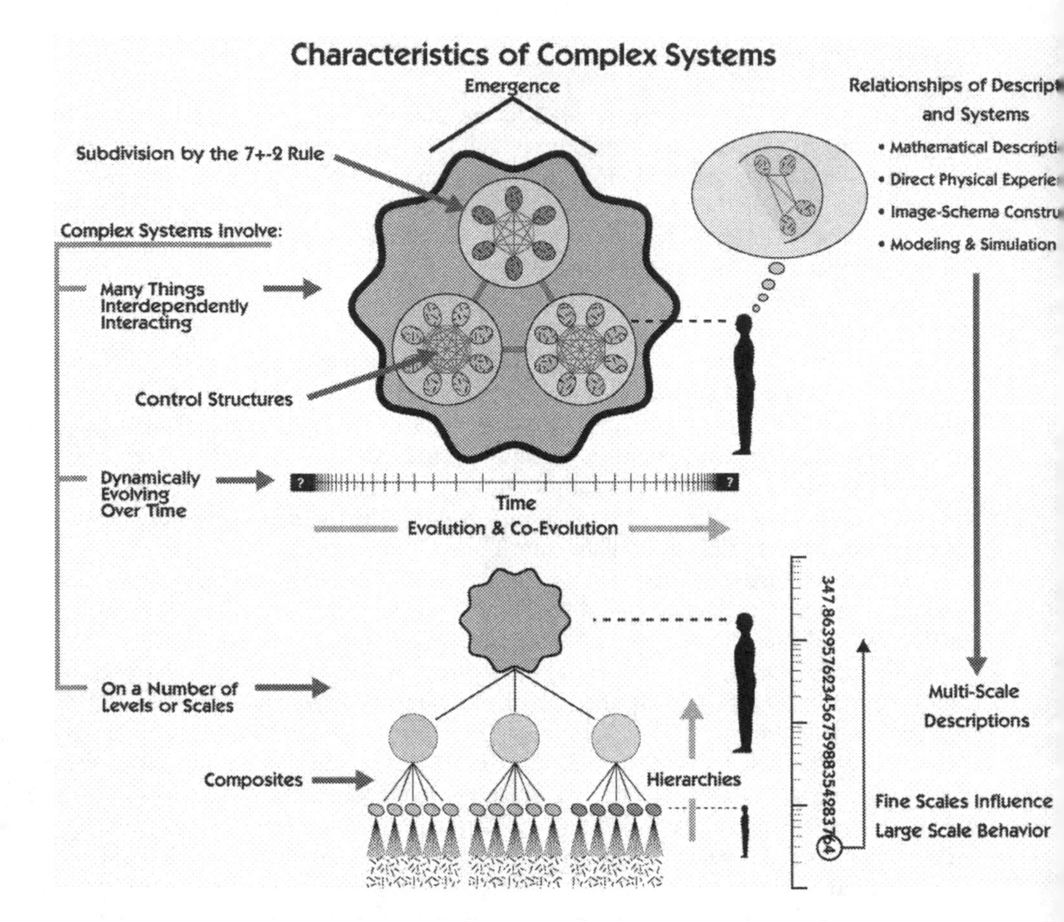

Figure 1. Example of a visual representation of core complex systems concepts.

4. Justifying and Visualizing Complexity

Presenter: Uri Wilensky

Wilensky presented an outline of a four-year computational modeling project sponsored by the National Science Foundation. The project, entitled "Making Sense of Complex Phenomena through Building Object-based Parallel Models" (http://www.ccl.tufts.edu/cm/), is developing modeling languages (such as StarLogoT) and tools for the study of complex systems and is studying students building models with these tools. In this presentation, Wilensky described research by his student Kenneth Reisman and himself in which students used the StarLogoT modeling language to explore the relations between different biological levels. (Note: For a more complete discussion of this research, see their paper elsewhere in the ICCS Proceedings.) In their research, students used the StarLogoT language to model the micro-rules that underlie the emergence of a phenomenon, and then observed the aggregate dynamics that resulted (see Figure 2). Two situations were discussed in the presentation in which this approach had been employed with students.

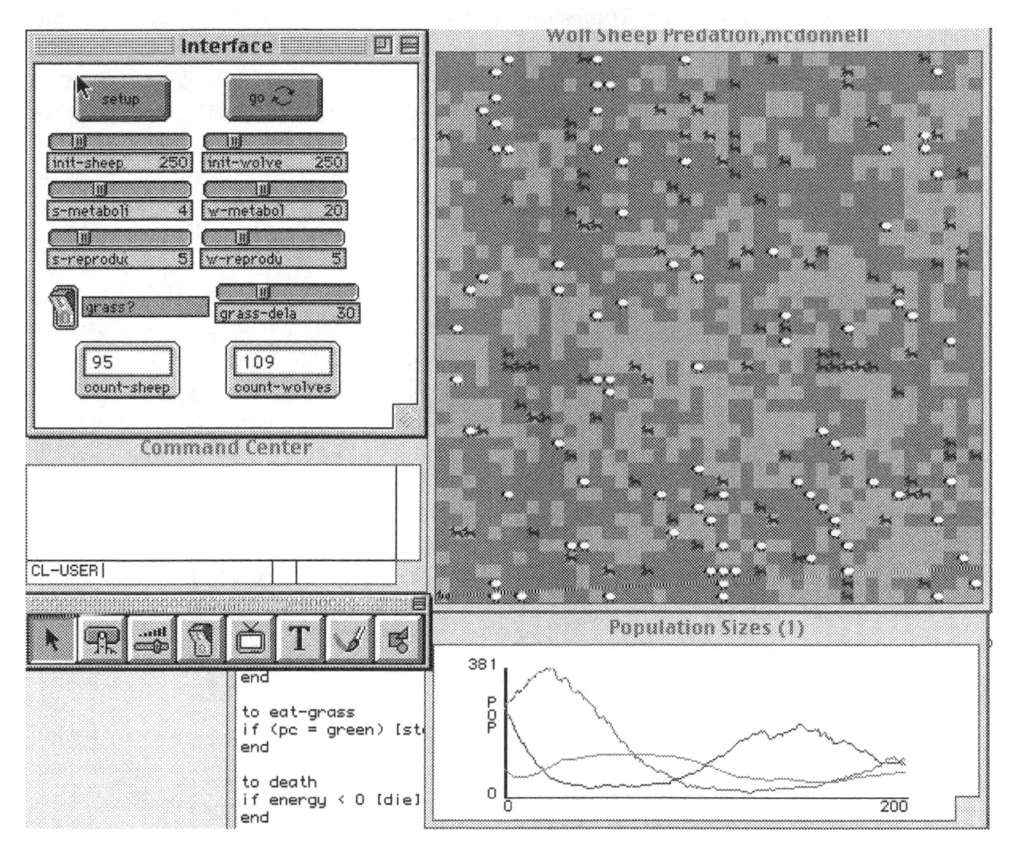

Figure 2. StarLogoT multi-agent model of a wolf-sheep-grass ecosystem.

5. On Growth and Form, the Random Universe, Patterns in Nature, and the Dance of Chance: Helping Students Learn How Order Comes Out of Chaos

Presenter: Kenneth Brecher

Kenneth Brecher discussed a series of NSF supported projects conducted during the 1990's by a collaborative team of university scientists, education specialists, and high school teachers that set out to explore how students might learn about a variety of complex systems. These projects had three main goals: (a) to bring forefront research topics into the classroom in order to provide students with the opportunity of dealing with current, rather than centuries old, ideas; (b) to test how new technologies, particularly computers, could best be used in education, particularly in the realm of simulations, but also in data gathering and analysis; and (c) to combine computer simulations with hands-on experiments to provide an inquiry based environment in which students construct their own knowledge effectively.

In the course of these projects, software was written with simple underlying rules that lead to the formation of complex patterns. Diffusion limited aggregation, percolation, and other current topics in complex systems could then be explored by students themselves. In addition, simple biological, chemical, and physical experiments were developed which showed how complex, but apparently ordered, patterns form from random physical processes. Techniques for determining the fractal dimensions of resulting patterns were also developed. Students were then placed in a position to compare theoretical models they could control with real physical phenomena that they themselves generated and measured.

This presentation displayed some of the products of this research and development efforts. It discussed some of the lessons learned in working with over 150 teachers during the development stages of the project. The presentation also described results from the classroom, particularly changes in the role of teachers and students; presented anecdotal stories concerning changes in students' understanding and attitude towards modern scientific research and the origin of complexity in nature. Materials from this project may be found at the following Web site: http://polymer.bu.edu/~trunfio/cpsproj-educ.html.

6. "Schoolifying" The Mathematical Side Of Complexity And Dynamical Systems: Conceptual And Curricular Integration

Presenter: Jim Kaput

Jim Kaput posed several questions and issues that must be addressed in order to incorporate Complexity and Dynamical Systems into core curriculum for mainstream students. For example: What does it mean to go beyond the pretty pictures? The answer involves both finding ways to make the ideas about complex and dynamical systems learnable, and finding ways to integrate and/or transform the relevant curricula now in place. Other issues include how to link with existing curricula, and the development of new curricula. On the mathematical side, education dealing with complexity and dynamical systems knowledge will need to integrate with the existing relevant school curricula involving notions of rate and ratio, periodicity, composition/iteration, randomness and probability, and so on. Also, there are topics that are usually associated with "high end" school curricula such as connections between rates and totals of varying quantities for different types of variation (e.g., linear, polynomial, exponential, logistic), limits, and continuity and approximation. In addition, ways must be found to deal with topics such as closeness, sensitivity, dimensionality, and so on that are not part of school curricula, as well as skills such as visualization, scaling, and "clever testing" of multivariate situations that are not part of explicit curricula at any level.

Working out answers to questions and issues such as these is likely to be a generation-long task, although the Kaput indicated he was predisposed to be highly optimistic regarding "learnability" issues. Indeed, Kaput asserted that many of the traditionally "hard" topics of science and mathematics may become more learnable within the rich and integrated approaches afforded by combinations of highly visual, student-controllable simulations, meaningful physical and social settings that tap into students' naturally occurring linguistic, kinesthetic, perceptual and cognitive sense-making powers, and local learning objectives that build on their interests and felt needs. The explorable modeling and simulation worlds that are now becoming available could be immensely more accessible and interesting to mainstream students than the highly idealized universe of classical science that was only available in the language of formal algebra—and which was put at the end of a long series of prerequisites that acted to filter out all but a tiny minority of students.

Kaput concluded his presentation by discussing selected conceptual and curricular issues in more detail based upon research with his colleagues Ricardo Nemirovsky, Walter Stroup, Jeremy Roschelle, and Helen Doerr as part of the ongoing NSF funded SimCalc Project. He expressed hope that projects such as the SimCalc Project may help lay the base for the next level of curricular innovation required to "schoolify" Complexity and Dynamical Systems.

7. Conclusion

The presentations in this session discussed a variety of issues related to the integration or use of complex systems knowledge in education and business. In particular, this work suggests that merely presenting or providing complex systems content to business executives or students—whether as concepts or as multi-agent modeling techniques—may not be enough. For example, Farrell observed that business executives may not have trust in complex systems modeling techniques since they are based on different assumptions than conventional top-down modeling approaches. Kaput poses a related question in education when he asks what does it mean to go beyond the "pretty pictures." Given the conceptual demands and counter-intuitive nature of many complex systems concepts and perspectives, helping adults in business or students in school to develop "trust" and understanding of complex systems conceptual and modeling approaches may prove to be significant challenges.

Fortunately, the education oriented presentations in this symposium discussed innovative representational, pedagogical, and learning approaches that show promise for meeting these challenges. These presentations discussed several ways that students may use powerful computer visualizations and modeling software as part of constructive science and mathematical learning experiences that often bring students into direct contact with cutting edge scientific issues and knowledge. Of course, much more work is necessary to expand and generalize approaches such as these for more widespread integration into the science and social science curricula our students are involved with. Still, the challenges and the work to be done notwithstanding, the importance of the projects described in these presentations should not be underestimated. They are showing us glimpses of the future of science, mathematics, and social science learning and of ways to use advanced knowledge in the "real world." These are very exciting glimpses.

Distributed Knowledge Networks

Vasant Honavar, Leslie Miller and Johnny Wong
Artificial Intelligence Research Laboratory
Department of Computer Science
226 Atanasoff Hall
Iowa State University
Ames, IA 50011. U.S.A.
http://www.cs.iastate.edu/~honavar/aigroup.html

Distributed Knowledge Networks (DKN) provide some of the key enabling technologies for translating recent advances in automated data acquisition, digital storage, computers and communications into fundamental advances in organizational decision support, data analysis, and related applications. DKN include computational tools for accessing, organizing, transforming, assimilating, and discovering knowledge from heterogeneous, distributed, possibly mobile, data and knowledge sources (including traditional databases, legacy systems, text repositories, sensors, image collections, mobile devices and specialized simulations) for distributed problem solving and decision making under tight time, resource, and performance constraints. DKN find applications in such diverse areas as organizational memory and decision support systems, internet and intranet based information systems, healthcare information systems, and agile distributed design and manufacturing environments. This paper summarizes the current state of our effort to design and implement Distributed Knowledge Networks.

1.1 Introduction

Organizational decision support (situation assessment, decision-making under uncertainty), distributed design and manufacturing, healthcare information systems advanced scientific research (e.g., the genome project), military applications (e.g., intelligence data handling, situation assessment, command and control), law enforcement (e.g., terrorism prevention), crisis management, power systems, communications infrastructure, offer numerous examples of scenarios which call for reactive as well as proactive decision making, often under tight time, resource, and performance constraints to accomplish the desired objectives. Recent advances in sensor, high throughput data acquisition, and digital information storage technologies, have made it possible to acquire and store large volumes of data in digital form. Advances in computers and communications, Internet, and mobile computing have made it possible for decision makers, to have at their disposal, at least in principle, large volumes of data as well as analysis and decision support tools residing on multiple, geographically distributed, heterogeneous, hardware and software platforms around the globe that are connected to the Internet. However, in order to translate the advances in our ability to acquire, store, and analyze data in increasing volumes and at increasing rates into gains in our understanding of the respective domains and new capabilities for effective decision-making, sophisticated tools are needed for information retrieval, knowledge discovery, and distributed problem solving and decision making.

The following examples motivate some of the problems that need to be addressed: Decision-makers in organizations, in order to be able to effectively perform their responsibilities, need to have critical information in a timely fashion. Day-to-day operations of organizations involve data gathering and analysis, situation monitoring and assessment, and search for potentially interesting patterns in data as it is being gathered to help in the decision making process. This information can be extremely valuable to decision makers in taking both proactive as well as reactive measures designed to ensure effective functioning of organizations.

Large organizations will benefit from digital repositories of transcripts of meetings, memos, performance data, key decisions, etc. that make up their *organizational memories*. Such organizational memories, if suitably structured and indexed in a form that lends itself to automated context-sensitive retrieval and automated or semi-automated analysis can significantly enhance the effectiveness of the decision making process.

In a distributed manufacturing scenario, organizations desirous of participating in a virtual enterprise, need to be able to access and effectively use information (e.g., specifications, availability, prices, production schedules) about parts produced by potential partners around the globe.

Healthcare providers, in order to provide effective and timely care for a patient in an emergency, need access to the patient's records regardless of the patient's place of residence. The patient's medical history and may be distributed in information systems autonomously owned and operated by healthcare providers around the country. Increasing the efficiency and quality of healthcare calls for constantly monitoring the healthcare system, gathering and analysing data, to discover what works and what does not (using automated or semi-automated knowledge discovery tools) to support the decision makers.

Development of sophisticated information systems to support precision agriculture calls for tools to access, gather, analyse, and use data and knowledge from multiple heterogeneous data and knowledge sources (geographical information systems, weather reports, economic data, etc.)

Scientists working on a complex problem (e.g., identification of the cause of a new disease) need to access and analyse a wide range of data sources (e.g., case histories of patients around the globe that report similar symptoms, databanks maintained by various national and international laboratories about known viruses, etc.).

Leaders of multinational businesses, in order to continually position their organization to succeed in a competitive, dynamic, and increasingly interdependent global market, need tools to utilize information from diverse sources (e.g., political developments, stock market data, labor market information, economic developments, technology news, etc.)

Having the right information at the right time at the disposal of the decision makers will often mean that potential hotspots can be dealt with and major crisis prevented. However, given the complex dynamics and increasingly interdependent nature of organizations, it is impossible to completely avoid unanticipated developments or crises. In a crisis, it is of utmost importance to make available to the decision makers, accurate information at the right time without overwhelming them with large volumes of irrelevant data. The availability of data in digital form also presents opportunities for discovery of otherwise hidden regularities or patterns in specific domains. The knowledge so gathered can be useful in the decision-making process.For example, businesses can discover and use knowledge to predict and hence respond to anticipated demands of specific products.

Translating recent advances in our ability to gather, store, and analyze a wide variety of data on multiple, geographically distributed, heterogeneous and often autonomously owned and operated hardware and software platforms into significant breakthroughs in distributed problem solving and decision making in industrial, scientific, business, and related applications presents significant challenges in several areas of computer and information science and technology. It is important to emphasize that no existing system comes close to meeting this need although researchers have begun to attack some of the technical challenges that need to be overcome.

We propose a modular and and scalable approach to addressing these challenges by building distributed knowledge networks (DKN). The rest of this paper is organized as follows: Section 1.2 briefly describes the key technical problems that need to be addressed in the design and implementation of DKN. Section 1.3 describes our approach to the design of DKN and briefly discusses the current state of the implementation and applications of DKN. Section 1.4 concludes the paper.

1.2 Technical Problems

Some of the key technical problems that need to be addressed in the design implementation, and successful deployment of distributed knowledge network tools for practical applications are summarized below.

In many instances, the data sources are geographically distributed. This calls for the use of *information assistants* or *software agents* for intelligent, selective, and context-sensitive data gathering and data assimilation prior to large scale data analysis. Hence, DKN include tools for monitoring different data sources and routing the appropriate information selectively to relevant sites or specific users. Since the information of interest is user and context-dependent, the tools are customizable to specific users and information contexts.

Given the large volumes of data involved, it is desirable to perform as much analysis as feasible at the sites where the data is located and transmit only the results of analysis rather than flooding the network with data. Hence, DKN use *mobile* software agents that can transport themselves to appropriate sites, carry out the computation on site, and return with useful results.

Since the data sources are often autonomously owned and operated, and reside on heterogeneous hardware and software platforms, their effective use requires a sufficient degree of interoperability among the different data sources (despite their heterogeneity). For example, military applications have to access and utilize data from multiple sensors, intelligence sources, etc. Hence, DKN make use of intelligent software agents to provide seamless access to such data sources.

The data sources contain multiple types of data (text, images, relational databases, sequence data, spectrograms, protein structures, etc.) Hence, DKN provide tools for extracting, transforming, and assimilating relevant information from heterogeneous data sources into a *data warehouse* where it can be further analyzed to facilitate knowledge discovery.

The data sources are dynamic (i.e., they change quite rapidly over time as data items are added or modified). Hence, DKN include software agents that detect and propagate the changes and trigger the necessary updates in the affected data and knowledge repositories. For example, in organizational decision support applications, interpretation of data from different units of the organization might be influenced by the contents of independently gathered financial reports and reports of political developments in the relevant regions.

The large volumes of data, the range of potentially relevant and useful complex interrelationships that need to be discovered, and the diversity of data sources challenge state-of-the-art approaches to *data mining* and *knowledge discovery*. Hence, DKN adapt and extend current statistical and artificial intelligence tools to support data-driven knowledge acquisition and incremental theory refinement from multiple heterogeneous, structured as well as semi-structured data and knowledge sources (including multiple types of sensor data, text, images, etc.)

Design of complex information systems in general, and knowledge networks in particular, in order to be feasible, often requires modular design which involves the decomposition of the overall task into more manageable subtasks. Hence, DKN consist of *multi-agent systems* that are made of multiple, more or less autonomous agents, each of which is responsible for a data source (e.g., an independently managed database) or analysis capability (e.g., a knowledge discovery tool). Modular design also lends itself to being adapted and extended for a broader class of knowledge network applications. In order to ensure satisfactory operation of multi-agent systems, DKN include mechanisms for coordination and control of collections of agents.

In addition to the technical problems listed above, additional issues that need to be addressed have to deal with reliability, fault-tolerance, performance, and security of the infrastructure that is needed to provide the necessary connectivity among the distributed data and knowledge sources and users. These issues are being addressed by multiple research groups working on the physical (hardware) aspects of the information infrastructure and are beyond the scope of our research.

1.3 Design of Distributed Knowledge Networks

Our current design of DKN consists of the following components: a mobile agent infrastructure; intelligent agents for information retrieval, information extraction, assimilation, and knowledge discovery; and coordination and control mechanisms for multi-agent systems. The system consists of modular and extensible, object-oriented toolkits for rapid design and prototyping of multi-agent systems for different applications.

Mobile agent technology [22], facilitated by recent advances in computers, communications, and artificial intelligence, provides an attractive framework for the design and implementation of *communicating applications* in general, and distributed networks in particular [22, 4]. A mobile agent is a named object which contains code, persistent state, data and a set of attributes (e.g., movement history, authentication keys) [22] and can move about or transport itself from one host to another as needed for accomplishing its tasks. Mobile agents provide a potentially efficient framework for performing computation in a distributed fashion at sites where the relevant data is available instead of expensive shipping of large volumes of data across the network.

There is considerable ongoing research on mobile agent infrastructures (MAI) [22, 24]. Most MAI designs consist of at least three components: agent servers, agent interface, and agent brokers (service directory). Agent servers support basic agent migration mechanisms, authentication, and sometimes provide other services. Agent brokers provide addresses of agent servers and support mechanisms for uniquely naming agents and agent servers. The agent interface is used by application programs to create and interact with agents. Recently, a consortium of several companies and research groups has proposed standards (MAF) for key aspects of mobile agent infrastructure to facilitate interoperability among different mobile agent systems (despite their different architecture, design, and implementation choices).

We have explored DKN prototypes using the commercially available Voyager mobile agent infrastructure. We have also designed and implemented an MAF-compliant, platform independent, MAI [24] in Java using *Common Object Request Broker Architecture* (CORBA) for managing distributed objects. Ongoing research seeks to build on this work using DKN designs implemented on different mobile agent platforms to evaluate the portability and interoperability of DKN across multiple MAIs as to extend the design of MAI with emphasis on reliability, scalability, and performance.

Intelligent agents – software entities that perform specific tasks on behalf of users with varying degrees of autonomy and intelligence [2, 4] – offer an attractive approach to the design of DKN components. Of particular interest are reactive agents which respond reactively to changes that they perceive in their environment, deliberative agents that plan and act in a goal-directed fashion, utility-driven agents that act in ways designed to maximize a suitable utility function, learning agents which modify their behavior as a function of experience, and agents that combine different modes of behavior [17, 4].

The prototype DKN systems that we have implemented include intelligent agents for customized information retrieval, information assimilation, and knowledge discovery functions [7, 27, 12, 25]. Customizable information retrieval agents acquire user preferences using machine learning techniques [11, 10, 27, 28] and have been successfully incorporated into mobile agents for selective retrieval of journal articles, news articles, email messages, etc. from remote sites [27, 28]. Our current research aims to extend the capability of customizable information assistants for monitoring multiple, heterogeneous data sources (e.g., traditional databases, image data, scientific databases, text, legacy systems, simulations) to address information retrieval problems that arise in large organizations, healthcare, manufacturing, and business applications.

Use of data from heterogeneous data sources residing on multiple hardware platforms and operating systems at different geographical locations requires a robust and flexible framework for interoperability between the various data sources and clients. Interoperability among multiple hardware and software platforms while using standard relational databases has become easier to manage with the availability of platform-independent tools such as the JDBC and CORBA. However, DKN have to be able to provide seamless access to data that is distributed over multiple relatively autonomous databases, as well as data that is heterogeneous in form and/or content.

Approaches to processing heterogeneous data sources can be broadly classified into two categories: *multidatabase systems* [19] and *mediator based* systems [23]. The multidatabase systems [19] apply traditional database techniques to bridge the mismatch between the underlying data sources. Our work has approached this task using object-oriented views [30] which exploits the underlying object structure for incorporating the rich semantics of the common data types. It creates a uniform interface to multiple databases to hide the heterogeneity and the distributed nature of the underlying data sources. Mediator based systems [23] offer an approach to bridging the mismatch between heterogeneous data sources. In a mediator-based system, new data sources can be added by simply formulating a set of rules that define the new data source. Although no current implementation offers the full range of capabilities envisioned in [23], some implementations are underway.

Our approach to the design of DKN takes a pragmatic approach to heterogeneous database interoperability. It borrows from both the multidatabase as well as the mediator-based approaches to design and implement an object-oriented *data warehouse* [9, 3] based on object-oriented views using knowledge-based software agents. Data source, domain, and application-specific knowledge is used to extract, transform, assimilate, and organize information from multiple heterogeneous data sources (including legacy systems) in one or more data warehouse(s) [12] in a form that is suitable for selecting information using sophisticated queries as well as further analysis (e.g., using data-driven knowledge acquisition and theory refinement). Current work is aimed at building on and extending this system to incorporate a broad range of tools for data transformation and information extraction from different types of data. These tools will include software agents for extracting information from different types of data including text (using domain-specific text analysis) and images (using domain-specific image analysis) and legacy systems.

Given the large volume, diversity, and variety of data to be analyzed in specific applications and the range of scientifically relevant but complex relations that are likely to exist among them, DKN have to include sophisticated tools for data-driven knowledge discovery. *Machine learning* currently offers one of the most practical and cost-effective approaches to automated or semi-automated data-driven knowledge discovery and theory refinement. A variety of approaches to machine learning [13, 5] including artificial neural networks, statistical methods syntactic methods, rule induction and evolutionary techniques are available. Examples of problems where machine learning has produced knowledge that is competitive with that of human experts include diagnosis, credit risk assessment, etc. The nature of the knowledge acquisition and discovery process and the choice of specific algorithms or tools to be used for the task depends on a number of factors [5, 13] such as: Overall objectives of the knowledge acquisition task (e.g., pattern classification, prediction, theory refinement, control); The nature and amount of *a priori* domain knowledge that is available; Choice of data (e.g., feature vectors, text, images) and knowledge representation (e.g., decision trees, neural networks, rules); The amount as well as the quality of data that is available; and Whether all of the necessary data is available for learning from data or whether there is a need for *incremental* knowledge refinement as new data is gathered.

Automated knowledge acquisition and discovery from a diverse collection of heterogeneous data sources such as the ones encountered in DKN applications, calls for a system that supports the use of multiple machine learning paradigms and algorithms as needed. Furthermore, the system must be modular and extensible so that the individual knowledge acquisition modules using specific machine learning algorithms can be easily added, modified, or replaced with minimal effect on the rest of the system. To address this need, our recent work has focused on the design, implementation, and evaluation of a modular, flexible, and extensible toolkit of machine learning algorithms. The toolbox includes several algorithms for: the design of artificial network pattern classifiers [15, 29, 26, 6]; automated feature subset selection to improve the accuracy of learned classifiers [26]; induction of finite state automata from examples [16]. These algorithms have been successfully applied for automated knowledge acquisition in a number of domains including diagnosis [1] and the design of customizable information assistants [28, 27]. More recently, such tools have More recently, such tools have been applied to the problem of data-driven refinement of available knowledge (in the form of rules provided by domain experts) in several bioinformatics applications [14]. This resulted in substantial increase in prediction accuracy on novel data (not used for training) over that obtained using using expert knowledge alone.

In order for largescale application of machine learning algorithms in bioinformatics applications to be feasible, tools for handling diverse data sources (including text images, relational data, legacy data), as well as effective means of encoding and using domain-specific knowledge (available through domain experts in a form that lends itself to use by automated techniques need to be developed. Furthermore, knowledge discovery in the real world often has to contend with partial or incomplete data, there is a need for *incremental* data-driven knowledge refinement algorithms. Hence, our current research seeks to adapt, extend, and apply machine learning tools for automated and semi-automated incremental knowledge discovery and theory refinement heterogeneous data sources of the sort encountered in DKN applications. These tools will be integrated into stationary and mobile software agents for knowledge discovery. Specific technical issues to be addressed include design of learning algorithms to acquire knowledge from heterogeneous semi-structured data (e.g., data from genome and protein databases), data preprocessing and data transformations needed to facilitate learning, techniques for encoding and using available domain-knowledge to aid learning, algorithms for cumulative learning from multiple data sources over time, and new tools to aid the extraction, transformation, and assimilation of the acquired knowledge into distributed knowledge networks.

Multi-agent systems are natural consequences of a modular approach to designing complex distributed knowledge networks. In such multi-agent systems, satisfactory completion of the tasks at hand depend critically on effective communication and coordination among the agents. In order to harvest the potential power of such systems in practical applications, especially when the individual agents might be autonomous (e.g., the agents associated with independently managed data sources as is often the case when dealing with distributed data sources in practical applications), it is essential that suitable mechanisms be devised to exercise adequate control over the behavior of such systems. In multi-agent systems, the notion of control suggests such functions as coordination among agents, synchronization among multiple agents, activation and deactivation of individual agents or groups of agents, selection among agents, creation of new agents when needed, elimination of agents that are no longer needed, adaptation of individual agents and agent populations to changes in the environments or task demands, learning (both at the individual as well as group levels) from experience, and (at a much slower timescale) evolution of agent populations toward more desirable behaviors. Both natural and artificial systems offer rich sources of examples of a wide variety of coordination and control mechanisms that can be beneficially incorporated into DKN [8, 21].

The current implementation of DKN includes a modular and extensible implementation of a general framework for inter-agent negotiation inspired by the *contract net protocol* (CNP) [20, 18, 25] which provides an attractive framework for negotiation and coordination among self-interested rational agents. Within this framework, each agent can announce tasks, make bids, evaluate bids made by other agents to complete the tasks, and offer contracts. Prototype DKN systems that utilize CNP have been implemented and tested on information retrieval and knowledge discovery applications. Systematic evaluation of CNP and related coordination mechanisms and design and implementation of other inter-agent coordination mechanisms (e.g., hierarchies and other structures inspired by biological and social organizations) in the context of specific DKN applications are topics of ongoing research.

1.4 Summary

In this paper, we have outlined some of the key technical challenges involved in translating recent advances in computers, communications, data acquisition and digital storage technlogies into fundamental gains in science, technology, and decision support applications. We have also sketched out the distributed knowledge networks approach to meeting these challenges. DKN provide sophisticated tools for automated information retrieval, information extraction, knowledge discovery, information and data organization and assimilation, distributed problemsolving and decision support tasks. DKN build on and extend recent advances in artifcial intelligence (intelligent agents and multi-agent systems, machine learning, distributed problemsolving), distributed computing (mobile agents), databases (data warehouses, multidatabases), and information retrieval to provide a moduilar, extensible, and scaleable solution to these problems. DKN are being applied in a variety of domains including bioinformatics (e.g., computational genomics and proteomics), organizational decision support systems, monitoring and control of complex systems (e.g. intrusion detection and countermeasures in computer networks).

Acknowledgements

This research was supported in part by grants to Vasant Honavar from the National Science Foundation (through grant IRI-9409580), the John Deere Foundation, and the Iowa State University Graduate College and to Vasant Honavar, Johnny Wong, and Les Miller from the National Security Agency. Jihoon Yang and Guy Helmer (both of whom are graduate students in Computer Science at Iowa State University) have made significant contributions to our current implementation of Distributed Knowledge Networks. Other graduate students who are currently involved in this project include Fajun Chen and Doina Caragea. Our work on Bioinformatics applications of Distributed Knowledge Networks has benefited from discussions with Professors Pat Schnable, Dan Voytas, and Drena Dobbs.

Bibliography

[1] BALAKRISHNAN, K., and V. HONAVAR, "Intelligent diagnosis systems", *International Journal of Intelligent Systems* (1998), In press.

[2] BRADSHAW, J., *Software Agents*, MIT Press Cambridge, MA (1997).

[3] GUPTA, H., "Selection of views to materialize in a data warehouse", *Proceedings of the International Conference on Database Theory*, (1997).

[4] HONAVAR, V., "Intelligent agents", *Encyclopedia of Information Technology*, (J. WILLIAMS AND K. SOCHATS eds.). Marcel Dekker New York (1998), To appear.

[5] HONAVAR, V., "Machine learning: Principles and applications", *Encyclopedia of Electrical and Electronics Engineering*, (J. WEBSTER ed.). Wiley New York (1998), To appear.

[6] HONAVAR, V., "Structural learning", *Encyclopedia of Electrical and Electronics Engineering*, (J. WEBSTER ed.). Wiley New York (1998), To appear.

[7] HONAVAR, V., L. MILLER, and J. WONG, "Distributed knowledge networks", *IEEE Information Technology Conference* (Syracuse, NY), (1998).

[8] HONAVAR, V., and L. UHR, "Coordination and control structures and processes: Possibilities for connectionist networks", *Journal of Experimental and Theoretical Artificial Intelligence* **2** (1990), 277–302.

[9] INMON, W., "The data warehouse and data mining", *Communications of the ACM* **39**, 11 (1996), 49–50.

[10] JOACHIMS, T., D. FREITAG, and T. MITCHELL, "Webwatcher: A tour guide for the world wide web", *IJCAI*, (1997).

[11] MAES, P., "Agents that reduce work and information overload", *Software Agents*, (J. BRADSHAW ed.). MIT Press Cambridge, MA (1997).

[12] MILLER, L., V. HONAVAR, and J. WONG, "Object-oriented data warehouses for information fusion from heterogeneous distributed data and knowledge sources", *IEEE Information Technology Conference* (Syracuse, NY), (1998).

[13] MITCHELL, T., *Machine Learning*, McGraw Hill New York (1997).

[14] PAREKH, R., and V. HONAVAR, "Constructive theory refinement in knowledge based neural networks", *Proceedings of the International Joint Conference on Neural Networks* (Anchorage, Alaska), (1998).

[15] PAREKH, R., J. YANG, and V. HONAVAR, "Constructive neural network learning algorithms for multi-category real-valued pattern classification", *Tech. Rep. no. ISU-CS-TR97-06*, Department of Computer Science, Iowa State University, (1997).

[16] PAREKH, R. G., and V. G. HONAVAR, "Learning dfa from simple examples", *Proceedings of the Eighth International Workshop on Algorithmic Learning Theory (ALT'97), Lecture Notes in Artificial Intelligence 1316* (Sendai, Japan,), Springer (1997), 116–131, Also presented at the *Workshop on Grammar Inference, Automata Induction, and Language Acquisition* (ICML'97), Nashville, TN. July 12, 1997.

[17] RUSSELL, S., and P. NORVIG, *Artificial Intelligence: A Modern Approach*, Prentice Hall Englewood Cliffs, NJ (1995).

[18] SANDHOLM, T., "Contract types for satisficing task allocation: I theoretical results", *AAAI 1998 Spring Symposium: Satisficing Models*, (1998).

[19] SHETH, A., and *et al.*, "Federated database systems for managing distributed, heterogeneous, and autonomous databases", *ACM Computing Survey* (1990).

[20] SMITH, R., "The contract net protocol: High-level communication and control in a distributed problem solver", *IEEE Transactions on Computers -29*, 12 (December 1980), 1104–1113.

[21] UHR, L., *Algorithm-Structured Computer Arrays and Networks: Architectures and Processes for Images, Percepts, Models, Information*, Academic Press New York (1984).

[22] WHITE, J., "Mobile agents", *Software Agents*, (J. BRADSHAW ed.). MIT Press Cambridge, MA (1997).

[23] WIEDERHOLD, G., "The conceptual basis for mediation services", *IEEE Expert* **12**, 5 (Sep-Oct 1997).

[24] WONG, J., V. HONAVAR, L. MILLER, and V. NAGANATHAN, "Design and implementation of mobile agent infrastructure based on mobile agent interoperability facilities (maf)" (1998), Submitted for publication.

[25] YANG, J., R. HAVALDAR, V. HONAVAR, L. MILLER, and J. WONG, "Co-ordination of distributed knowledge networks using contract net protocol", *IEEE Information Technology Conference* (Syracuse, NY), (1998).

[26] YANG, J., and V. HONAVAR, "Feature subset selection using a genetic algorithm", *IEEE Expert (Special Issue on Feature Transformation and Subset Selection)* (1998).

[27] YANG, J., V. HONAVAR, L. MILLER, and J. WONG, "Intelligent mobile agents for information retrieval and knowledge discovery from distributed data and knowledge sources", *IEEE Information Technology Conference* (Syracuse, NY), (1998).

[28] YANG, J., P. PAI, V. HONAVAR, and L. MILLER, "Mobile intelligent agents for document classification and retrieval: A machine learning approach", *14th European Meeting on Cybernetics and Systems Research. Symposium on Agent Theory to Agent Implementation* (Vienna, Austria,), (1998).

[29] YANG, J., R. PAREKH, and V. HONAVAR, "DistAl: An inter-pattern distance-based constructive learning algorithm", *Intelligent Data Analysis* (1999), To appear.

[30] YEN, C., and L. MILLER, "An extensible view system for multidatabase integration and interoperation", *Integrated Computer-Aided Engineering* **2**, 2 (1995), 97–123.

Complexity of Allocation Processes: Chaos and Path Dependence

David A. Meyer
Department of Mathematics
University of California/San Diego
La Jolla, CA 92093-0112
dmeyer@chonji.ucsd.edu
and
Center for Social Computation
Institute for Physical Sciences
Los Alamos, NM

Allocation processes—the division of some commodity among multiple agents—are fundamental to social interactions in various arenas. Examples include wealth/income distribution in populations, natural resource exploitation, market share for competing corporations, satellite bandwidth division among many users, and CPU time usage by multiple software agents running simultaneously. In the case where each agent prefers more to less of the commodity—as in these examples—preference, or Condorcet, cycles are inevitable. We determine the consequences of this fact on an analytically tractable process of allocation subject to random external perturbations. This is a complex system: under majority rule the process is chaotic, while under weighted majority rule the system self-organizes to produce a path-dependent majority owner/dictator/monopolist.

Key Words: Condorcet cycle, allocation process, entropy, path-dependence, Brownian motion.

1 Introduction

EXAMPLE 1. Suppose Mom has baked an apple pie and is faced with the quintessential problem (no more American than are mothers or pies!) of allocating the pie among her children: Alice, Bob and Charlie. For reasons of her own she cuts a $\frac{1}{3}$ wedge of the pie and suggests to her children that she give it to Charlie and leave the remaining $\frac{2}{3}$ of the pie for Alice, with Bob getting none. Bob complains vociferously, so she gives the three children an alternative: either stick with this initial allocation $a_1 = (\frac{2}{3}, 0, \frac{1}{3})$, or change to the allocation $a_2 = (0, \frac{1}{3}, \frac{2}{3})$, where these ordered triples indicate the amounts allocated to Alice, Bob and Charlie, respectively. Assuming each of the children prefers more to less of the pie, Bob and Charlie will vote to change to a_2; only Alice is worse off this way. But then Mom suggests the allocation $a_3 = (\frac{1}{3}, \frac{2}{3}, 0)$ and now Alice and Bob prefer this to a_2 so the result of a (majority) vote is to switch again. Losing patience, Mom suggests $a_1 = (\frac{2}{3}, 0, \frac{1}{3})$ for the second time, and

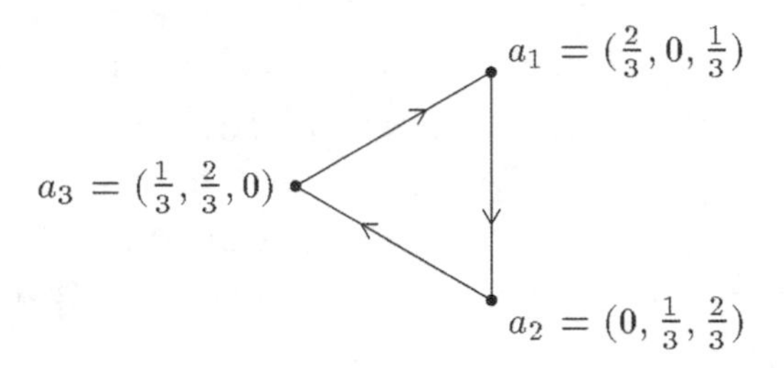

Figure 1: The directed graph f_p representing majority rule aggregation of the agents' more-is-better preferences contains a Condorcet cycle. The edges connecting each vertex to itself are omitted.

despite the fact that the children voted to switch from a_1 to a_2 and then to a_3, now they vote again by 2:1 to return to a_1, completing a cycle. Figure 1 illustrates this preference (Condorcet [1]) cycle: The three vertices represent the a_i while the directed edges indicate how decisions between two alternatives go when decided by majority vote. Under the very strict assumptions implicit in our story (*e.g.*, there are no side agreements between the children) this situation exemplifies a social system in which we attribute rationality to the individual agents (Alice, Bob and Charlie) in the form of totally ordered preferences [2] for allocations giving them bigger over smaller pie slices, yet cannot conclude that any equilibrium is achieved when these preferences are aggregated. In this paper we will explain the sense in which the absence of equilibria in allocation processes indicates that they are complex systems. Let us begin by recalling a pragmatic definition of complexity [3]: A system is *complex* if it is represented

SYSTEM	LOCAL MODEL	GLOBAL MODEL
fluid flow	scattering molecules	interacting vortices
population biology	interacting organisms	cycling ecology
market economics	agent utilities	market w/o stable equilibrium
iterated social choice	agent preferences	Condorcet cycles

Table 1: Four systems which can be complex ... when the global topology is nontrivial, producing the phenomena listed in the third column.

efficiently by different models at different scales. That is, if a system can be simulated with bounded error and with less computational effort at larger scales in terms of macroscopic variables than in terms of microscopic variables, and the macroscopic variables differ qualitatively from the microscopic ones, then the system is complex. Table 1 summarizes some examples of complex systems for which the qualitative difference of the macroscopic model derives from nontrivial global topology [3]. The second and third systems in this table are essentially similar: there is a finite number of continuous valued variables in the local models—relative populations in the second and prices of some finite number of goods in the third. The first system in the table, fluid flow, differs in having a continuum of continuous valued variables, the velocities $v(x)$, while the fourth, iterated social choice, has a finite number of discrete valued variables. In Example 1, these are the three pie allocations Mom proposes. Our goal is to bridge the gap between this situation and the one exemplified by market economics and population dynamics—by making the natural generalization to processes where a continuum of allocations is allowed [5]. To varying degrees, this is a good model for social systems such as wealth/income distribution in populations, natural resource exploitation, market share for competing corporations, satellite bandwidth division among many users, and CPU time usage by multiple software agents running simultaneously. Specifically, we want to extend the techniques developed to quantify the complexity of iterated social choice [4] to continuous variables. We do so in Section 3, after reviewing the analysis of the discrete problem in Section 2. In Section 4 we consider a natural modification of the allocation process and then conclude with a discussion in Section 5.

2 Iterated discrete choice

The general setting for the discrete choice problem of Example 1 is a finite set of alternatives A and a finite number of agents, each of whom has a *preference order* on A. The usual model for the preference order of a rational agent is a relation, denoted $\geq$, which is *complete* ($a, b \in A \Rightarrow a \geq b$ or $b \geq a$) and *transitive* ($a, b, c \in A$ and $a \geq b$, $b \geq c \Rightarrow a \geq c$) [2]. We formalize voting or aggregation by maps f from preference profiles p (a list of agent preference orders) to directed graphs f_p. As in Figure 1, the vertices of f_p correspond to alternatives in A and a directed edge $a \leftarrow b$ in f_p indicates that for profile p the map f chooses alternative a over alternative b. We call f a *voting rule* if for all profiles p, f_p is *complete* ($a, b \in A \Rightarrow a \leftarrow b$ or $b \leftarrow a$ in f_p) and *unanimous* ($a, b \in A$ and $a \geq b$ in each preference order in $p \Rightarrow a \leftarrow b$ in f_p). Notice that in Figure 1 we have omitted the directed edges—present in every f_p—connecting each vertex to itself. The directed graph f_p defines a *symbolic dynamical system*: Suppose the agents are presented with a sequence of alternatives. The results of successive pairwise votes between each new alternative and the current one form a sequence of symbols representing the chosen alternatives. The possible sequences are exactly the directed paths in f_p. These paths, together with the *shift map* (deletion of the first symbol of a sequence) form a dynamical system—a (*one-sided*) *subshift of finite type* [6]. To enumerate the paths we define the *transition matrix* F_p by $(F_p)_{ab} = 1$ if $a \leftarrow b$ in f_p and $(F_p)_{ab} = 0$ otherwise; then the number of paths of length N from b to a is $(F_p^N)_{ab}$. The *topological entropy* [7] of the symbolic dynamical system defined by f_p is

$$S[f_p] := \lim_{N\to\infty} \frac{1}{N} \log \operatorname{Tr} F_p^N = \log(\text{largest eigenvalue of } F_p). \qquad (1)$$

In [4] we showed that the topological entropy is positive exactly when there is a Condorcet cycle and observed that this makes the dynamical system chaotic [8]. In Example 1 there is a preference cycle, the transition matrix is

$$F_p = \begin{pmatrix} 1 & 0 & 1 \\ 1 & 1 & 0 \\ 0 & 1 & 1 \end{pmatrix},$$

and $S[f_p] = 2$. We can observe that the system is chaotic since, for example, arbitrarily close paths diverge like 2^t under t iterations of the shift dynamics. By contrast, suppose that in our example only Alice's preferences matter: since she prefers $a_1 > a_3 > a_2$, the direction on the edge between a_1 and a_2 in Figure 1 would be reversed. For this *dictatorial* voting rule f' we have an acyclic f'_p, the transition matrix is

$$F'_p = \begin{pmatrix} 1 & 1 & 1 \\ 0 & 1 & 0 \\ 0 & 1 & 1 \end{pmatrix},$$

and $S[F'_p] = 0$. This system is *not* chaotic: the set of paths converging to a_1 for large t has measure 1; neither is it complex: the whole system can be described by Alice's preference order—there is no difference between efficient models at the local (individual) and global (group) scales. The original system, however, is efficiently described at the global scale by f_p—a directed graph containing a nontrivial cycle, which is qualitatively different than an individual preference order. These examples illustrate that the fundamental distinction between iterated discrete choice systems is *topological*—the presence or absence of preference cycles. Their existence was observed by Condorcet [1]; Arrow analyzed conditions under which they are inevitable [2]; Chichilnisky recognized their topological importance [9]; and we have noted that the topological entropy quantifies the amount of cyclicity, distinguishing simple from complex discrete choice systems [3], and suggested that it be used to quantify complexity in this context [4]. Now we will extend this approach to a situation with a continuum of alternatives—continuous allocation processes.

3 Continuous allocation processes

EXAMPLE 2. Suppose Mom is able to cut the pie into any three portions x_1, x_2 and x_3 for Alice, Bob and Charlie, respectively. Since there is only one pie, each allocation is described by a vector $x = x_1\hat{e}_1 + x_2\hat{e}_2 + x_3\hat{e}_3 \in \boldsymbol{R}^3$ with $x_1+x_2+x_3 = 1$ and $x_i \geq 0$. The x_i are barycentric coordinates on the *allocation simplex*—the two dimensional simplex bounded by the triangle with vertices at $\{\hat{e}_i\}$ ($\hat{e}_1 := (1,0,0)$, *etc.*). We supposed that each child prefers more pie to less, so the ith child prefers allocation $y = (y_1, y_2, y_3)$ to $x = (x_1, x_2, x_3)$ iff $y_i > x_i$. This defines a profile for three agents on the space of allocations. Let us suppose, as in Example 1, that the three preference orders in this profile are aggregated by majority rule, *i.e.*, the group prefers allocation y to x iff

$$\sum_i \operatorname{sign}(y_i - x_i) \geq 0, \tag{2}$$

in which case we write $y \leftarrow x$. [1] The consequences of this voting rule are captured by the lightcone-like diagram shown in Figure 2: The point labelled x in the allocation simplex is preferred under majority rule to all points in the shaded regions, while all points in the unshaded regions, *e.g.*, a_1, are majority preferred to x. Possible sequences of outcomes in an iterated voting procedure then, consist of points each of which lies in 'the future lightcone'—the unshaded region—of its predecessor. Unlike the (usual) situation in Lorentzian geometry, however, cycles are prevalent: *e.g.*, the $a_1 \rightarrow a_2 \rightarrow a_3 \rightarrow a_1$ cycle of Example 1 exists also in this continuous generalization. Our goal is to measure the complexity of this system by extending the topological entropy analysis we described for discrete choice systems [4] in Section 3. One might attempt to do so by *discretizing* Example 2 to some finite number V of allocations—such as the vertices of the triangular lattice shown in Figure 2—and then taking the limit as $V \rightarrow \infty$. The voting rule (Eq. 2) defines a directed graph on these vertices and we can compute the topological entropy (Eq. 1) of the associated symbolic dynamical system. As we refine the discretization to more vertices, however, the largest eigenvalue of the transition matrix, $\Lambda_V \sim V$ as $V \rightarrow \infty$, so to get a continuum limit we must rescale to Λ_V/V. The *relative entropy* [11] is the logarithm of the rescaled largest eigenvalue, *i.e.*,

$$S := \lim_{V\rightarrow\infty} \log \Lambda_V - \log V. \tag{3}$$

[1] With this voting rule the situation is equivalent to the general three-person zero-sum game considered originally by von Neumann and Morgenstern [10].

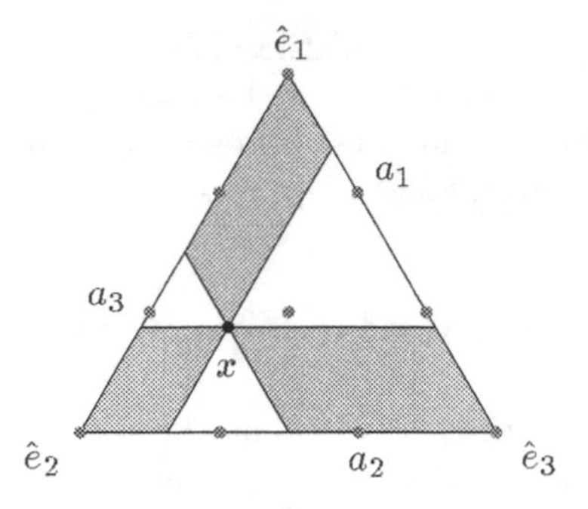

Figure 2: Aggregated preferences on the allocation simplex: unshaded points are majority prefered to x. The triangular lattice including alternatives a_1, a_2 and a_3 is a discrete approximation to the continuous allocation simplex.

Since $\Lambda_V \leq V$, the relative entropy is nonpositive, vanishing for the maximally cyclic case of complete directed graphs. The relative entropy, therefore, measures how far the system is from being *maximally* chaotic. To compute the relative entropy, however, it is more convenient—as well as aesthetically pleasing—to work directly in the continuum, where a transition matrix becomes a transition *operator*, or Green's function. Here it is defined by

$$T(y,x) := \begin{cases} 1 & \text{if } y \leftarrow x; \\ 0 & \text{otherwise.} \end{cases}$$

Matrix multiplication is replaced by integration: y can be reached from x in *two* steps provided

$$T^2(y,x) := \int_\triangle T(y,z)T(z,x)\mathrm{d}z > 0,$$

and in N steps if

$$T^N(y,x) := \int_\triangle T^{N-1}(y,z)T(z,x)\mathrm{d}z > 0,$$

where $\triangle$ indicates that the variable z is integrated over the allocation simplex. Just as in the discrete situation, the 'number' of cyclic paths of length N is given by the trace of this operator, which is again defined by integration. Thus the relative entropy of Eq. 3 can equivalently be defined as:

$$S := \lim_{N\to\infty} \frac{1}{N} \log \int_\triangle T^N(x,x)\mathrm{d}x - \log \text{area}(\triangle), \tag{4}$$

i.e., as the logarithm of the largest eigenvalue of the operator T, rescaled by the area of the allocation simplex. The eigenvalue problem for such an integral operator is

$$\lambda f(y) = \int_\triangle T(y,x)f(x)\mathrm{d}x. \tag{5}$$

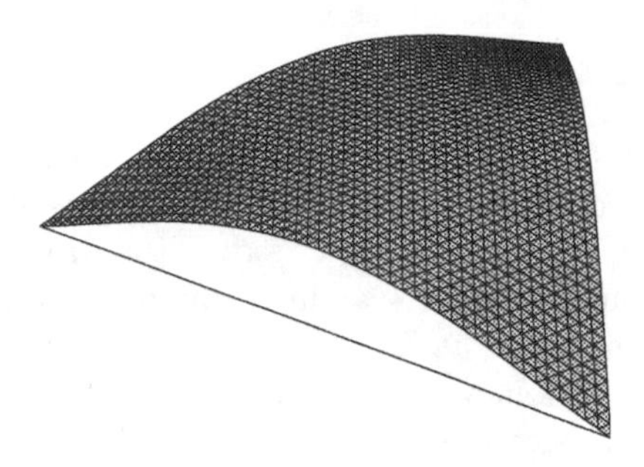

Figure 3: The approximate eigenfunction of T with largest norm eigenvalue $\Lambda \approx 0.28494$.

Clearly $f(x) \equiv 0$ solves this equation; the problem is to determine the *eigenvalues* λ_i (in our case we mostly care about the one, $\Lambda := \lambda_1$, with the largest norm) for which Eq. 5 has nontrivial solutions, the *eigenfunctions*. Leaving the calculational details for another paper [12], Figure 3 shows the eigenfunction of T with the largest norm eigenvalue, $\Lambda \approx 0.28494$. Notice that $\Lambda < \frac{1}{4}\sqrt{6} = \text{area}(\triangle)$ so that the relative entropy (Eq. 4) $S := \log \Lambda - \log \text{area}(\triangle) < 0$. That is, there is a lower density of Condorcet cycles than there would be if $T(y,x) \equiv 1$ (in which case the largest norm eigenvalue solving Eq. 5 is $\frac{1}{4}\sqrt{6}$ and the corresponding eigenfunction is constant), but it is high enough that the relative entropy is finite. As in the previous section, we can compare with a *simple* system: Suppose again that only Alice's preference matter so that there are no preference cycles. Then the transition operator is

$$T'(y,x) := \begin{cases} 1 & \text{if } y_1 \geq x_1; \\ 0 & \text{otherwise,} \end{cases}$$

and it is clear that the only nontrivial eigenfunction of this operator is concentrated at $\hat{e}_1$. Thus the eigenfunction is proportional to $\delta(\hat{e}_1 - x)$ and according to Eq. 5, the corresponding eigenvalue is 0. In this case the relative entropy is $-\infty$. These examples demonstrate that the finiteness of the relative entropy (Eq. 3 or Eq. 4) for iterated continuous choice systems is the criterion corresponding to positivity of the entropy (Eq. 1) for iterated discrete choice systems. Each reflects the presence of topologically nontrivial paths in the space of alternatives [2] and identifies the system as complex.

[2]It is amusing to note that if we extend our model to continuous *time* choices, there are continuous periodic paths in the allocation simplex, as well as continuous paths from any point to any point other than the vertices $\hat{e}_i$. The discrete time version of this statement was proved by Ward [5].

4 Inequity

The eigenfunction of T shown in Figure 3 also reflects the presence of Condorcet cycles—by not being concentrated at a single point as is the eigenfunction of the transition operator T' modelling Alice's dictatorship. We may consider both the discrete and continuous choice systems probabilistically: at each timestep a new alternative y is presented at random—uniformly from all the alternatives. This probability distribution, weighted by $T(y, x)$, defines a transition probability from x to y. The eigenfunction is a projectively stationary distribution for this non-probability preserving process. For many allocation processes, how-

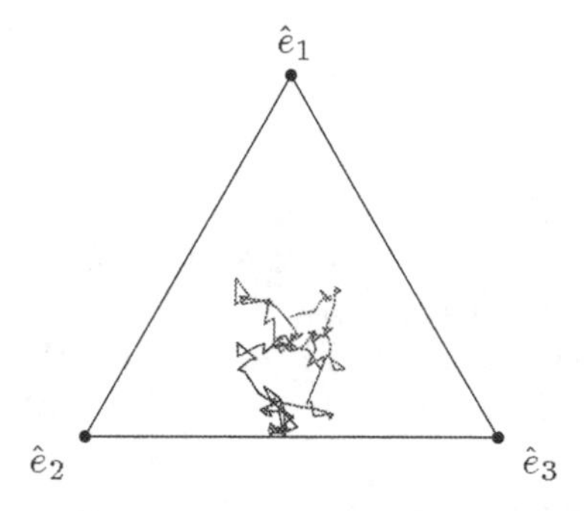

Figure 4: The first 400 steps of a path starting at the equal allocation $(\frac{1}{3}, \frac{1}{3}, \frac{1}{3})$ and driven by a Gaussian random process with $\sigma = 0.03$.

ever, this is not the most realistic probabilistic model. In the case of wealth distribution, for example, it seems likely that a possible change in the allocation will be fairly small—just a perturbation of the *status quo* implemented by, say, a small change in the tax code. Rather than a uniform distribution for the new alternative, therefore, we might consider a Gaussian distribution centered at the *status quo*. The voting rule (Eq. 2) still determines which new alternatives are preferred. Figure 4 shows the first 400 steps of a path starting at the equal allocation $(\frac{1}{3}, \frac{1}{3}, \frac{1}{3})$ and generated by a Gaussian with width $\sigma = 0.03$. Notice the cycles in the path, and the apparently larger probability for allocations near the center of the allocation simplex, both phenomena we derived in the previous section. We must also remark that majority rule (Eq. 2) may be an unrealistic idealization for some allocation processes. In the case of wealth distribution, for example, it seems likely that the wealthier agents will have a greater say in the decision about a new allocation. Similarly, changes in market share may be more heavily influenced by corporations which dominate the market. To model unequal influences we need only weight the corresponding terms in Eq. 2, defining the aggregate preference $y \leftarrow x$ iff

$$\sum_i x_i \mathrm{sign}(y_i - x_i) \geq 0. \tag{6}$$

This voting rule weights the preference of each agent proportionaly to his/her fraction in the current allocation. Figures 5 and 6 illustrate the consequences of this inequitable voting rule with sample paths generated by Gaussian distributions. Each path starts at the equal allocation $(\frac{1}{3}, \frac{1}{3}, \frac{1}{3})$. Figure 5 shows the first 100 steps with a Gaussian of width $\sigma = 0.03$, just as in Figure 4. Notice that while there are still cycles, the path converges rapidly to $\hat{e}_3$. Figure 6 shows the first 400 steps with a narrower Gaussian, $\sigma = 0.02$. In this case the allocation process takes longer to leave the center of the allocation simplex, and squeezes agent 1's fraction down near 0 before converging to $\hat{e}_2$. [3] These examples illus-

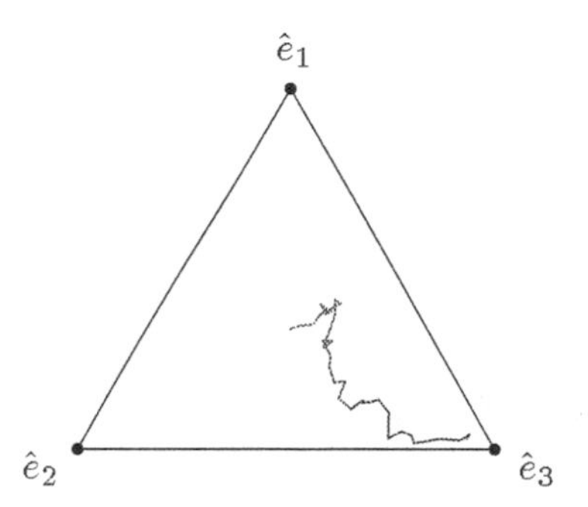

Figure 5: The first 100 steps for the inequitable rule (Eq. 6) driven by Gaussian fluctuations with $\sigma = 0.03$. There are cycles, but the path converges to $\hat{e}_3$.

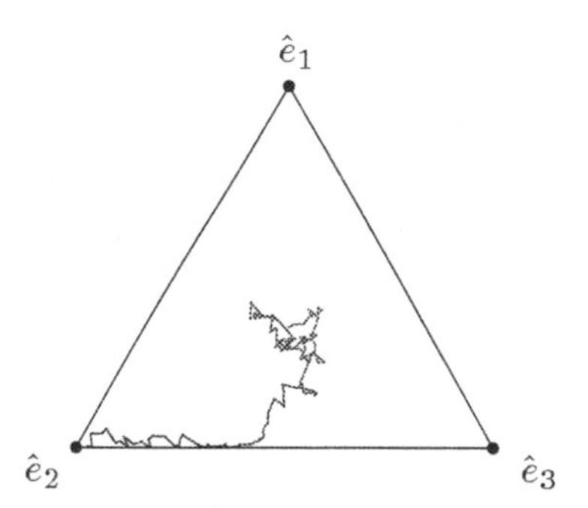

Figure 6: The first 400 steps for the same process driven by a narrower Gaussian ($\sigma = 0.02$). The path takes x_1 near zero and then converges to $\hat{e}_2$.

trate that although the voting rule (Eq. 6) is symmetric (under permutation of the agents [10]), the fact that it is inequitable leads to *symmetry breaking* [13]: Almost all paths converge to one of the $\hat{e}_i$, but to *which* is a consequence of the Gaussian random process—deviations from the equal distribution are amplified

[3] Log normal (*i.e.*, Gaussian for the log of a multiplicative factor) distributions for the x_i, rescaled to keep the total 1, give similar—although not identical—results.

and eventually 'frozen in' by the inequity of the voting rule (Eq. 6). So complex allocation processes can include the venerable economics phenomenon of increasing returns [14] and demonstrate the path-dependent consequences [15].

5 Discussion

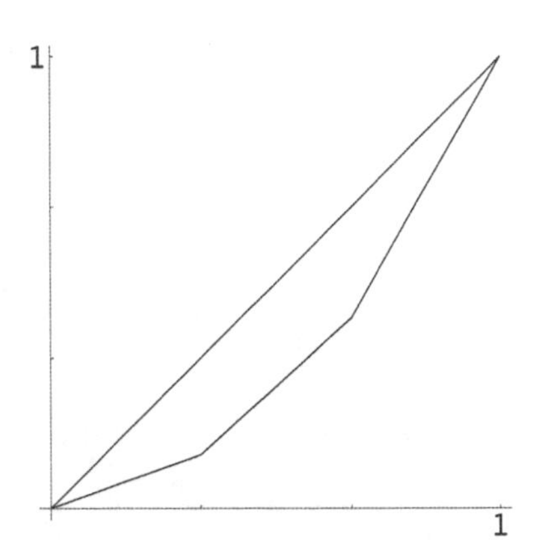

Figure 7: The Lorenz curve for the expected allocation of the eigenfunction shown in Figure 3 lies below the Lorenz curve for an equal distribution (the diagonal line). Fraction of the total is graphed as a function of 'poorest' population fraction.

We have generalized the entropy measure of complexity for iterated discrete choice systems to the economics setting of continuous allocation processes. The most complex system we have considered is the majority rule allocation process; in Section 3 we showed that the relative entropy can be computed precisely in this case. Furthermore, the density of Condorcet cycles makes the equal allocation $(\frac{1}{3}, \frac{1}{3}, \frac{1}{3})$ only the most probable; even with this equitable voting rule the expected values satisfy

$$\langle \text{max. fraction} \rangle > \frac{1}{3} > \langle \text{min. fraction} \rangle,$$

implying inequality in the distribution. In Figure 7 we illustrate this inequality with a Lorenz curve [16] which plots the cumulative allocated fraction as a function of the 'poorest' fraction of the population. For the distribution given by the eigenfunction in Figure 3, the Lorenz curve lies below the diagonal Lorenz curve for the equal allocation. In Section 4 we considered allocation processes driven away from an initial equal allocation by Gaussian fluctuations. The expected inequality of these random processes increases in each case, stabilizing at some intermediate value for majority rule (Eq. 2), but converging to the maximum for the inequitable rule (Eq. 6). The latter situation exemplifies the evolution of a complex system into a simple one!

Acknowledgements

I thank Randy Bank, Thad Brown, Peter Doyle, Bruce Driver and Mike Freedman for useful discussions about various aspects of this project, and Sun Microsystems for providing computational support.

Bibliography

[1] M. J. A. N. de Caritat, Marquis de Condorcet, *Essai sur l'application de l'analyse à la probabilité des décisions rendues à la pluralité des voix* (Paris: *l'Imprimèrie Royale* 1785).

[2] K. J. Arrow, *Social Choice and Individual Values* (New York: Wiley 1951).

[3] D. A. Meyer, "Towards the global: complexity, topology and chaos in modelling, simulation and computation", chao-dyn/9710005, *InterJournal Complex Systems*, Article [123], http://dynamics.bu.edu/InterJournal/.

[4] D. A. Meyer and T. A. Brown, "Statistical mechanics of voting", cond-mat/9806359, *Phys. Rev. Lett.* **81** (1998) 1718–1721.

[5] B. Ward, "Majority rule and allocation", *J. Conflict Resolution* **5** (1961) 379–389.

[6] W. H. Gottshalk and G. A. Hedlund, *Topological Dynamics*, AMS colloquium publications, vol. 36 (Providence, RI: AMS 1955).

[7] C. E. Shannon, "A mathematical theory of communication", *Bell System Tech. J.* **27** (1948) 379–423; 623–656;
W. Parry, "Intrinsic Markov chains", *Trans. Amer. Math. Soc.* **112** (1964) 55–66;
R. L. Adler, A. G. Konheim and M. H. McAndrew, "Topological entropy", *Trans. Amer. Math. Soc.* **114** (1965) 309–319.

[8] T. Li and J. Yorke, "Period three implies chaos", *Amer. Math. Monthly* **82** (1975) 985–992;
M. Misiurewicz, "Horseshoes for continuous mappings of an interval", in C. Marchioro, ed., *Dynamical Systems*, proceedings of the CIME session, Bressanone, Italy, 19–27 June 1978 (Napoli, Italy: *Liguori Editore* 1980) 125–135.

[9] G. Chichilnisky, "The topological equivalence of the Pareto condition and the existence of a dictator", *J. Math. Econom.* **9** (1982) 223–233.

[10] J. von Neumann and O. Morgenstern, *Theory of Games and Economic Behavior*, third edition (Princeton: Princeton University Press 1953).

[11] S. Kullback and R. A. Leibler, "On information and sufficiency", *Ann. Math. Statist.* **22** (1951) 79–86.

[12] D. A. Meyer, in preparation.

[13] J. Goldstone, "Field theories with 'superconductor' solutions", *Nuovo Cimento (10)* **19** (1961) 154–164;
Y. Nambu and G. Jona-Lasinio, "Dynamical model of elementary particles based on an analogy with superconductivity. I; II", *Phys. Rev.* **122** (1961) 345–358; **124** (1961) 246–254.

[14] A. Marshall, *Principles of Economics* (London: Macmillan and Co. 1890).

[15] For some recent work, see:
E. Helpman and P. R. Krugman, *Market Structure and Foreign Trade: Increasing Returns, Imperfect Competition, and the International Economy* (Cambridge, MA: MIT Press 1985);
W. B. Arthur, *Increasing Returns and Path Dependence in the Economy*, with a foreword by K. J. Arrow (Ann Arbor: University of Michigan Press 1994);
and references therein.

[16] M. O. Lorenz, "Methods of measuring the concentration of wealth", *J. Amer. Stat. Assoc.* **9** (1905) 209–219.

Caution! Agent Based Systems in Operation

T. Middelkoop **A. Deshmukh**
The University of Massachusetts at Amherst
Department of Mechanical and Industrial Engineering
mtim, deshmukh@farm.ecs.umass.edu

1 Introduction

As agent based systems move out of the controlled laboratory and simulated environment into real world applications, they face significant design and operational challenges. This paper attempts to highlight some of the major issues involved in the design and implementation of agent based applications. Our intent is not to prescribe specific solutions or actions, which will depend on the application domain, scope and resources available, but to make the reader aware of the issues which need serious consideration when using agents to tackle real world problems. The following sections each address a specific issue related to design, implementation or operational efficiency of multi–agent systems.

2 When to use Agents?

The foremost decision a system designer has to make is when to use agents to solve a problem. It is important to note that not all problems benefit from agent based approaches. Thus, it is crucial to identify the key characteristics which make a problem suitable for agent solutions. Agent based system closely represent how natural systems work by distributing a problem among a number of autonomous entities[7, 6]. Thus, if the problem cannot be effectively divided into a series of interacting subproblems or subgoals, an agent approach may not be successful. The agency approach offers the power of problem decomposition and parallelism for tackling complex problems. The autonomous nature of agents also offers flexibility in dealing with uncertain situations and localized failures. However, agent based systems require significant life–support and monitoring overhead.

For example, one application domain where agent based systems offer significant advantages is distributed system monitoring. In this domain, agents process large amounts of local data and only communicate meta–level results to other agents. This vastly reduces the bandwidth requirements of the entire system. Applications in this domain include telephone, power and computer network monitoring. On the other hand, the additional overhead may not be justified for using agents to transfer data between two applications, where the data format and conversion requirements are known and fixed.

3 Agent Functionality

As mentioned in the previous section, problem decomposition and parallelism are key advantages of the agency approach. Hence, the level of functionality assigned to each agent is an important characteristic of the system. On one hand, an agent containing too much functionality can defeat the advantages of agency and parallelism. Whereas, if each agent has limited functionality, the system can be overwhelmed by large number of agents and excessive communications. In many cases the agent functionality is determined by the problem decomposition process. The decisions made by each agent are determined by their respective goals, tasks and available resources. Every agent in the system does not need to have the same level of functionality. The use of agent hierarchies can result in efficient organizations. Alternatively, higher lever organization can emerge out of the agent interactions, where certain agents act as information and decision consolidators, in order to reduce the interactions among agents and coordinate tasks.

The optimal distribution of functionality among various agents in the system depends on various factors, such as available communication bandwidth, processing capability of individual agents, real–time decision requirements, level of fault tolerance desired and the nature of the task to be performed.

4 Incorporating Values Into Agents

The representation of goals and values in individual agents requires translation of overall goals and problem requirements into an internal representation for each agent. The agents have to be able to evaluate the value of each action from a local perspective and also consider its impact on societal objectives.

Specification of utility functions or value functions is commonly used to compare different options available. Appropriate translation of the cost of uncertainty and risk associated with each option needs to be done by the system designer in order to guide the overall system performance to desirable levels. For example, the importance placed on risk may be different in making mission critical decisions as compared to certain backup or redundant operations. In case utility function cannot be specified *a priori*, the agents could be equipped with learning mechanisms to judge the utility of their decisions and improve as the system evolves. In most cases individual agents are myopic in their decision making. However, it is important to realize that the collection of these myopic decisions leads to system level objectives. This interplay must be examined to realize the potential effect that a single agent may have on the overall system.

Another important aspect which needs careful examination is the difference between actual and perceived utility. Consider an example of the role of advertising and promotions on customers. If a product selection decision was solely based on quantitative functional characteristics of the product, then advertising would not be effective in promoting product sales. However, the perceived utility of a product can be altered by changing the emphasis placed by customers on different product qualities using advertisements. Such methods can be employed by agents, especially when the information available to agents involved in a negotiation is asymmetric. The system designers also have to be aware of conflicting goals of different agents and devise conflict resolution methods to handle these situations. For example, a resource agent on the shop floor may use maximizing its utilization as a local goal, whereas from a customer's perspective it may desirable to minimize the time spent in the system. Cooperative or environmentally friendly agents can be developed to alleviate some of the issues discussed before. These agents understand the notion of sharing resources, such as bandwidth and computations, as well as follow appropriate protocols to coordinate tasks with other agents.

5 Incorporating Intelligence into Agents

It is important to balance learned and programmed behavior in agents. A strictly programmed agent will be rigid and inflexible, whereas a pure learned behavior may be time consuming and will not guarantee desired results. For example, an agent that uses search engines having an entirely programmed behavior will need to be reprogrammed every time the search engine interface is updated. Alternatively, if the agent used a purely learning approach it may not deliver any useful information until it learns for a significant period of time. There are variety of learning techniques that agents can employ, such as neural networks, regression, retrospection and temporal differencing[4]. A agent based system developer needs to determine the appropriate level of learning and suitable learning mechanism to be incorporated in agents to meet the application needs.

Another form of intelligence is the domain knowledge. As with learning, there exists a balance between domain knowledge and generic information. One approach to this problem in multi–agent systems is to distribute the domain knowledge among agents. Domain knowledge can be expressed in the form of expert systems, procedural reasoning systems, first order logic and Bayesian decision trees. The domain knowledge can then be traded between agents as and when required.

In general, agent intelligence and learning issues have to deal with balancing requirements and resources, such as solution quality, memory, computation time, communication and interruptability (anytime algorithms).

6 Context and Ontological Issues

In order to achieve the overall goals, agents must communicate their goals and intents with other agents. These communications need to be unambiguous and structured so as to allow agents to interpret each others objectives. Specification of a context plays an important part in the agent communications. For example, an agent inquiring about the *load* of an electro–mechanical object needs to specify the context: whether the load is structural or electrical. Similar examples of contextual and ontological mismatches can be found while using currently available Internet search engines.

The common vocabulary and context allows the agents to work together and have a shared intent. It is the expression of this shared intent that facilitates agents working towards system wide goals. For example, in order for the Internet to work correctly individual autonomous routers, or agents, must communicate with specifically defined protocols. The shared context allows the coordination of a large system, such as the Internet, to work effectively.

One way of maintaining a common context is to use existing standard meta languages, such as KQML[2], XML[13], and KIF[3]. These languages provide both a well defined meta context, a library of programming tools and the ability to interact with agents using similar standards. However, specific ontologies have to be developed for each domain by the system designers.

7 When Agents Make Mistakes

Dealing with errors, misinformation and rogue information is an important issue with agents, especially with the use of agents in electronic commerce. System designers need to pay close attention to the authentication and data consistency issues, and incorporate evaluation mechanisms in the agent architecture. For example, the information gathered by an agent from an Internet site may be verified from other sources before it is used in a decision process. Assessing the importance of information can help in determining appropriate verification processes. In a resource constrained environment, it is not always possible nor desirable to validate all information accurately.

Even though agents may not maliciously present false data, errors can occur in data transmission or translation. In such cases, apportioning the blame becomes an interesting legal issue. The legal and ethical issues related to errors, intended or unintended, due to agents in operational settings need to be resolved before implementation. Agents may need to securely log important transactions in order to maintain records and perform self diagnostics.

8 Convergence and Negotiations

Many applications require solutions, of a desired quality, in a specified amount of time. In some cases, it might be desirable to obtain optimal or near optimal solutions. Agent based systems, especially when iterative negotiations are used, do not always guarantee convergence to a solution in a specific time period. Moreover, the optimality or robustness of the solutions are not assured. It is important to understand these limitations and plan appropriate tests to guarantee quality of the results generated by agent based systems.

The quality of the solutions generated by agent based systems can be evaluated by comparing the results to theoretical optimal solutions generated by simplified models or lower bounds obtained by relaxations of the actual formulations, or by basing the agent systems on well studied analogies, such economic or financial markets. Wellman *et al.* draw analogies from the financial world into agent negotiations[14]. Their decentralized scheduling system is based on free markets utilizing bidding protocols. This technique represents utility functions in terms of currency and the agents *bid* for processing resources. The system uses a generalized Vickery auction to evaluate and allocate bids. Since the characteristics of the Vickery auction are well known, the convergence and solution quality of the decentralized scheduling system can be easily extrapolated. Systems based on financial models often use bidding, auctions, negotiations and argumentation techniques for agent interactions[8, 12, 1]. These techniques and their associated attributes are based on interacting autonomous entities, which can be easily translated into agent based settings.

This approach however requires strong assumptions to be made about the system in order to fit a known model or obtain solution bounds. For example the game theoretic framework assumes each agent has a utility function and that it is known. Many times this utility function is not known and can be very hard to determine explicitly. Hence, in order to use game theoretic constructs one needs to assume certain utility values for agents, which may not be realistic[9].

9 System Level Issues

Agent based systems, like other asynchronous distributed environment, are faced with the system level issues, such as stalling, deadlock, cycling, etc. Arguably, a centralized observer or controller is the antithesis of the distributed decision making qualities of agents. However, most real world agent applications need some level of system monitoring in order to ensure error free operations.

For example, consider the deadlock problem that can arise due to information dependencies among agents. Agent A could be waiting on information from agent B which is waiting on information from agent C which is waiting on information A. Can this be detected? Can it be eliminated? Questions like these must be answered in order to have a robust system. The agent based systems also need to be monitored for resource consumption at the system level. For example, the communication bus is shared by all the agents in the system. Although each agent uses only a fraction of the bandwidth, the level of negotiations and number of agents involved may result in a communications bottleneck. Another example of system level parameters which need to be monitored can be given using an analogy from the national economy. The price inflation is closely monitored by the Federal Reserve in order to maintain the balance in the economy. Similar inflationary pressures can exist in agent based systems, where price cascades can result in virtual shutdown of activities in the system.

The ability to observe the system from both an aggregate view and an agent view is paramount. Agent systems can be hard to debug due to their distributed asynchronous nature. Hence special attention needs to given to diagnostic tools for agent based systems before they are deployed in the real world.

10 Implementing Agent Systems

Agent based systems need a set of basic tools and services in order to implement them, such as birth/death services, communication channels, interaction protocols, etc. If the agents are mobile over heterogeneous systems they require additional capabilities for process migration. A large body of standard tools and protocols is available to accomplish these tasks. It is advisable to use these standards where possible in order to reduce the implementation overheads and ensure interoperability among different agent systems. It is important to note that the final goal is to develop a solution to an application and implementation itself is simply a means of achieving that end[15].

One of the most popular programming languages for agents is Java[11]. This network aware platform independent language provides the core functionality that an agent system needs by isolating the operating system and hardware details from the agent. Other commonly used languages for agents include AgentTCL, lisp, and prolog. These languages only provide the rudimentary capabilities that an agent system requires. The agents need an environment to work in, called a Multi Agent System (MAS)[10]. The MAS provides an environment for the agents to exist in and perform their activities. Several MAS have been developed commercially, most of them based around the Java language. Such systems include Voyager by Objectspace, Odssey by General Magic and Aglets by IBM. Other software techniques for facilitating agent interaction on a lower level include the Common Object Request Broker Architecture (CORBA)[5] and Distributed Component Object Model (DCOM).

In order to effectively communicate with each other, agents need a shared structured communication protocol. Knowledge Query and Manipulation Language (KQML)[2] is an example of an agent communication language which provides a standard way for agents to interact. This language provides basics tools for exchanging information. However, in order have a meaningful dialogue the agents must have a shared context. The Knowledge Interchange Format (KIF)[3] is an example of a formal mechanism for agents to exchange knowledge in a structured format.

11 Designing Optimal Agent Systems

The question of designing best agents for a particular task still remains unanswered. Moreover, the design of a society of agents which is ideally suited for solving a complex problem requires the use of prescriptive tools for analyzing distributed agent systems. Traditionally, simulation has been used to analyze the performance of agent based systems under different operating conditions before they are implemented in the real world. Simulation is, however, only a descriptive tool. The computational expense of exhaustive testing of all operational scenarios may be significantly high, if at all possible.

The difficulty in developing prescriptive analytical tools for agent based systems is due to the non–equilibrium conditions under which they operate, the difference in the information available to each agent for decision making and, more practically, the sheer size of some of the real world agent system applications.

Several problem or goal decomposition techniques can be used to design individual agents. Although, the solutions generated by these systems may not be optimal, we can guarantee the quality of the solutions or the deviation from the theoretical optimal that can be achieved by such decomposition procedures. Several negotiation protocols also have been studied by researchers and their long range characteristics are known. Hence, prescriptive methods for design of agent based systems may be available in the near future.

12 Summary

Agent systems offer significant advantages for modeling complex adaptive systems. However, they are not the solution to all complex problems. It is important to note the strengths, weaknesses and implementation issues related to agent based systems. The issues discussed in this paper need to be studied for specific application domains to determine the appropriate strategies. Although, much remains to be learned about the emergent behavior of agent systems and how best to design them, this approach has the potential of revolutionizing the system design field.

Bibliography

[1] CLIFF, Dave, and Janet BRUTEN, "Zero is not enough: On the lower limit of agent intelligence for continuous double auction markets", *Tech. Rep. no. HPL-97-141*, Hewlett-Packard Laboratories, (1997).

[2] FININ, Tim, Rich FRITZSON, and Don MCKAY, "A language and protocol to support intelligent agent interoperability", *Proceedings of the CE & CALS Washington '92 Conference*, (June 1992).

[3] GENESERETH, Michael R., and Richard E. FIKES, *Knowledge Interchange Format Version 3.0 Reference Manual*, Computer Science Department Stanford University Stanford (1994).

[4] GREEN, Shaw, Leon HURST, and Brenda NANGLE, "Software agents: A review", *Tech. Rep. no. TCD-CS-1997-06*, Intelligent Agents Group (IAG), (May 1997).

[5] GROUP, Object Management, "Common object request broker architecture (corba)", http://www.corba.org (1998).

[6] JENNINGS, N. R., K. P. SYCARA, and Michael WOOLDRIDGE, "A roadmap of agent research and development", *Journal of Autonomous Agents and Multi-Agent Systems* **1**, 1 (July 1998), 7–36.

[7] JENNINGS, Nicholas R., and Michael J. WOOLDRIDGE, *Agent Technology: Foundations, Applications, and Markets*, Springer-Verlag (March 1998).

[8] KARUS, Sarit, "Negotiation and cooperation in multi-agent environments", *Artificial Intelligence journal, Special Issue on Economic Principles of Multi-Agent Systems* **94**, 1-2 (1997), 79–98.

[9] KRAUS, Sarit, and Daniel LEHMANN, "Designing and building a negotiating automated agent", *Computational Intelligence* **11**, 1 (1995), 132–171.

[10] LEVY, Renato, Kutluhan EROL, and Howell MITCHELL, "A study of infrastructure requirements and software platforms for autonomous agents", *Proceedings of ICSE-96*, (July 1996).

[11] MICROSYSTEMS, Sun, "Java programming language", http://java.sun.com (1998).

[12] SIERRA, Carles, Nick R. JENNINGS, Pablo NORIEGA, and Simon PARSONS, "A framework for argumentation-based negotiation", *Proc. Fourth Int. Workshop on Agent Theories Architectures and Languages (ATAL-97)* (Rode Island, USA,), (1997), 167–182.

[13] WC3, "Extensible markup language", http://www.w3.org/XML/ (1998).

[14] WELLMAN, Michael P., William E. WALSH, Peter R. WURMAN, and Jeffrey K. MACKIE-MASON, "Auction protocols for decentralized scheduling", *In submission to Games and Economic Behavior* (July 1998).

[15] WOOLDRIDGE, Michael, and N. R. JENNINGS, "Pitfalls of agent-oriented development", *Proceedings of the Second International Conference on Autonomous Agents* (K. P. SYCARA AND M. WOOLDRIDGE eds.), ACM Press (May 1998).

Entropy estimation from finite samples

Dirk Holste and Hanspeter Herzel
Humboldt University, Institute for Biology, Berlin
Ivo Grosse
Free University, Institute for Molecular Biology, Berlin
{d.holste,i.grosse,h.herzel}@itb.biologie.hu-berlin.de

Entropy analysis is a common method to search for statistical patterns hidden in experimental symbol sequences or time series. In order to obtain a more comprehensive description, generalized Rényi (H_q) and Tsallis (H'_q) entropies have been introduced. A practical problem lies in the numerical estimation of H_q and H'_q from finite samples, which can lead to systematic and statistical errors. We focus on the problem of estimating H_q and H'_q from finite samples and derive the Bayesian estimators $\widehat{H}_q$ and $\widehat{H}'_q$. We compare $\widehat{H}_q$ and $\widehat{H}'_q$ with the standard frequency-counts estimators of H_q and H'_q and find by numerical simulations that $\widehat{H}_q$ and $\widehat{H}'_q$ reduce statistical errors for higher-order Markov processes.

1 Introduction

Measuring the complexity of experimental symbol sequences or time series is a central issue that occurs in many fields of sciences such as in computational physics [1, 15] and biology [3, 6], dynamical systems theory [10], and linguistics [4]. While most approaches are based on the assumption of infinite sequences, in many circumstances one has to cope with finite samples. As such, for instance, biosequences range from 10^2 (the characteristic length of protein sequences) to 10^9 (as in the human genome) symbols, and excerpts from natural languages contain typically $10^3 \ldots 10^6$ characters. The Shannon entropy (H) has been proposed to characterize complex behavior in model and real systems, and later on several generalizations of H have been proposed [11, 12, 15]. Since H is formally defined as an average value, the idea underlying most generalizations is to replace the average of logarithms by an average of powers [17]. This gives rise to generalized Rényi (H_q) and Tsallis (H'_q) entropies, which are used to characterize complex systems and (multi-)fractal structures in general. Here, the parameter q describes the inhomogeneous structure of the underlying probability distribution.

We address the problem of estimating H_q and H'_q from finite samples. The estimation of H_q and H'_q is problematic, since they are based on probabilities which are not known for finite experimental data. Commonly, the sampled relative frequencies f_i are substituted for probabilities p_i, but experience shows that replacing p_i by f_i can produce large statistical and systematic deviations of the entropy estimate $\widehat{H}_q$ ($\widehat{H}'_q$) from the true value H_q (H'_q). This problem becomes severe when the number of data points is of the order of magnitude of the number of different states which can occur. Then the choice of an estimator with small deviations from the true value becomes important. Over the last decades, several different estimators of H and corrections to them have been derived [7, 8, 13, 14]. Different estimators for H_q and for the generalized dimensions related to them have also been derived [9]. Here, we derive Bayesian estimators for H_q and H'_q, and we show that for finite samples of higher-order Markov models that the fluctuation of H_q and H'_q are much smaller than the naive estimate.

2 Generalized entropies

We consider a sample of size N, where each of the N data points is sampled from M possible different outcomes, and a random variable A that can assume M discrete values a_i ($i = 1, \ldots, M$). The outcomes $(a_1, \ldots, a_M)$ are associated with a probability vector $\mathbf{p} = (p_1, \ldots, p_M)$ with components $p_i \equiv \mathrm{Prob}(A = a_i)$. Then, the Shannon entropy (H) of A is defined by [12]

$$H(\mathbf{p}) \equiv -\sum_{i=1}^{M} p_i \log_2 p_i. \tag{1.1}$$

H measures the average uncertainty of the outcome of A and is usually measured in units of bits. It follows from Eqn. (1.1) that events a_i having either a very high or low p_i do not contribute much to H. In order to weight particular components p_i, one can consider the function

$$Z_q(\mathbf{p}) \equiv \sum_{i=1}^{M} p_i^q, \tag{1.2}$$

where the parameter q is a real number. In the framework of statistical mechanics, by identifying $E_i = -\log_2 p_i$ as the energy of state a_i and q as the inverse temperature, Z_q plays the role of the partition function. For $q = 0$, we obtain that $Z_0 = M$ (the number of states a_i with $p_i > 0$) and that $Z_1 = 1$ (due to normalization of p_i). By using Z_q, the Rényi entropy is defined by [11]

$$H_q(\mathbf{p}) \equiv \frac{1}{1-q} \log_2 Z_q(\mathbf{p}). \quad (1.3)$$

Similarly, the Tsallis entropy is defined by [15]

$$H'_q(\mathbf{p}) \equiv \frac{1}{1-q} \frac{Z_q(\mathbf{p}) - 1}{\ln 2}. \quad (1.4)$$

We find that for fixed q, H_q and H'_q are monotonic functions of each other, since they are functionally related through $H_q(\mathbf{p}) = \log_2[1+(1-q)\ln 2\, H'_q(\mathbf{p})]/(1-q)$.

3 Bayesian entropy estimators

We derive an estimator of H_q and H'_q by using a Bayesian approach[1]. The Bayesian estimator has the optimal property to minimize the mean-squared deviation of the estimate from the true value under a given prior hypothesis. We derive the Bayesian estimators $\widehat{H}_q$ and $\widehat{H}'_q$ for the class of Dirichlet distributions.

For a given sample of N data points, we define the vector $\mathbf{f} = (f_1, \ldots, f_M)$ and calculate for each $f_i = n_i/N$ the number n_i of events a_i, where $N = \sum_{i=1}^{M} n_i$. Then, under the assumption of identical independent-distributed sampling, the frequency-counts vector $\mathbf{n} = (n_1, \ldots, n_M)$ is multinomially distributed:

$$P(\mathbf{n}|\mathbf{p}) \equiv \frac{N!}{n_1! \cdots n_M!} p_1^{n_1} \cdots p_M^{n_M} = \frac{1}{W(\mathbf{n})} p_1^{n_1} \cdots p_M^{n_M}, \quad (1.5)$$

where we define the normalization constant $W(\mathbf{n}) = n_1! \cdots n_M!/N!$. For some experiment under consideration, the individual components p_i always assume specific values, and so one uses a prior distribution $P(\mathbf{p}|\lambda\mathbf{u})$ to regularize the counts n_i. Here, we choose the Dirichlet probability density function (p.d.f.) $P(\mathbf{p}|\lambda\mathbf{u})$ for the conditional variable $\mathbf{p}|\lambda\mathbf{u}$, since it is conjugate to $P(\mathbf{n}|\mathbf{p})$. It is defined by [3]

$$P(\mathbf{p}|\lambda\mathbf{u}) \equiv \frac{1}{W(\lambda\mathbf{u})} p_1^{\lambda u_1 - 1} \cdots p_M^{\lambda u_M - 1}, \quad (1.6)$$

with the normalization constant $W(\lambda\mathbf{u}) = \Gamma(\lambda u_1) \cdots \Gamma(\lambda u_M)/\Gamma(\lambda)$ and the parameter λ. Γ is the gamma function, which simplifies to a factorial in case of an integer argument. The vector $\mathbf{u} = (u_1, \ldots, u_M)$ is normalized ($\sum_i u_i = 1$) such that $\int d\mathbf{p} P(\mathbf{p}|\lambda\mathbf{u})\mathbf{p} = \mathbf{u}$ can be interpreted as the expectation value of $\mathbf{p}$. The role of λ is that it measures how different we expect typical $\mathbf{p}$ to be from $\mathbf{u}$. The larger the value of λ, the closer the sample values of p_i tend to u_i. In the case of $\lambda u_i = 1$, Eqn. (1.6) becomes independent of p_i.

[1]In comparison, frequency-counts (or maximum-likelihood) estimators of H_q and H'_q are defined by $\widetilde{H}_q = [\log_2 Z_q(\mathbf{f})]/(1-q)$ and $\widetilde{H}'_q = [Z_q(\mathbf{f}) - 1]/(1-q)/\ln 2$.

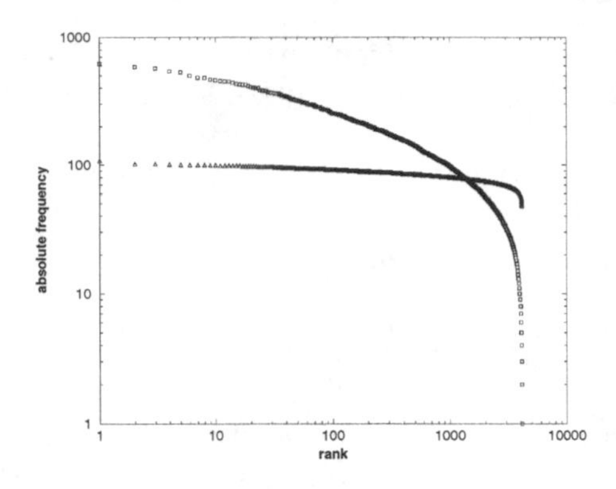

Figure 1: Double-logarithmic representation of the histogram of all hexamers (6-letter words) of the complete DNA sequence of *H. influenzae* [5] as a function of their rank (□). We plot the frequency of occurrence versus its rank, starting with the most abundant one. For a comparison, we display the histogram of hexamers sampled from a corresponding uncorrelated random sequence (△) and find that □ contains many over- and underrepresented hexamers as compared to △. Fig. 1 can be regarded as a typical example representing hexamer histograms in bacterial DNA sequences.

Having the likelihood $P(\mathbf{n}|\mathbf{p})$ and the prior p.d.f. $P(\mathbf{p}|\lambda\mathbf{u})$, we can write down the conditional probability of $\mathbf{p}$ given the data $\mathbf{n}$

$$P(\mathbf{p}|\mathbf{n},\lambda\mathbf{u}) = \frac{P(\mathbf{n}|\mathbf{p})P(\mathbf{p}|\lambda\mathbf{u})}{P(\mathbf{n}|\lambda\mathbf{u})} = \frac{1}{W(\mathbf{n})W(\lambda\mathbf{u})}\frac{\prod_i^M p_i^{(n_i+\lambda u_i-1)}}{P(\mathbf{n}|\lambda\mathbf{u})}. \quad (1.7)$$

Using $P(\mathbf{n}|\lambda\mathbf{u}) = \int \mathrm{d}\mathbf{p}\ P(\mathbf{n}|\mathbf{p})P(\mathbf{p}|\lambda\mathbf{u}) = \int \prod_j^M \mathrm{d}p_j\ p_j^{(n_j+\lambda u_j-1)}/W(\mathbf{n})/W(\lambda\mathbf{u})$, we cancel out the nomalization factors in Eqn. (1.7) and find the posterior p.d.f.

$$P(\mathbf{p}|\mathbf{n},\lambda\mathbf{u}) = \frac{\prod_i^M p_i^{(n_i+\lambda u_i-1)}}{\int \prod_j^M \mathrm{d}p_j p_j^{(n_j+\lambda u_j-1)}}. \quad (1.8)$$

Eqn. (1.8) states the probability of $\mathbf{p}$ given both the hypothesis $(P(\mathbf{p}|\lambda\mathbf{u}))$ of $\mathbf{p}$ before and the likelihood $(P(\mathbf{n}|\mathbf{p}))$ of $\mathbf{n}$ after the sampling. One calculates the Bayesian estimator [2] as the expectation value of a random variable over its posterior distribution. If we thus write for H'_q

$$\begin{aligned}\widehat{H}'_q(\mathbf{n},\lambda\mathbf{u}) &= \int \mathrm{d}\mathbf{p}\ H'_q(\mathbf{p})P(\mathbf{p}|\mathbf{n},\lambda\mathbf{u}) \\ &= \frac{1}{1-q}\frac{\int \mathrm{d}\mathbf{p}\ Z_q(\mathbf{p})P(\mathbf{p}|\mathbf{n},\lambda\mathbf{u})-1}{\ln 2}, \end{aligned} \quad (1.9)$$

we note that in order to obtain $\widehat{H}'_q$, we have to calculate the Bayesian estimator of $Z_q(\mathbf{p})$. We can obtain the estimator $\widehat{Z}_q(\mathbf{n},\lambda\mathbf{u})$ analytically [9] and we obtain

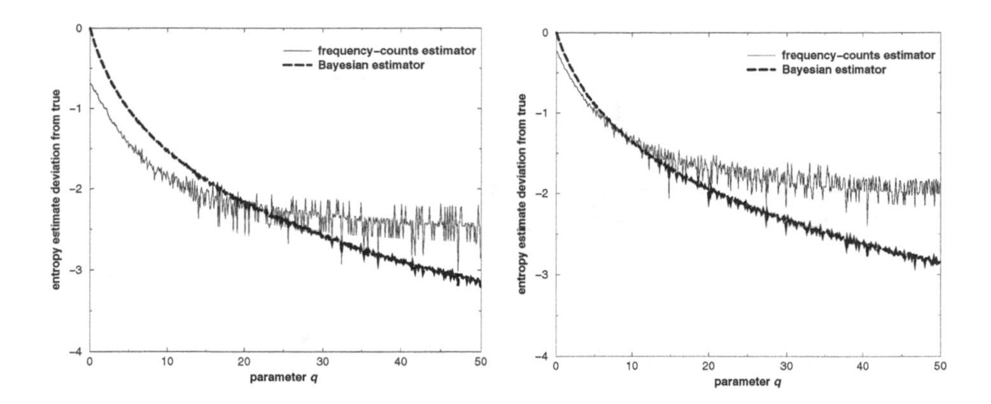

Figure 2: Comparison of different estimators of H_q for the parameter set $M = 4096$, $p_i = 1/M$, $\lambda u_i = 1$, and $q \in (0, 50)$ for a single realization. **Left**. We simulate for $N = 4000$ an equidistributed Markov process of 0th order and find a lower variability of the Bayesian estimator as compared to the frequency-counts estimator. **Right**. We repeat the simulation by doubling the sample size and find that the Bayesian estimator shows for $N = 8000$ a lower variability as compared to the frequency-counts estimator.

$$\begin{aligned}\widehat{Z}_q(\mathbf{n}, \lambda\mathbf{u}) &= \sum_i^M \frac{\int \prod_j^M \mathrm{d}p_j \; p_j^{(n_j+\lambda u_j-1)} p_i^q}{\int \prod_j^M \mathrm{d}p_j \; p_j^{(n_j+\lambda u_j-1)}} \\ &= \frac{\Gamma(N+\lambda)}{\Gamma(N+\lambda+q)} \times \sum_i^M \left\{ \frac{\Gamma(n_i+\lambda u_i+q)}{\Gamma(n_i+\lambda u_i)} \right\} \end{aligned} \quad (1.10)$$

for the parameter interval $q \in (-\lambda u_{\min}, \infty)$, where $u_{\min} = \min_{\forall i}\{u_i\}$. We note in passing that $\widehat{Z}_0 = M$ and $\widehat{Z}_1 = \sum_i (n_i + \lambda u_i)/(N + \lambda) = 1$, and hence for $\lambda = 1$ the Bayesian estimator of p_i is given by $\widehat{p}_i = (N_i + u_i)/(N+1)$. According to Schürmann & Grassberger [8], for $u_i = 1/M$ this is the estimator that yields 'best' entropy estimates when inserted as p_i. So, the 'best' entropy estimator of H corresponds to choosing a uniform prior p.d.f. over the simplex of $\mathbf{p}$.

Since H_q' is a generalization of H, we expect $\widehat{H}_q'$ to be a generalization of the Bayesian Shannon entropy estimator $\widehat{H}$. We calculate for any integer λu_i the limit $q \to 1$ and obtain

$$\widehat{H}(\mathbf{n}, \lambda\mathbf{u}) = \frac{1}{\ln 2} \sum_i^M \frac{n_i + \lambda u_i}{N + \lambda} \left(\sum_{k=n_i+\lambda u_i+1}^{N+\lambda} \frac{1}{k} \right) \quad (1.11)$$

which is equivalent [8] for $\lambda u_i = 1$.

We turn to the Bayesian Rényi entropy estimator $\widehat{H}_q$. In analogy to (1.9), the derivation of $\widehat{H}_q$ requires the calculation of the integral

$$\widehat{H}_q(\mathbf{n}, \lambda\mathbf{u}) \propto \int \mathrm{d}\mathbf{p} \; \log_2 Z_q(\mathbf{p}) P(\mathbf{n}|\mathbf{p}) P(\mathbf{p}|\lambda\mathbf{u}). \quad (1.12)$$

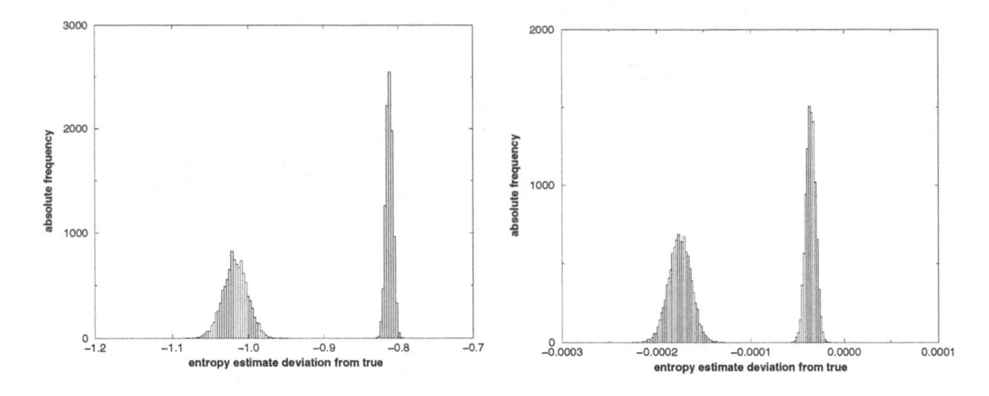

Figure 3: Comparison of different estimators of H_q and H'_q for the parameter set $M = 4096$, p_i taken from *H. influenzae*, $\lambda u_i = 1$, and $q = 2$. We estimate the sample variance of estimators by generating an ensemble of 10000 sequences. **Left**. We simulate for $N = 4000$ a Markov process of 0th order and observe the smaller variance of $\widehat{H}_q$ (right) as compared to the frequency-counts estimator (left). **Right**. We simulate for $N = 8000$ a Markov process of 5th order and observe the clearly smaller width of $\widehat{H}'_q$ (right) as compared to the frequency-counts estimator (left). While the systematic error can partially be corrected, the reduction of the variance requires a different estimator.

Even in the binary case $(p, 1-p)$, we cannot derive an analytical expression. Hence, we use the following strategy to estimate $\widehat{H}_q$. We use the relationship between H'_q and H_q and hence $\widehat{Z}_q$ to define the (indirect) Bayesian entropy estimator of H_q

$$\widehat{H}_q(\mathbf{n}, \lambda\mathbf{u}) \equiv \frac{1}{1-q} \log_2 \widehat{Z}_q(\mathbf{n}, \lambda\mathbf{u}). \tag{1.13}$$

Clearly, in the limit $q \to 1$, we obtain again the Bayes estimator of the Shannon entropy $\widehat{H}$.

4 Numerical tests

To elucidate the application of $\widehat{H}_q$ and $\widehat{H}'_q$, we test the Bayesian estimation of H_q and H'_q from DNA sequences. By means of Monte-Carlo simulations, we simulate DNA sequences and investigate the sample variance of the entropy estimators for the cases $N = 4000$, $N = 8000$, and $M = 4^6 = 4096$.

In Fig 2, we simulate equidistributed Markov processes of 0th order. We vary q in the parameter range $q \in (0, 50)$, incrementing q by $\Delta q = 0.1$, and for each q we estimate for a single realization H_q by using both the Bayesian estimator $\widehat{H}_q$ and the frequency-counts estimator. In Fig. 3, we simulate Markov processes of 0th and 5th order. In order to simulate DNA sequences which are biologically relevant, we take the transition probabilities from the complete $1,830,240$ base pairs long DNA sequence of the bacterial genome of *Haemophilus influenzae* Rd (cf. Fig. 1). We generate 10,000 sequences from which we estimate both H_q and H'_q for $q = 2$ by using the Bayesian and the frequency-counts estimators, and we compare the estimators to the true H_q and H'_q. We denote the difference between the estimated and the theoretical values as entropy estimate deviation from true.

5 Concluding remarks

We derived the Bayesian estimators $\widehat{H}_q$ and $\widehat{H}'_q$ for the class of Dirichlet distributions. We obtain both $\widehat{H}_q$ and $\widehat{H}'_q$ from the estimator of the partition function $\widehat{Z}_q$. A test of the reliability by which $\widehat{H}_q$ and $\widehat{H}'_q$ as well as the corresponding frequency-counts estimators estimate H_q and H'_q from Markov chains of 0th and 5th order (cf. Fig. 2 and Fig. 3) shows that for $q \in (0, 50)$ both $\widehat{H}_q$ and $\widehat{H}'_q$ have a smaller variance than the frequency-counts estimators. Thus, Fig. 2 and Fig. 3 suggest to prefer the Bayesian entropy estimator over the frequency-counts estimator in cases where the sample size is small and the underlying rank-order distribution corresponds to Fig. 1. Clearly, the regularization $\mathbf{u}$ is application-dependent. Given no prior information, one can consider all $\mathbf{p}$ equiprobable and then a uniform distribution constitutes the least biased guess.

Bibliography

[1] BECK, C. and F. SCHLÖGL, *Thermodynamics of Chaotic Systems*, Cambridge University Press, Cambridge (1993).

[2] BERGER, O., *Statistical Decision Theory and Bayesian Analysis*, Springer, New York (1985).

[3] DURBIN, R., S. EDDY, A. KROGH, and G. MITCHINSON, *Biological Sequence Analysis*, Cambridge University Press, Cambridge (1998).

[4] EBELING, W. and R. FEISTEL *Physik der Selbstorganisation und Evolution*, Akademie-Verlag, Berlin, (1982).

[5] FLEISCHMANN, R. *et al.*, *Science* **269** 496–512 (1995).

[6] GATLIN, L.L., *Information Theory and the Living System*, Columbia University Press, New York (1972).

[7] HUTCHESON, K. and L.R. SHENTON, *Comm. Stat.* **3** 89–94 (1974); COOK, G.W., D.F. KERRIDGE, and J.D. PRYCE, *Indian J. Stat.* **36** 443–448 (1974); HARRIS, B., *16. Topics in Information Theory* **16** 323–355 (1975); LEVITIN, L.B. and Z. REINGOLD, *Annu. Conf. Israel Stat. Assoc.*, Tel Aviv (1978);

[8] W. Li, *J. Stat. Phys.* **60** 823–837 (1990); Schmitt, A.O., H. Herzel, and W. Ebeling, *Europhys. Lett.* **23** 303–309 (1993); Pöschel, T., W. Ebeling, and H. Rosé, *J. Stat. Phys.* **80** 1443 (1995); Ebeling, W., T. Pöschel, and K.-F. Albrecht, *Int. J. Bif. & Chaos* **5** 51–61 (1995); Wolpert D.H. and D.R. Wolf, *Phys. Rev. E* **52** 6841–6854 (1995); Treves, A., and S. Panzeri, *Neural Comput.* **7** 399–407 (1995); Schürmann, T. and P. Grassberger, *CHAOS* **6** 414–427 (1996); Grosse, I., in *Dynamik, Evolution, Strukturen*, ed. J.A. Freund, Verlag Köster, Berlin (1996); Roulston, M.S., *Physica D* **125** 285–294 (1999).

[9] Pawelzik, K. and H.G. Schuster, *Phys. Rev. A* **35** 481–484 (1987); Grassberger, P., *Physics Lett.* **A 128** 369–373 (1988); Holste, D., I. Grosse, and H. Herzel, *J. Phys. A* **31** 2551–2566 (1998).

[10] Pompe, B., *J. Stat. Phys.* **73** 587–610 (1993).

[11] Rényi, A., *Probability Theory*, Amsterdam, North Holland (1970).

[12] Shannon, C.E., *Bell Syst. Tech. J.* **27** 379–423 (1948);

[13] Shannon, C.E., *Bell Syst. Tech. J.* **27** 50–64 (1951); Miller, G.H., *Information Theory in Psychology*, ed. H. Quastler, Free Press, Glencoe (1955); Basharin, G.P., *Theory Prob. Appl.* **4** 333–339 (1959);

[14] Fraser, A. and H. Swinney, *Phys. Rev. A* **33** 1134–1140 (1986); Herzel, H., *Syst. Anal. Mod. Sim.* **5** 435–444 (1988);

[15] Tsallis, C., *J. Stat. Phys.* **52** 479–487 (1988).

[16] http://tsallis.cat.cbpf.br/~biblio.htm.

[17] Wehrl, A., *Rev. Mod. Phys.* **50** 221 (1978).

In the Looking Glass: The Self Modeling of Social Systems

Christina Stoica, Jürgen Klüver & Jörn Schmidt
Center of Research in Higher Education
University of Essen, Germany
christina.stoica@uni-essen.de

Summary

The hybrid system SOCAIN is described as a formal model of the self modeling of social systems. The system consists of a stochastic cellular automaton (CA) coupled with a genetic algorithm (GA) and an interactive net (IN), also coupled with a GA. The IN is understood as the self-model of the CA. It can be shown that the adaptive success of the whole system SOCAIN is much greater than the successes of both subsystems alone. Reasons for this are given and the logic of the system is discussed both logically and sociologically.

1. Introduction

Hybrid systems are simulation programs which consist of two - or even more - different subprograms; these are coupled together to fulfill tasks which neither of them could do alone.

Goonatilake and Khebbal (1995,7) proposed a general classification scheme for hybrid systems: They differentiate between (a) "function replacing hybrids" and (b) "intercommunicating hybrids" (we skip their third category, because that is no hybrid at all). For reasons we cannot discuss here we prefer the terms "vertically coupled systems" and "horizontally coupled systems": Vertically coupled systems are combinations of programs where one program is operating upon the rules and/or parameters of the second. In such cases, according to the concepts of logical semantics

(Tarski 1956), we speak of the first system as the metasystem and of the second as the base system. An example for a vertically coupled system which is described below is the hybrid system SOZION, a combination of a cellular automaton (CA) as the base system and a genetic algorithm (GA) as the metasystem. The GA modifies the rules, especially certain parameters according to certain evaluation criteria. Neural nets that are combined with GAs are also vertically coupled systems.

Horizontally coupled systems operate in a sort of division of labor insofar as one system does one part of a job, transferring its results to the second system, and the second system performs another part of the task. The two systems are operating on the same logical level, in contrast to the case of vertical coupling. An example of a horizontally coupled system is THEOPRO (Klüver 1996) which consists of a knowledge based system and a neural net. The knowledge based system constructs a special net, consisting of concepts of a theory of society, and a weight matrix; the neural net performs simulations of social processes according to the theory and gives its results back to the knowledge based system. Of course, with horizontally coupled systems the problem arises whether the results of one system are compatible with the results of the second, as the two systems normally operate with very different rule sets. We shall return to this problem in section 4.

We have now intended to create higher order hybrids by combining hybrids of these two types: The program SOCAIN (see below) contains a CA, horizontally coupled with an interactive neural net (IN), and GAs operating on both base systems. In this case we have a horizontally-vertically coupled system.

Hybrid systems offer important possibilities for the modeling and simulating of social systems. These shall be demonstrated with the problem of the self modeling of social systems.

2. The Problem

Since Luhmann (1984) it has become common to speak of social systems as "self referential systems". It is not possible here to define this and related concepts precisely (see e.g. Roth 1987); apparently, "self-reference" contains several different aspects among which the concept of "self-modeling" is a salient feature. This means that complex systems like language, mind, or society are able to construct models of themselves as parts of their different states: one can speak about language by using language and construct models of it, the mind can model itself by thinking about thinking, and social systems build models of themselves by mapping their structures onto parts of them. Two famous examples shall illustrate this:

(a) As part of his ethnological studies, Geertz described the institution of the Balinese cockfight as a simulation of the "social matrix" of the Balinese society in the sense that the important structural aspects of the society are mapped - mirrored - onto the institution of the cockfight. Among these are class distinctions, the inequality of wealth and the hierarchy between the sexes. The cockfight is an important part of the Balinese culture and as a part it constitutes a model of the whole society - "it is a story they tell themselves about themselves" (Geertz 1972, p.26)).

(b) In their studies about the French higher education system, and especially the French universities, Bourdieu and Passeron (1964) analyze the education system as a "reproduction" of the society in general. With "reproduction" they imply two aspects: On the one hand, the structures of society, especially the structures of inequality, are mirrored in the subsystem of education; on the other hand, the educational process, determined by the structures of the educational system, reproduces the structures of inequality as a function for the whole society. Of course, the educational system can only do this because its structures are reflections of societal structures in general and are originated this way. Thus, the educational system as a part of society is also a model of society. The importance of social reproduction via modeling the society as a whole by a part of society has always been stressed by theorists who are influenced by Marx (e.g. Habermas 1981), though not in these terms.

Self-reference can easily lead into logical paradoxes, as is well known since the logical and set theoretical antinomies. That is why self-reference in formal models is inadmissible in logical semantics (Tarski 1956). But, as we have seen, complex systems like societies possess self-reference and especially self-modeling as two of their most fundamental aspects; we can assume that the main difference between the systems of the natural sciences on the one hand and systems like society, mind, or language - the topics of the humanities - on the other hand lies exactly in these features. Of course, physical or biological systems are very often complex too. Social systems just have an additional dimension of complexity. If formal models cannot capture these aspects then the construction of formal models is indeed, as many theoretical sociologists believe, not suited for the analysis of social systems. Therefore, the most important task is to show that it is possible despite the cautions of the logical semanticists to construct consistent models and corresponding computer programs with the ability of self modeling.

3. The Program SOCAIN

As mentioned above, SOCAIN is a horizontally-vertically coupled system which consists of two hybrid systems. The base systems are a CA and an IN, the metasystem in both systems is a GA.

The CA is rather simple because it belongs to a formal analogue of Wolfram class I (Wolfram 1986); that means that it will reach an attractor state after a limited number of time steps and will remain there (a point attractor). Its rules are "totalistic": The transition of a cell is determined by computing the mean value of the cells of the neighborhood (a Moore neighborhood). In contrast to most CAs, the rules are not deterministic but stochastic; changing of a state value only occurs with a certain probability. The state of the cells can take one of nine values - which one depends on the neighborhood and on the probability values which define the relations between the different states. The different values of the cell states symbolize different social classes.

The CA was constructed to simulate social differentiation, namely the differentiation of a homogenous society into a class society (see Klüver 1996a). In the original version the CA was coupled with a GA which modified the probability values and decided on switching any of the rules on or off.

The IN is a recurrent network. Each unit is connected with all others, but the weight values of the network are not changed during the runs. This makes INs suitable for simulations of special social processes (Stoica 1999), though INs are not able to learn by themselves. Our IN consists of nine units and, accordingly, a weight matrix of 9 * 9 = 81 weight components. A run of the IN begins with an "external activation" of one ore more units; the run ends when the activation values of the units stabilize. The IN uses the standard activation rule

$$A_i = \Sigma_j A_j * w_{ij} \qquad (1),$$

where A_i is the activation value of the receiving unit i, A_j is the activation value of a sending unit j and w_{ij} is the weight value of the relation between the units j and i. Other activation rules are possible, especially nonlinear ones, but we wanted to keep the whole system as simple as possible.

The GA is also standardized. It contains the usual genetic operators of crossover and mutation and is non elitistic. That means that the best results from a GA-run are not preserved, though elitistic GAs often do better than non elitistic ones. Its evaluation function, i.e. the algorithm for choosing the "best" solution, is described below. The GA operates on vectors in which those rules of the base systems are represented that are to be changed; they use simple numerical codes, i.e. integers for the CA-rules and real numbers for the IN (they are the values of the weight matrix).

The two systems are coupled in the following way: The transformation of cell states within the CA is determined by a so called "mobility matrix", i.e. the 9 * 9 matrix containing the transition probabilities from each of the nine possible cell states to any other. The GA of this first hybrid system operates on the transition *rules*, allowing or forbidding certain transitions, as well as on the transition *probabilities*, decreasing or increasing those values. The CA always runs three steps; then the "success" of the development of the CA is evaluated. Afterwards the GA starts its changing operations.

Cell transformation in the CA proceeds according to the following rules:

1. A cell with index value i **increases** its index value (ascendence),
   i. if the average of cell indices in its Moore neighborhood is greater or equal i (i ($u \geq i$) **and**
   ii. if the corresponding component in the antecedens part of the rule vector has the value 1 or 3 (neutral element) **and**
   iii. if the corresponding component of the action part of the rule vector is 1 **and**
   iv. if the probability of ascendence according to the procedure defined below is given.
2. Correspondingly a cell **decreases** its index value (descendence),
   i. if the average of cell indices in its Moore neighborhood is less than i ($u < i$) **and**
   ii. if the corresponding component in the antecedence part of the rule vector has the value 0 or 3 (neutral element) **and**
   iii. if the corresponding component of the action part of the rule vector is 1 **and**
   iv. if the probability of descendence according to the procedure defined below is given.
3. Otherwise the cell does not transform.

To calculate the transformation probability the elements of the rows i of the upper and lower triangle of the mobility matrix **A** (representing the probabilities for ascendence and descendence) separately are consecutively added according to

$$c_{ij} = \Sigma a_{ik} \qquad (2),$$

where k runs from i+1 to j (upper triangle) or from j to i-1 (lower triangle), forming the cumulated mobility matrix **C** (c_{ij}=0).
A random number ($0.0 \leq r < 1.0$) is generated and the cell index is transformed to the first index j - beginning from the main diagonal of the cumulated mobility matrix - with c_{ij} greater than the random number; if no element c_{ij} fulfills the condition, the cell is not transformed.

After 10 runs (CA+GA) the mobility matrix is taken over by the second hybrid system as the weight matrix of the IN. In addition, the relative numbers of cells in each cell state of the actual CA serve as external activations of the IN. The GA of the second system optimizes the weight matrix with reference to its distance to the "target vector", which is the very same as that of the first system (see below). The timing of this system is: the IN runs several time steps (usually 50-80) until it reaches an attractor state. Then the GA starts its operations. The optimized weight matrix - and only this - is then fed back to the first system, and so on.

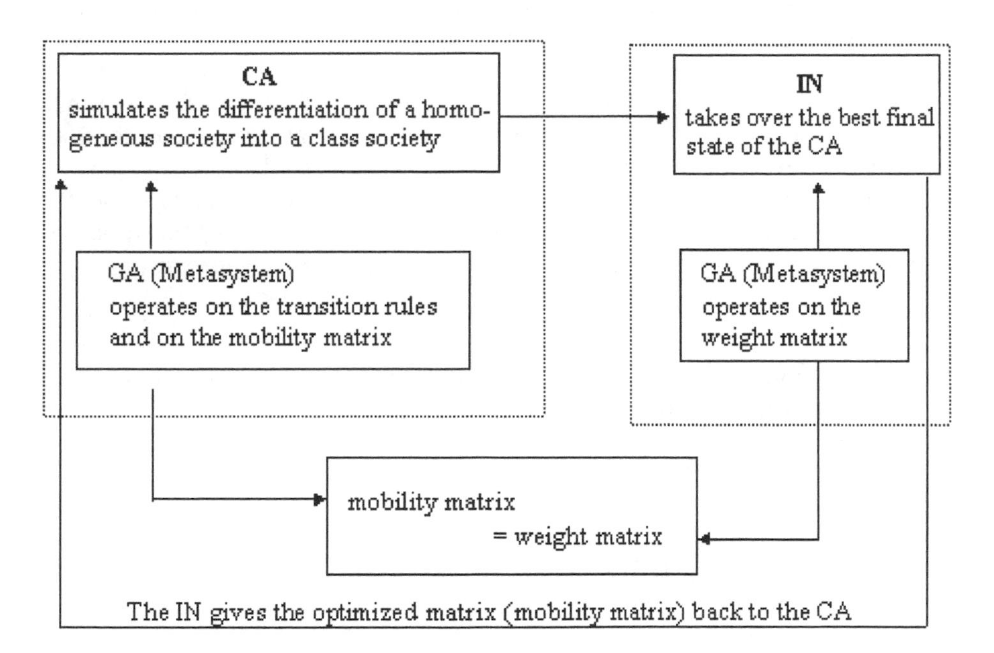

Picture 1 The structure of the hybrid system SOCAIN

In more detail, SOCAIN - SOcial Cellular Automaton with an Interactive Network - always starts with the CA that symbolizes the "real" and whole society. It has one initial state, namely a homogenous state with only the lowest cell state representing rural people. The GA uses 20 "rule vectors" that correspond to 20 rule sets for the CA, or, briefly expressed, to 20 different CA. The selection of the best of the different CAs is based on the evaluation vector

$$V = (\Sigma_i A_i * B_{i1} * F_1 , \ldots\ldots., \Sigma_i A_i * B_{i6} * F_6) \qquad (3),$$

where V is the "value" of the social system, A_i is the number of cells in the state i and therefore the number of members of the class i, B_{ij} is the contribution of the class i to the system function j, and F_j is a weight factor representing the value that the function j has to the whole system (this is explained in more detail in Klüver 1996a). The selection criterion for the GA is the smallest possible distance d of the resulting vector V to a "target vector" or "environment vector" E:

$$d = | E - V | \qquad (4)$$

After the GA has operated a certain number of times on the CA, the CA maps itself onto an IN in the way described above. The initial state of the IN, that is the initial activation values of the units, is given by the best final state of the CA before the mapping CA → IN. Then the IN starts its runs, gets optimized by the corresponding GA, and gives a modified matrix back to the CA which starts again as before. This procedure is repeated until a sufficiently low value of d is achieved for the whole system. The IN is a model of the CA insofar as the mobility matrix of the CA and a global measure of its last state are directly transformed into the IN. Thus the IN reproduces the structure of the CA. But it is a reductive model because on the one hand, the important geometry of the CA has no counterpart in the IN - it operates only on a macro level from the view of the CA; the CA operates on a micro level and gets its macro level as a kind of aggregation effect. The GA of the second system, on the other hand, can only change the weight values of the IN-matrix and does not affect the transition rules as in the case of the CA. Moreover, it has to be pointed out that the basic algorithms of the two base systems CA and IN are totally different.

4. Results

SOCAIN was compared to the original – SOZION, where there is no coupling to the IN - with reference to its effectiveness to minimize d. The results are quite remarkable:

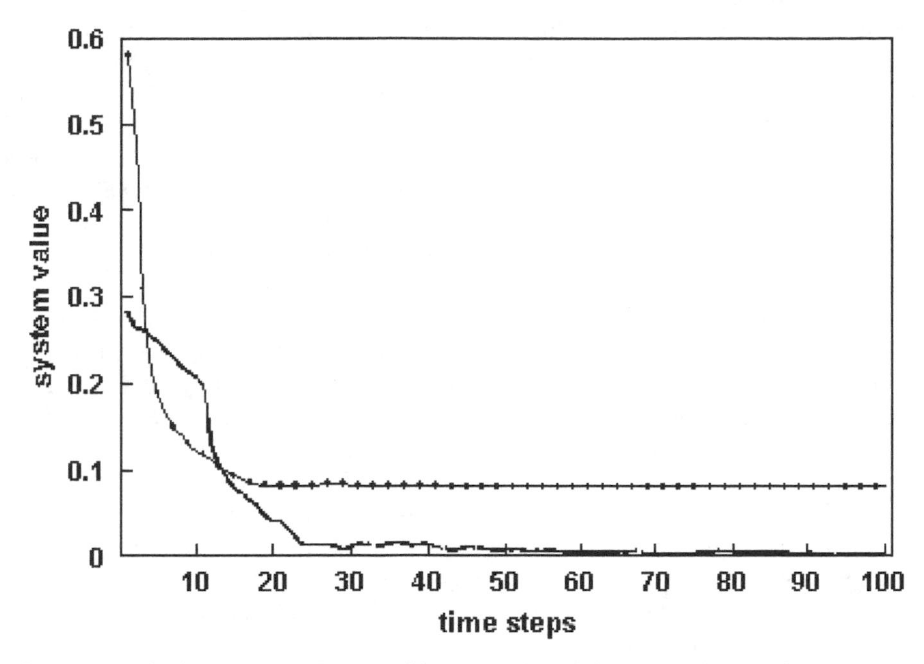

Picture 2 Optimization with IN and without IN (dotted) in 2 typical runs of SOCAIN.

SOZION typically reaches optimal distances of about 0.08 after 50 to 100 time steps, a result which is not improved by further optimization. The reason is that the set of rule vectors of the GA tends to become nearly homogenous, so that only higher mutation rates can lead to variation.

Combining SOZION with the second system - SOCAIN - leads to much lower distances - typically lower than 0.01 - within fewer time steps. Surprisingly, the optimization process within the second system, i.e. the GA operating on the IN, is in itself much less effective; it rarely achieves distances below 0.2. Nevertheless, these sub-optimal weight matrices are capable of considerably improving the optimization process within the first system when taken over by this. Apparently, the IN/GA system is a very suitable model for the CA/GA system, improving its operation by "pushing" the values of the mobility matrix.

These results are rather astonishing at first sight since CA and IN operate with very different kinds of rules. This has to be explained in connection with particular control parameters which we call "meta parameters":

Logically speaking, adaptive systems do not have one set of rules on one logical level but at least two different sets of rules: The one set of rules constitutes the *base system,* i.e. the whole of the rule governed interactions and actors of the "real" system. The second set contains the *meta rules* of the system, i.e. the rules by which the base rules are changed. That is why we call adaptive systems *hybrid systems* in a logical sense: Their rule system is composed of two ore more different rule sets.

If one tries to classify adaptive, that is, hybrid systems by introducing control parameters then one has to look for *meta parameters,* i.e. parameters on the meta level of the whole rule system (meaning base rules and meta rules together). These meta parameters "control" the adaptive success of an adaptive system: in other words, specific values of meta parameters decide whether an adaptive system reaches its targets fast or not, with a sufficient accuracy or not, and in a sufficient time. Meta parameters describe the manner of changing the base rules in a quantitative way and are thus a feature of the meta rules (for a systematic analysis of meta parameters see Klüver forthcoming). They can be looked upon as the parallel of the well known control parameters of the rules systems of CAs or Boolean nets (Kauffman 1993).
When distinguishing between base rules and meta rules, an obvious criterion for the variability of an adaptive system is the ruthlessness (ρ) by which the meta rules operate on the base rules. The ρ-parameter measures the quantity of changing a specific rule. In the case of SOCAIN the GA operates on the CA in the way that the values of the mobility matrix are changed only rather softly, i.e. by raising or lowering them by a constant factor of approx. 0.05. So a mobility parameter is changed as a consequence of a GA-operation for example from 0.3 to 0.35. The corresponding weight matrix of the IN is changed much more radically: A weight value in the matrix of the IN can be changed by one GA-operation from 0.3 to 0.9 or even from 0 to 0.9. In the case of the CA $\rho = 0.05$, in the case of the IN $\rho = 0.9$. When the CA gets its mobility matrix from the IN with the more radical changing, the CA gets much better results in optimizing itself to a specific environment than with its "own" matrix, i.e. the matrix which resulted from the more prudent changes. So it seems that high ruthlessness gets better results, but things are not that simple: The hybrid IN with high ruthlessness is often not as good as the hybrid CA with low ρ. The best results are obtained when the hybrid CA operates with low ρ, i.e. in a soft way, with the results it gets from the operations of the IN with high ρ. Therefore the best way seems to be to operate with *changing* ρ. The high ruthlessness of IN/GA is the way for CA/GA to get a fresh start when its own way of changing is too subtle. The difference of ρ in the two particular hybrid systems CA/GA and IN/GA explains the whole behavior of SOCAIN as well as the not very successful adaptive behavior of the IN.

5. Discussion

1. The GA of the CA only operates on the mobility matrix according to the idea that a social system can change the probability values of its rules more easily than the rules in general. If we take a look at political decisions, we can see that politicians first try to change the "probability" structure of the system, for example the possibility to choose a profession or to get a better education. Changing social rules is a sensitive point, and it is the last possibility in the event of the probability restructure not having the successful effect.
2. In a formal sense the system IN/GA can be interpreted as the self mapping of the system CA/GA. So the operations of IN/GA are to be understood as virtual

experiments with the own future of the whole system - a sort of mathematical gedankenexperiment. That is why the ruthlessness of the operations of IN/GA can be much greater than in the case of CA/GA. Virtual experiments can be done with no consideration for the costs. If they are successful, the system can take over the results without having to pay for failures.

3. Sociologically speaking, social systems can achieve the ability of modeling themselves in rather different ways, as was demonstrated with the examples in section 2. Another possibility is a political one: The CA is a "bottom up" system, i.e. it can be understood as the model of a society where the social processes are local interactions between different social roles. The IN, on the other hand, is a "top down" system insofar as only macro aspects of the society - social classes - are taken into consideration. Thus a social system can model its bottom up processes by a top down model and improve its results; metaphorically speaking, the system is able to make politics "from above" through the process of self modeling. This is the old dream of social planners and dictatorial utopians which often yielded disastrous results. The Killing Fields of Cambodia will not be the last example, we fear. All these cases have got one thing in common: wishful thinking, based on some ideology, had designed a model of a "real" society and tried to carry it through by force. Neither means nor ends were clearly analyzed.

Our results hint at the possibility that democratic societies can use their own capability of self modeling in a "soft" way, namely by modeling a hypothetical future and discussing the advantages and disadvantages of the results. These modeling processes could be done in the way and by the use of computer systems like SOCAIN. Presently, however, this is rather science fiction than science.

References:

Bourdieu, P. & Passeron, J.C. (1964). Les Héritiers. Les Etudiants et la Culture. Paris: Editions de la Minuit.

Geertz, C. (1972). Deep Play: Notes on the Balinese Cockfight. In: *Daedalus* 101, 1-37.

Goonatilake, S. & Khebbal, S. (1995). Intelligent Hybrid Systems. Issues, Classifications and Future Directions. *In: Goonatilake, S. & Khebbal, S. (eds.) (1995). Intelligent Hybrid Systems. Chichester-New York: John Wiley.*

Habermas, J. (1981). Theorie des kommunikativen Handelns. Frankfurt: Suhrkamp.

Kauffman, S., 1993: The Origins of Order. Oxford: Oxford University Press

Klüver, J. (1996). Sociological Discourses in Virtual Reality. In: *Social Science Computer Review* 14,3, 280 - 292.

Klüver, J. (1996a). Simulations of Selforganizing Social Systems. In: Faulbaum, F. & Bandilla, W. (eds.): *SoftStat 95. Advances in Statistical Software.* Stuttgart: Lucius

Klüver, J. (2000): Dynamics and Evolution of Social Systems. New Foundations of a Mathematical Sociology. Dordrecht: Kluwer Academics Publishers,

Luhmann, N. (1984). Soziale Systeme. Frankfurt: Suhrkamp

Roth, G. (1987). Autopoiese und Kognition: Die Theorie H.R. Maturanas und die Notwendigkeit ihrer Weiterentwicklung. In: Schmidt, S. J. (ed.): *Der Diskurs des Radikalen Konstruktivismus*. Frankfurt: Suhrkamp

Stoica, C. (2000): Die Vernetzung sozialer Einheiten. Hybride Interaktive Neuronale Netzwerke in den Kommunikations- und Sozialwissenschaften. (Networks of Social Units. Hybrid Interactive Neural Nets in the Social Sciences). Wiesbaden: DUV

Tarski, A. (1956). Logics, Semantics, Metamathematics. Oxford: Oxford University Press

Wolfram, S. (1986). Universality and Complexity in Cellular Automata. In: Wolfram, S.: *Theory and Applications of Cellular Automata.* Singapore: World Scientific

Harnessing Evolution For Organizational Management

Jim Hines & Jody House
MIT, Sloan School of Management
E53-309
77 Massachusetts Avenue
Cambridge, MA 02139-4307
JimHines@Interserv.Com, Jhouse@MIT.EDU

Synopsis. Organizations are too complex to "solve". Consequently, corporate improvement programs are as likely to be deleterious as beneficial. Evolutionary mechanisms offer an alternative approach to organizational improvement. Organizational policies correspond to biological genes; policy innovation corresponds to genetic mutation; and learning corresponds to genetic recombination. These correspondences form a foundation for creating mechanisms that will foster more rapid organizational evolution in the direction managers desire. Computer simulation offers a laboratory environment for exploring issues in organizational evolution including learning drift, hierarchy, and team learning.

1. Introduction: The importance of organizational evolution

Managers hope to improve their companies. Ultimately, they seek higher profitability for shareholders, more satisfaction for employees, and better service for customers. Unfortunately, managers who improve one part of a complicated business system, run a significant risk of creating problems in another part.

Kofman, Repenning and Sterman (1997) recently investigated a case of improvement gone wrong at Analog Devices. Analog's TQM program succeeded in slashing variable manufacturing costs. However, people in marketing continued to set prices using the traditional policy of multiplying variable costs by 2 to arrive at the price[1]. The old pricing formula, applied to the new lower variable costs, triggered price reductions which no longer covered fixed costs. The company's performance slipped and then slid as Analog touched off a price war with competitors.

Fortunately, Analog has recovered. The example, though, is telling: An existing trait – fixed costs low relative to variable costs – worked adequately well at the local, manufacturing level and was well integrated into other and widely separate areas of the company and industry. An improvement effort inadvertently modified the trait, resulting in lower performance for the company as a whole.

Once revealed connections between one part of an organization and another can seem either subtle or obvious. In either case, however, these connections are often only revealed when something goes wrong.[2] Connections that knit local activities together into an entire company form an invisible web. Like unmapped, long-buried gas mains in a city, hidden organizational connections make improvement efforts hazardous.

Organizations are too complex to map. But this does not mean organizations cannot be improved. Evolution is a process that in the biological realm has consistently improved extraordinarily complex organizations (organisms) – sometimes in remarkably short periods of time as the unexpected emergence of drug-resistant bacteria attests. Computer simulations of evolutionary processes, strongly suggest that biological evolution is only one example, albeit the archetypal example, of more general processes that can improve complex systems (Holland 1995, Koza 1992, Goldberg 1989). There is every reason to believe that evolutionary processes can be harnessed by managers to create organizations that are ever better.

2. The Elements of Organizational Evolution.

Understanding evolution – whether organizational or biological -- means understanding three things: (1) The "genetic material" of evolution, (2) how novel genetic material can arise and (3) how genetic material can be manipulated so that "children" can surpass "parents".

Genes and Policies: the Genetic Material of Evolution. A significant achievement this century in genetics was pinning down precisely what fundamentally is changing during

---

[1] The factor of two was gleaned from the history of past prices and variable costs. In an interview an Analog manager said they multiplied by 315%. The researchers believe that perhaps the manager meant 315% of labor costs. This probably translates to about twice variable costs. (Nelson Repenning, 1997, personal communication).

[2] In the case of Analog the key link between the TQM program and the pricing policy was not revealed until it was clear that something had gone drastically wrong. Even today, the link between the TQM program and the financial problems remains highly controversial at the company.

evolution. The answer is genes, meaningful segments of DNA molecules.[3] A gene forms a template which is used to create a stream of proteins in the cell[4]. These proteins form cellular structure, raw materials and enzymes that speed reactions. A gene produces a continuing flow of activity in the cell.

The organizational counterpart of a gene is an *idea.* Evolutionary biologist Richard Dawkins uses the term "meme" for an idea that can evolve and be passed from person to person (Dawkins 1990). Today, the new field of "memetics" is concerned with all such ideas from little tuneful commercial jingles to the idea of a wheel. We do not have to be quite this inclusive, since we are concerned only with ideas that control organizations. Jay Forrester has called these ideas "policies" (Forrester 1961).

By "policy" Forrester means any rule – implicit or explicit – that produces decisions. For example, a pricing policy might be: "Multiply variable costs by 2". A manufacturing policy might be: "Decrease production when inventories rise". Some policies are explicit, perhaps even written in policy manuals, but most policies are implicit, rarely if ever articulated and stored only in the heads of a company's employees. Policies guide pricing, marketing, budgeting, accounting, production, research, development, construction, acquisition and every other category of activity undertaken by an organization. Policies, like genes, produce a continuing stream of action. Policies, like genes, are a fulcrum on which evolution can operate. If policies evolve, so will the company.

Mutation and Innovation. In the popular press and even among some evolutionary biologists, mutation is considered the key to evolution. In fact, most mutations are deleterious. As a consequence, even primitive cells possess sophisticated "error checking" processes that make mutations in genes exceedingly rare. Estimates vary, but roughly a mutation occurs only once in every 1 million to 100 million cell divisions (Brown 1995, Smith 1989, Nei 1987).

The problem with biological mutation is that it not only can create a slightly better gene, it can also produce a much worse gene. This problem becomes more severe as the fitness of the gene increases. Evolutionist John Holland has recently offered a homey illustration of this point: Randomly changing an ingredient of your favorite recipe is unlikely to improve the dish (Holland 1995).

Turning to human organizations, consider the business policy: "When inventories are low raise prices". A random mutation might be to replace the word "low" with the word "yellow": The policy "When inventories are yellow raise prices" is unlikely to improve matters, unless, perhaps, the business happens to involve selling ripening bananas. Most changes to policies are either misunderstandings or improvement attempts. Changing "low" to "yellow" might result from a mistake or misunderstanding and is clearly similar to a biological mutation. Less obviously, perhaps, improvement attempts, too, are similar to mutations: Because we don't understand our complex organizations,

---

[3] Some organisms, particularly certain viruses, carry their genes in RNA. Even in these cases, however, RNA is usually converted to DNA within the host cell so that it can be transcribed using the usual machinery.

[4] Arguably the term "gene" could also include control regions on a DNA molecule.

"improvements" have unpredictable – that is random – effects with respect to the functioning of the business as a whole. Improvement attempts usually "make sense" – such as Analog's TQM program. This means that of all possible "random" changes, some are more likely to be pursued, but still are no more likely to succeed. The case is similar for biological mutation where certain kinds of mutations are also more likely than others.

If innovations in policy correspond to mutations in genes it is likely that organizations need to control innovation for the same reason that organisms control mutations, and for the same reason that you stick to the recipe: A policy innovation is more likely to disrupt a well functioning business policy than it is to improve it.

Recombination and Learning. Compared to their miserly attitude toward mutation, cells are profligate with recombination. Consider your own cells. Half the chromosomes in each of your cells comes from your mother and half from your father. In recombination, a chromosome from your mother crosses with a similar chromosome from your father to create a chromosome which is part Mom's and part Dad's. You package these combo-chromosomes into your sperm or your eggs (depending on obvious parameters) and pass them on to your children. In the process you might just produce a child who is better than either of your parents, just as combining the best part of eggs Benedict with the best part of a sandwich produces an Egg McMuffin, which is better than either "parent", at least if you happen to be driving in the morning.

A numerical example might help. Say we have two "chromosomes" made of 5 digits each *4,987* and *7,329*. A recombination event at the coma will produce two children: *7,987* and *4,329*. Assuming that bigger numbers are better numbers, one of the children is better than either parent (or course, the other one is not as good). The process of recombination can speed up evolution by combining good parts of different parents to produce offspring that are even better. Holland (Holland 1992, Goldberg 1989) has termed this the "building block" hypothesis.

The organizational analog of recombining chromosomes or culinary dishes is the recombination of policies. "Policy recombination" goes under the more common name "learning". Say that my pricing policy is "increase prices when inventories are low", and your pricing policy is "set prices at a margin of 200% over variable costs." My policy is good because it responds well when demand exceeds supply causing inventories to fall. Your policy is good because you will never price below costs. Perhaps you will learn from me, combining my idea with your own. Your policy afterwards might be "set prices at a margin over variable costs, but raise that margin when inventories are low". Your new policy still ensures that prices never drop below costs, but it also responds to supply and demand.

3. Simulations and insights

Recombination is learning. It occurs whenever one person adopts a policy or part of a policy from another person; and under the right conditions, it can be the workhorse of organizational evolution.

Unfortunately, organizational evolution is difficult to track in the "wild": Learning is difficult to measure, policies difficult to explicitly articulate, and the process occurs over a period of months or years. Computer simulations make it possible to study evolutionary learning under "laboratory conditions" where the pace can be quickened, policies can be explicitly specified, and learning can be closely monitored. As we shall see learning by itself is not sufficient to guarantee improvement.

Learning drift. We will simulate a company that is somewhat like a Microsoft or a Lotus Development: The simulated company has a number of different products – say, word processor, spread sheet, data base, presentation, and organizer. The company develops and releases a sequence of new versions of each product. Each release of each product is modeled as a simple system dynamics project model (Abdel-Hamid and Madnick 1991, Cooper 1980, 1993). As programmers code, they empty the stock of unwritten code, producing correctly written code. Unfortunately, the programmers also produce bugs, which eventually are discovered and then need to be rewritten. Mathematically,

$$\textit{Code to write} = \int (\textit{Discovering bugs - Writing code})dt$$

$$\textit{Writing code} = \textit{Programmers} * \textit{Productivity}$$

$$\textit{Programmers} = \{\textit{determined by agent-managers}\}$$
$$\textit{Productivity} = 2$$

$$\textit{Discovering Bugs} = \textit{Undiscovered bugs/Bug discovery time}$$

$$\textit{Bug Discovery Time} = 0.25$$

$$\textit{Undiscovered bugs} = \int (\textit{Creating bugs - Discovering bugs})dt$$

$$\textit{Creating Bugs} = \textit{Writing code} * (1 - \textit{Quality})$$

$$\textit{Quality} = 0.7$$

$$\textit{Writing code correctly} = \textit{Writing code} * \textit{Quality}$$

$$\textit{Correct code} = \int (\textit{Writing code correctly})dt$$

The dynamics of a release are simple: The stock of correctly written code increases until it reaches (almost) 100%. The more programmers, the faster the project is completed. After the version is released, programmers begin work on a new release.

As the third equation might suggest, the number of programmers is determined by managers (agents) which learn from one another through recombination. Each manager has in mind a number between 0 and 15 which specifies the number of programmers he would like to hire. A manager learns in a process of "plasmid-type" recombination: The learner recombines his pre-existing policy (expressed as a base-2 integer) with a random portion of his teacher's policy. This recombination is similar to plasmid dissemination and recombination in that (i) the donor's plasmid (policy) is unaffected, (ii) both the recipient and the donor survive the process, and (iii) no children are created (Summers

1996, *cf.* Banzhaf 1998). To phrase this more operationally, two managers might engage in conversation and one of them might come away with a modified idea.

If we are interested in shipping as quickly as possible it is intuitive that the highest number of programmers (fifteen in our simulation) is best. However, we do not let the managers in the model have this intuition. What we want to know is whether they converge on the right policy by learning from one another. A typical simulation appears as Figure 1.

Two observations can be made: (1) Managers are not tending toward the optimum number of programmers (i.e. fifteen), but (2) the managers do converge.

These results can be explained as follows. Learning is not moving managers toward the proper answer, because learning is undirected. The simulated managers are in the position of real managers, who operate in a complexity significantly beyond their comprehension. In the simulation managers copy parts of each other's ideas, but without direction and without foreknowledge of what is optimal. Which ideas they copy is random with respect to the performance of the company.

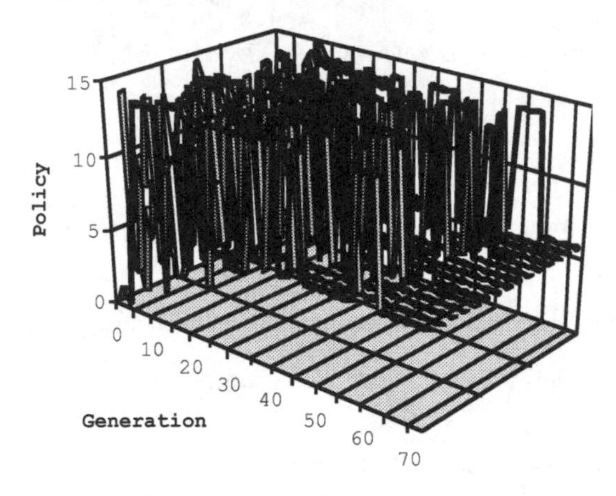

Figure 1: Learning drift: Each line represents the evolution of one manager's beliefs about the proper number of programmers over the course of 50 learning generations.

Convergence is a mathematical property of the simulation[5]. But convergence is also an organizational property. Put a group of managers together in a room to decide something and a consensus is sure to emerge eventually, as long as the managers have some common ground. Consensus is dependable enough that consultants can make a living from it by guaranteeing that they will bring a group of managers to agreement.

[5] Lost alleles are not re-introduced (there is no mutation in the simulation), and there is a finite probability that an allele in any given position will be lost in each generation. When all alleles except one are lost for a particular position, recombination simply continues to reproduce the surviving allele in each learner. The probability reaches 100% that eventually only one allele will survive in each position.

The nature of the consensus (and whether it is good or bad) is unpredictable, but consensus is assured.

Evolutionary systems operating without knowledge or direction are subject to random convergence. Geneticists call the biological case "genetic drift". We label the organizational case "learning drift". A pure learning company drifts because the company is too complex for people to know what they should be learning.

Pointing and Pushing. How can evolution create a better company if no one can be relied upon to specify what people should be learning? Consider nature's answer.

Nature never specifies *what* to inherit; it specifies *who* to inherit from. Nature does not tell baby tigers which genes to choose, it tells them *whose* genes to choose – namely, its parents'. More conventionally, nature does not tell an amorous adult tiger *which* tiger to (re)combine with, instead it limits the choice to those other tigers who happen to be alive, those who have been fit enough to survive. Once Nature selects with *whom* tigers should should team, it *pushes* team-formation (so to speak) via the sex drive.

Similarly, organizations do not need to specify *what* to learn, but only *who* to learn from. Having pointed out the good teachers, the organization then only needs to push learners to learn. We call organizational features that promote learning from selected teachers "pointing and pushing mechanisms". These mechanisms are the organizational counterparts of natural selection and the sex drive.

Organizations must somehow find or invent pointing and pushing mechanisms that will single out people who have been successful and will encourage others to learn from (or imitate) them. As one example of a potentially powerful pointing and pushing device, consider position in the management hierarchy. Managers expend an enormous amount of time trying to figure out why their bosses do the things they do, trying to understand what their bosses think, and what attitudes they have. People seem *pushed* to imitate or learn from people *pointed-out* by high position. People want higher position and they believe they can get it by imitating (or learning from) the people who have it. If the people who have position attained it because they were successful, we have the makings of an effective evolutionary system.

We can add position to our simulation: Initially, a random positions is assigned to each manager. As the simulation proceeds, position changes based on performance, measured by how long it would take that manager to ship a product. The manager with the best performance (i.e. the manager who wants the most programmers) is promoted by a factor p^1, and the manager with the worst performance is demoted by a factor of p^{-1}. Managers performing in between these two extremes are promoted or demoted by a distribution of powers of p ranging between 1 and -1. In the following simulation $p=4$. Hence, a manager with position of 2 who performed best would be promoted to a new position of $8 = 2*4^1$. Had that same manager performed the worst, his new position would have been $0.5 = 2*4^{-1}$

In each product generation, each manager has an opportunity to learn from another. A learner chooses a teacher probabilistically based on the teacher's position. The probability of a manager learning from fellow-manager j is:

$$p(learning\ from\ j) = \frac{position_j}{\sum_i position_i}$$

where i ranges over all managers.

A typical simulation run now shows both convergence and improvement. In Figure 2, managers converge at the maximum number of programmers, fifteen, which is optimal in this situation.

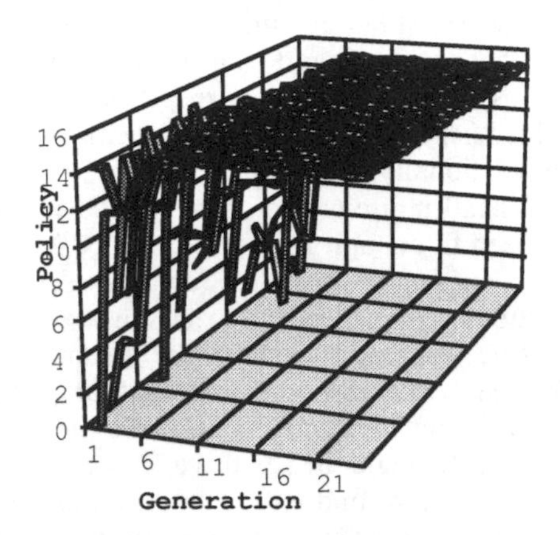

Figure 2: When managers tend to learn from a superior and when promotions are based on successful performance, improvement results. The graph indicates the policy of twenty-five managers in the simulation over a number of generations.

We do not mean to suggest that hierarchy is the only or even a desirable pointing and pushing device. We *do* mean to suggest that *a* pointing and pushing device is necessary for organizational evolution.

Defining success and team learning. A pointing and pushing device depends on the ability to tell whether someone has succeeded or not. The definition of success is up to the organization and could include such elements as personal improvement and happiness as well as the more traditional profit-based criteria.

For evolution to work organizations must point to people who are successful. In most organizations, individual success is difficult to gauge. Most people's performance depends on the performance of others. Individuals in organizations work in teams.

One of the more hopeful insights of our simulations to date is that team-based pointing and pushing mechanisms can be as effective as individual-based ones. Consider one last simulation: Instead of having individual managers managing each product, we will let teams manage the products. Promotions will now be based on team performance: All individuals on the best performing team will be promoted to a position which is a factor of p greater than their prior position. And, all individuals on the worst performing team will be demoted by a factor of p^{-1}. At the end of each product generation, teams will be mixed randomly. Even though wise and foolish policies on the same team receive the same (factor) promotion; mixing teams provides enough discrimination to pick out the better policies, as shown in the following simulation.

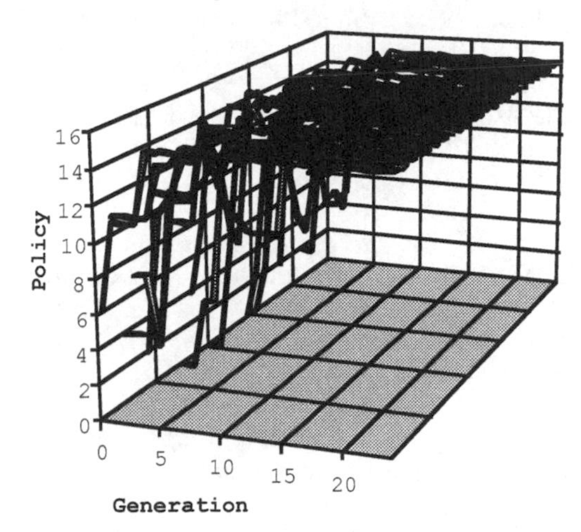

Figure 3. Convergence and accuracy are similar whether the pointing and pushing mechanism is individual-based or team-based.

4. Conclusion: Evolutionary management

In this paper we have laid a foundation that utilizes advances in evolutionary algorithms to understand what is possible in organizational evolution. Much remains to be done, however, these early results suggest that managers can create mechanisms which will make their companies evolve quickly in whatever direction desired. We term this activity "evolutionary management". Evolutionary management replaces the impossible task of understanding complex organizations with the merely difficult task of understanding organizational evolution.

References

Abdel-Hamid, Tarik and Stuart Madnick. 1991. *Software Project Dynamics: An Integrated Approach.* Englewood Cliffs, NJ: Prentice Hall.

Banzhaf, Wolfgang, Peter Nordin, Robert E. Keller, Frank D. Francone. 1998. *Genetic Programming*. San Francisco, CA: Morgan Kaufmann Publishers.

Brown, TA. 1992. *Genetics: A Molecular Approach.* New York: Chapman & Hall.

Cooper, Kenneth G. 1993, "The Rework Cycle: Vital Insights Into Managing Projects." IEE Engineering Management Review. Vol. 21, Number 3, pp. 4-12.

Cooper, Kenneth. 1980. "Naval Ship Production: a Claim Settled and a Framework Built". *Interfaces*. Vol. 10, No. 6.

Dawkins, Richard. 1990. *The Selfish Gene.* Oxford: University of Oxford Press.

Forrester, Jay W. 1961. *Industrial Dynamics.* Portland, OR: Productivity Press.

Goldberg, David E. 1989. *Genetic Algorithms in Search, Optimization and Machine Learning.* Reading, MA: Addison_Wesley Publishing Company, Inc.

Holland, John H. 1992. *Adaptation in Natural and Artificial Systems , Second Edition.* Cambridge, MA: The MIT Press.

Holland, John H. 1995. *Hidden Order: How Adaptation Builds Complexity.* Reading, Ma: Addison-Wesley Publishing Company, Inc.

Kofman, Fred, John D. Sterman, and Nelson P. Repenning. 1997. "Unanticipated Side Effects of Successful Quality Programs: Exploring a Paradox of Organizational Improvement". *Management Science.* Vol. 43, No. 4: 503-521.

Koza, John R. 1992. *Genetic Programming.* Cambridge, MA: The MIT Press.

Nei, Mastoshi. 1987. *Molecular Evolutionary Genetics.* New York: Columbia University Press.

Smith, John Maynard. 1989. *Evolutionary Genetics.* New York: Oxford University Press.

Summers, David K. 1996. *The Biology of Plasmids.* Oxford, U.K.: Blackwell Science Ltd.

AUTHORS INDEX

CPSIA information can be obtained
at www.ICGtesting.com
Printed in the USA
JSHW010827201219
3108JS00001B/49